Science and Art

The Contemporary Painted Surface

Science and Art
The Contemporary Painted Surface

Edited by

Antonio Sgamellotti
Accademia Nazionale dei Lincei, University of Perugia, Italy
Email: sgamellotti.antonio@gmail.com

Brunetto Giovanni Brunetti
INSTM, University of Perugia, Italy
Email: brunetto.brunetti@unipg.it

and

Costanza Miliani
CNR-Istituto per le Scienze del Patrimonio Culturale, Naples, Italy
Email: costanza.miliani@cnr.it

ROYAL SOCIETY
OF **CHEMISTRY**

Print ISBN: 978-1-78801-469-4
PDF ISBN: 978-1-78801-638-4
EPUB ISBN: 978-1-78801-976-7

A catalogue record for this book is available from the British Library

The Royal Society of Chemistry is a charity, registered in England and Wales, Number 207890, and a company incorporated in England by Royal Charter (Registered No. RC000524), registered office: Burlington House, Piccadilly, London W1J 0BA, UK, Telephone: +44 (0) 20 7437 8656.

Visit our website at www.rsc.org/books

Printed in the United Kingdom by CPI Group (UK) Ltd, Croydon, CR0 4YY, UK

Preface

Science and art are increasingly interconnected and their natural meeting ground is heritage science, a recent field devoted to the study, preservation and enhancement of cultural heritage. This encounter is particularly relevant in contemporary art, where, in an interdisciplinary context, different knowledge, tools, techniques, methodologies and experiences are combined to better understand the esthetical and physical complexity of the artworks made up of natural and synthetic organic, inorganic and hybrid materials, pure and in mixture, intentionally assembled by artists in the most diverse conformations.

This volume, "Science and Art. The contemporary painted surface", in continuity with the previous volume published by the Royal Society of Chemistry in 2014, "Science and Art. The painted surfaces", consists of a series of chapters, mostly co-written by scientists, art historians, conservators/curators, with the peculiarity of often including authorship of, and interviews with, the artists themselves. Topics covered include execution techniques, diagnostics, conservation, conceptual aspects and languages, described in such a way as to interest specialists, but also to be accessible to students and art lovers in general.

The book is composed of twenty-three chapters. It opens with a focus on artists already historicized, such as Jackson Pollock, Piero Manzoni, Lucio Fontana, Enrico Baj and Joseph Albers. These masters can be considered as *avant-garde* of the contemporaneity with their use of unconventional materials vehiculating innovative languages. Then, the core of the volume is made up of examinations and discussions on works of contemporary artists, considering the sixties as the

Science and Art: The Contemporary Painted Surface
Edited by Antonio Sgamellotti, Brunetto Giovanni Brunetti and Costanza Miliani

Published by the Royal Society of Chemistry, www.rsc.org

starting point of contemporaneity. These painters and sculptors, all protagonists of the development of new ideas and techniques, belong to thirteen different countries and have been selected to offer a wide vision of how innovations in contemporary art have been conjugated in various regions and cultures. Materials and practices of masters such as David Hockney, Gary Hume, Avinash Chandra, Cai Guo-Qiang, Jan Fabre, Eduardo Kac, Miquel Barceló, Sarah Lucas, Fabrizio Plessi, Loris Cecchini, Tim Otto Roth, Ângelo de Sousa, Julião Sarmento, Lourdes Castro, Juan Melé, Gregorio Vardánega, Raúl Lozza, Judith Lauand, Lina Bo Bardi, Gunter Gerzso, José Gutiérrez and Alina Szapocznikow are presented and discussed, corresponding to artists who worked or are currently working in countries including the United States, Great Britain, India, China, Germany, Belgium, Spain, Italy, Brazil, Argentina, Mexico, Portugal, France and Poland.

The last three chapters are dedicated to contemporary industrial design, net-art and the painted surface in contemporary architecture.

Each contemporary artist chooses a personal way of expression among virtually infinite numbers of possibilities and combinations, on the basis of preferences on technical aspects or other predilections, adopting a variety of practices that are generally related to traditions, memories or individual inspirations. These choices are the common thread that runs through all the chapters.

In some of them it is shown how a large variety of oil-containing paints, composed of semi- or non-drying oils mixed with linseed oil, or of synthetic resins (alkydic, vynilic and acrylic) in different formulations, have been employed by contemporary artists with the aim of achieving specific effects on polychrome surfaces. In several cases these materials, intentionally prepared or used as ready-made industrial paints, are proven to be linked to the conservation problems exhibited by the pictorial layers. Actually, in contemporary art the unprecedented richness of choices available to artists finds a counterpart in alterations and changes in appearance that can be more chemically complex than in traditional paintings.

Singular and fascinating is the use of gunpowder as a medium for drawings, or of coloured gunpowder, commonly also used to create spectacular pyrotechnic displays, to paint unique polychrome works on canvas. Accurate studies provided chemical information on used gunpowder, nature of products after deflagrations and their likely changes with time. Unusual applications include the exploitation of polymers, as for example poly-methyl-methacrylates, polyesters and polyurethanes for the creation of sculptures, taking advantage of their chemical properties, colour, and transparency.

Some of the chapters deal with themes elaborated on by artists in a conceptual way through multiple techniques and supports, including film, video-performances and video-installations, discussed in relation to the conservation of the works. Other chapters describe the exploration of effects and perception of light and colour in space, or even the assemblage of forms and structures reminiscent of chemistry and biology that dialogue with the environment. All these works are explorations of territories where art and science, natural complexity and artificial systems, subject and language interweave together.

A significant group of chapters reports on reflections and experiences, from different perspectives, on the topic of restoration of contemporary art, showing how science can contribute towards disentangling crucial problems regarding properties and complexity of employed materials and technologies, whose knowledge is essential in restoration to plan and develop decisive interventions. Restoration is also strongly linked to the protection of the identity of the work and of the rights of the artist. From this point of view, in contemporary art two fundamental questions must be addressed: that of the restoration which intervenes on permanent works, and that of the restoration which involves works destined for decay by intention of the author. As is shown in the book, in both cases there are potentially problematic situations that must be tackled and solved before any intervention: the first in relation to interventions that must not change the meaning of the work according to the intention of the artist; the other in relation to operations that, against the artist's purpose, affect the self-destructive nature of the artefact, as happens in works where the fundamental subject is natural decomposition or the ephemeral of the present.

This book, while obviously not exhaustive, aims to offer a general look at how the latest technological advances and the advent of new ideas and concepts have been translated by contemporary artists into material choices, executive practices and languages. Within this framework, it also shows that holistic studies that integrate scientific, technical and historical-artistic research on contemporary works of art, can open the way to new perceptions and unprecedented understanding of the creative process. In particular, when these studies are further combined with discussions with artists, they can provide additional information, otherwise impossible to attain, about details of unusual procedures or hidden intentions and meanings. Interviews can also reveal fundamental key aspects on the attitudes and thoughts of each artist towards the ineluctable problems of aging and conservation.

Antonio Sgamellotti
Brunetto Giovanni Brunetti
Costanza Miliani

Contents

Science and Art: The Contemporary Painted Surface
Edited by Antonio Sgamellotti, Brunetto Giovanni Brunetti and Costanza Miliani

Published by the Royal Society of Chemistry, www.rsc.org

CHAPTER 1

Jackson Pollock's Drip Paintings: Tracing the Introduction of Alkyds Through Non-invasive Analysis of Mid-1940s Paintings

FRANCESCA ROSI[a], COSTANZA MILIANI[b], JOHN DELANEY[c], KATHRYN DOOLEY[c], LENA STRINGARI[d], GRAZINA SUBELYTE[e] AND LUCIANO PENSABENE BUEMI*[e]

[a]CNR-Istituto di Scienze e Tecnologie Chimiche (SCITEC), Perugia, Italy; [b]CNR-Istituto di Scienze del Patrimonio Culturale (ISPC), Naples, Italy; [c]National Gallery of Art, Washington DC, USA; [d]Solomon R. Guggenheim Foundation, New York, USA; [e]Peggy Guggenheim Collection, Venice, Italy
*E-mail: lpensabene@guggenheim-venice.it

1.1 Introduction

From an art historical perspective, the term 'dripping' is generally associated with the visual idiom of the American Abstract Expressionist Jackson Pollock (1912–56), one of the most influential painters of the twentieth century. From a technical perspective, Pollock's materials and tools were used in relatively unconventional ways. The artist laid the canvas on the floor and divested the brush of its familiar function, using the brush and other tools to drip, pour and splatter paint onto a horizontal plane from above.

Science and Art: The Contemporary Painted Surface
Edited by Antonio Sgamellotti, Brunetto Giovanni Brunetti and Costanza Miliani

Published by the Royal Society of Chemistry, www.rsc.org

The exact development of the drip technique has been a matter of long and inconclusive debates, but signs appeared already in the mid-1940s. This was a seminal period of experimentation which marked Pollock's abandonment of the symbolic imagery of his earlier pictures and his adoption of more abstract means of expression. It was an innovative period also in term of materials, since he began to use the materials that had become available at the time: industrial household paints and metallic pigments.

The technical investigation of several poured paintings from 1949 to 1952 demonstrated the extensive use of alkyd-based paint.[1,2] Exploiting the advanced rheological properties and rapid drying time of alkyd resin, Pollock could experiment in innovative ways, using various tools to fling and guide fluid paint. Alkyd was a perfect medium to allow Pollock to develop his new kind of visual expression, marked by sinuous and liquid lines.

Consequently, the objective of this technical study is to correlate the change in Pollock's use of materials with the evolution of his visual vocabulary, *i.e.* Pollock's abandonment of a more figurative subject matter for a completely abstract one, achieved with the dripping technique. During this period, Pollock shifted from artists' oil paints to a greater reliance on commercial alkyd-based paints. Using a non-invasive analytical approach, the authors investigated eleven works from the Peggy Guggenheim Collection (PGC), Venice, ranging from 1942 to 1947. These were the crucial years that marked Pollock's experimentation, which would eventually lead to a fully mature drip technique.

1.2 Jackson Pollock's Search for New Methods of Expression

As noted, the 1940s were a watershed decade for Pollock, since his practice underwent a considerable evolution with regard to both his technique and materials. During those years, he abandoned figurative and symbolic subject matter in favour of a non-representational network of gestural drips and splashes. Any figurative connotations observable in Pollock's earlier works disappear after 1947. The influence of Cubism—Pablo Picasso in particular—had endured in Pollock's earlier work, but his interest in Navajo sand paintings and the experimental processes used by the Surrealists was an important force in his exploration of new techniques and subject matter. Pollock's transition was gradual, even halting: he progressed from figuration to images concealed under paint densely applied with fingers, palette knife or paint tubes, to more abstract paintings made solely by pouring and splattering paint on the canvas. The unprecedented spontaneity and liberty of gestures in his 'drip' works altered and expanded the definition of painting. Pollock worked from all sides of the canvas, which was placed horizontally on the floor. He felt that this positioning enabled him to control the application of paint and be most connected to the act of painting.

Such an innovative artistic process was at least partially based on the flow of imagination and plain creative impulse, or creative accident, as Pollock tested various methods of paint application that challenged convention. In considering the spontaneous element of chance in Pollock's compositions, one should reflect on his interest in the Jungian notion of the collective unconscious and the Surrealist concept of automatism, when unpremeditated actions are used to express the force of the unconscious. Such an experimental process was an important force in Pollock's search for a new, American[†] subject matter and for the technical means to give it visual expression. However, it is also important to note that Pollock would return to his paintings after some time in order to finish them, thus moving back and forth between them. This allowed him to exert control and develop his compositions, working on several paintings simultaneously. In fact, Pollock denied the spontaneous and impulsive character of his gestures. In a handwritten statement, he emphasized that '*new needs demand new techniques—total control—denial of the accident —states of order—organic intensity—energy and motion made visible*'.[5] Yet, in an interview with Selden Rodman in the spring of 1956, shortly before his tragic death, when asked whether he had "any preconceived visual image in mind, or [whether it was] the result of a wholly spontaneous, something that happens in the process of paintings", Pollock replied hesitantly: "How do I know? I have and I haven't." Nevertheless, he then pointed out: "Something in me knows where I'm going," thus ultimately and undeniably asserting the intentionality of his working process[6].[‡]

Each painting created by Pollock was idiosyncratic and varied considerably in terms of paint application, thus proving his mastery. In his statement, Pollock implicitly admitted that he had a specific vision for each painting. Indeed, he declared that his paintings were not disordered, essentially affirming that he looked at the liberated dripping technique and the lines he laid on the canvas, even if in constant tension and dialogue with his unconscious urge to paint freely, as controlled and not accidental. Additionally, although his poured works appear on the verge of complete abstraction, Pollock suggested that "he wasn't just throwing paint, he was delineating some object, some real thing, from a distance above the canvas".[8] *He believed that he was drawing not on canvas, but in the air*. Lee Krasner described Pollock as "working in the air", creating "aerial form[s] which then landed".[8] As Naifeh and White Smith concluded, Pollock "was a figurative painter in hiding. In fact, he never stopped being a figurative painter, according to [Krasner], but his figures were ephemeral

---

[†]It must be acknowledged that Pollock did not advocate the idea of a purely American art, claiming that "the idea of an isolated American painting...seems absurd to me... An American is an American and his painting would naturally be qualified by that fact, whether he wills it or not. But the basic problems of contemporary painting are independent of any one country."[3] Yet Lee Krasner posited that "of course he was very aware of European art, but what he identified with was about as American as apple pie".[4]

[‡]An extensive key study on the intentionality in Pollock's work is in ref. 7.

creations, 'airy nothings' that existed only momentarily in mid-air loops of paint, then disappeared, leaving behind their vacated 'skins' on the canvas".[8]

1.3 The Evolution of Jackson Pollock's Methods Reflected in the Works of the Peggy Guggenheim Collection, Venice

The evolution of Pollock's techniques and the *crescendo* of his ever more abstract imagery are perfectly reflected in the eleven works in the holdings of the Peggy Guggenheim Collection (Solomon R. Guggenheim Foundation), which date from 1942 to 1947. These works encompass Pollock's earlier abstract-figurative paintings, often marked by mythological subject matter, as in *The Moon Woman* (1942), painted with a traditional easel painting technique of oil paint applied with a brush. The composition, palette and execution all attest to Pollock's indebtedness at the time to Pablo Picasso's visual idiom, in particular to his painting *Girl before a Mirror* (1932; Museum of Modern Art, New York). *Circumcision* (1946) belongs to the transitional period of Pollock's evolving visual idiom. The artist then moved towards the abstract and heavily impastoed paintings of the *Sounds in the Grass* series, such as *Croaking Movement* (1946) and *Eyes in the Heat* (1946), both of which he executed by squeezing paint from tubes directly onto a canvas. The group of paintings Peggy Guggenheim decided to keep for herself[§] reaches a culmination with Pollock's signature examples of his drip paintings, *Alchemy* (1947) and *Enchanted Forest* (1947), which he created with the canvas placed on the floor. *Alchemy* is a complex work painted with a variety of materials applied in various ways: squeezed directly from the tube, applied with a paintbrush or a spatula or dripped and splattered on the canvas.

1.4 Dripping as a Revolutionary Material Choice

Pollock was introduced to the use of liquid paint in 1936 at an experimental workshop in New York City taught by the Mexican muralist David Alfaro Siqueiros. In the early 1940s, Pollock started to pour colour directly from the can or with the help of unconventional tools. In October 1945, Pollock moved to Springs, East Hampton, at the east end of Long Island, New York, in a house currently known as the Pollock-Krasner House and Studio. There he began to paint on unstretched canvases laid on the studio floor, approaching the canvas from all sides, using sticks, old hardened brushes and knives to collect the paint from the jars and then drop it onto the canvas, developing

---

[§]Peggy Guggenheim consolidated Pollock's reputation through the donations of at least fifteen works by him to museums worldwide, including the Stedelijk Museum in Amsterdam, the Museum of Modern Art in New York, the Galleria Nazionale d'Arte Moderna in Rome, the Museum of Art at the Rhode Island School of Design, the University of Iowa Museum of Art, the Art Museum of Seattle and the Tel Aviv Museum of Art.

what would be later called his *drip* technique. His mature dripping technique appears in paintings he created after 1948. Scientific data[1] suggest that he regularly used the relatively new, synthetic resin-based paint called alkyd enamel. Photographs of Pollock in his studio[9] and interviews with his wife Krasner[1] indicate that he used paints produced by the DuPont Company, a leader in the manufacture of alkyd resins.[10]

Alkyd resins, also called glyptal,[10] were synthesized at the beginning of the twentieth century by combining oil or oil-derived fatty acids into a polyester polymer structure. Chemically, an alkyd resin is an oil-modified polyester made by condensation polymerization of three components: di-carboxylic acid (phthalic anhydride), polyhydric alcohol and triglyceride lipids. Glycerol, trimethylolpropane and pentaerythritol are the polyalcohols commonly employed.

The development of alkyd resins completely revolutionized the paint and coating industry. Alkyd-based house paints are commonly found in works of art. During World War II (1939–45), these gloss enamel paints were more readily available than artists' oil paints and less expensive. However, the enhanced properties of these newly produced synthetic resins were also most attractive when compared to traditional artists' oil paints. The fast drying time (on the order of hours) was appreciated by artists accustomed to waiting weeks for the drying of traditional artists' oil paints. In addition, alkyds' glossy appearance appealed to artists wishing to juxtapose matte and glossy passages. Overall, the improved rheological properties of alkyd resins allowed Pollock to adopt unorthodox techniques such as dripping and pouring, enabling the creation of a range of unusual effects and textures that could not be achieved using traditional artists' oils.

1.5 Non-invasive Identification of Alkyd Resins

Among the scientific methods available to heritage scientists, those based on mass spectrometry allow the most detailed information on the molecular composition of the synthetic binding media. As for alkyd resins, by combining Gas Chromatography–Mass Spectrometry (GC-MS), High-Performance Liquid Chromatography (HPLC)–MS and flow injection analysis (FIA)–MS it is possible to obtain from a minute sample of the paint the fatty acid profile and aromatic fraction of the resin, as well as identifying pentaerythritol, phthalic acid ester and the triglyceride residues of the synthesis process.[11]

The past decade has witnessed significant changes in heritage science as paint sampling has increasingly given way to non-invasive, analytical techniques carried out with mobile instruments that can be brought to the paintings for *in situ* analysis.[12,13]

Among the non-invasive methods available today, Fourier Transform Infrared (FT-IR) reflection spectroscopy (in the mid 4000–400 cm^{-1} and near 12 000–4000 cm^{-1} infrared range) is most suitable for binding media analysis. The analysis of the spectral properties of paint reconstructions made of different types of synthetic (and natural) polymeric paints allowed Rosi *et al.*[14] to demonstrate that, despite the spectral distortion occurring in reflection mode, the main differences among the vibrational profiles of different polymeric backbones are still appreciable and exploitable for their non-invasive identification. Figure 1.1 reports a comparison of the FT-IR spectral profiles of different binding media that are of interest for this study. In particular, the alkyd resin profile (grey line), recorded from a model painted with white barium sulphate in the wide infrared range (6000–400 cm^{-1}), is compared with the spectra recorded on the same white pigment painted with oil (black line), acrylic (blue line) and vinyl resin (magenta line). In the mid-IR range the marker vibrational mode of the alkyd resin is the C–O–C stretching,[14,15] appearing as a derivative-like feature at about 1270 cm^{-1}, being absent or very weak in polymers like oil and acrylic or upshifted with respect to the corresponding mode in vinyl resin. This band falls in a range relatively free from overlapping, which, combined with its strong/medium intensity, makes its visualization possible also in complex paint mixtures.[14] The other portion of the spectral profile is rather similar to that of oil, including the carbonyl stretching (νC=O) at about 1740 cm^{-1} and the CH stretching mode (νCH) at about 3000–2800 cm^{-1}. In the near-IR range, except for the combination ($\nu + \boldsymbol{\delta}$) of the CH functional group which is very similar to that of the oil binder, alkyd resin can be distinguished from oil by observing a band at 4684 cm^{-1}, possibly due to the combination of the CH and C=O stretchings (νCH + νC=O),[14] but also tentatively assigned to combination bands of the aromatic component of alkyds.[16] This last feature is clearly present also in vinyl and acrylic resins and cannot be used for the alkyd identification (Figure 1.1).

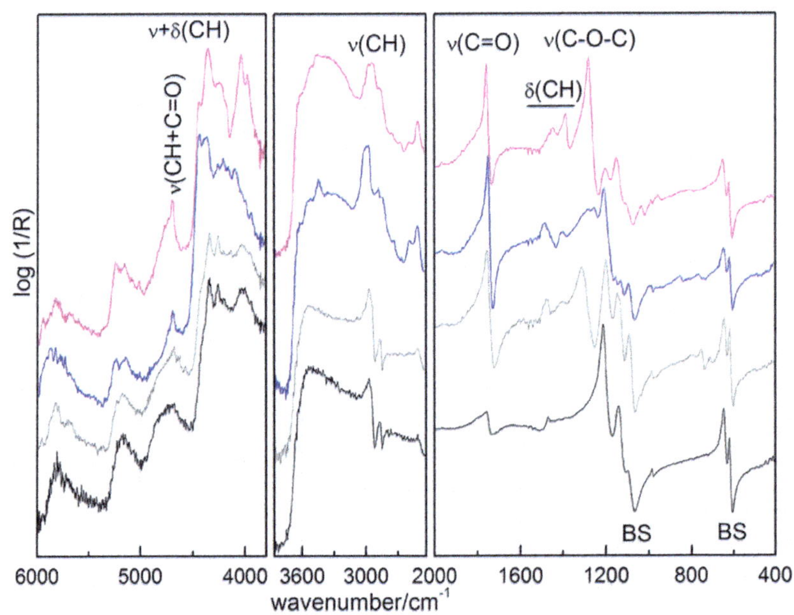

Figure 1.1 Reflection mode FT-IR spectra recorded on laboratory models based on barium sulphate painted with vinyl resin (magenta line), acrylic resin (blue line), alkyd resin (grey line) and oil (black line). © The Pollock-Krasner Foundation ARS, NY and DACS, London 2019.

Although the mid-IR range is more specific for the alkyd identification with respect to the near-IR window, the latter is presently also accessible through hyperspectral imaging instruments that allow the chemical identification and spatial localization at the macro-scale of the material composition.[16,17]

To the best of our knowledge, the results of micro-destructive MS analysis of binding media were published for fourteen paintings by Pollock (see Table 1.1).[1,2,18,19] Overall, the collection of MS data reveals that alkyd appeared first in 1947, then consistently in all the later paintings, as already reported by Lake *et al.*[1]

The exploitation of non-invasive FT-IR reflection spectroscopy allowed for an extension of the information focusing on paintings realized in 1942 to 1947 belonging to the PGC (data are collected in Table 1.1).

In agreement with previous data,[1,18,20] our analysis clearly evidenced the use of alkyd resins by Pollock in 1947 as binder of the glossy and matte black paints (possibly made of vegetal carbon black) poured in *Alchemy*[20] and *Enchanted Forest* (Figure 1.2). The identification in both paintings was achieved by evaluating the mid-IR range because strong absorption by the black pigments in the near-IR range resulted in little-to-no reflected near-IR light being collected.

Table 1.1 List of the paintings in chronological order subjected to non-invasive (NI) or destructive (D) investigation indicating the identified organic binder (oil or alkyd) and the texture/paint application.¶

Painting title	Date	Organic binder	Texture/application
Moon Woman	1942	Oil (NI)	Brushstrokes
Composition With Pouring II[1]	1943	Oil (D)	Brushstrokes, squeezed, dripped
Mural[19]	1943	Oil (D)	Brushstrokes, splatter, smear, dab
Untitled 1944	1944	Oil (NI)	Brushstrokes
Two	1943–45	Oil (NI)	Brushstrokes, smear, dab
Direction	1945	Oil (NI)	Brushstrokes, dab, dripped
There Were Seven in Eight[1]	1945	Oil (D)	Brushstrokes and dripped
Bird Effort	1946	Oil (NI)	Brushstrokes, dab
Water Bull[21]	1946	Oil (NI)	Brushstrokes

¶The table intends to trace temporally the use of the oil-modified alkyd binder with respect to drying oil; it is not inclusive of all the other possible organic components of the paint formulations. The 'Texture/application' column refers to the main visual aspects of the alkyd and drying oil paints under consideration. It does not detail the different paint applications of each paint, but it is aimed at evidencing the moment unconventional/complex applications entered Pollock's painting.

(*continued*)

Table 1.1 *(continued)*

Painting title	Date	Organic binder	Texture/application
Croaking Movement	1946	Oil (NI)	Brushstrokes, squeezed, smear
Circumcision	1946	Oil (NI)	Brushstrokes, dripped
Shimmering Substance[1]	1946	Oil (D)	Squeezed, stiff brush or palette knife application
Eyes in the Heat	1946	Oil (NI) Oil-modified alkyd (NI) – orange	Squeezed, brush-strokes, smear
Untitled 1946	1946	Oil (NI) Oil-modified alkyd (NI) – black and yellow	Brushstrokes, dripped Dripped
Alchemy[20]	1947	Oil (NI) Oil-modified alkyd (NI) – black matte and glossy	Brushstrokes, squeezed, dripped Dripped
Full Fathom Five[1]	1947	Oil (D) Oil-modified alkyd (D) – dark blue, dark orange	Dripped Dripped
Enchanted Forest	1947	Oil (NI) Oil-modified alkyd (NI) – black	Dripped Dripped
Reflection of the Big Dipper[18]	1947	Oil (D) Oil-modified alkyd (D) –black	Dripped, squeezed Dripped
Number 1A[1]	1948	Oil (D) – white, oil (NI) – white, turquoise, red Oil-modified alkyd (D, NI) –cream	Dripped, squeezed Dripped
Summertime: Number 9A[22]	1948	Oil (D) Oil-modified alkyd (D) – black and silver	Dab Poured, dripped and sprayed
Number 3, 1949: Tiger[1]	1949	Oil (D) Oil-modified alkyd (D) –orange, yellow, rust, green, white, black	Dripped Dripped
Number 12[1]	1949	Oil-modified alkyd (D) – all paints	Dripped
Number 10[1]	1949	Oil (D) Oil-modified alkyd (D)	Dripped Dripped
One: Number 31[1]	1950	Oil-modified alkyd (D)	Dripped
Number 1, 1950 (Lavender Mist)[16]	1950	Oil (NI) – white Oil-modified alkyd (NI) – white and blue	Squeezed Dripped
Number 14[23]	1951	Oil-modified alkyd (D)	Dripped
Yellow Island[2]	1952	Oil (D) Oil-modified alkyd (D) – yellow, orange, silver, white	Dripped Dripped

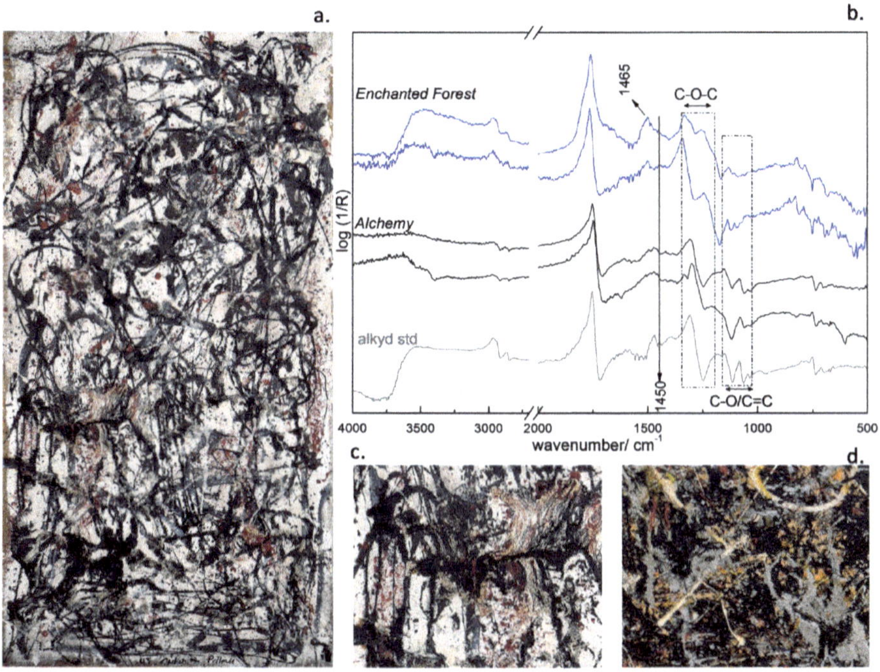

Figure 1.2 (a) *Enchanted Forest* (1947; Peggy Guggenheim Collection, Venice); (b) reflection mode FT-IR spectra recorded on black paints of *Alchemy* and *Enchanted Forest* compared with the spectrum of a reference standard of an alkyd resin; (c) and (d) details of the dripping paintings *Enchanted Forest* and *Alchemy*, respectively. © The Pollock-Krasner Foundation ARS, NY and DACS, London 2019.

More surprisingly, we also observed the presence of alkyd in two out of the seven paintings dated 1946. In *Eyes in the Heat* (late 1946; a detail is shown in Figure 1.3a), the poured orange paint is composed of chrome orange (as indicated by Raman spectroscopy; Figure 1.3b), lead white and barium sulphate clearly bound with an alkyd binder (Figure 1.3c). In *Untitled 1946* (a 1946 work on paper), some of the black and yellow poured paints were possibly made of an alkyd medium. In Figure 1.4 the spectra recorded in some of the glossy black and transparent yellow areas of the painting are reported.

Finally, leaving aside the works in the Peggy Guggenheim Collection and looking at those in the National Gallery of Art, Washington, Figure 1.5 depicts the macro-scale spatial distribution of the oil and oil-modified alkyd binders in the painting *Number 1, 1950 (Lavender Mist)*, as obtained by hyperspectral near-IR reflectance imaging already published in ref. 16. In order to enhance the spectral differences of oil and alkyd, a first-derivative transformation was applied to the hyperspectral cube and the resulting processed images clearly unravel the complex paint texture and layering executed with the two oil-based binders.

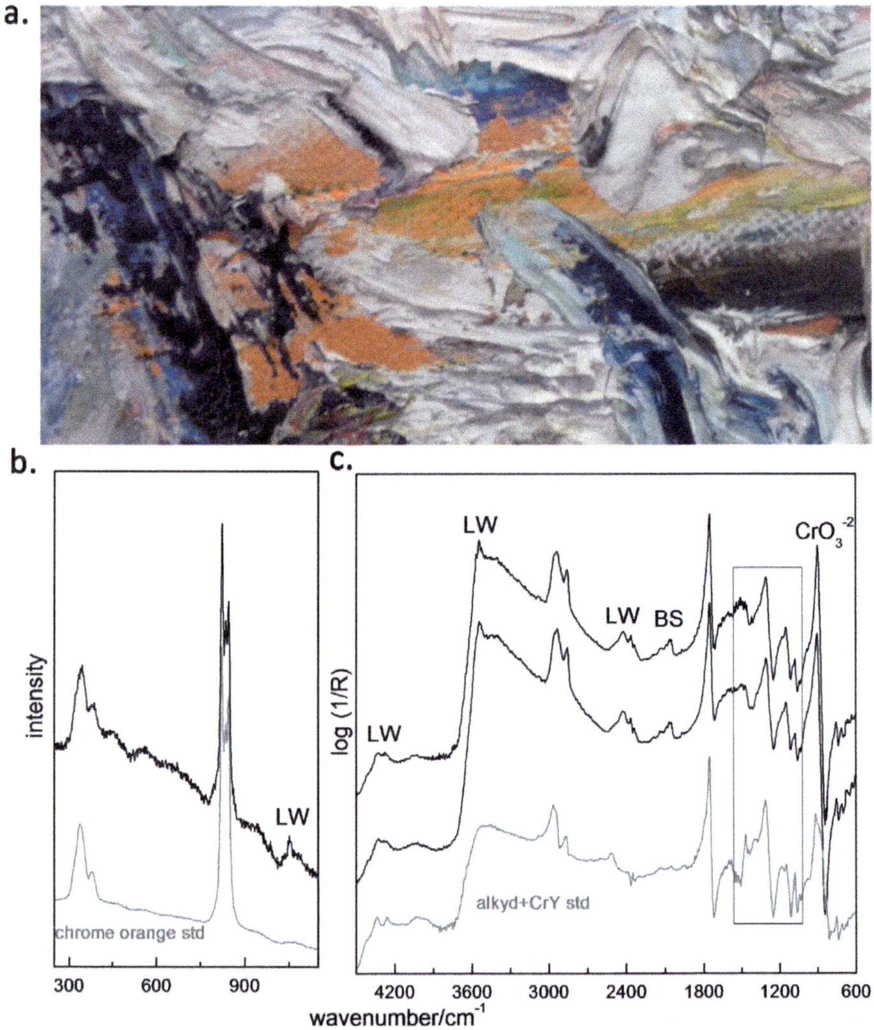

Figure 1.3 (a) Detail of the orange paints in *Eyes in the Heat* (1946); (b) Raman spectra recorded on the orange paint compared with the spectrum acquired from a reference standard of chrome orange; (c) reflection mode FT-IR spectra recorded on the orange paints compared with the spectrum acquired from a reference standard paint model based on alkyd resin with chrome yellow (CrY). Outlined box highlights the identifying bands of alkyd in this specific example. LW = lead white, BS = barium sulphate. © The Pollock-Krasner Foundation ARS, NY and DACS, London 2019.

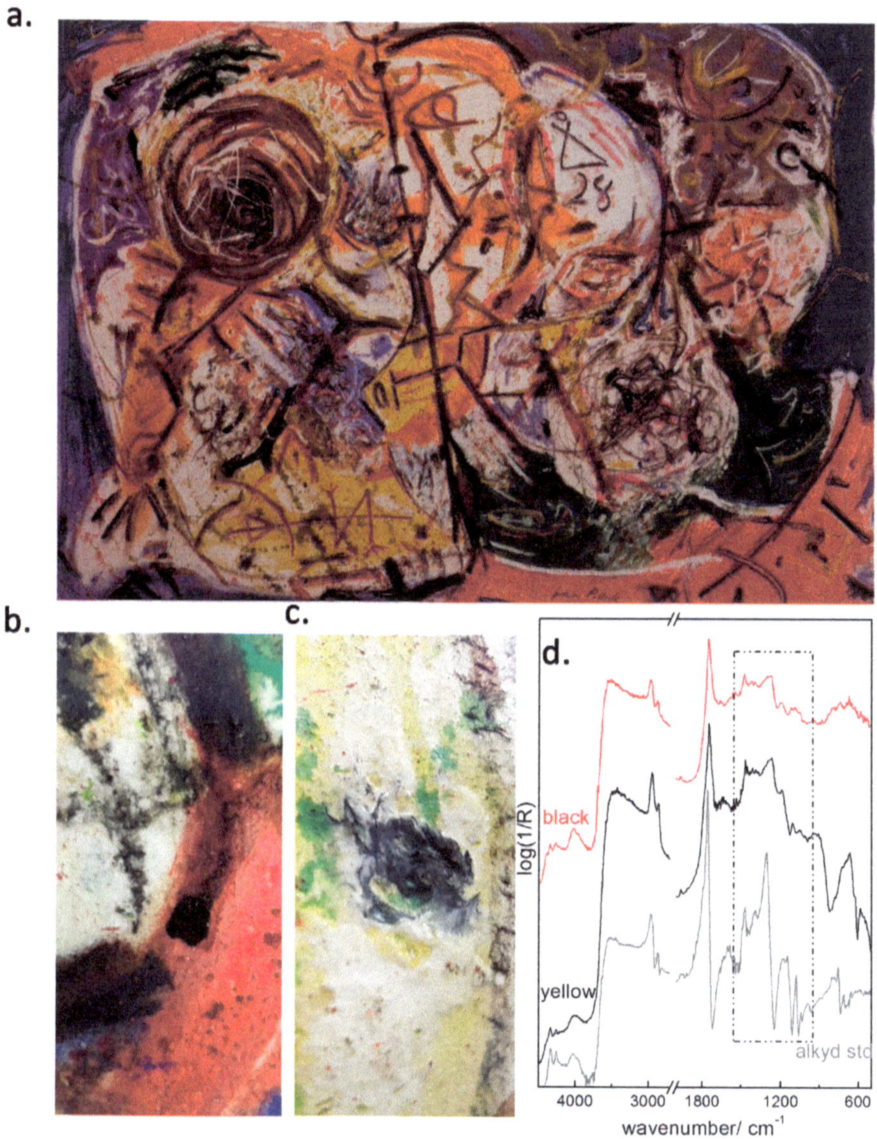

Figure 1.4 (a) *Untitled 1946*; (b) and (c) detail of the black glossy and transparent yellow paints possibly composed of alkyd resins respectively, as indicated by reflection FT-IR shown in (d), where the corresponding spectra are compared with a spectrum of an alkyd reference. Outlined box highlights the identifying bands of alkyd in this specific example. © The Pollock-Krasner Foundation ARS, NY and DACS, London 2019.

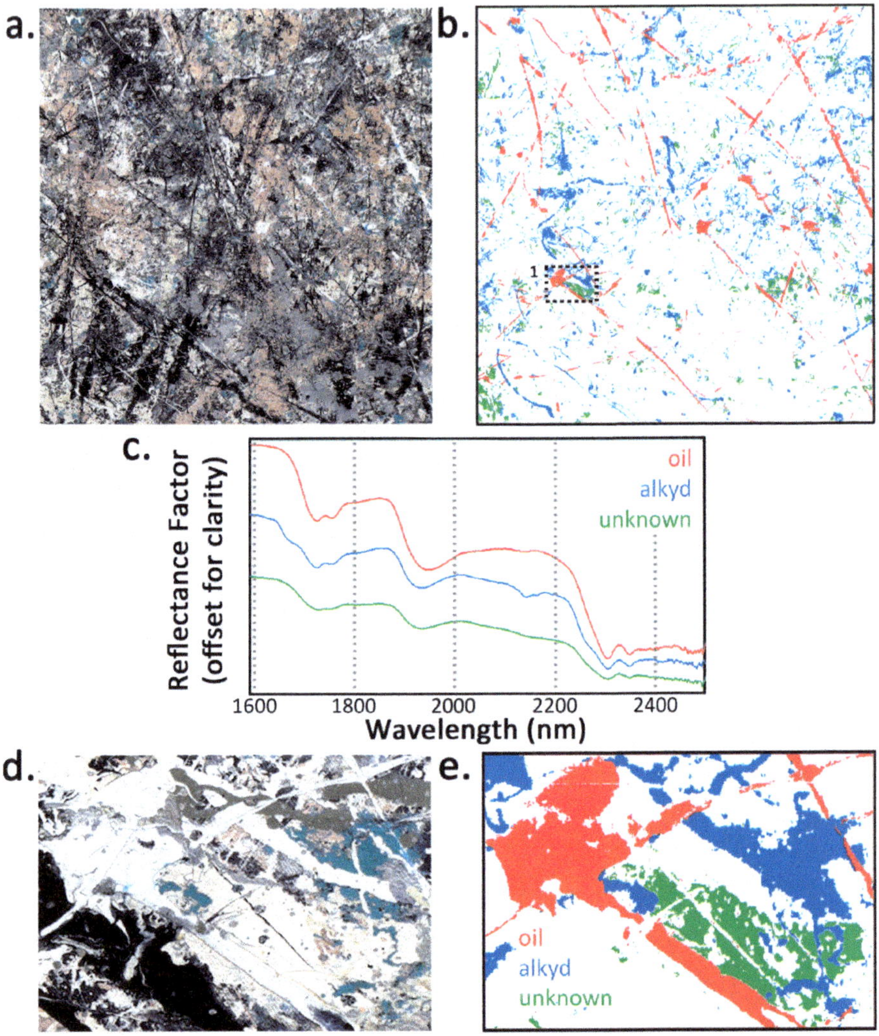

Figure 1.5 (a) Detail of ~1 × 1 m² area of *Number 1, 1950 (Lavender Mist)* (Ailsa Mellon Bruce Fund, National Gallery of Art, Washington); (b) material map showing the locations that contain oil (red), alkyd (blue) and an unknown binder (green); (c) point spectra of the three binding media from region #1 denoted with dashed lines in (b); (d) colour detail; and (e) binder map of detail #1. Reproduced from ref. 16 with permission from The Royal Society of Chemistry. Artwork © The Pollock-Krasner Foundation ARS, NY and DACS, London 2019.

1.6 Discussion

The comparison of scientific analyses on binding media of twenty-seven paintings by Pollock, with related textural effects, offers interesting observations (Table 1.1).

Oil binders (both as artists' materials and/or house paint commercial products) have been mainly identified in the paintings of the first half of the 1940s, where the more traditional paint brush applications prevail over limited unconventional squeezed and/or dripped paint (Table 1.1). Pollock's earlier abstract-figurative paintings are still painted with a traditional brush and oil paint on a primed and stretched canvas placed on an easel, as observed in two works in the Peggy Guggenheim Collection: *The Moon Woman* and *Two* (1943–45). Although Pollock was still using relatively conventional painting methods, during this period he began introducing other, more intricate techniques, as observed in such works as *Composition with Pouring II* (1943). This work reveals a certain complexity, as Pollock used a variety of techniques ranging from brush, squeezed and splattered paints to poured and dripped applications. He mainly used artists' paints in tempera, oil and oleoresin media, and house paint in the glossy dripped black, one of the last paints applied, although not composed of oil-modified alkyd resins.[1]

Surprisingly, the first documented occurrence of alkyd resin, so far dated to 1947, in the present study is now anticipated to be 1946.[1,18,20]

Indeed, it has been detected in the orange paints of *Eyes in the Heat* and possibly in the yellow and glossy black paints of the work on paper *Untitled*, both dated 1946.

Indeed, 1946 is a transitional period. *Circumcision*, defined by figurative references and a relatively organized pictorial field, reveals a looser application of paint, albeit highly controlled. The animated line applied with evident vigour becomes the protagonist of the scene. Nonetheless, in the same year Pollock painted the *Sounds in the Grass* series, such as *Croaking Movement* (1946), *Eyes in the Heat* (1946) and *Shimmering Substance* (1946)[1]. In these paintings he manipulated the pigments on the surface, spreading and pushing them with the help of utensils, and purposefully creating an emphatic, painterly texture. *Eyes in the Heat*, for instance, is covered with thick and vigorously worked paint, leaving no part untouched. However, Pollock still seemed to concentrate on the space as delimited by the edges of the canvas, rather than allowing the paint to potentially extend beyond the pictorial field. His ceaseless, active lines twist and turn towards the margins, yet do not surpass the edges. The dense, rhythmic web of white strokes seems to conceal the bodies of the creatures to which the 'eyes' of the title belong. Partially obliterated, thick, black lines—outlining what may be a 'primitive' figure—provide structure to the composition. The viewer's eye finds no repose, and this induces the sensations of intensity and heat inferred in the title. It is key to note that here Pollock used alkyd paints even prior to placing his canvases on the floor and developing his later drip compositions.

As a matter of fact, historical records report that the winter of 1946–47 marked a decisive turning point in Pollock's style: he began experimenting with painting by placing the canvas on the floor of his barn-studio in Springs. This watershed moment was accompanied by the permanent appearance of alkyd resins in Pollock's art.

Among the four investigated paintings, those dated to 1947 have all been executed with at least one alkyd-based paint. Carbon black alkyd media were found in *Alchemy, Enchanted Forest* and *Reflection of the Big Dipper* (carbon black-containing green paint) (Stedelijk Museum, Amsterdam, Netherlands),[18] whereas in *Full Fathom Five* (Museum of Modern Art, New York), dark blue and dark orange alkyd paints are splattered with oil-based artist's binders and oleoresin commercial house paints.[1]

Although the dripping experimentation started earlier (Table 1.1), paintings of 1947 present a radical departure from figurative work that relies on a Cubist grid and symbolic subject matter, in favour of a non-representational network of gestural drips and splashes. In other words, Pollock's so-called 'drip period' started and marked the beginning of his mature style.

Among the paintings of this period, one should not discount the possibility that *Alchemy* (1947) might have been Pollock's first highly experimental 'drip' work. The photographs taken by Herbert Matter in the summer of 1947 are the seminal documentary evidence remaining of Pollock at work on *Alchemy* in his studio in Springs. In one of them, Pollock crouches on the floor, and the canvas is attached to a quilting frame that belonged to his mother Stella, who was a skilled seamstress and weaver.[‖]

In *Alchemy*, Pollock envisioned a precise compositional order created through his controlled drips of paint, and through an arrangement of counterpoints and symmetries, in which straight lines play against curved lines, brilliant colours against opaque colours, black with silver and blue with red.[24] From the outset, Pollock had a general framework in mind and proceeded with his guiding hand to direct the composition. Therefore, *Alchemy* seems to represent the various steps of the unfolding of Pollock's subconscious on the canvas surface through a predetermined working method, a process akin to jazz improvisations.

Alchemy itself is the prime example demonstrating the fact that Pollock did not apply all layers of paint at once. The material surface of the painting is dense and heavy, with fifteen different colours applied. Historical photographs depict *Alchemy* essentially finished in his studio, while four smaller dripped or poured paintings and at least one earlier work are stacked around the walls on the barn-studio. The paintings—*Composition with Black Pouring*, *Galaxy*, *Phosphorescence* and *Reflection of the Big Dipper*—are not all finished at the time the photographs were taken, suggesting that Pollock must have returned to complete some of his works after some time. Most likely, he put a layer of paint on the canvas, and while waiting for it to dry he proceeded to work on his other paintings.

It may also be assumed that the experimentation with *Alchemy* made him understand that a canvas with a thick, rich surface could crack when put on a stretcher, generating losses. The losses and the paint cracking that he may have witnessed in *Alchemy* might have given him the impetus to change his painting style in order to make it more fluid, resulting in a flatter surface. If this were true, then *Enchanted Forest*, characterized by a markedly

‖The cloth on which *Alchemy* was painted most probably also belonged to Pollock's mother Stella. It was cut from what seems to have been a cotton curtain or tablecloth, decorated with pale red roses and vases, which are still visible on the verso.

flatter surface, would have been created after *Alchemy*. *Enchanted Forest* has characteristics of many of Pollock's mature compositions: a reduced range of colour, space for black and white to breathe and find their balance on the canvas. Line, laid on the canvas in a way that resembles script, embodies the only form the eye can trace: it neither implies space through perspective nor defines shape through contour. Its constant mobility allows no rest for the viewer's eye, even if the palette is subdued. Small patches of red and gold protrude, making the composition ever more vivid. There is no central focus here, and we are denied the safe ground offered by the traditional subordination of technique to subject matter. Here, colour and line are themselves the subjects.

In subsequent years, Pollock began to use even more diluted paint that flowed better and allowed him to have even more control over his works. In a handwritten statement in *ca.* 1950, he noted: "Most of the paint I use is liquid, flowing/the brushes are used more as sticks and do not touch the surface/I'm able to be more free and move about with greater ease/it seems to be possible to control the flow of paint to a great extent/I deny the accident...".[5]

It is not possible to demonstrate that Pollock's satisfaction expressed by this statement could be connected to the availability of the new oil-modified alkyd resin offering him the possibility of technically expressing his art. However, based on scientific examinations (Table 1.1), after the first introduction of alkyds, the new resins appear frequently in the later paintings, clearly indicating their extensive use by Pollock.

According to ref. 1 and 2, destructive investigation of a number of samples revealed that: in painting *Number 12* (1949) although not specifying the total number, all the samples are alkyds; in *Number 10* (1949), six samples out of seven are alkyds; in *One: Number 31* (1950) three samples out of three are alkyds; and in *Yellow Island* (1953) six samples out of seven are alkyds. This results in a percentage of oil-modified alkyd presence in the range of 85–100%. The extensive use of the oil-modified alkyd binder can be clearly appreciated from the material maps obtained by near-IR reflectance hyperspectral imaging of *Number 1, 1950 (Lavender Mist)* (Figure 1.5) where the texture of squeezed artists' oil paint is complemented by poured alkyd resin house paint.[16]

1.7 Conclusions and Perspectives

The present non-invasive study revealed that oil-modified alkyds were introduced in Pollock's materials in 1946, updating previous records referring to 1947. The identification of the synthetic resin was carried out by non-invasive FT-IR reflection spectroscopy on Pollock's works at the Peggy Guggenheim Collection in Venice, which holds eleven paintings from the period 1942–47. Combined with previous scientific examinations carried out on fourteen paintings through destructive methods on samples, the study enabled the evaluation of the occurrence of the oil-modified alkyd resin in the paintings dated 1942–52. The dripping paints appeared before the use of alkyds (in agreement with the previous study,[1] Pollock in fact started experimenting with both artists' and house paints). The introduction of the oil-modified alkyd offered Pollock the suitable binder for expansively expressing his *action painting* in the paintings after 1947.

From a methodological point of view, it should be emphasized that the overall considerations resulted from analysis of tiny samples that are not fully representative of the entire material composition of a painting. The results are improved by using non-invasive methods such as single-point FT-IR reflection spectroscopy, enabling the non-invasive collection of a theoretically infinite number of points. As a matter of fact, when investigating complex paintings such as those by Pollock, a non-invasive single-point technique also presents a limited representation. Presently, the use of hyperspectral reflectance imaging methods can overcome this limitation by maintaining the spectral specificity of single-point techniques while offering the macro-scale chemical localization of the binders, thus enabling both a better representation of the composition overall and visualization of complex layering and textures as demonstrated for *Number 1, 1950 (Lavender Mist)*.

References

1. S. Lake, E. Ordonez and M. Schilling, A technical investigation of paints used by Jackson Pollock in his drip or poured paintings, *Stud. Conserv.*, 2004, **49**, 137.
2. F. Cappitelli, THM-GCMS and FTIR for the study of binding media in *Yellow Islands* by Jackson Pollock and *Break Point* by Fiona Banner, *J. Anal. Appl. Pyrolysis*, 2004, **71**, 405.
3. N. Jachec, *Jackson Pollock: Works, Writings and Interviews*, Ediciones Poligrafia, Barcelona, 2011, p. 134.
4. B. Rose, Jackson Pollock at work: An interview with Lee Krasner, *Partis. Rev.*, 1980, **47**(1), 82–92.
5. P. Jackson, Handwritten Statement, *ca.* 1950, in *Jackson Pollock: Works, Writings and Interviews*, ed. N. Jachec, Ediciones Poligrafia, Barcelona, 2011.
6. S. Rodman, Interview with Jackson Pollock, Spring 1956, in *Jackson Pollock: Works, Writings and Interviews*, ed. N. Jachec, Ediciones Poligrafia, Barcelona, 2011, p. 148.
7. J. Coddington, No Chaos Damn It, in *Jackson Pollock. New Approaches*, ed. K. Varnedoe and P. Karmel, Museum of Modern Art, New York, 1999, p. 101.
8. S. Naifeh and G. W. Smith, *Jackson Pollock: An American Saga*, Woodward/White, Aiken, 1989, p. 539.
9. H. Matter, *Jackson Pollock and Lee Krasner Papers*, Archives of American Art, Smithsonian Institution, 1947.

10. H. Standeven, The Development of Decorative Gloss Paints in Britain and the United States C, *J. Am. Inst. Conserv.*, 2006, **45**, 51.
11. J. La Nasa, I. Degano, F. Modugno and M. P. Colombini, Industrial alkyd resins: characterization of pentaerythritol and phthalic acid esters using integrated mass spectrometry, *Rapid Commun. Mass Spectrom.*, 2015, **29**, 225.
12. C. Miliani, F. Rosi, B. G. Brunetti and A. Sgamellotti, In Situ Non-invasive Study of Artworks: The MOLAB Multi-technique Approach, *Acc. Chem. Res.*, 2010, **43**(6), 728.
13. G. B. Brunetti, C. Miliani, F. Rosi, B. Doherty, L. Monico, A. Romani and A. Sgamellotti, Non-invasive Investigations of Paintings by Portable Instrumentation: The MOLAB Experience, *Top. Curr. Chem.*, 2016, **374**, 10.
14. F. Rosi, A. Daveri, P. Moretti, B. G. Brunetti and C. Miliani, Interpretation of mid and near-infrared reflection properties of synthetic polymer paints for the non-invasive assessment of binding media in twentieth-century pictorial artworks, *Microchem. J.*, 2016, **124**, 898.
15. R. Ploeger, O. Chiantore, D. Scalarone and T. Poli, Mid-Infrared Fiber-Optic Reflection Spectroscopy (FORS) Analysis of Artists' Alkyd Paints on Different Supports, *Appl. Spectrosc.*, 2011, **65**(4), 429.
16. K. A. Dooley, J. Coddington, J. Krueger, D. M. Conover, M. Loew and J. K. Delaney, Standoff chemical imaging finds evidence for Jackson Pollock's selective use of alkyd and oil binding media in a famous 'drip' painting, *Anal. Methods*, 2017, **9**, 28.
17. K. A. Dooley, D. M. Conover, L. D. Glinsman and J. K. Delaney, Complementary standoff chemical imaging to map and identify artist materials in an early Italian Renaissance panel painting, *Angew. Chem., Int. Ed.*, 2014, **53**(50), 13775.
18. L. Wijnberg and K. J. van den Berg, Reflection of the Big Dipper (1947), in *Problemi di Conservazione e Restauro*, Edifir, 2019, p. 52.
19. Y. Szafran, L. Rivers, A. Phenix, T. Learner, E. G. Landau and S. Martin, *Jackson Pollock's "Mural": The Transitional Moment*, J. Paul Getty Museum, Los Angeles, 2014.
20. F. Rosi, C. Grazia, R. Fontana, F. Gabrieli, L. Pensabene Buemi, E. Pampaloni, A. Romani, C. Stringari and C. Miliani, Disclosing Jackson Pollock's palette in Alchemy (1947) by non-invasive spectroscopies, *Heritage Sci.*, 2016, **4**, 18.
21. *Jackson Pollock. La Figura Della furia*, exhibition, Florence, Palazzo Vecchio, 16 April-27 July 2014. Analyses were executed by MOLAB (CNR-ISTM&UNIPG-SMAArt), Report: C. Miliani et al., Non-invasive Investigations on the Water Bull by Jackson Pollock, 06 June 2015, p. 7.
22. O. Fuesers and S. Zumbühl, The Influence of Organic Solvents on the Mechanical Properties of Alkyd and Oil Paint, *9th International Conference on NDT of Art*, Jerusalem Israel, 25–30 May 2008 and references therein.

23. H. Standeven, Cover the Earth: A History of the Manufacture of Household Gloss Paints in Britain and the United States from the 1920s to the 1950s, in *Modern Paints Uncovered: Proceedings from the Modern Paints Uncovered Symposium*, Getty Conservation Institute, Tate Modern London, 16–19 May 2006, and references therein.
24. L. Pensabene Buemi, F. Bettini and R. Bellucci, 'Alchimia' di Jackson Pollock. Viaggio all'interno della materia, *Problemi di Conservazione e Restauro*, Edifir, 2019, p. 52.

CHAPTER 2

Piero Manzoni, Achromes: *An Initial Investigation into the Chemical and Technical Characterization of the Artist's Materials*†

ROSALIA PASQUALINO DI MARINEO*[a], LUCA BOCHICCHIO[b] AND LUISA MENSI[c,d]

[a]Fondazione Piero Manzoni, Milan, Italy; [b]University of Genova – MuDA/ Asger Jorn House Museum, Albissola, Italy; [c]IUAV University, Venice, Italy; [d]School of Restoration of the Brera Academy of Fine Arts, Milan, Italy
*E-mail: fondazionepieromanzoni@gmail.com

2.1 *Achrome*: What Lies Underneath

This essay presents the first exclusive and wide exploration of a number of issues associated with the characterization of the materials used in the *Achromes*: one of the most distinctive series of works in the extraordinary, albeit brief, life and career of Piero Manzoni (Soncino 1933–Milan 1963).[1,2]

†The production of this chapter was coordinated by Rosalia Pasqualino di Marineo for the Fondazione Piero Manzoni, Milan. The first section was written by Luca Bochicchio. The second section was written by Luisa Mensi.

Science and Art: The Contemporary Painted Surface
Edited by Antonio Sgamellotti, Brunetto Giovanni Brunetti and Costanza Miliani

Published by the Royal Society of Chemistry, www.rsc.org

According to a well-known document written by Manzoni himself,[3] the first works of this kind date from 1957 and the artist explored the innumerable possibilities with diverse materials and techniques through to 1963, the year of his sudden death.[4–9] While Manzoni did not initially ascribe the term to these works, the *Achromes* represent the artistic turning point in his career: the first expression of a wholly independent and fully defined aesthetics, even taking into account the nonetheless excellent work that, between 1955 and 1956, had placed the artist alongside the *Movimento Arte Nucleare* (Nuclear Art Movement).[10]

In order to identify briefly the distinctive characteristics of Piero Manzoni's *Achromes* (and as an introduction to the following paragraphs), it may be useful to observe them from two different yet complementary points of view, as essential to one another as the two faces of the same coin: the philosophical or symbolic and the technical or procedural perspectives.

In philosophical-symbolic terms, the *Achrome* corresponds to a search for absolute values, of which the work of art can no longer be a passive and secondary carrier with respect to the image painted by the artist (whether

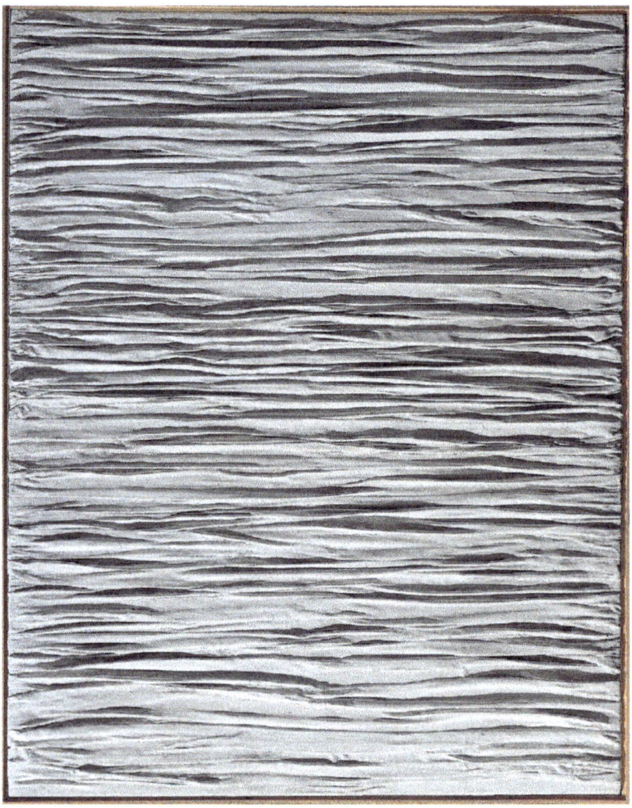

Figure 2.1 *Achrome*, 1958–1959, creased canvas and kaolin, 100 × 80 cm^2. Image © Fondazione Piero Manzoni, Milan. Reproduced with permission from Elena Manzoni, on behalf of The Estate of Piero Manzoni.

Figure 2.2 *Achrome*, 1958–1959, canvas in square and kaolin, 80 × 60 cm^2. Image © Fondazione Piero Manzoni, Milan. Reproduced with permission from Elena Manzoni, on behalf of The Estate of Piero Manzoni.

it be expressionist, figurative, abstract or otherwise). The question of the universality of the image and the medium is essential here, "foundative" in fact[‡]. Manzoni intellectually supports the radical need to free the surface of any subjective intervention, which would inevitably lead to the exclusion of the infinite possibilities of the image in favour of that sole option chosen and imposed by the artist on the canvas through the processing of a constructed or, rather, painted image. The technical verification of such theoretical hypotheses occurs by way of achromatism, that is to say, an even more advanced degree of monochromatism obtained not by "painting" the surface of the canvas with the "colour" white, but in many cases by immersing the canvas in liquid solutions of inert substances tending to white, such as for example gesso and kaolin (Figures 2.1 and 2.2). The difference is fundamental. As emphasized in the historiographical and critical literature, the authorial process is present here but reduced to the bare minimum. In stretching the canvas previously

[‡]The concept of "foundative" art underpins Manzoni's theory and is cited in his text *Oggi il concetto di quadro... [prolegomena]* (1957) and reprised in the subsequent texts.[11]

immersed in the gesso or the kaolin on the frame, Manzoni indulges the physical tendency of the support in its process of drying and consolidation. The folds obtained through this process become visual and plastic elements that are only partially casual. Visually speaking, the observer's gaze can free itself on the achromatic surface obtained through a carefully calibrated process that was quite innovative for the time: the achromatism embodies the key concepts of "the discovery of a zone of images" and "free dimensions" expressed by Manzoni in his theoretical writings.[12,13]

The reduction of the language to an empty primary field to be completed and the reduction of the visual sign to monads or minimal traces (in the case of the *Alphabets*, for example, or the *Lines* series) reflects a whole series of related international situations of which Manzoni showed he was an aware and attentive critic (Figures 2.3 and 2.4).

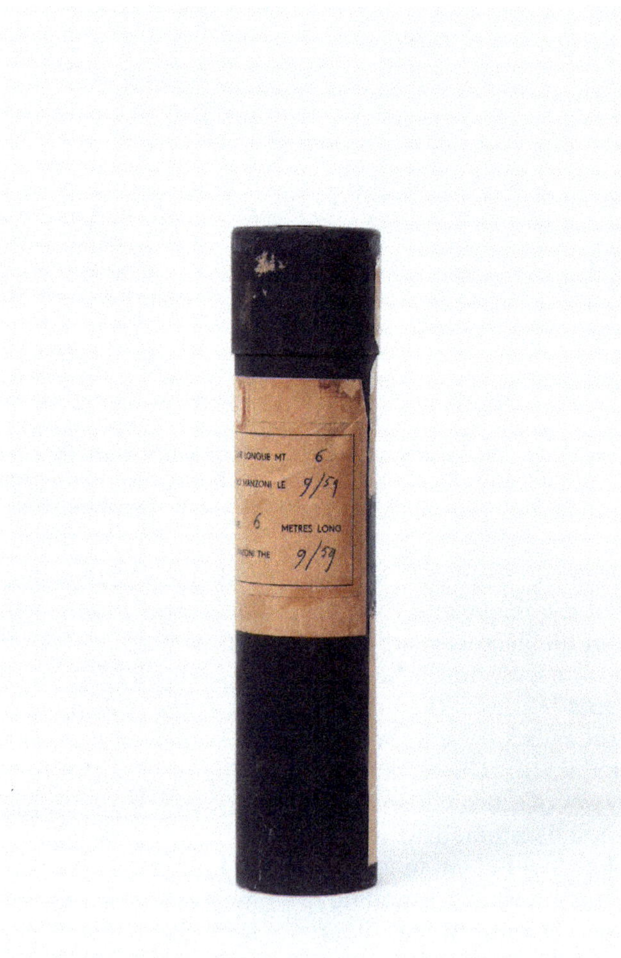

Figure 2.3 *Linea lunga m 6*, September 1959, ink on paper, cardboard tube, 25.5 cm × ø 6 cm. Image © Fondazione Piero Manzoni, Milan. Reproduced with permission from Elena Manzoni, on behalf of The Estate of Piero Manzoni.

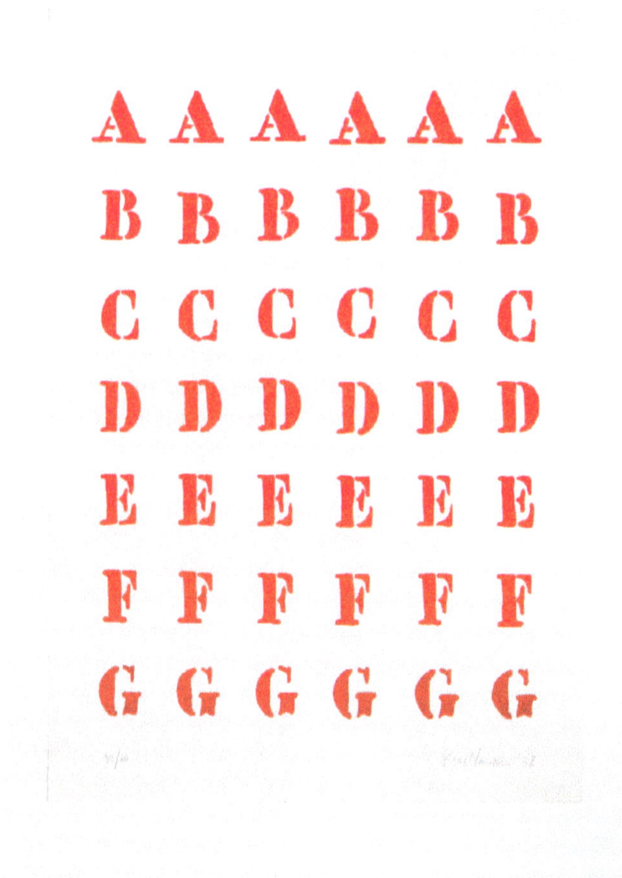

Figure 2.4 *Alfabeto*, 1958, photolithography on paper, 50 × 35 cm^2 (from 8 *Tavole di Accertamento*, 1962). Image © Fondazione Piero Manzoni, Milan. Reproduced with permission from Elena Manzoni, on behalf of The Estate of Piero Manzoni.

First and foremost was his attention to Zen philosophy, new interpreters of which were emerging not only within the Japanese *avant-garde*, such as the Gutai group, but also in Europe and the United States, with an abstract bridge linking Fluxus to New Dada.[5,14,15] Then there was his affinity with the trends towards the annulment of language present in Northern Europe,[16] but also the legacy of Marcel Duchamp, the impact of which was beginning to be evaluated in those years.[14] As well as all this, Manzoni's *Achromes* established dialogues at various levels with the diverse and contemporary experimentation of Lucio Fontana, Enrico Baj, Alberto Burri, Jean Tinguely and Yves Klein.

Having verified the possibility of expressing his philosophical principles initially through the procedure of canvases soaked in gesso or kaolin, as Manzoni saw it the technical and phenomenological potential of the

Achromes became infinite: the canvases could be portions of machine-sewn fabric or single pieces of wrinkled canvas, the kaolin and the gesso may cover bread rolls and fishing hooks, the canvas may be replaced with synthetic (polystyrene, fibres) or natural (stones, straw) materials (Figures 2.5–2.7).

Naturally, the artist's biography influenced both the aesthetic and procedural aspects of the *Achromes*. From the point of view of his philosophical-intellectual research, Manzoni's studies and readings[6,17,18] without doubt had a primary influence, but so did his frequentation of a certain artistic milieu.[19,20] From the procedural point of view, apart from the fundamental comparison with the most innovative solutions to the use of materials and techniques proposed by the aforementioned international *avant-garde* movements, mention should also be made of Manzoni's frequentation of very specific environments such as Albisola: a small town on the western coast of Liguria, a few kilometres from Genoa and close to Turin and Milan, the centuries-old and renowned ceramics industry which had in part been subverted by the Futurist revolution of the early decades of the century. From

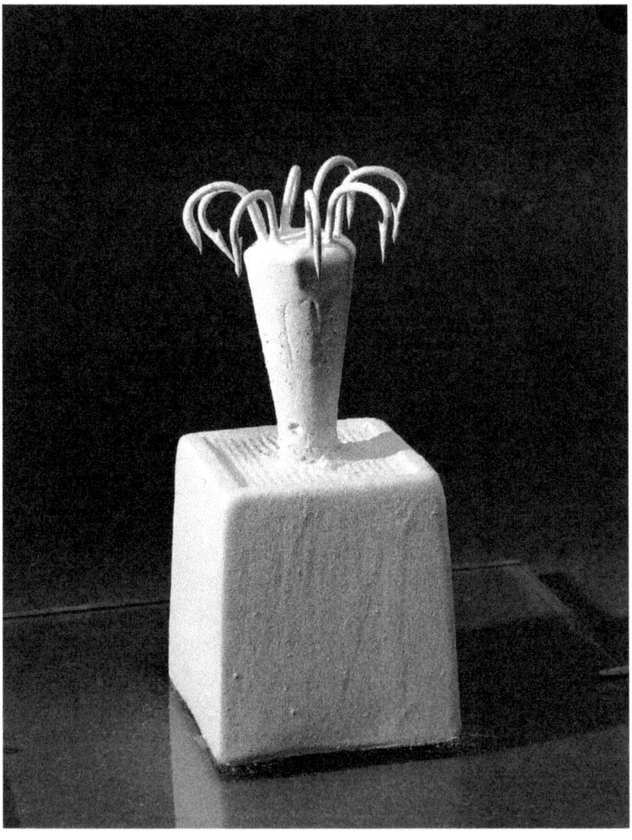

Figure 2.5 *Achrome*, *ca.* 1960, wood base, fishing hooks and kaolin, $4 \times 4 \times 10$ cm^3. Image © Fondazione Piero Manzoni, Milan. Reproduced with permission from Elena Manzoni, on behalf of The Estate of Piero Manzoni.

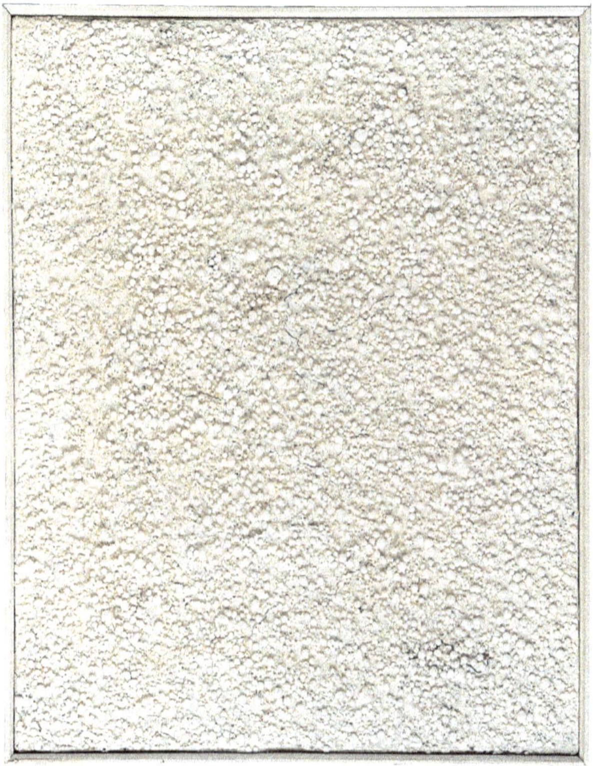

Figure 2.6 *Achrome*, 1962–1963, polystyrene balls and kaolin, 45 × 35 cm^2. Image © Fondazione Piero Manzoni, Milan. Reproduced with permission from Elena Manzoni, on behalf of The Estate of Piero Manzoni.

the mid-1920s, in fact, the Futurists had established a prolific collaboration between craftsmen and outside artists at a number of ceramic manufacturers at Albisola, with many being open to experimentation.[21,22] In the second half of the 1950s, Manzoni (who inhabited the small Mediterranean town with the confidence of a boy who had grown up there, spending annual summer holidays in the area[6,23]) found himself continuing his research in a context rich in cultural stimuli, artists, dealers and writers and on a par with Milan, a city that had intimate ties to Albisola (Figure 2.8).

Such was the pervasiveness of the craft workshops and ceramic factories, Manzoni had the opportunity to frequent the spaces and meet the workers despite the fact that he had no interest in the working of ceramics in a plastic, sculptural, decorative or pictorial sense. In particular, Manzoni established relationships with one of the leading firms, the historic Mazzotti Giuseppe Albisola (MGA), where under the influence of Tullio d'Albisola (1899–1971) the second-generation Futurists of the 1920s–1930s had taken root. From 1936, Lucio Fontana was a virtually constant presence at MGA, as were Aligi Sassu, Agenore Fabbri, Franco Garelli, Emilio Scanavino and, from 1954, Asger Jorn and Enrico Baj (to mention but a

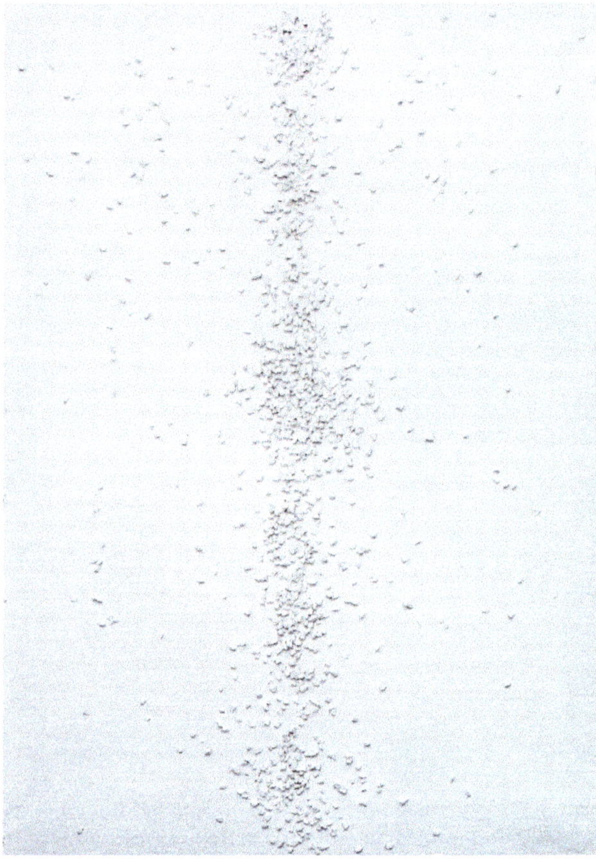

Figure 2.7 *Achrome, ca.* 1962, pebbles and kaolin, 50 × 70 cm^2. Image © Fondazione Piero Manzoni, Milan. Reproduced with permission from Elena Manzoni, on behalf of The Estate of Piero Manzoni.

few). Although an initial contact between Manzoni and MGA may have come through Fontana, the relationships with the family and the Mazzotti firm are documented from 1957, when the artist established a friendship with Tullio d'Albisola's granddaughter, of his own age, Teresa Baldantoni Mazzotti, known as Esa (Albisola Superiore 1932–Albissola Marina 2005), and her husband, Rinaldo Rossello (Savona 1925–1971). In 1959, MGA was divided into two separate companies:[§] Tullio d'Albisola, together with his sister Vittoria Mazzotti and her husband, Giuseppe Mariano Baldantoni (an expert in the on-glaze third firing technique beloved of Fontana among others), created Vittoria Mazzotti Albisola (VMA). It was within the context of VMA that Esa (a ceramist and jewellery designer) and Rinaldo (a building contractor) consolidated their roles as representatives and mediators for the new generations of the *avant-garde*, above all in Milan, a city they

[§]In 1959 the three offspring of Giuseppe Bausin Mazzotti (founder of MGA), Torido, Tullio and Vittoria, founded respectively Giuseppe Mazzotti 1903 (under the leadership of Torido) and Vittoria Mazzotti (under the leadership of Tullio and Vittoria).

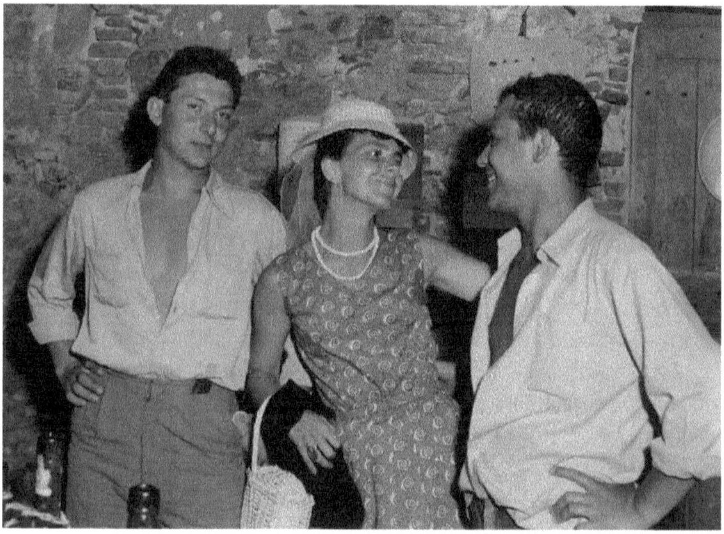

Figure 2.8 Angelo Verga and Piero Manzoni in front of Fontana's studio at Pozzo Garitta, Albissola Marina, summer 1957. Image © Fondazione Piero Manzoni, Milan. Reproduced with permission from Elena Manzoni, on behalf of The Estate of Piero Manzoni.

frequented with a certain constancy. It was in fact thanks to this couple that an exponent of this generation such as Piero Manzoni could progress from being an occasional visitor to the factory (between 1956 and 1957) to being a habitué, albeit without being influenced by the appeal of the sculptural "trade", as had been the case firstly for the Futurists and then for Fontana and Jorn. Despite the lack of interest Manzoni demonstrated in ceramics, he established a friendship with Esa and Rinaldo's family and the older Tullio d'Albisola that led him to take a further step in the development of the *Achrome*. When visiting the factory workshops, Manzoni in fact witnessed the daily ritual of the immersion of the biscuit ware (a term indicating the clay pieces after the first firing at around 900°) in the bath of white majolica glaze (a liquid, tin- or lead-based glaze). He will therefore have seen, set on long trolleys, dozens of white pieces left to dry before being hand-painted by the decorators with diverse traditional and modern ornamental motifs. When it dries and before being fired again, the majolica takes on a flat, opaque white appearance and the surface seems almost dusty. As an alternative to the majolica glaze, the ceramic industry frequently covers the biscuit ware with a slip, a layer of white or coloured clay having a covering function in preparation for further painted decoration. When the slip is white it is based on kaolin, diluted in water.

Manzoni might well have noticed – but here we are in the realm of speculation – the achromatic purity of the ware prepared with a majolica glaze or slip. What he claims for his *Achrome* works ("to free the surface of the picture", preventing the infinite possibilities of images being blocked by painting and colour), finds perfect correspondence in the production of majolica pottery; just as in the pictures the inert and colourless layers of gesso, kaolin and glue evidently refer to the traditional priming of the canvas ready to be painted, in the pottery the glaze primes the biscuit ware to receive its surface decoration. What do kaolin and gesso on the one hand and majolica on the other have in common? The both underlie painting, are covered by it and above all are achromatic white and therefore free and open to every possibility of image, at least until they are covered.

Further works came out of the collaboration between VMA and Manzoni, with the development between 1960 and 1961 of the *Achrome* concept and more. Among the sculptures realized on the basis of the ready-made process was the one composed of a squid hook fixed to a small base and covered with a layer of white, achromatic matter (which could in fact be a layer of slip). The hook was given to Manzoni by Rinaldo Rossello (who actually used it for fishing). The fact that Manzoni also realized a number of "sewn canvases" using Esa's sewing machine shows the extent to which he was at ease with the Mazzotti–Baldantoni–Rossello family, feeling free to experiment with new forms of *Achromes*. Manzoni produced the majority of the successive works in this series in his studio in Milan; however, his friendship with Rinaldo Rossello and contacts with Vittoria and Tullio's factory were to continue, as shown by the only ceramic work ever made by Manzoni: a majolica plate with a horizontal black line made at VMA in 1961, at the insistence of Tullio d'Albisola himself (Figure 2.9).

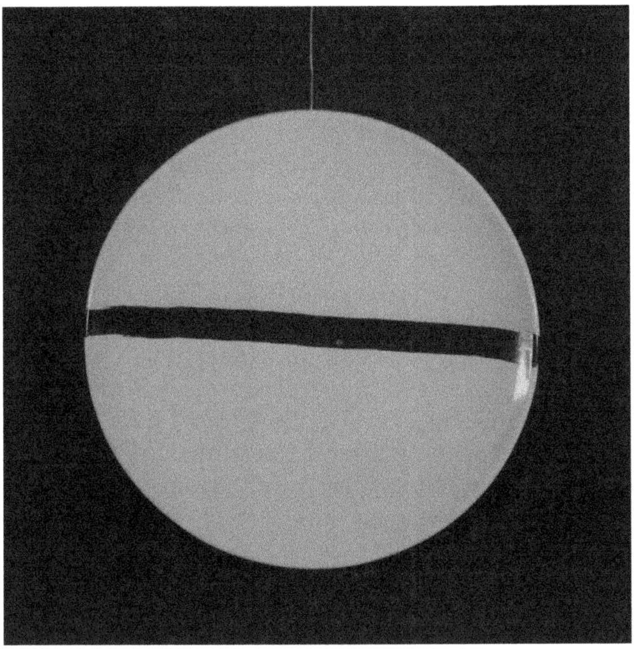

Figure 2.9 *Linea*, 1961, majolica plate and colour for majolica, ø 40 cm. Image © Fondazione Piero Manzoni, Milan. Reproduced with permission from Elena Manzoni, on behalf of The Estate of Piero Manzoni.

2.2 *Achrome*: White Is Not a Colour

It is well known that most artists do not like to tell about the artistic media or techniques employed in the making of their works of art, because of reserve, reluctance or lack of interest in these specific details. In fact, the technical aspect and the materials used remain for many artists the means (the medium) to an end – the work of art – and not a subject worthy of extensive debate.

Although Piero Manzoni will not dwell in his writings and letters on the technical details, he affirms with authority to have used different materials in different works; materials that are by themselves carriers of aesthetic qualities and therefore of an artistic message.

His own education plays a role in explaining the thought and procedure of the artist: he never attended institutional art courses – neither historical nor practical – apart from drawing and watercolour lessons taken privately and the two weeks during which he frequented the life drawing class at the Brera Academy of Fine Arts.[24] His educational training is unusual, enrolling first on a law course and turning in a second stage to philosophy, though even this was suspended after a short time.

He consequently demonstrated, on the one hand, an uncommon freedom compared with other contemporary artists in the choice and use of materials and, on the other, an intolerance for using them at length, in the continuous, incessant and immense attempt to comprehend – in the dual sense of the term, to investigate and to embrace – the entire universe. Further, the fact that he had only attended theoretical courses did not encourage him to refine techniques or practical procedures, but rather to seek in nature and trade for all that was needed to make explicit his idea.[1]

Manzoni has been among the first artists to give a conceptual meaning to the materials of a work of art. They are expressly chosen to render explicit and phenomenal the concept of *achrome* and put it into practice: *achrome* is, in fact, an abstract and non-commensurable concept, defined negatively (alpha privative + chrome), which having been purged of all the subjective aspects has claims to universality.

While many artists with academic backgrounds refined their techniques and procedures, Piero Manzoni tested an abstract concept through materials, for which any compound or product could be adapted to the purpose. Therefore, there are a great number, potentially infinite, of "sensitive objects that occupy a spatial entity"[25] which he uses to create art, always – programmatically – without colour (Figures 2.10 and 2.11).

In 1963 he wrote in the magazine *Evoluzione delle lettere e delle arti*:

> My first Achromes were born in '57: in canvas soaked in kaolin and glue … from '59 the raster of the "achrome" was sewn by machine.

Figure 2.10 *Achrome*, 1957–1958, canvas and gypsum, 130 × 97 cm^2, Rira Collection, Köln. Image © Fondazione Piero Manzoni, Milan. Reproduced with permission from Elena Manzoni, on behalf of The Estate of Piero Manzoni.

In '60 I made others in cotton wool, in expanded polystyrene, I experimented with phosphorescent and others soaked in cobalt chloride which change colour with variations in the weather. In '61 I continued with others in straw and plastic and with a series of pictures, always white, in cotton wool balls and then hairy ones, with clouds, in natural or artificial fibres.

I also made a sculpture in rabbit fur ...[26]

It should be pointed out how materials and techniques used in his works are mentioned as expressions of his art, as well as how everything that surrounds us, belonging to the real world, can acquire the status of a work of art.

We are thus able to explain the egg, (Figure 2.12) the breath, the shit, the polystyrene, the polyamide fibres, but also the kaolin, the gesso, the

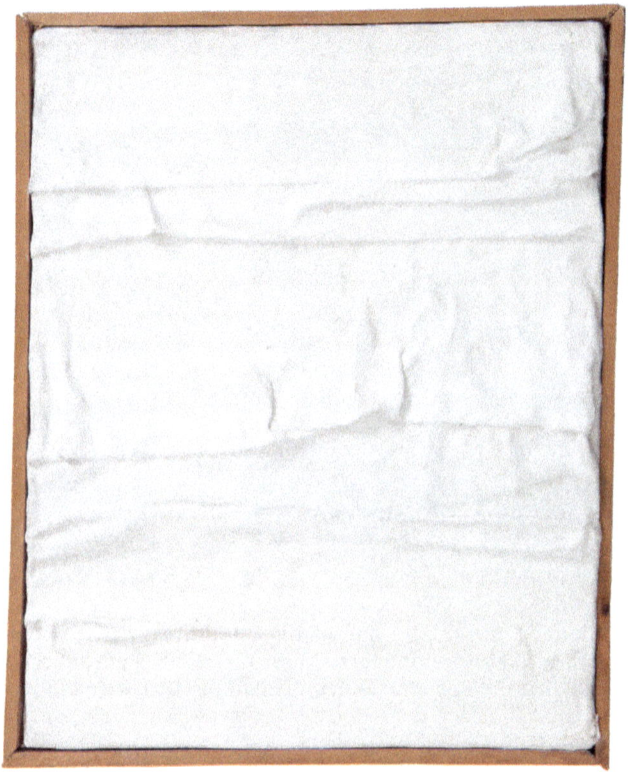

Figure 2.11 *Achrome*, 1958, creased canvas and kaolin, 30 × 24.5 cm^2. Image © Fondazione Piero Manzoni, Milan. Reproduced with permission from Elena Manzoni, on behalf of The Estate of Piero Manzoni.

household gloss paint, the tar and then the model, the world, hoisted onto its base and measured in its circumference, and all those who, seriously or as a joke, have climbed onto the Magic Base or have been gratified with a certificate of authenticity.

In relatively recent years, some diagnostic campaigns have been conducted on a number of works by Manzoni, generally with the aim of identifying the presence of areas of overpainting and their scale, but a truly systematic technical-scientific research program on execution techniques has yet to be established.

A number of reasons explain why these studies are still at an early stage.

First and foremost, as is well-known, the artist died before he was 30 years old and his output, while widely admired by his peers and still today of seminal importance to the younger generations, is much more restricted than that of other artists who lived and worked longer.

Second, this relatively small number of works of art is distributed in collections throughout the world, a fact that has strongly reduced the opportunities of conducting extensive diagnostic campaigns through the years.

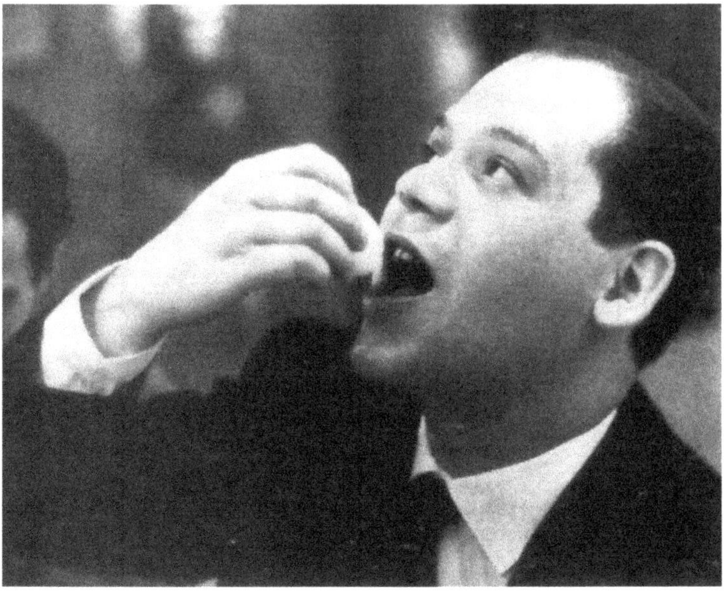

Figure 2.12 Piero Manzoni eats one of his sculptures, *Egg*, Copenhagen, 1960. Image © Fondazione Piero Manzoni, Milan. Reproduced with permission from Elena Manzoni, on behalf of The Estate of Piero Manzoni.

Further, many works have a layer of white impasto only on the front of the canvas with no possibility of taking samples with a significant information content on the tacking margins or on the lateral edges. This implies a mandatory use of non-invasive techniques, even if not all cases should give exhaustive information.[27]

What we will intend to carry out is a series of investigations concerning Manzoni's widest production throughout his career, exploiting the most advanced non-invasive techniques, starting with the series of *Achromes* that represent in a rather wide way the first phase of his production.

In fact, *Achrome* is the title of two thirds of the works included in the general catalogue of the artist. These works, despite having the same title, were created with diverse techniques and materials. The current survey is aimed at around 350 examples of '*Achromes*', where the white impasto covers the creased, sewn or square canvases, and the waved, so called, lasagne (Figure 2.13).[5]

Within the general context of understanding which materials and techniques have been used by Manzoni in the *Achromes* executed between 1957 and 1958, we mention some results obtained recently in a preliminary examination of the work corresponding to number 433 in the latest General Catalogue edited by Germano Celant and published in 2004.

While this *Achrome* is a typical example of Manzoni's initial production, it does feature certain differences in terms of the aesthetic result. It presents a monochromatic white surface with a band of horizontal creases in the centre (Figure 2.14).

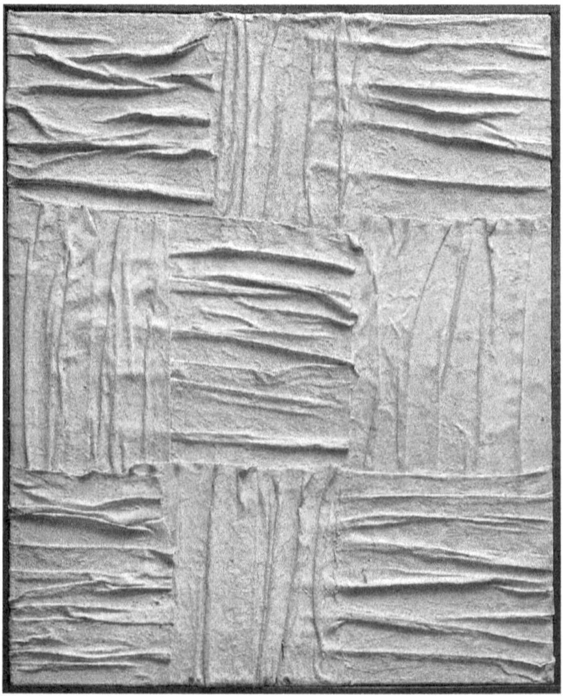

Figure 2.13 *Achrome, ca.* 1959, canvas in creased square and kaolin, 73 × 61 cm^2. Image © Fondazione Piero Manzoni, Milan. Reproduced with permission from Elena Manzoni, on behalf of The Estate of Piero Manzoni.

The work was executed by stretching a burlap canvas over a wooden strainer; it is composed of four members and a horizontal crossbar. A second canvas with a series of creases in the centre was attached to the stretcher over the first; the creased canvas was covered with a layer of smooth, levelled white impasto with no brushstrokes.

The burlap canvas, a coarse fabric with a loose, bristly weave, is functional as a secondary support as, among the natural textiles, it is the most impermeable thanks to its high lignin content. These characteristics are useful in the work of this artist, given that the scarce elasticity of the fabric together with the aforementioned strength and impermeability allow it to support the weight of the other canvas, which shall hitherto be indicated as the primary support.[28]

This primary support is a cotton canvas, the natural yarn of which is obtained from the seeds of the plant of the same name; the fibre obtained is around 98% pure cellulose. The fabric is elastic, light in colour and highly absorbent: it is useful to the artist for "modelling" the creases and is impregnated with white impasto.

A feature of Piero Manzoni's practice that is very evident in this work is the absence of brushstrokes. The work was not in fact painted, with paint

Figure 2.14 *Achrome*, *ca.* 1958, creased canvas and gypsum, 55 × 45 cm^2. Image © Fondazione Piero Manzoni, Milan. Reproduced with permission from Elena Manzoni, on behalf of The Estate of Piero Manzoni.

being laid in layers with a brush, but was created by pouring a whitish liquid onto the picture which was set horizontally and then rocked to spread the impasto evenly. This procedure clearly reminds us of the practice of pottery making, notably the step called *engobe*, which is a clay slip coating applied to the "leather-hard" partially dried impasto and which remains opaque even after firing. It may be covered or not with glaze, the transparent or coloured layer found on many ceramic manufactures.[29]

This execution procedure has become more evident over time due to the ageing of the work, as the traces of water and protein-based binder, already visible in older photographs, have become darker. Further evidence of this technique is provided by the fact that the traces do not run in a single direction, but rather in straight lines in all four directions (forward, backward, right, left).

From a visual examination of the work and of its ageing it is also very probable that the artist added a notable quantity of binder to make the colloidal impasto less liquid and more workable and viscous. As testified by the images found in the Fondazione archive, a drying crack immediately formed along the central crease, caused by the shrinking of the protein binder during the evaporation of the aqueous diluent.

Another feature that characterizes all these *Achromes* by Manzoni that we might define as material and not pictorial is that the canvas, while present as a support of creases and undulations, is never exposed: the weave is never left in evidence. The subtle weave of the canvas was instead in evidence in the series of *Achromes*, not covered with white impasto, just showing lines of stitching on the cotton fabric and hence becoming known as "sewn canvases" (Figure 2.15).

The preliminary investigation carried out on a small quantity of material collected by scraping the right tacking margin towards the lower right corner of *Achrome* (catalogue number 433) has indicated that the impasto, at least in this region, is mainly composed by a protein binder (most likely animal glue) and gypsum.

Figure 2.15 *Achrome*, 1959, canvas sewn in squares, 24 × 18 cm^2. Image © Fondazione Piero Manzoni, Milan. Reproduced with permission from Elena Manzoni, on behalf of The Estate of Piero Manzoni.

The analyses entailed the execution of a series of preliminary micro-chemical tests, followed by measurements with two instrumental analytical techniques applied in the lab on the scratched powder. The micro-analytical tests were designed to identify proteinaceous substances (typical of animal glues, casein and egg tempera), saponifiable compounds (attributable to the use of drying oils, certain waxes and lipids in general) and natural resins. Fourier Transform Infrared Spectroscopy (FT-IR) was then exploited to better characterize organic and inorganic components.

All together, the measurements revealed the presence of proteins and of a large quantity of *calcium sulphate* (the gypsum), together with a minor quantity of other compounds as calcium carbonate and silica.[30] These results have, however, to be considered only partial, since they regard only material taken off in a very marginal part of the work. To achieve more accurate results, a careful and extended study *via* non-invasive spectroscopic measurements is needed, especially focused on the organic and inorganic components of the impasto and preparation, which characterize the whole frontal area of the work. These measurements have already been planned and will be carried out in the near future.

2.3 Conclusion

It is now accepted that Manzoni, through the absence of colour in his work, was seeking an absolute.

In order to limit the artist's emotiveness and achieve the absence of subjectivity, he stopped his work a layer before the subjective layer *par excellence*, that of the painting.

He stopped at the preparatory layer, presenting all the possibilities with a variety of materials, flat, light, covering, opaque and absorbent; even when they have character, the preparatory layers never dominate the "paint" above them, rather they accompany it, that is to say, they are neutral.

Regarding the execution technique, in his writings Manzoni lists the materials he has used in such an assertive and clear way that nobody questioned what he has affirmed.

Traditionally, the *Achromes* are considered works that feature a white impasto in one of the tones which are catalogued to correspond to gesso and glue or kaolin and Vinavil® (the Italian trade name for polyvinyl acetate, also called *white glue*). So, from this point of view, the *Achrome* number 433 examined in this work would belong to the first group.

However, only very few studies, mainly on the occasions of maintenances or restorations, have been carried out to date on the materials effectively present in the works of Manzoni, with results that have not yet led to satisfactory answers on a series of questions which still remain open. These questions regard: the intimate nature and composition of the various substances or mixtures used by the artist (starting from the *Achromes*); the modality of their application, including the use or not of pigments apart from kaolin or gesso; the reliability of the convention in which the gesso–glue or kaolin–Vinavil® pairings are adequate to define and comprehend these works; the modality of development of techniques through the years of the artist's short career.

Such responses may only be provided by as extensive an examination as possible of groups of works, through the synergy of diverse instrumental analyses, following the mandatory protocol for the study of works of art, that implies first the exploitation of all the possible non-invasive techniques and then, only in a second instance, and whenever possible to take a micro-sample, an accurate examination to solve the problems still open, especially those related to stratigraphies.

Acknowledgements

The authors would like to thank Stefano Volpin, Gallerie dell'Accademia, Venice and Marco Realini and Antonio Sansonetti of CNR (Istituto per la Conservazione e la Valorizzazione dei Beni Culturali), Milano Bicocca, for preliminary analyses.

References

1. L. Mensi, Naturale, artificiale, sintetico: l'universo dei materiali di Piero Manzoni, in *Piero Manzoni. Nuovi Studi*, ed. R. Pasqualino di Marineo, Carlo Cambi Editore, Poggibonsi, 2017.
2. L. Mensi, Without Color: The Matter of the Achromes, in *Piero Manzoni. Materials of His Time*, ed. R. Pasqualino di Marineo, Hauser & Wirth Publisher, Zurich, 2019, pp. 40–48.
3. P. Manzoni, I miei primi "Achromes" sono del '57... [alcune realizzazioni – alcuni esperimenti – alcuni progetti] (1963, written in 1962), in *Piero Manzoni. Scritti Sull'arte*, ed. G. L. Marcone, Abscondita, Milan, 2013, pp. 51–52; 86–88.
4. G. Celant, *Piero Manzoni. Catalogo Generale*, Skira, Milan, 2004, vol. 1, p. 53.
5. G. Celant, *Piero Manzoni. Catalogo Generale*, Skira, Milan, 2004.
6. F. Gualdoni, *Piero Manzoni. Vita d'artista*, Johan & Levi, Milan, 2013.
7. *Piero Manzoni. Achrome*, ed. C. Kazarian and C. Lévêque-Claudet, Musée Cantonal des Beaux-Arts, Lausanne, 2016.
8. *Piero Manzoni. Achromes: Linea Infinita*, ed. G. L. Marcone, Carlo Cambi Editore, Poggibonsi, 2016.
9. *Piero Manzoni. Scritti Sull'arte*, ed. G. L. Marcone, Abscondita, Milan, 2013.
10. T. Sauvage, *Arte Nucleare*, Galleria Schwarz, Milan, 1962.
11. *Piero Manzoni. Scritti Sull'arte*, ed. G. L. Marcone, Abscondita, Milan, 2013, pp. 22–24.

12. P. Manzoni, Per la scoperta di una zona di immagini (1958), in *Piero Manzoni. Scritti Sull'arte*, ed. G. L. Marcone, Abscondita, Milan, 2013, pp. 25–27.
13. P. Manzoni, Libera dimensione (1960), in *Piero Manzoni. Scritti Sull'arte*, ed. G. L. Marcone, Abscondita, Milan, 2013, pp. 34–38.
14. L. Bochicchio, The Line comes first: Sign and Myth in Piero Manzoni's Line, in *Piero Manzoni. Lines*, ed. R. Pasqualino di Marineo, Hauser & Wirth Publisher, Zurich, 2019, pp. 16–29.
15. G. Celant, *Su Piero Manzoni*, Abscondita, Milan, 2014.
16. F. Pola, *Piero Manzoni e ZERO. Una regione creativa europea*, Electa, Milano, 2014.
17. E. Grazioli, *Piero Manzoni*, Bollati e Boringheri, Turin, 2007.
18. G. L. Marcone, *Piero Manzoni. Diario*, Electa, Milan, 2013.
19. R. Perna, *Piero Manzoni e Roma*, Electa, Milan, 2017.
20. *Azimuth. Continuità e nuovo*, ed. L. M. Barbero, Marsilio Editore, Venice, 2014.
21. M. Fochessati, Ceramics, in *Handbook of International Futurism*, ed. G. Berghaus, Walter de Gruyter, Berlin, Boston, 2018, pp. 88–100.
22. E. Crispolti, *La ceramica futurista da Balla a Tullio d'Albisola*, Centro Di, Florence, 1982.
23. F. Pola, *Una visione internazionale. Piero Manzoni e Albisola*, Electa, Milan, 2013.
24. F. Gualdoni, *Piero Manzoni. Vita d'artista*, Johan & Levi, Milan, 2013, pp. 27, 30, 45, 218.
25. Material (and ff), in *Dizionario Enciclopedico Italiano*, ed. G. Treccani, Istituto dell'Enciclopedia Italiana, Rome, 1955.
26. P. Manzoni, Alcune realizzazioni – Alcuni esperimenti – Alcuni progetti (1962), in *Piero Manzoni. Scritti Sull'arte*, ed. G. L. Marcone, Abscondita, Milan, 2013.
27. B. G. Brunetti, C. Miliani, F. Rosi, B. Doherty, L. Monico, A. Romani and A. Sgamellotti, Non-invasive investigations of paintings by portable instrumentation: the MOLAB experience, *Top. Curr. Chem.*, 2016, **374**, 41–76.
28. S. Lorusso and L. Gallotti, *Caratterizzazione, Tecnologia e Conservazione dei Manufatti Tessili*, Pitagora, Bologna, 2007, pp. 97, 112.
29. L. Campanella, *et al.*, *Chimica per l'arte*, Zanichelli, Bologna, 2007, p. 254.
30. S. Volpin, *Scientific Analyses of Two Micro-samples of Material Taken From a Work by Piero Manzoni*, January 2019, unpublished report.

CHAPTER 3

1949–1968 Concetti spaziali *by Lucio Fontana: A Historical-artistic and Technical Study*

BARBARA FERRIANI*[a], LUCA MASSIMO BARBERO[b] AND FRANCESCA CATERINA IZZO[c]

[a]Barbara Ferriani srl, Via Vettabbia 1, 20122, Milan, Italy; [b]Associate Curator Peggy Guggenheim Collection Venice, Italy; [c]Cà Foscari University of Venice, Department of Environmental Science, Informatics and Statistics, Via Torino 155/b, 30174, Venezia Mestre, Italy
*E-mail: info@barbaraferrianisrl.com

3.1 Introduction

This study deals with Lucio Fontana's artworks, *Concetti spaziali*, produced between 1949 and 1968, and reports the results of a larger research project on his technique carried out in collaboration with the Fondazione Fontana by international, multidisciplinary work groups in the last decade,[1–6] with a special focus on a selection of artworks from the series *Olii* (Oils) series.

Fontana's activity was characterized by constant experimental research within a context formally spanning figurative and abstract expression. Although Fontana's versatility and propensity for innovative techniques and materials is well known – from the use of concrete and metal in the 1930s to the application of new media such as light, neon, "black light",

Science and Art: The Contemporary Painted Surface
Edited by Antonio Sgamellotti, Brunetto Giovanni Brunetti and Costanza Miliani

Published by the Royal Society of Chemistry, www.rsc.org

and fluorescent colours in the 1940s and 1950s – the painting medium itself became an experimental field for the artist. Throughout his very long career, Lucio Fontana used many different synthetic paints that were entering the market at that time – nitrocellulose, alkyd colour, PVA (polyvinyl acetate)-based house paint (called *Idropittura*), and PVAc–acrylic mixtures – without ever neglecting traditional oil paint. As he mixed these materials in different layers, and also together, to create complex works of art, the identification of binding media and additives was essential to an understanding of painting techniques and their ageing, as both of these aspects can play a role in the planning of preventive and sustainable conservation procedures.

3.2 Lucio Fontana Upon his Return From Argentina at the End of the 1940s

Fontana returned to Italy from Argentina in 1947. His sojourn in Argentina, from 1940 to 1946, was a period of "stock-taking", of distancing himself from the creative European vitality which he had experienced to the full as a leading player in the years preceding his departure.

Italy was going through a very particular period after the terrible drama of World War II, and there was an urgent need for reconstruction and for the recovery of social life and of the urban context, as well as for a new life, of which Milan appeared to be a leading and active example. Right from the outset, this was a period of new projects for Fontana, with a wealth of activities and thoughts, characterized by his constant syncretism of highly inventive figurative research and a constant, more purely abstract approach.

He brought with him to Milan the *Manifesto Blanco* (White Manifesto), which had inspired the art pupils and fellow artists in Argentina when he was teaching at the Altamira Academy. As Crispolti writes, "[...] the ideas of the *Manifesto Blanco*, inspired by Fontana, did not come about from his sculpture of the first half of the 1940s, but rather from his experience of the highly advanced, pure research that he carried out in the 1930s."[7] From the beginning of the 1930s, Fontana resolutely embarked on anticipation of the use of signs and materials that would not be practiced until twenty years later by the European *Informel*. If Fontana's non-figurative sculpture in reinforced concrete and sheet metal established a new dialectic of spatial graphic signs, the synthetically figurative ceramic sculptures developed a further mode of expansive spatial connection through a plastic vitalism exalted in the play of possible luministic dissolves, thanks to the bright expansive colouring of the ceramics.

However, he brought some fundamental drawings dated to 1946 that constitute the first indications of the new spatial research conducted by the artist, which anticipate the *Concetti spaziali* (Spatial Concepts) cycle. Drawing proved to be fundamental in Fontana's work. Indeed, it identifies the sign with the birth of the idea and of its "functionality", or rather with the central role it plays in creative development. The sign is identified with the action on and in the material to be sculpted, and painted. It is this that Enrico Crispolti points to when he writes that, from the outset, for Fontana "[...] the sculptural and painted image [...] is primarily one of drawing in its structural form. It is image as drawing, and in a certain sense one might say it is basically a drawn image. It can then be given concrete form by carving clay, creating outlines with figures or pure signs or by engraving metal, or even, as a painter, by lacerating or slashing the canvas. It is by affixing signs in this drawing operation that in Fontana's view a brutally primeval, telluric material [...] acquires the power of meaning from the presence of a conceptual vision of the image."[8] It was through his use of paper and the sign (in other words, his personal way of "conducting the sign" with regard to material) that Fontana achieved what was to become one of the features of his process, indeed something that made its mark, or "sign", on every future approach to contemporary art: the hole.

The holes not only represented space, but also recalled it in its fundamentally infinite dimension. They are fully part of an original investigation of a new idea of space and of a profound identification of the cosmos as a possible new subject, quite probably as unknown as it is "abstract". As Fontana wrote: "The true conquest of space made by man is the detachment from Earth, from the horizontal line that was for millenniums the basis of its aesthetics and proportion. Having created the 4th dimension, volume is now contained in space in all its dimensions."[9] The holes of the first series of the Spatial Concepts cycle were originally thought of as futuristic "screens" through which space dynamically marks its own passage and its emergence as an effective sign in time. These extraordinarily concentrated and succinct works, which continued until 1952, use the hole in its purest conceptual intention. It is entirely geared towards the creation of the spatial event. This means it is not a sign that is distributed graphically, creating simply allusive bodies and figures, but rather a further spatial element, new and symbolically infinite.

Crispolti again: "[...] in conceptual measure, these physical signs, which open up to an infinite (cosmic) dimension of space, are a sort of plan for this further spatial dimension, a sort of imaginative symbolisation of it, but by no means its "representation". [...] They are a sort of spatial project, for they are exactly the *Concetto spaziale* that Fontana was to use constantly after that, giving it a more precise, intentional meaning in light of the experience of Conceptual Art in the second half of the 1960s and the early 1970s."[10]

3.3 The *Concetto Spaziale* Cycle

3.3.1 The *Buchi* (Holes), *Pietre* (Stones), and *Barocchi* Series

The first works of the *Concetto spaziale* cycle were created by perforating the support, initially on canvas-backed paper and on raw canvases, from both the front and the back, using awls of different shape and sizes. If at the beginning the surfaces are unpainted or monochromatic, with paint layers applied on industrially prepared canvas, later he would gradually introduce areas of colour as well as material additions in the form of sequins, marble dust, and sand. For these works, realized at the beginning of the 1950s, he mainly used oil-based paint, probably linseed oil, as evidenced by analyses carried out on a few works of this cycle (51 B 26),[4] (60 B 5).[11]

In the new cycles, *Pietre* (1952–1958) and *Barocchi* (1954–1957), he added to the holes a carefully designed, entirely invented distribution of coloured pieces of Murano glass (stones), tracing out their position in relation to the presence of the holes. The altered spatial dimension, which is also suggested by the stones, was soon enriched with new materials, including sand, oil spread in dense ribs, and, especially, as the umpteenth perceptive distraction, the shifting, radiant luminosity of sequins that took these works to an almost "baroque" and material dimension. The surfaces of the paintings are enriched by different superimposed layers of paint and the puncturing and holes are at times obscured by the paint, while the materials, fragments, and glass, at times also partially painted, create ulterior spatial effects on the surface.

As revealed by analyses conducted on some works of this series (52 P 5, 52 P 7, 53 P 6),[4] the artist abandoned the traditional oil paint and mainly used alkyd paints, on which in some works he overlaid a new PVA-based paint recently introduced into the market for industrial or household purposes and not yet employed within the artistic milieu. He used it not only in his paint formulation called *Idropittura*, but also in the transparent adhesive formula called "Vinavil®" to embed sand, marble powder, sequins, and stones, as well as transparent painting material, which he sometimes left visible and sometimes covered with other coloured layers. Although the first experiments date back to the 1940s, only at the beginning of the 1950s did the company, Rhodiatoce, founded by Montecatini and Rhône-Poulenc, begin industrial production of emulsified polyvinyl acetate. Thus was born Vinavil®, the famous transparent adhesive still used today, whose name is the acronym for VINyl Acetate and VILladossola, the name of a town near Ravenna where its production was launched. This was a new resin that was no longer emulsified, had innovative characteristics, and could be diluted in water instead of in toxic organic solvents. It can be used as an adhesive but also as a medium for pigments. It has completely innovative properties: it dries very fast and, although dilutable in water, once dry it cannot be dissolved in that same medium. This property makes it possible to place a new layer of paint on top without risking damage to the underlying layers. Its texture, which is already quite fluid, can be varied: it can be diluted in water or thickened with aggregates (56 BA 13),[4] (56 BA 8, 56 BA 12).[11] Once dry, it demonstrates good adhesive properties and considerable elasticity. Contrary to oil and alkyd paints, house paints have no acid ingredients and can be applied directly to the canvas without requiring an underlying preparation layer (56 BA 13).[4] Thus, it is possible to obtain even opaque surfaces devoid of brushstrokes, which allow the texture of the underlying canvas to show through. Thanks to these features, Fontana used them in his new series, *Tagli* (Cuts), which he began at the end of the 1950s (1958–1968).

3.3.2 The *Tagli* (Cuts) Series

In this period, the mature Fontana lost none of his inventiveness but, on the contrary, revealed a wealth of inventive vigour that led him to take all his previous experiences into new experimental forms that gave new life and dynamism to his creativity. This new energy and experimental potential can truly be seen in the research involved in the *Tagli*. This started as early as 1957, when some canvas-lined papers had holes and serried, distributed cuts made with gestural randomness. These became increasingly longer, until they turned into openings and lacerations that were gradually refined using increasingly thin, sharp blades. In early 1958 they turned into swarms of cuts that slashed and sheared the surface, leading towards a purification of what appears to come with the advance of monochromy on the canvas. The monochrome background takes them to a new dimension, with a new role. From the objectifying concept of the monochrome surface and thus from the "body" adopted by the canvas, the *Tagli* increasingly appear as a carefully thought-out and ordered structure, to the point that they can even cut through the surface a number of times along diagonal, tilted, or even slightly arched lines. As Crispolti writes: "[...] we begin to see a decisive characteristic use of monochrome in the new cycle of works, together with a desire to give order to the cut or cuts as primary elementary structures that are exquisitely calibrated".[12]

The new use of monochrome, which recalls the original one of the *Buchi* in the late 1940s, removes the material quality of the informal canvases and becomes an ideal ground also from a conceptual point of view. For these works, the artist used *Idropittura* applied directly to the unprimed side of the canvas, while the industrially primed side, mainly an alkyd-based preparation such as Cementite, was turned to the rear (60 T 85, 64 T 26),[4] (59 T 32, 65 T 139, 67 T 52, 68 T 42, 68 T 92).[11]

He applied one or two layers of colour, which differed slightly in terms of tonality, and the paint overflowed onto the border of the canvas, along the edges, and onto the back of the stretcher. Fontana then executed the cuts. These cuts have clean, un-frayed edges that show the deliberate action of the cutter's blade, and are backed on the rear by strips of black gauze adhered to the support by means of the transparent adhesive Vinavil®. It is interesting to note how some of these works reveal a pictorial composition containing PVAc–acrylic mixtures, and not only the PVAc medium of "house paint". Indeed, the analyses show the presence of methyl methacrylate (MMA) or ethyl acrylate (EA), the most common copolymers in acrylic paint (64 T 149, 64/65 T 79, 65 T 7, 68 T 12),[4] (64 T 15, 65 T 139, 68 T 42, 68 T 143).[11] It is not yet proven that the artist, in order to obtain more or less opaque surfaces, made these mixed formulas himself or used industrially prepared paints, as many manufactures were beginning to add a percentage of acrylic resins to house paint to improve performance. In the initial phase of the elaboration of this series, Fontana used not only rectangular but also shaped canvases, an invention that anticipates the succession of such works in America by at least five years, once again illustrating Fontana's almost pioneering abilities. Indeed, he had already started studying and creating shaped canvases in his works on paper and in the series *Barocchi* (56 BA 13, 56 BA 14) at the beginning of the second half of the 1950s, a period of rare, extraordinarily intense inventiveness. In the space of just a few years, along with the Cuts came the series of *Gessi* (1954–1958), the advent of *Inchiostri* (1956–1959), as well as the *Quanta* (1959–60). The conclusive *Teatrini* series (1964–1966) also forms a part of this natural, ongoing research.

3.3.3 The *Teatrini* (Little Theatres) Series

This series was succinctly linked to the possible symbolic figurative reduction of a new world and of a new nature that we might dare call "synthetic", artificial, and cosmic. While the sign of the *Tagli* extends right through to the final test pieces with an almost neoclassical and architectural solemnity, Fontana himself said that the absolute direction of the *Teatrini* is a form of "Realist Spatialism: new shapes, a projecting frame of often lacquered organic forms surrounds a canvas in the background, often run through by a series of holes arranged in a regular orbital manner." Mario Portalupi wrote about these misunderstood creations: "At first sight, they are synthetic projections of landscape skylines with trees, above which is the sky: the holes appear like flocks of birds in flight [...]."[13] The critics immediately wrote about this "natural/artificial" aspect, for they were aroused by the strong colours and the often shrill, contemporary juxtapositions of colours that were not only far removed from the trend towards "objectification" in the *avant-garde* at the time, but also conceptually close to the very lively experiments in new forms of industrial design being carried out in the early 1960s. For the frames of these *Teatrini*, often characterized by acid colours and huge contrast, Fontana actually used the same industrial paints used for automobiles, nitrocellulose-based lacquer coating, and he commissioned autobody professionals to carry out the painting. The backgrounds, however, were carried out using prepared canvases mounted on stretchers and painted with both oils and mixtures of PVA and acrylic co-polymers (65 TE 51),[4] (64–65 TE 1).[11]

3.3.4 The *Olii* (Oils) Series

In 1960, the artist began a new series of works, classified in the General Catalogue as *Olii* (Oils). And it was again holes and cuts that traversed this series, the coaxingly baroque surfaces of the golds and silvers on the canvases of its *Venezie* (Venice) sub-series, and the shaped canvases of its *Fine di Dio* (The End of God) series (1963–1964), which constituted the highest moment of the *Olii*. In a letter to Jef Verhein dated 1961, Fontana writes: "At the moment I am working a lot with oil, it is a novel, interesting experience, but I need to work much more in order to obtain results."[14]

Oil, the medium *par excellence* used by artists since the fifteenth century and by Fontana from the beginning of his work, became a new research area for the artist. Oil paint was traditionally composed of finely ground pigments mixed with linseed, walnut, and/or poppy seed oil, drying oils that made it possible to create a paint "impasto" that could be worked with over very long periods. Depending on the oils and pigments, oil paints can take from several weeks to several months to become dry to the touch, while the complete process of drying and polymerization can take from years to decades. In this series, however, the artist was exploring new possibilities for this medium. The colour was applied with a thick impasto that had body, using a brush as well as his fingers and spatulas of different sizes that allowed him to shape the paint. The presence of a build-up of colour along the border of the canvases and along the frame indicates that the artist initially placed the canvas horizontally to avoid leakage that might have caused unexpected plastic effects.

Before setting the work vertically on an easel and continuing his work, Fontana waited until the colour began to dry and could be shaped and manipulated. Only then was it possible for him to cut through the material with incisions and scratches, embed stones and hard inorganic matter, and penetrate the material with cuts, holes, and tears. Depending on how dry the colour was, he was able to achieve completely different plastic effects. When the paint was still fluid, the edges of the incisions were sinuous and enveloping, and the painting material moved inside, penetrating to the back of the canvas. Little by little, as the paint began to dry, the material began to be less malleable and became sharper and more uneven. Fontana generally used monochrome paint: dark, earth-like hues, bright greens, or brilliant pinks, morbidly sensual colours, deep and dazzling, or tender as flesh. In some artworks, the texture of the paint is enhanced by gold and silver finishes which were sometimes embedded in the paint itself and sometimes applied in successive layers, thereby making the most of the underlying chromatic layers, as the old masters did. The artifice of metallic colour, its mimetic, symbolic, reflective quality, fascinated Fontana: metal, the way light reflects from and at the same time penetrates it, revealing its plasticity, had always represented a challenge for him. "Gold is as beautiful as the Sun," he wrote on the back of one of his *Olii* (64 O 11). To amplify the effects of light on surfaces, particularly in the *Fine di Dio* series, the artist applied sequins to the surfaces, embedding them in the oil paint or, in a second stage, mixing them into a PVAc resin (63 FD 30). As can be seen by the tubes of paint found in his studio and visible in the historical photographs, he used oil paints manufactured by the most famous brands of the time: *Colori ad olio* by Maimeri™, *Mussini* by Schmincke®, *Rembrandt* by Talens™, *Académie* by Bodson & Nélis, *Viktoria* by Schutzman, and also local, less known manufacturers, such as *Longo* oil paints.

The paint layers of the works belonging to this series are normally compact and well preserved, and seldom present conservation problems. As is normal, over time the pictorial surfaces have begun to show the effects of the drying process and ageing – which generally appear in the form of craquelure – but rarely do we find actual losses in the paint layer, deformation, or its detachment from the canvas. Despite the apparent consistency, however, some works belonging to the above-mentioned series have revealed discrepancies in terms of both the physical-mechanical behaviour of the pictorial film and alterations resulting from ageing and degradation processes.[15]

3.3.5 The *Venezie* Sub-series

One example is the sub-series of the *Olii*, entitled *Venezie*, where, together with the first group of *Metals, New York*, he included a non-abstract place in the subtitle. Fontana's intentions were never illustrative nor informative. He worked, as always, by analogy, and the basis for his images continued to be profoundly conceptual, though with accents specific to each of the cities: one, the island of architectural wonders, Byzantine and Baroque; the other, the metropolis of the future. Fontana carried out the *Venezie* – 22 works, all of whose stretchers measured 1.5 square meters – in just a few months, beginning in the Autumn of 1960 in preparation for the *Arte e Contemplazione* exhibition, which was scheduled to take place at the Palazzo Grassi in Venice in the Summer of 1961 and in New York City at the Martha Jackson Gallery immediately afterwards.

The surfaces of some of these works, made with a thick impasto of viscous and malleable material, showed conformation that was much more rigid than that of the other *Olii*, as well as completely anomalous deformations and detachments. Although an article entitled, "I Concetti spaziali di Lucio Fontana" (Spatial Concepts by Lucio Fontana), which was published in 1972 in the magazine *Materiali plastici ed elastomeri* (Plastic materials and elastomers), reported that "[...] the use of polyester in some works of the 1959–1961 period made it possible for Fontana to create highly pictorial effects of plastic and *chiaroscuro* softness, without neglecting the theme of the slashes [...]",[16] this detail did not appear in the artist's General Catalogue. Instead, as shown by diagnostic investigations carried out on seven works of this cycle, Fontana experimented with polyester resins and/or short-oil styrenated alkyd paints, mostly used in the nautical and building sectors at the time, since he could not afford the time required for the drying process of traditional oils (61 O 39, 61 O 47, 61 O 48, 61 O 53, 61 O 58)[4] (61 O 33, 61 O 43).[11] As a more recent and ongoing study of these works has shown, he used polyester-based paints, in particular alkyd-based paints modified with drying and semi-drying oils (60 O 82, 61 O 30, 61 O 52, 61 O 58).[17]

As is well known, alkyd paints were characterized by fast drying, like almost all colours introduced at the beginning of the 1900s, while still maintaining malleability: they had a natural tendency to level out and could be applied vertically due to their high thixotropy. Indeed, to decrease the drying time, manufacturers sometimes also added other ingredients to oil paints: small percentages of polyester or short-chain oils used in the alkyd formulas, as is shown by analyses conducted on *Longo* oil paint and on the work 63 FD 9.[11] But these new polyester paints, having been formulated predominantly for industrial use, behave differently from alkyd paints used by artists over time. It is interesting to note that Lucio Fontana, as stated in a personal communication by President Nini Laurini of the Fondazione Lucio Fontana and confirmed by analytical results, sometimes added a small percentage of a PVA-based emulsion, likely Vinavil®, to oil paints (60 O 10)[11] to speed up the drying process.

3.4 A Technical Study on Selected *Concetti Spaziali* From the *Olii* Series

Besides the particular physical-mechanical behaviour of the pictorial film and of alterations resulting from the degradation processes of the *Venice* sub-series, several other *Concetti spaziali* artworks from the *Olii* series were very interesting for the peculiar and anomalous degradation phenomena observed.

How can one explain the atypical behaviour of the 1960 "oil" painting, *Concetto spaziale* (60 O 81) (Figures 3.1 and 3.2), in which widespread craquelure on the pictorial surface created "islands" that tended to detach from the fabric support? What could have caused the light stains that appeared

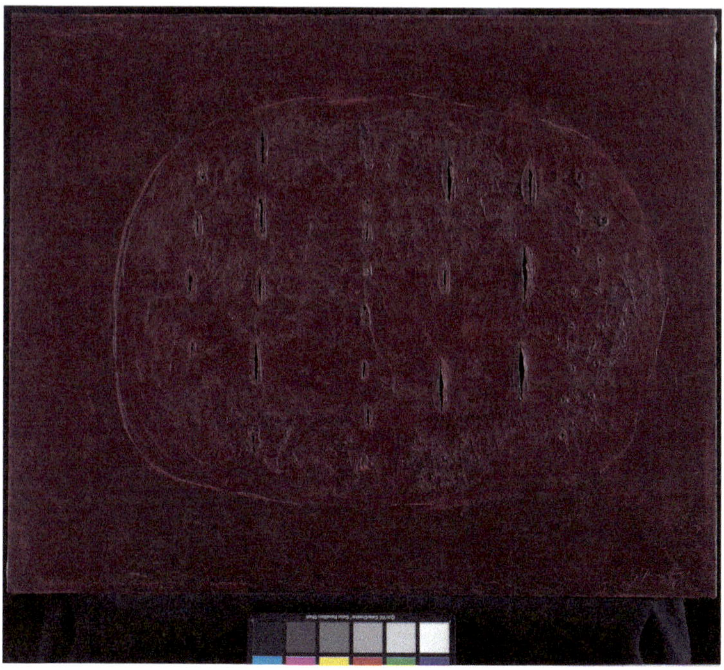

Figure 3.1 Lucio Fontana, *Concetto spaziale*, 60 O 81. Reproduced with permission from Fondazione Lucio Fontana.

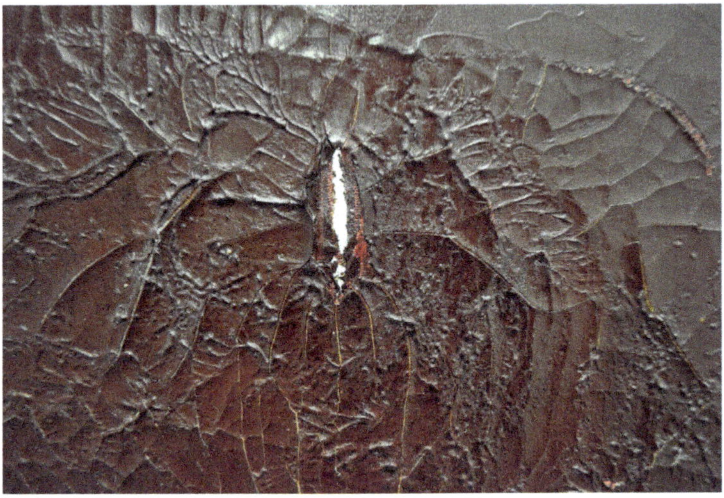

Figure 3.2 Lucio Fontana, *Concetto spaziale*, 60 O 81, detail. Reproduced with permission from Fondazione Lucio Fontana.

as superficial alterations on the pictorial surfaces of the pink *Olii* from this period? And what was the cause of the unusual mechanical behaviour of the punctures in some of the works from the *Fine di Dio* series?

We could have interpreted these anomalies as the results of ageing, humidity, heat, radiation, or previous restoration, but instead we decided to investigate the possibility that these anomalies resulted from the nature of the pictorial medium itself – *i.e.* pigments and oils with differing compositions or containing varying additives.

The opportunity to investigate the above-mentioned works by Fontana from the *Olii* and *Fine di Dio* series occurred when these paintings required restoration, and samples were provided thanks to the generosity of the works' owners and the Fondazione Lucio Fontana. Minute sample fragments of the respective pictorial layers were taken from the back sections or lateral margins of the paintings, some of which also contained the preparation layer of the canvas. These samples were found to be composed of different layers of monochromatic paints and, in some cases, contained sequins. A multi-analytical approach – involving optical microscopy, X-ray fluorescence (XRF), Fourier-transform infrared spectroscopy (FTIR), and gas chromatography–mass spectrometry (GC-MS) – was adopted to study and identify the materials employed by the artist and the products of their degradation processes.

3.4.1 *Concetto Spaziale* (60 O 81)

The first work to be examined was a 1960 painting entitled *Concetto spaziale* (60 O 81), belonging to the *Olii* series, whose rectangular canvas is coated with a monochromatic red–brown paint and the pictorial surface shows punctures, cuts, and scratched incisions (Figure 3.1). The reddish-brown pictorial layer was wrinkled and a widespread craquelure had created "islands"

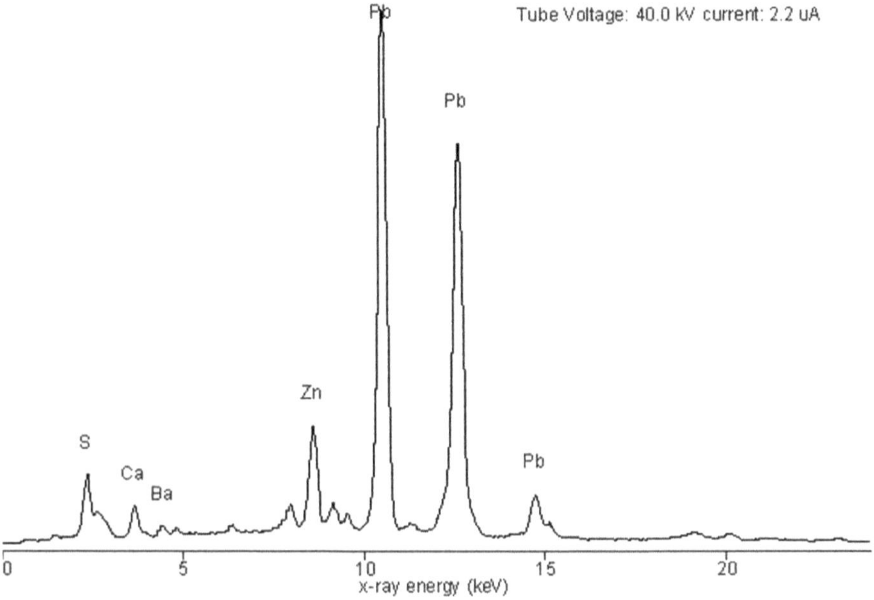

Figure 3.3 XRF spectrum of the ground layer from Lucio Fontana, *Concetto spaziale*, 60 O 81.

that tended to detach from the fabric support. In some places, this phenomenon was so advanced that it had caused losses in the paint layer (Figure 3.2). In some parts, the paint layer appeared to be extremely brittle and fragile, while in others it was still sticky and only partially dry – but why?

After separating and isolating the ground from the pictorial layer, a paint sample from this painting was carefully analysed. XRF results suggested that the white ground layer is mainly composed of lead white ($2PbCO_3{\cdot}Pb(OH)_2$), although other whitening materials were also detected as traces of zinc (likely zinc oxide, ZnO), calcium (calcium carbonate, $CaCO_3$), and barium (probably the corresponding sulphate, $BaSO_4$) (Figure 3.3). The presence of a lead soap was observed at about 1530 cm^{-1} in the FTIR–ATR spectrum (Figure 3.4): this was most likely due to the absorption of lead carboxylates formed as products after the reaction of the oil binder with the lead pigment.[18,19]

The red–brown colouration of the paint is attributable to the presence of alizarin, an organic red dye mainly composed of 1,2-dihydroxyanthraquinone, as is shown by comparison of the recorded infrared spectrum with the reference spectrum of natural alizarin (Figure 3.4). The medium, after transesterification and GC-MS analysis, was identified as an oil: in fact, the chromatogram (Figure 3.5) presents all the typical peaks referring to an oil binder. The paint sample contains methyl esters: glycerol, monocarboxylic fatty acids (such as palmitic, stearic, arachidic, and behenic), dicarboxylic fatty acids (such as suberic, azelaic, and sebacic), and unsaturated fatty acids (such as oleic and linoleic). In addition, the presence of ricinoleic acid, such as methyl ricinoleate and methyl methoxy ricinoleate, was detected. This monounsaturated C18

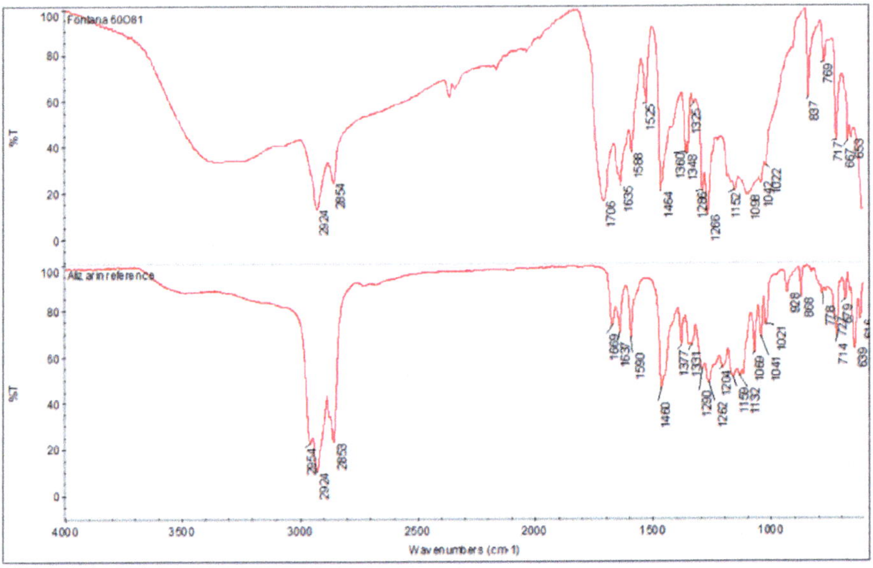

Figure 3.4 FTIR spectra of the red sample from Lucio Fontana, *Concetto spaziale*, 60 O 81 (top) and the reference of natural alizarin (bottom).

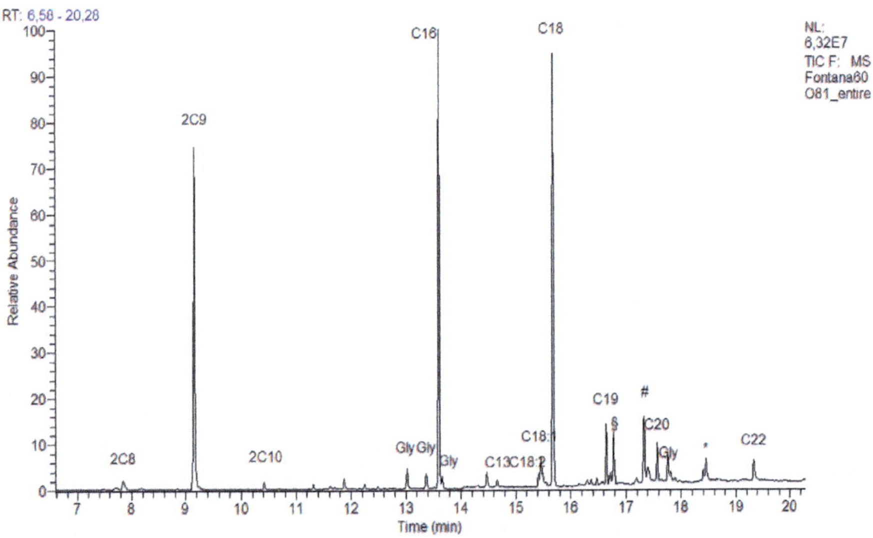

Figure 3.5 Total ion current chromatogram of red sample from Lucio Fontana, *Concetto spaziale*. 60 O 81 after transesterification and GC-MS analysis (Gly = glycerol, 2C8 = suberic acid, 2C9 = azelaic acid, 2C10 = sebacic acid, C16 = palmitic acid, C13 = tridecanoid acid, C18 : 1 = oleic acid, C18 = stearic acid, C18:2 = linoleic acid, C19 = nonadecanoic acid, internal standard, C20 = arachidic acid, C22 = behenic acid, # = ricinoleic acid, § = ricinoleic acid, methoxy, * = 9,10-dihydroxy octadecanoic acid).

fatty acid is a bio-marker that commonly implies the presence of castor oil. Castor oil is a vegetable oil obtained from castor seeds and is considered to be a slow/non-drying oil as it is composed of ricinoleic (90%), oleic, and linoleic acids (less than 8%). It is likely that the red paint in this work contained largely castor oil, as suggested by the P:S (palmitic:stearic) acid ratio. In fact, the 1:1 ratio confirms the presence of this non-drying oil and after 50 years the painting does not appear to have dried completely.

The presence of castor oil may explain some of the signs of degradation evident in the pictorial layers, and the wrinkles that have formed on the surface might be linked to the use of this oil in the thick, bodied pictorial layer – an application technique used by Fontana with the aim of bringing his painting closer to sculpture. Over time, the evaporation of volatile compounds (such as alcohols, aldehydes, and ketones) occurred due to oxidative cross-linking reactions, leading to weight losses.[20,21] As the drying speed is slowed down by the presence of a non-drying oil, it is possible that evaporation also occurred slowly, causing the formation of more visible, deeper wrinkles.[22]

It is not clear whether Fontana added the castor oil himself or if he bought a ready-made paint containing a linseed and castor oil mix. Indeed, little is known about the brands of the materials he used, and we do not know if Fontana had direct contact with industrial paint manufacturers or if he simply bought commercial paints from local hardware suppliers. Even though it is known, as mentioned above, that in some works Fontana sought paint formulations that would allow the pictorial film to dry more rapidly, we cannot exclude that for other works he purposefully used binders that dried very slowly.

3.4.2 The Pink Oil: *Concetto Spaziale* (62 O 66), *Concetto Spaziale* (1737/92), *Concetto Spaziale, La Fine di Dio* (63 FD 14)

Thus, given that the results show the presence of a variety of oils and the implication that these oils could play a role in the degradation processes observed, we decided to analyse other works from the *Olii* and *Fine di Dio* series – Fontana's "pink" oils.

When confronted with one of these works for the first time, strange, lighter stains in the form of drops and puddles were noted. Initially it was presumed that these stains indicated accidental damage caused by contact with external elements such as detergents or condensation on the glass of a frame. However, with time these same stains were observed in other "pink" works and could obviously not be simplistically attributed to accidental damage or unsuitable conservation condition, but had to be the result of the paint composition itself. In order to verify the origin of these alterations, three monochromatic pink works that presented the same kind of alteration were analysed: two *Concetti spaziali* belonging to the *Olii* series (62 O 66, 1737/92) (Figures 3.6, 3.7, 3.8, 3.9) and a painting from the *Fine di Dio* series (63 FD 14) (Figures 3.10 and 3.11).

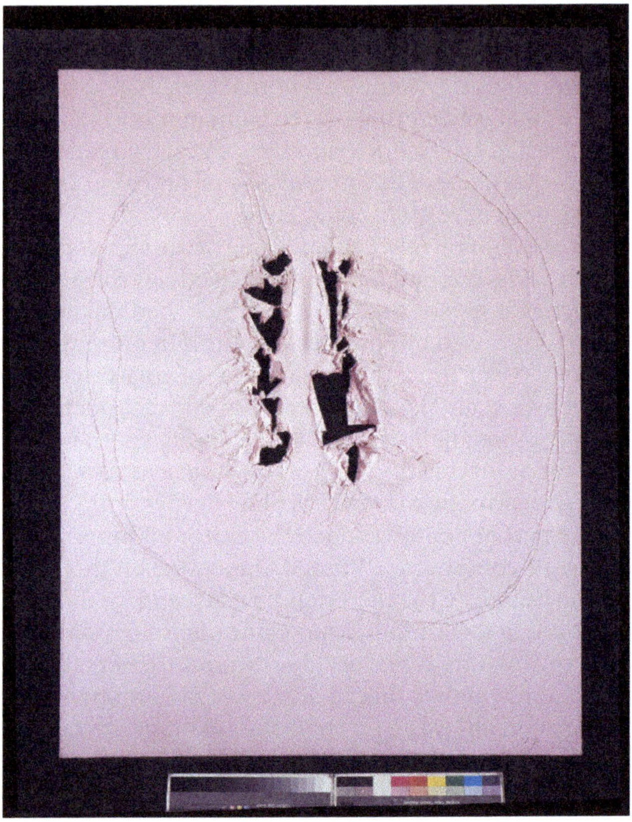

Figure 3.6 Lucio Fontana, *Concetto spaziale*, 62 O 66. Reproduced with permission from Fondazione Lucio Fontana.

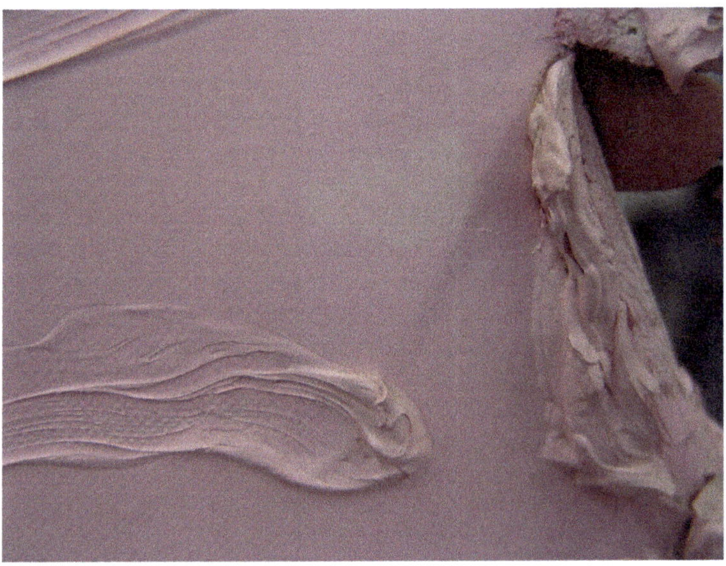

Figure 3.7 Lucio Fontana, *Concetto spaziale*, 62 O 66, detail. Reproduced with permission from Fondazione Lucio Fontana.

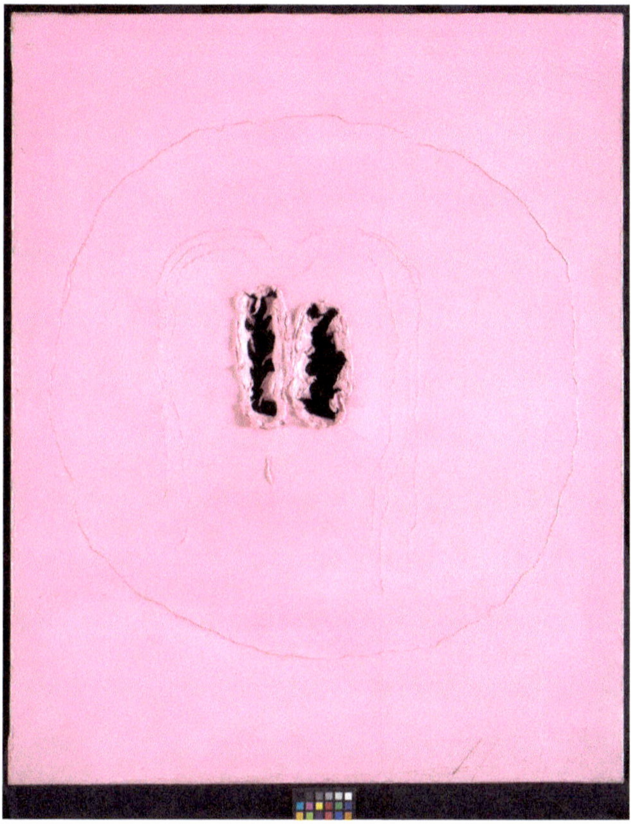

Figure 3.8 Lucio Fontana, *Concetto spaziale*, 1737/92. Reproduced with permission from Fondazione Lucio Fontana.

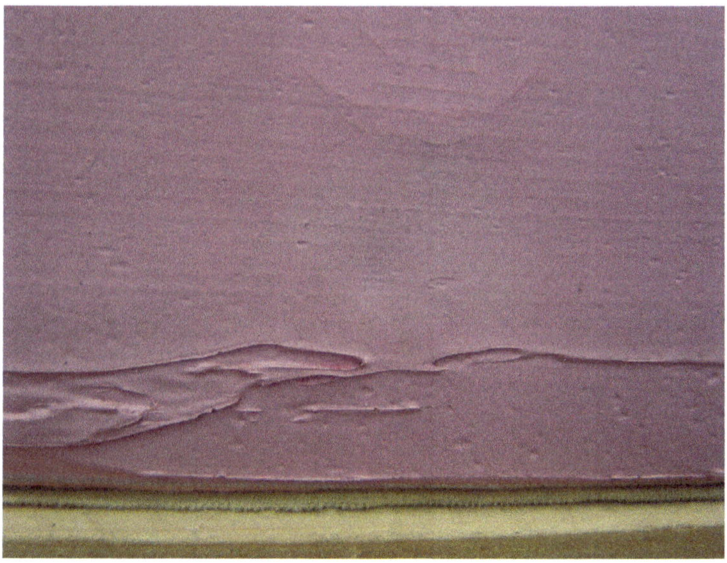

Figure 3.9 Lucio Fontana, *Concetto spaziale*, 1737/92, detail. Reproduced with permission from Fondazione Lucio Fontana.

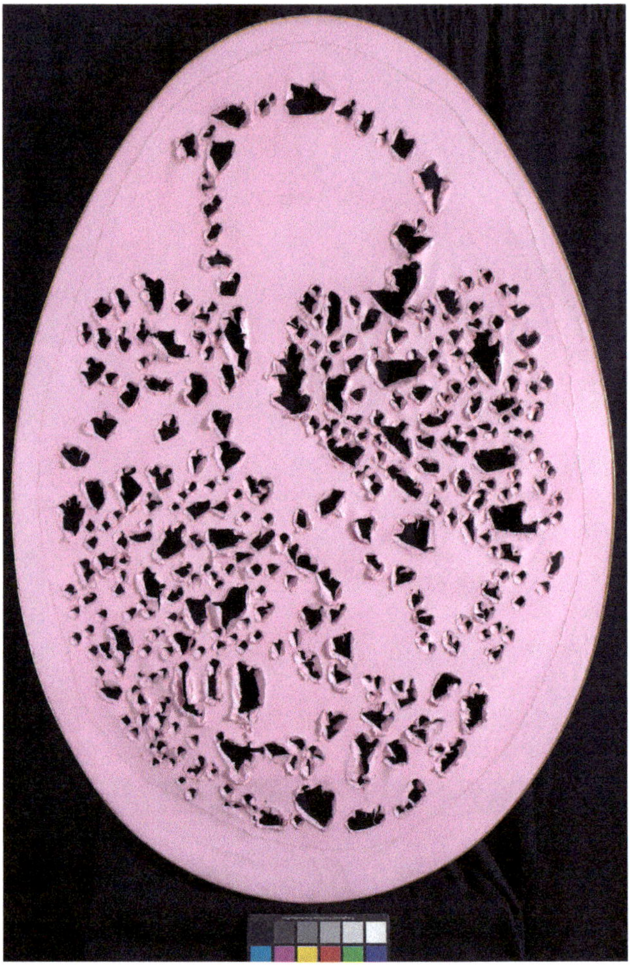

Figure 3.10 Lucio Fontana, *Concetto spaziale*, 63 FD 14. Reproduced with permission from Fondazione Lucio Fontana.

With regard to the colouration of all three of the works analysed, XRF analysis shows the presence of zinc in all of the paints examined, suggesting the use of zinc white (ZnO) mixed with an organic red dyestuff. FTIR-ATR analyses performed on the pink painted layers (Figure 3.12) show the presence of the typical absorption peaks of drying oils at 3340, 2919, 2851, 1738, 1300–1090 and 722 cm^{-1}. In all the FTIR spectra one notices a clear, broad carboxylate absorption around 1570–1540 cm^{-1}, which consists of several separate peaks (1620, 1588, 1540 cm^{-1}). These peaks are possibly the result of the influence of zinc white in the drying and ageing of the oil, as they could refer to the formation of metal carboxylates (soaps). Metal soaps can be formed

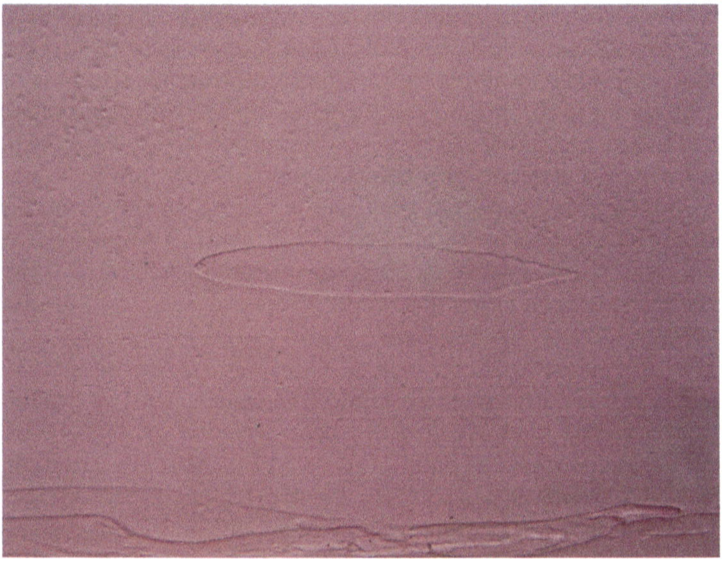

Figure 3.11 Lucio Fontana, *Concetto spaziale*, 63 FD 14, detail. Reproduced with permission from Fondazione Lucio Fontana.

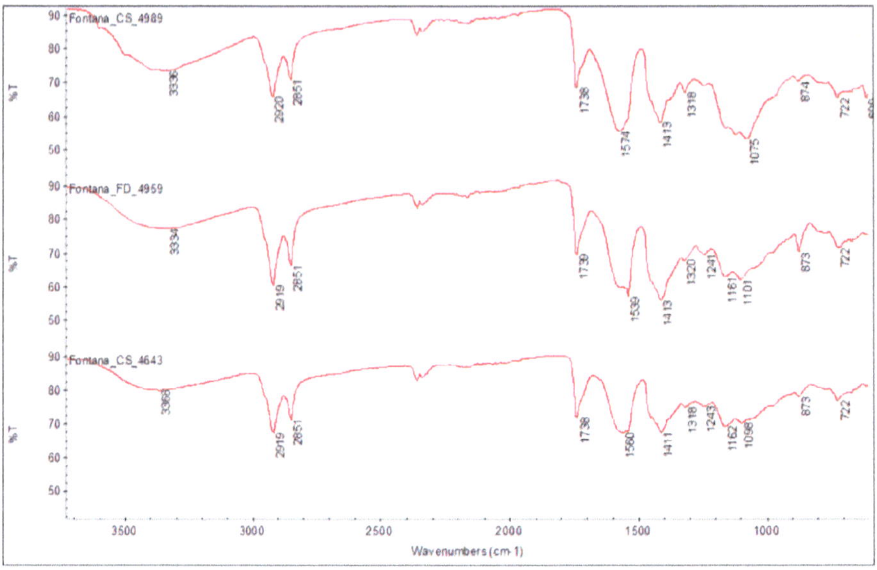

Figure 3.12 Comparison of the FTIR spectra of the pink *Olii* by Lucio Fontana, *Concetto spaziale* (17 37/92) (top), *Concetto spaziale, La Fine di Dio* (63 FD 14) (middle) and *Concetto spaziale* (62 O 66) (bottom).

after hydrolysis of the fatty acid molecules in the triglycerides of drying oils with metal-based pigments. The presence of added stearates in the paint formulation cannot be excluded.[23,24] A minor addition of calcium carbonate, probably as an inert compound, is suggested by the carbonatic absorption peaks at 1413 and 873 cm^{-1}.

GC-MS analyses carried out on the pink paints show similar results in all three artworks and confirm that they are oil paints. In the chromatogram, obtained after transesterification and GC-MS analysis (Figure 3.13), saturated (lauric, myristic, palmitic, stearic), unsaturated (oleic), and dicarboxylic (suberic, azelaic, sebacic) fatty acids and glycerol derivatives were recorded (as methyl esters).

Amongst the characteristic drying oil compounds, two uncommon fatty acids were detected in the three pink *Olii*: erucic acid (13-docosenoic) and gondoic acid (11-eicosenoic) and their oxidation products (13,14-dihydroxydocosanoic acid and 11,12-dihydroxyeicosanoic acid, respectively).

These unsaturated fatty acids are considered to be bio-markers, as they are commonly found in abundance in oils obtained from the seed of *Brassicaceae*, such as rapeseed oil.[25] Rapeseed oil is a semi-drying oil that was introduced in the twentieth century as a lubricant and is now used in the production of biodiesel.

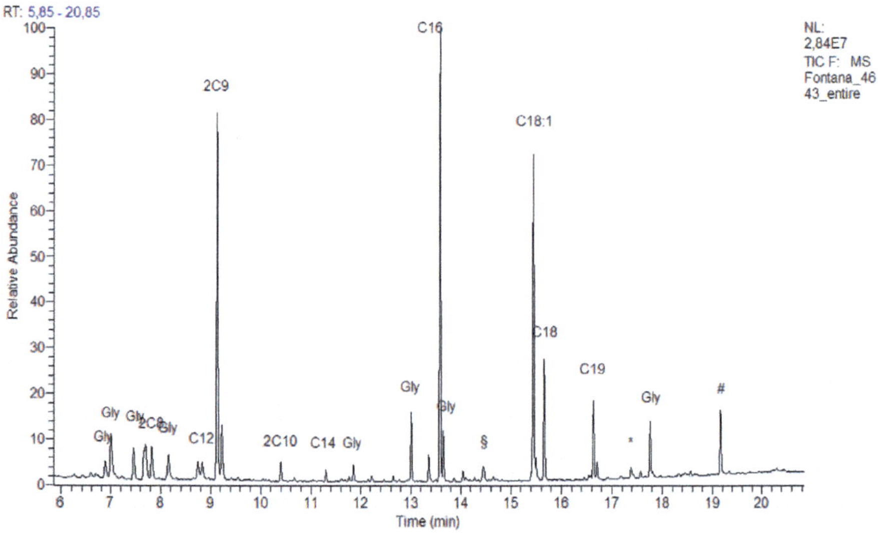

Figure 3.13 Total ion current chromatogram of pink paint from Lucio Fontana, *Concetto spaziale, La Fine di Dio* (63 FD 14), after transesterification and GC-MS analysis (Gly = glycerol, 2C8 = suberic acid, C12 = lauric acid, 2C9 = azelaic acid, 2C10 = sebacic acid, C14 = myristic acid, C16 = palmitic acid, C18 : 1 = oleic acid, C18 = stearic acid, C19 = nonadecanoic acid, internal standard, * = gondoic acid, # = erucic acid).

According to artists' handbooks, olive oil or rapeseed oil and glycerol were added to avoid industrially primed grounds of commercial canvases from drying too fast, and these compounds functioned to keep the ground elastic.[26] High levels of glycerol were detected in the case studies, but it is not clear whether the commercially primed canvases of the pink *Olii* were prepared with the addition of rapeseed oil or if the rapeseed oil was mixed (probably with linseed oil) as a binding medium.

The high content of oleic acid and the consequentially high degree of unsaturation is most likely to be a consequence of the presence of a slow-drying oil. However, the significant amount of oleic acid, which has also been noted in other artworks, may be linked to the presence of zinc oxide, because in an oil-based medium zinc oxide forms a packed structure that is able to trap oleic acid in the pictorial layer. This effect can also occur years after the oxidation process ends.

The P : S ratio is 3.5 for *Concetto spaziale* (1737/92), 3.7 for *Concetto spaziale* (62 O 66), and 3.4 for *Concetto spaziale, La fine di Dio* (63 FD 14). These values are relatively similar and may refer to the presence of rapeseed oil (whose P : S is around 3.5–3.8).

The unusual degradation phenomena observed in all of these paintings (surface stains in the form of drops and puddles) might be the result of the presence of a mixture of linseed oil with a slow-drying oil (rapeseed oil). These mixtures contain a large proportion of unsaturated acids (oleic, erucic, and gondoic were detected after almost 50 years) whose triglycerides reacted with zinc white from the painted layers to form metal soaps. As FTIR analyses suggest, metal soaps are present in the form of zinc carboxylates, and hence it is probable that zinc soaps are present as formed and/or added soaps and that they migrated and conglomerated on the surface, forming the white spots.

However, given that the alterations observed in the pink *Fine di Dio* (63 FD 14) were limited to the light stains present on the pictorial surface, we questioned what could have provoked the strange physical-mechanical behaviour of the pictorial film in other works from the *Fine di Dio* series.

3.4.3 *Concetto Spaziale, La Fine Di Dio* (63 FD 23), *Concetto Spaziale, La Fine Di Dio* (64 FD 5)

At this point, two other works from the *Fine di Dio* series were analysed in a comparative study: a 1963 black painting with punctures, slashes, and sequins glued on the surface (63 FD 23) (Figures 3.14 and 3.15) and a 1964 yellow work from the same series (64 FD 5) (Figures 3.16 and 3.17).

Despite the fact that all preceding analyses carried out on other works belonging to the *Fine di Dio* series showed that the artist had used oil paints and all presented similar physical-mechanical characteristics, the black *Concetto spaziale, La fine di Dio* (63 FD 23) analysed here showed completely different physical behaviour. In the majority of *Fine di Dio* works, the pictorial

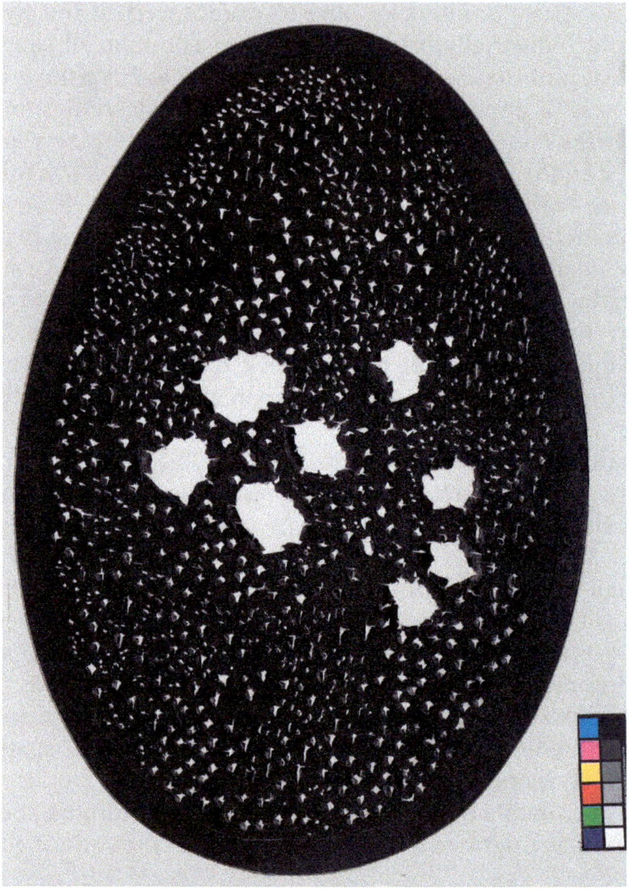

Figure 3.14 Lucio Fontana, *Concetto spaziale*, 63 FD 23. Reproduced with permission from Fondazione Lucio Fontana.

layers are compact and the oil flowed into the internal edges of the holes, which, in turn, maintained their original position. In contrast, in this black *Concetto spaziale, La fine di Dio* (63 FD 23), the edges of the slashes appeared to be deformed and, instead of protruding onto the rear, appeared to have contracted and pushed through onto the front of the canvas (Figure 3.15). What could have caused this anomalous physical-mechanical behaviour?

Could this discrepancy in the alterations of the different works within this series be related to the formulation of the medium used – which at first glance appeared to be a synthetic medium, but was later proven to be an oil – or was it simply a result of the extremely thin pictorial film found in this specific case?

In order to attempt to clarify what could have caused the anomalous deformation of the edges of these slashes, another work from the same series was examined. Thus *Concetto spaziale, La Fine di Dio* (64 FD 5) – a monochromatic yellow painting on an oval canvas with punctures, slashes, and scratches dating to 1964 – was analysed, as it did not present any unusual alterations and therefore represented an ideal reference standard (Figure 3.17).

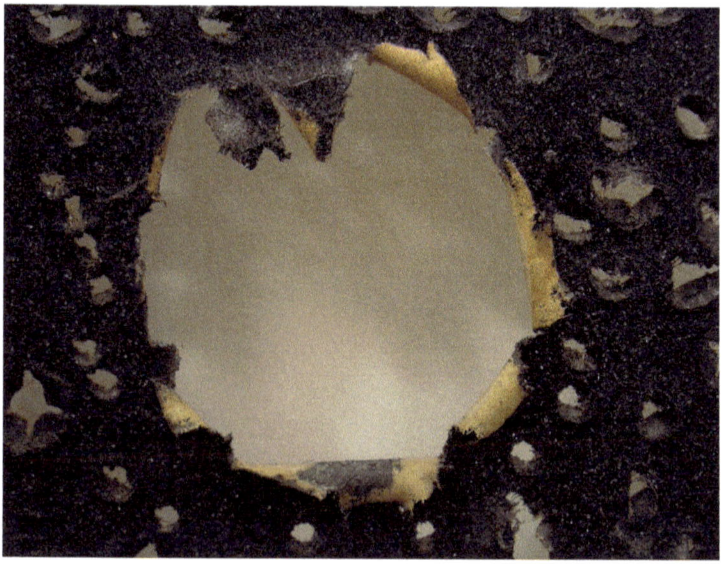

Figure 3.15 Lucio Fontana, *Concetto spaziale*, 63 FD 23, detail. Reproduced with permission from Fondazione Lucio Fontana.

The elemental analysis of the yellow paint revealed the presence of zinc (Zn), chrome (Cr), and potassium (K) (Figure 3.18): the yellow pigment used is possibly zinc yellow, whose composition is mainly $ZnCrO_4$ plus K_2CrO_4. Accurate reading of FTIR–ATR spectra (not reported here), analogous to what was observed for the pink series, suggested the presence of carbonatic materials (absorption peaks at 1413 and 873 cm^{-1}), which are most likely whitening additives, while the peaks at 3425, 2963, 2853, 1736, and 720 cm^{-1} indicated the use of drying oils. As observed in the pink *Olii*, the broad peak at 1574 cm^{-1} was explained by the presence of zinc soaps formed and/or added in the paint formulation. GC-MS chromatograms (not reported) were also comparable with those of the pink series. Besides the presence of common fatty acids (palmitic, stearic, and azelaic) and glycerol derivatives, the yellow paint exhibited high amounts of unsaturated oleic, gondoic, and erucic acids. The presence of rapeseed oil was also suggested in this case. The abundant oleic acid content might also be related to the action of zinc pigments.

Having verified these results, we then moved on to tackle the case of *Concetto spaziale, La Fine di Dio* (63 FD 23). This monochromatic black work presented punctures and sequins glued to the surface and, as already mentioned, during restoration treatments an unusual alteration was noted: the canvas was very flexible and the punctures protruded outwards instead of inwards.[15] From samples taken of the pictorial layer (including areas in which the sequins were present), it was ascertained *via* GC-MS that the components of the black paint include: glycerol derivatives, monounsaturated fatty acids (such as palmitic and stearic acids), dicarboxylic acids (such as

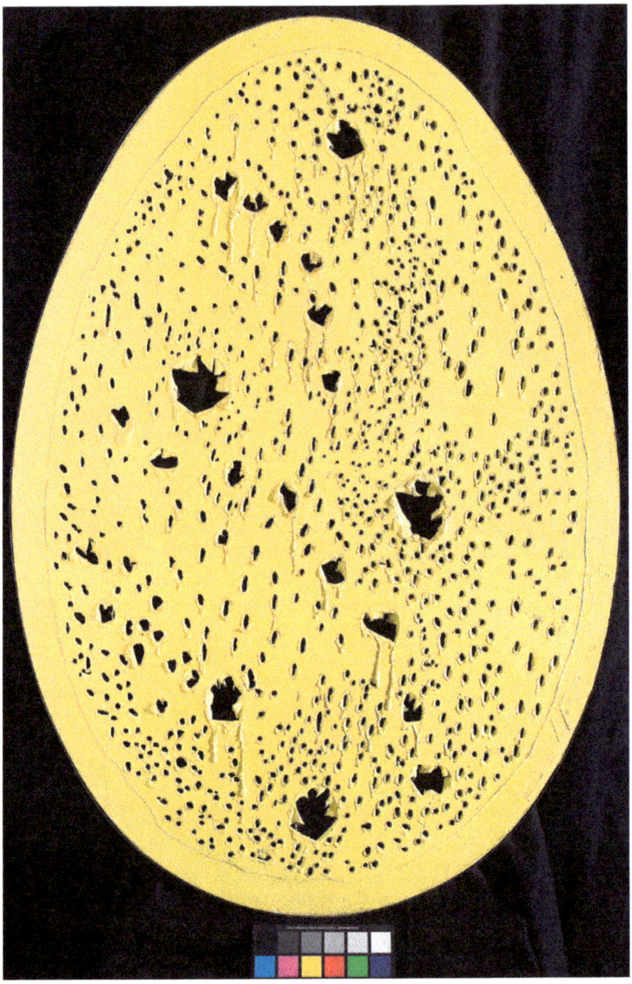

Figure 3.16 Lucio Fontana, *Concetto spaziale*, 64 FD 5. Reproduced with permission from Fondazione Lucio Fontana.

suberic, azelaic, and sebacic acids), and unsaturated acid (Figure 3.19). The high degree of unsaturation – also seen in *Concetto spaziale* (60 O 81) – associated with the high content of oleic acid, suggests either that the paint, even 50 years after its application to the canvas, is still immature or that a non-drying oil was added to the paint. The P:S ratio is 1.3, a value that conventionally refers to linseed oil; nevertheless, the use of this traditional binder is rather improbable, due to the different behaviour shown by the pictorial layer under examination. It is likely that other lipidic binders, exhibiting a similar P/S ratio, were employed. In this sample, the sequins appeared not only to be embedded in the pictorial film, but in some places were also embedded in a superficial layer of PVA-based adhesive (most likely Vinavil). This is illustrated by the FTIR spectrum (Figure 3.20) in which all the characteristic

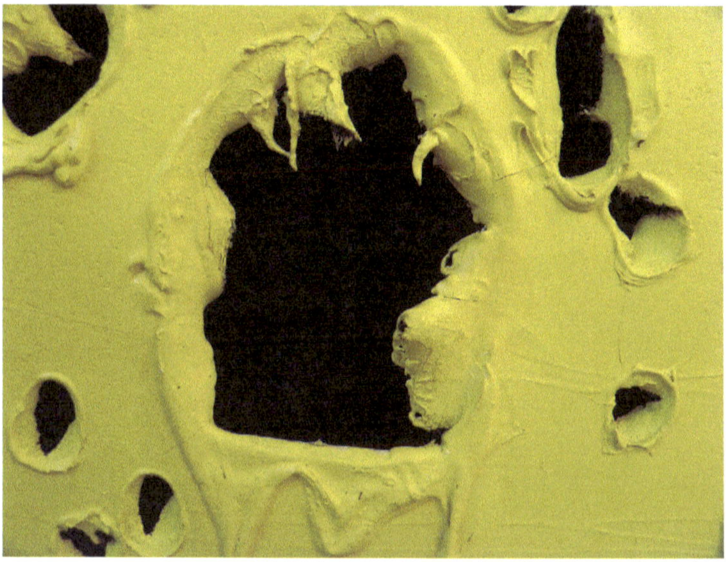

Figure 3.17 Lucio Fontana, *Concetto spaziale*, 64 FD 5, detail. Reproduced with permission from Fondazione Lucio Fontana.

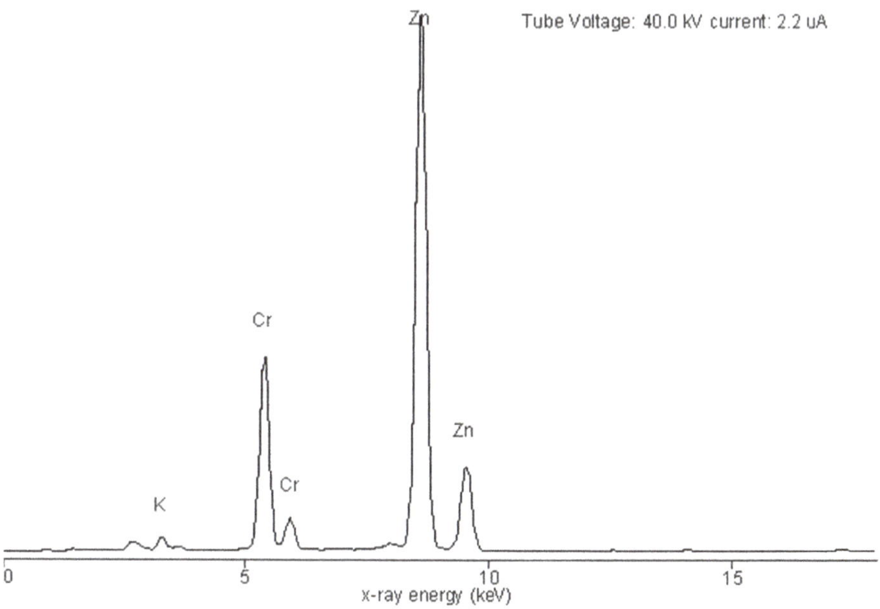

Figure 3.18 XRF spectrum of Lucio Fontana, *Concetto spaziale, La Fine di Dio* (64 FD 5).

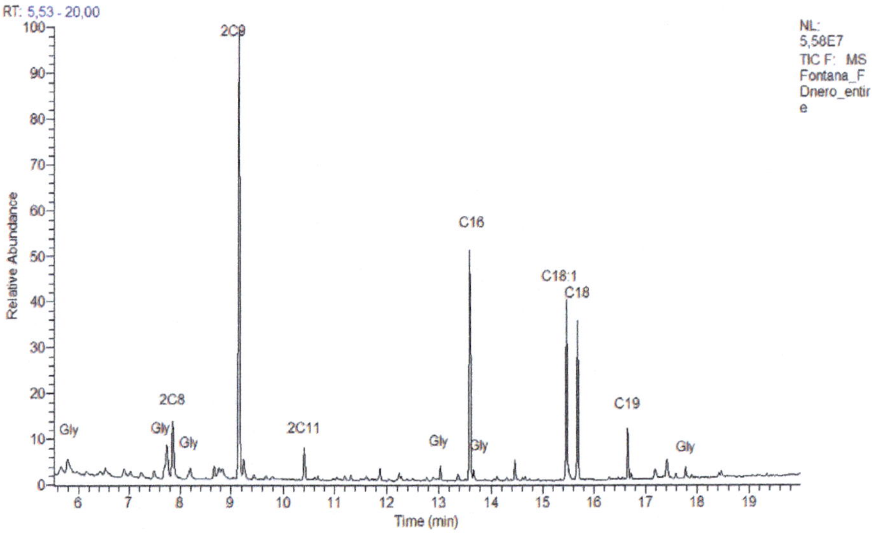

Figure 3.19 Total ion current chromatogram of the black paint from Lucio Fontana, *Concetto spaziale, La Fine di Dio* (63 FD 23), after transesterification and GC-MS analysis (Gly = glycerol, 2C8 = suberic acid, 2C9 = azelaic acid, 2C10 = sebacic acid, 2C11 = undecanedioic acid, C14 = myristic acid, C16 = palmitic acid, C18:1 = oleic acid, C18 = stearic acid, C19 = nonadecanoic acid, internal standard).

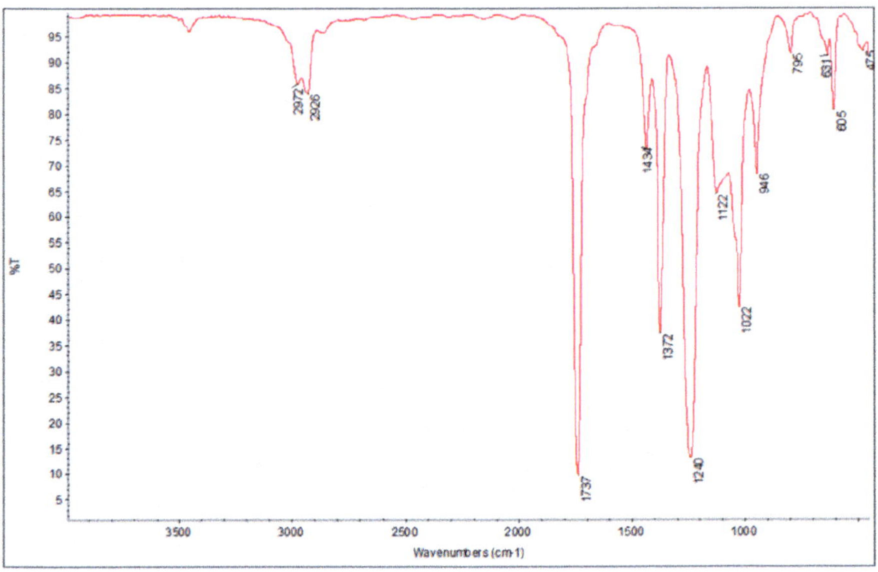

Figure 3.20 FTIR spectrum of the adhesive used to glue sequins in Lucio Fontana, *Concetto spaziale, La Fine di Dio* (63 FD 23).

peaks of the polymer are present. Although not proven, this glue could have played a role in the creation of tensions and the deformation of the pictorial layers, however it is also possible that this alteration could be caused by the extreme subtlety of the pictorial film (as mentioned in the previous examples).

The analyses carried out on numerous works in this series actually revealed different material compositions and in some cases the use of paints that, besides traditional drying oils, also contained semi-drying oils (such as rapeseed oil) or non-drying oils (such as castor oil), probably added to lower the price of the product.

3.5 Conclusion

This study provided us with important information regarding the works of the *Olii* series, created by Lucio Fontana between 1960 and 1964. During the same period, the artist experimented with several different and innovative pictorial binders, but still tended towards more traditional binding materials. Nevertheless, as demonstrated here, Fontana used oil paints containing different lipidic binders: a non-drying oil such as castor oil, a semi-drying oil such as rapeseed oil, and the traditional siccative binder, linseed oil.

It is not clear if Fontana wished to experiment with different kinds of oils in order to achieve different effects or if he simply used ready-made paints without knowing their composition. In any case, the materials and techniques he used are here proven to be closely linked to the conservation problems exhibited by the pictorial layers.

Further, these unexpected but useful findings have not only provided us with a new insight into the nature of the constituent materials of the examined works, but also facilitated planning of suitable and effective conservation methods.

Without these investigations, it is probable that the advanced state of degradation observed in *Concetto spaziale* (60 O 81) would have been attributed to extreme or erroneous conservation conditions rather than to the nature of the binders used in the paint's formulation. As a result, conservation measures would probably have proved ineffective overtime, if not harmful. The same can be said of the unusual alterations seen on the surfaces of the pink *Olii*.

However, it still remains to ascertain the real cause of the anomalous behaviour observed in the black *La Fine di Dio* (63 FD 23). In the past, this work had been partially examined on the occasion of a maintenance intervention carried out for an exhibition, but at that time it was possible to analyse only a tiny sample taken from the edge of the painting.[4] From this analysis, the presence of a lipidic binder, most likely linseed oil, was hypothesized, but a few months later, thanks to the opportunity provided by the necessity for further conservation treatments, new examinations were carried out following micro-sampling from different parts of the painting. Initially, the identification of a PVA-based adhesive used in addition to the pictorial medium to affix the sequins to some areas of the surface – which could also have been applied at a later stage in order to reinforce the adhesion or to add more sequins – led us to believe that the presence of this synthetic binder could have been the cause of the anomalous behaviour of the slash edges. However, the appearance of another black *La Fine di Dio* painting which presented the same alteration but showed neither sequins nor polyvinyl acetate in its composition, raised doubts about this interpretation and led us to believe that the most probable cause is the extreme subtlety of the pictorial layer.

From an analytical point of view, this study has reinforced previously stated conclusions that modern and contemporary artists used, and still use, several varying forms of the traditional oily medium – with the generic label "oils" – without any specific regard to their composition and properties. The uninformed use of these oils demands that each case study be considered separately and carefully analysed in order to develop suitable and effective conservation strategies.

3.6 Details of the Experimental Method

XRF was carried out using a Bruker® Tracer III–V portable XRF instrument, equipped with a low power Rh-tube operating at 40 kV and 2.2 uA and a Peltier® cooled Si-PIN detector.

FTIR–ATR analyses were performed with a Thermo Nicolet® Nexus 670 FTIR spectrophotometer combined with a Smart Orbit Single Reflection Diamond ATR accessory from 4000 to 400 cm^{-1} for 128 scans with 4 cm^{-1} resolution. Spectra were elaborated with Omnic® 8.0.

A Trace GC™ 2000 instrument with a capillary Supelco® Equity-5 column, 30 m, 0.25 mm, 0.5 µm interfaced with a Trace MS 2000 was employed for GC-MS analysis. The temperature programme was set from 80 °C to 300 °C 10 °C min^{-1}. The MS was run in Full Scan mode (m/z 40–600), 1.9 scans s^{-1}. The transfer line was at 280 °C and the source temperature was 150 °C. Electron ionization energy was 70 eV. Quantitative GC-MS analysis was performed using nonadecanoic acid as the internal standard. The oil samples were transesterified using (trifluoromethylphenyl) trimethylammonium hydroxide, overnight reaction.[5,27] The molar ratios amongst the most significant fatty acids were calculated: palmitic to stearic acid (P:S), azelaic to palmitic acid (A:P), oleic to stearic acid (O:S) ratios were considered and compared to previous results obtained by the authors on twentieth-century artists' oil paints.[5,28]

Acknowledgements

The authors would like to thank Nini Laurini, Paolo Laurini, Silvia Ardemagni, and Maria Villa of the Fondazione Lucio Fontana, Milan, without whose collaboration these studies would not have been possible.

References

1. B. Ferriani, Lucio Fontana: I materiali 1959–1961, in *Lucio Fontana, Venezia, New York*, ed. L. M. Barbero, Peggy Guggenheim Collection, 2006, pp. 219–224.
2. B. Ferriani, presented in part at *La conservazione del presente: leganti sintetici, plastiche, video e installazioni, Giornate di studio sulla conservazione e il restauro dell'arte contemporanea*, Pinacoteca di Brera, Milan, 31 March 2008.
3. B. Ferriani, R. Ploeger, T. Poli and O. Chiantore, Lucio Fontana Between Tradition and Innovation, in *Art d'aujourd'hui patrimoine de demain – Conservation et restauration des oeuvres contemporaines*, SFIIC, Champs-sur-Marne, 2009, pp. 179–185.
4. O. Chiantore, R. Ploeger, T. Poli and B. Ferriani, Materials and techniques in the pictorial oeuvre of Lucio Fontana, *Stud. Conserv.*, 2012, **57**(2), 92–105.
5. F. C. Izzo, *20th Century Artists' Oil Paints: A Chemical–Physical Survey*, Università Ca' Foscari, Venice, 2010.
6. F. C. Izzo, B. Ferriani, K. van den Berg, H. van Keulen and E. Zendri, 20th Century Artists' Oil Paints: The Case of the Olii by Lucio Fontana, *J. Cult. Herit.*, 2014, **15**, 557–563.
7. V. Minardi, E. Crispolti and L. Fontana, *Drawings and Graphic Works – Exhibition Catalogue*, ed. V. Minardi, Istituto Italo-Latino-Americano, Rome, 2 March–15 April 1972.
8. E. Crispolti, Il disegno di Fontana, manualità concettuale, in *Disegni di Fontana anni trenta-quaranta – Exhibition catalogue*, Milan, Galleria d'Arte Moderna Farsetti, October–November 1989, Firenzelibri, Collana FarsettiArte, Florence, 1989, p. 5.
9. L. Fontana, *First Congress of Proportions at Triennale, 1951*, 9th Triennale of Milano, unedited manuscript/draft for Manifesto Tecnico Spaziale, Transcription of page 51/text 41, Fondazione Lucio Fontana, Milan, 1951.
10. E. Crispolti, L'avventura creative di Fontana nell'arte del XX secolo, in *Lucio Fontana. Catalogue Raisonné of Sculptures, Paintings, Environments*, ed. E. Crispolti, N. Laurini and V. Ernesti, Skira, 2006, vol. I, p. 64.
11. P. Gottschaller, N. Khandekar, L. F. Lee and D. P. Kirby, The evolution of Lucio Fontana's painting materials, *Stud. Conserv.*, 2012, **57**(2), 76–91.
12. E. Crispolti, *Catalogue Raisonné of Sculptures, Paintings, Environments*, Skira, 2006, vol. I, p. 76.
13. M. Portalupi, Le ombre cinesi di Lucio Fontana, in *La Notte*, Milan, 28 October 1965.
14. P. Campiglio, *Lucio Fontana, Lettere 1919–1968*, Skira, Milan, 1999.
15. B. Ferriani, unpublished results.
16. I Concetti spaziali di Lucio Fontana, in *Materie Plastiche Ed Elastomeri*, Milan, July 1972, pp. 584–586.

17. F. C. Izzo, B. Ferriani and E. Balliana, *A New Insight into Fontana Oil Paints*, 2019, unpublished.
18. R. J. Meilunas, J. G. Bentsen and A. Steinberg, Analysis of aged paint binders by FTIR spectroscopy, *Stud. Conserv.*, 1990, **35**, 33–51.
19. J. van der Weerd, A. van Loon and J. J. Boon, FTIR studies of the effects of pigments on the ageing of oil, *Stud. Conserv.*, 2005, **50**, 3–22.
20. H. Wexler, Polymerization of drying oils, *Chem. Rev.*, 1964, **64**, 591–611.
21. J. D. J. van den Berg, K. J. van den Berg and J. J. Boon, Chemical changes in curing and ageing oil paints, in *ICOM Committee for Conservation Preprints*, 1999, vol. 1, p. 253.
22. D. Erhardt, S. C. Tumosa and M. F. Mecklenburg, Long-term chemical and physical processes in oil paint films, *Stud. Conserv.*, 2005, **50**, 143–150.
23. C. S. Tumosa, A brief history of aluminum stearate as a component of paint, *WAAC Newsletter*, 2001, vol. 23, pp. 10–11.
24. A. Burnstock, K. J. van der Berg and S. de Groot, *et al.*, An investigation of water-sensitive oil paints in 20th century paintings, in *Modern Paints Uncovered*, ed. T. J. S. Learner, P. Smithen, J. Krueger, *et al.*, The Getty Conservation Institute, Los Angeles, 2007, pp. 177–188.
25. M. P. Colombini and R. Modugno, *Organic Mass Spectrometry in Art and Archaeology*, Wiley & Sons, New York, 2009.
26. M. Doerner, *The Materials of Artists and Their Use in Painting*, Harvest House, San Diego, New York, London, 1984.
27. F. C. Izzo, K. B. van der Berg, H. van Keulen, B. Ferriani, and E. Zendri, Modern oil paints – formulations, organic additives and degradation: some case studies, in *Issues in Contemporary Oil Paint*, ed. K. J. van den Berg, A. Burnstock, M. de Keijzer, J. Krueger, T. Learner, A. de Tagle and G. Heydenreich, Springer International Publishing, Switzerland, 2016, pp. 75–104.
28. K. R. Sutherland, Derivatisation using m-(trifluoromthyl) phenyltrimethylammonium hydroxide of organic materials in artworks for analysis by gas chromatography-mass spectrometry: unusual reaction products with alcohols, *J. Chromatogr. A*, 2007, **1149**, 30–37.

CHAPTER 4

Josef Albers' Use of 20th Century Pigments: A Non-invasive Analytical Approach

GIANLUCA POLDI*[a], CHIARA ANSELMI[b], ALESSIA DAVERI[c] AND MANUELA VAGNINI[c]

[a]Università degli Studi di Bergamo, Visual Art Centre, Via Pignolo 123, 24121, Bergamo, Italy; [b]CNR-IRET, via G. Marconi 2, 05010, Porano (TR), Italy; [c]Laboratorio di Diagnostica Beni Culturali, Spoleto, Italy
*E-mail: g.poldi@gmail.com

4.1 Introduction and Aims

Josef Albers (1888–1976) was a German-born, Bauhaus-educated designer who in 1934 emigrated to the United States, where he became a globally influential artist and teacher. As a teacher, after Bauhaus (1924–1933) he worked first at the Black Mountain College, North Carolina, from 1933 to 1949, and then at Yale University, until 1960. His life-long interest in visual perception and interaction of colors culminated in the series of paintings *Homage to the Square*, for which he is best known. The *Catalogue Raisonné of the Paintings of Josef Albers, 1914–1976*, edited by Jeannette Redensek and The Josef and Anni Albers Foundation (New Haven, Connecticut), currently in progress, is documenting some 2500 paintings made between 1914 and 1976.

Science and Art: The Contemporary Painted Surface
Edited by Antonio Sgamellotti, Brunetto Giovanni Brunetti and Costanza Miliani

Published by the Royal Society of Chemistry, www.rsc.org

Albers used manufactured oil pigments, mainly unmixed and directly from the tube, on prepared wood composition boards, making great use of the palette knife. He made careful records of the materials he used, recording in studio notebooks and inscribing on the reverse of paintings the grounds, colors, manufacturers, varnishes and sometimes even the dates the color tubes were purchased. There are color notes for about 90% of Albers' paintings.

There are also a large number of color studies made on white paper[1] (Figures 4.1 and 4.2A) – typically thick, soft blotting paper often used in printmaking – and the shelves of paints, palette knives, brushes and other tools remaining in the artist's studio at the time of his death.

With this extraordinarily detailed corpus of data about the materials, any analysis or study of them seems unnecessary. Nevertheless, a deeper knowledge of Albers' technique can be useful for assessing some conservation issues, characterizing materials where no records appear (like in some of his earlier paintings), studying how he prepared the support (ground,

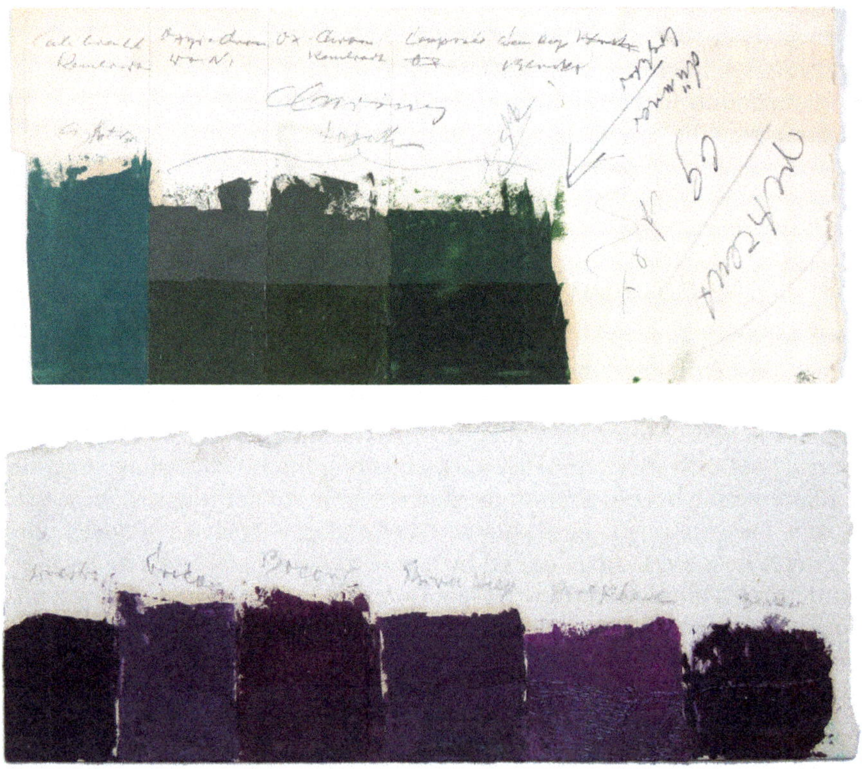

Figure 4.1 Two of Josef Albers' color tests on paper, from *The Josef and Anni Albers Foundation*, Bethany: greens (top, inv. 1976-2-1432) and violets (bottom, inv. 1976-2-1381). Photos: Gianluca Poldi. Courtesy of the Josef and Anni Albers Foundation. Reproduced with permission.

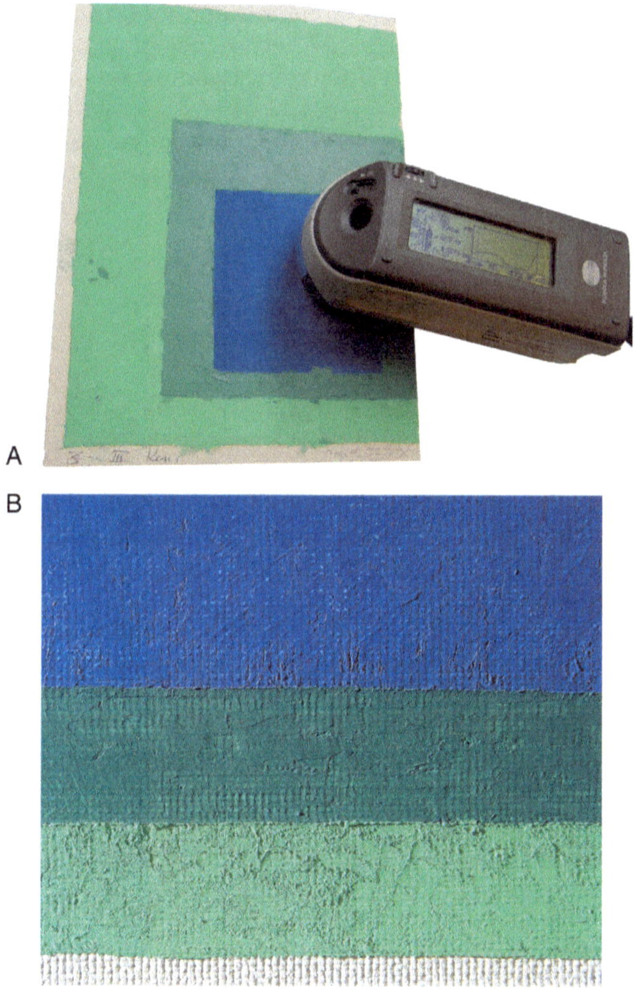

Figure 4.2 Josef Albers' color study on paper for *Homage to the Square* (inv. 1976-2-1410) during vis-RS analyses (A) and a raking light detail (B) of the relative painting *Homage to the Square*, 1976 (1976-1-524), both belonging to *The Josef and Anni Albers Foundation*, Bethany. Photos: Gianluca Poldi. Courtesy of the Josef and Anni Albers Foundation. Reproduced with permission.

underdrawing, *etc.*) and understanding which pigment or mixture of pigments constitutes the single color named on the tubes and recorded by the painter on the *verso* of his works. From the perspective of Conservation Science, this latter point is a particularly important contribution, considering the high number of colors and producers he used during his career and the possible alterations or preservation problems concerning some of them.

The study of Albers' paintings on Masonite and paper favored the acquisition of a large database regarding the composition of 20th-century color tubes and the possible alterations of organic and inorganic components. It can clarify the efficacy of non-invasive techniques in characterizing them. Besides, it is worthwhile to verify if scientific examinations can help – within Albers' work – to discover fakes.

The findings are supposed to contribute toward a greater understanding of Albers' technique and the creation of a large database of 20th-century pigments, from the 1930s to 1970s.

4.2 On Albers' Technique

Important accounts of Albers's practice were published on a few occasions.[2]

Josef Albers' major activity as a painter started fundamentally after 1934, when he was 46 years old and after his arrival in the United States. There, he started using compressed wood fiberboard panels (also called Masonite, referring to one of the manufacturers) with a smooth and a raw side. He painted on the smooth side until the mid-1940s, when he started experimenting also with the raw side, which in the 1950s became the preferred one, providing a more stable support for the paint. He occasionally used other supports.

Apart for some works painted on unprepared supports, Albers prepared his paintings with a white ground. In his early American paintings he used many layers of zinc white and chalk in an egg tempera and linseed oil emulsion, with shellac. Then he started using commercially manufactured interior house paints, firstly using Sherwin-Williams®' Kem-Tone® and the United States Gypsum Company™'s Texolite® (oil-and-water mixtures, emulsified with casein) and then the brand Luminall, another casein-based interior paint. By the early 1950s he turned to DuPont® flat white and semi-gloss interior paints, based on alkyd resins. After 1959, he exclusively employed the Liquitex® brand, applying it in five to eight thin coats, in order to create a brilliant light surface for pigments.

Concerning color layers, Albers preferred oil paints but also used casein ones in the early 1950s and acrylic paints on a few 1960s works. He used paints from many manufacturers, but favored the pigments produced by Grumbacher®, Permanent Pigments™, Winsor & Newton™, Shiva™, Lucien Lefebvre, Blockx® and Bocour™ (see Table 4.1).

He didn't use extra drying oil or turpentine, preferring "to keep the nature of the pigments [...] the characteristics of the paint", being opaque or more transparent, thinner or thicker, depending on the particle size and the density of the binder.

As mentioned above, at least from the mid-1930s Josef Albers kept careful records of the grounds, paints (name and manufacturer), possible mixtures of pigments, occasional additives like drying oils and thickeners and the varnishes he used to make each work.[2] He inscribed this information using pen or pencil on the backs of the paintings, and he frequently recorded the materials in his studio notebooks alongside titles, dates and dimensions. The studio notebooks are currently held in the artist's archive in Bethany, Connecticut.

Table 4.1 Albers' paints examined by a large set of non-invasive spectroscopic analyses, with their names as recorded by Albers on the *verso* of his paintings. Pretested belongs to Grumbacher manufacturer.

Color	Paint and manufacturer
Violet	Cobalt Violet (Winsor & Newton™), Cobalt Violet (Shiva™), Thio Violet (Pretested®)
Blue	Cerulean Blue Shade (Pretested®), Monastral™ Blue Light (Greco®), Ultramarine No. 2 (Winsor & Newton™)
Green	Cadmium Green (Shiva™), Cadmium Green (Winsor & Newton™), Chrome Oxide Green Light (Shiva™), Cobalt Green (Winsor & Newton™), Cobalt Green Deep (Greco®), Emerald Green (Malfa), Permanent Bright Green (Grumbacher® F.), Permanent Green Light (Pretested®)
Yellow	Aureolin (Rembrandt®), Barium Yellow (Permanent Pigments™), Cadmium Yellow Deep (Greco®), Cadmium Yellow Deep (Blockx®), Cadmium Bar. Yellow Deep (Permanent Pigments™), Cadmium Yellow Medium 3 (Lefebvre™), Pretested Cadmium Light (Schmincke®), Yellow Ochre Light, Yellow Ochre (Winsor & Newton™), Naples Yellow (Shiva™), Naples Yellow Reddish (Lukas™), Scheveningen Yellow Lt (Old Holland®)
Orange	Cadmium Orange (Pretested®), Cadmium Orange (Rembrandt®), Cadmium Orange (Winsor & Newton™)
Red	Alizarin Crimson (Pretested®), Cadmium Red Deep (Pretested®), Cadmium Red Medium (Pretested®), Cadmium Red Light (Greco®), Cadmium Red Light (Permanent Pigments™), Cadmium Red Purple (Shiva™), Carmin (Winsor & Newton™), Harrison Red (Winsor & Newton™), Venetian Red (Sargent®)
Brown	Burnt Umber (Winsor & Newton™), Burnt Umber Pret (Winsor & Newton™), Gold Ochre (Utrecht®), Light Red (Shiva™), Terra Rosa (Winsor & Newton™), Venetian Red (Greco®)
Gray	Chapin Neutral Gray I (Shiva™), Chapin Neutral Gray II (Shiva™), Chapin Neutral Gray III (Shiva™), Illustrators gray #2 (Grumbacher® F.), Illustrators gray #3 (Grumbacher® F.), Illustrators gray #4 (Grumbacher® F.), Payne's Gray (Shiva™), Reilly's Gray #3 (Grumbacher®), Reilly's Gray #4 (Grumbacher®), Reilly's Gray #7 (Grumbacher®), Reilly's Gray #8 (Grumbacher®), Reilly's Gray #9 (Grumbacher®), Schminke Gray 3 (Grumbacher®), Schminke Gray 4 (Grumbacher®)
Black	Ivory Black (Pretested®), Mars Black (Winsor & Newton™)
White	Optical Gray #0 warm (Marabu®), Permalba (Weber®), Titanium White (Utrecht®), Yacht White (DuPont®), Zinc Weiss (Pretested®), Zinc White (Winsor & Newton™)

Paint tubes were sometimes dated, because he noticed some variation in colors among different batches of similar tubes from the same manufacturer. In fact, Albers usually used paints directly from the tubes, and only on rare occasions mixed them to reach specific tones, and adapted the palette knife technique according to the specific characteristics of the paint. The quality of texture and the paint thickness can be appreciated through a close viewing, especially by raking light (Figure 4.2B).

When he made his major pictorial cycle, *Homage to the Square*, from 1950 to 1976, Albers followed a specific routine. After the ground was applied, he drew lines with a ruler and a very hard pencil, 7H,[3] and the panel was laid flat on a table in his studio, where he sat under fluorescent lights. "With palette knife in one hand and a tube of paint in the other (Figure 4.3), he applied the pigment from the center squares outward. The paint was laid down freehand, without any masking aid such as tape or a straightedge. The slow-drying character of oils meant that he could paint only one area at a time. To keep the boundaries crisp and to prevent the paint from smearing required a pause of days or even weeks to let the paint in one square dry enough before proceeding to the next square. This meant that Albers worked on several pictures at once."[4].

The signature of his paintings was typically incised, including the letter 'A' and the last two ciphers of the date.

As Garland observes,[5] many of his early paintings were varnished with a Grumbacher® matte varnish (PVA) mixed with wax, or 'M' varnish, as Albers would note it. He later abandoned this, because he found, as described

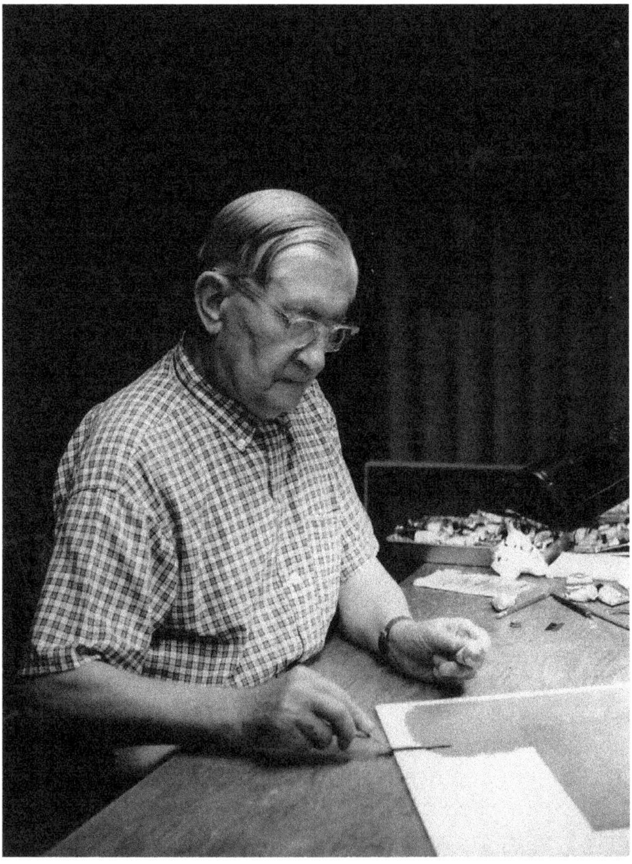

Figure 4.3 Josef Albers in his studio, New Haven, Connecticut, 1961. Photo: Alan Ludwig. Courtesy of the Josef and Anni Albers Foundation. Reproduced with permission.

in some of his correspondence, that a bloom would appear over certain colors, particularly Cobalt Blue. For perhaps the longest period he used *N*-butyl methacrylate (Hypalon™ P-4, Lucite®-44). Albers was never really pleased with the glossiness of this varnish and continually looked for alternatives, including combining the two mentioned varnishes in alternating layers. The paintings varnished with 'M' varnish exhibited highly discolored surfaces.[5]

Among the varnishes indicated on the *verso* were: Mat Varnish (Weber®), Preserving Varnish (Pratt + Lambert®), Damar Varnish (Grumbacher®), Methacrylate resin solved in Xylene, Poly Vinyl Acetate in Toluene. Eonite™ was used at least until 1959. In the early 1960s, Albers frequently let a few collaborators varnish his paintings.

In some color tests on paper we can find half-varnished color samples, in order to foresee how the colors might be altered in the final painting.

Albers was careful also in mounting and framing his works.[6]

4.3 The Performed Analyses

Firstly, a selected group of paintings were examined in Perugia, Italy, during the exhibition *Josef Albers: Spirituality and Rigor*[7] using non-invasive tools by the University of Bergamo, the University of Perugia, CNR-ISTM and the Laboratorio di Diagnostica Beni Culturali di Spoleto.

More than 30 paintings were studied by means of Infrared Reflectography (IRR), False Color IR (IRC), Macrophotography and a group of spectroscopic techniques useful for detecting pigments and other organic and inorganic materials: Reflectance Spectroscopy in the UV–visible–IR range (FORS or vis-RS), UV–visible Fluorescence (FOFS), X-ray Fluorescence (XRF), Raman Spectroscopy (Raman), Mid-IR and Near-IR Spectroscopy (MIR and NIR). Among them, the 18 paintings studied in depth are listed in Tables 4.2 and 4.3. In particular, Table 4.3 is focused on the *Homage to the Square* compositions (henceforth *HTS*) from 1951 to 1976, while Table 4.2 is related to some of his earlier paintings, with different characteristics, dating from 1937 to about 1952.

Successively, some significant paintings were examined *in situ* by Gianluca Poldi at the Albers Foundation in Bethany, Connecticut, only by means of IRR and vis-RS (Figure 4.2A). Colorimetric data of the measured points were acquired, too. A few paintings belonging to private collections were also examined.

4.4 Results

4.4.1 IR Imaging: Underdrawing and Other Technical Aspects

Imaging exams proved to be an important tool in investigating Albers' paintings, for technical as well as conservation issues.

Table 4.2 Josef Albers' paintings examined, dating 1937—*ca.* 1952, with the main results of the spectroscopic analyses. The last line of results refers to the white ground and to binders or varnishes, if detected. All images courtesy of the Josef and Anni Albers Foundation. Reproduced with permission.

Painting	***Proto Form A*** 1937 Oil on masonite 23 × 24 inches Inv. 1976.1.100	***White Cross*** 1937 Oil on masonite 18 × 20 inches Inv. 1976.1.1006	***Slanting Cross (on green)*** 1938 Oil on masonite 17 3/8 × 22 3/4 inches Inv. 1938.1.6	***Related I (red)*** 1938–1943 Oil on masonite 24 × 18 inches Inv. 1976.1.1015	***Variant of Related*** ca.1943 Oil on masonite 16 × 13 inches Inv. 1976.1.1052
	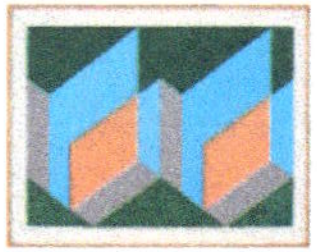	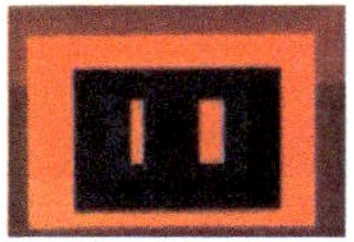	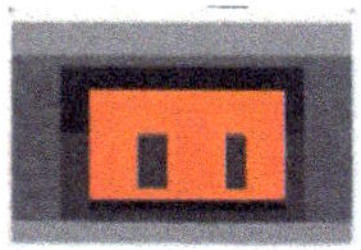	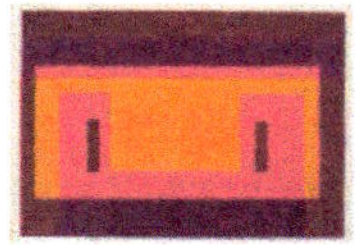	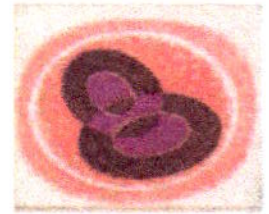
Albers' color notes on verso	**Gründerung:** hinten: 1 X Leimwasser 2 X flat paint vorn: Leimwasser Spachtel mit Wehltegrund 2 X Leimwasser 1 X formalin 1 X Preserving Varnish (Pratt + Lambert) **Malerei:** reine ölfarbe ohne Malermittel Pretested Cadmium Light (Schminke) (Untermalung) darüber: Cadmium (_____farbe -- Neisch) Cobalt Violett (Winsor + Newton) Zinc Weiss (Pretested) mit Harrison Red (Winsor + Newton)	No color notes on *verso*	No color notes on *verso*	**Ground:** front: 3 X Texolite, sandpapered [...] **Painting:** Center: Cadmium Red Light (Greco) outer frame: Harrison Red (Winsor + Newton) Background: Upper part: Cadmium Red Deep (Pretested) lower part: half and half mixture of Cadmium Red Deep and Medium (Pretested) all without painting medium Design + front first painted 1937 painted anew 1943 **Varnished** with Weber Matvar January 1944	After sanding off a former painting 3 coats of Texolite (Amer. Gypsum Co.) [...] **Painting:** figures: Center: Cadmium Red Light (Permanent Pigment) Outer parts: Harrison Red (Winsor + Newton) Rectangle: Cadmium Red Deep (Pretested) Center: On top of Cad. R.D. into wet ground Carmin (Winsor + Newton) Background: Ultramarine [...] No. 2 (Winsor + Newton) All colors without any paint medium put on with a palette knife **Varnished** with Weber Mat.

Main results	**White:** Zinc oxide **Pink:** Organic dye, zinc oxide **Purple:** Co–Mg arsenate **Red:** Red organic dye over strontium chromate **Yellow:** Strontium chromate Kaolin, oil	**White:** Zinc oxide **Purple:** Top: Co–Mg arsenate, zinc oxide Bottom: Co–Mg arsenate **Black:** Little iron oxides Gypsum, oil	**Light green edge:** Cu-based green, little zinc oxide **Green:** Cu-arsenate green, zinc **Dark green:** Hydrated chromium oxide (viridian), little zinc oxide **Blue:** Cobalt blue (with zinc oxide in lighter hues)**White** (along the lines): Zinc oxideCa-carbonate ($CaCO_3$), oil	**Dark Red** (upper and lower), **Red** (rectangular frame) and **Orange** (inner shapes): Different Cd-based reds, zinc **Brilliant red** (frames of the inner shapes): Organic red, zinc Oil, kaolin, barium sulfate	**Blue:** Ultramarine blue, little cobalt-blue (probably belonging to the previous version), zinc **Dark Red** (inner rectangle): Organic dye (on cadmium red), zinc **Red** (rectangular frame) and **Orange** (inner shapes):Different Cd-based reds**Brilliant red** (frames of the inner shapes): Organic red Kaolin, oil, barium sulfate

Painting	Tautonym B 1944 Oil on masonite 16 × 24 inches Inv. 1976.1.1036	Variant/Adobe, N. M. Black - Pink [New Mexico Black Pink] 1947 Oil on masonite 14 × 27 inches Inv. 2010.1.1	Variant/Adobe, Pink Orange Surrounded by 4 Grays 1947–1952 Oil on masonite 15 × 27 inches Inv. 1976.1.1380	Variant/Adobe, Familiar Front 1948–1952 Oil on masonite 13 × 21 inches Inv. 1976.1.1383
	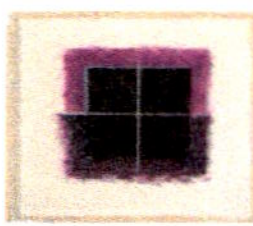	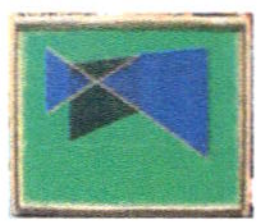	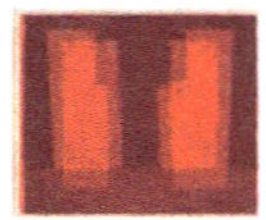	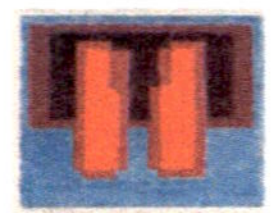

(*continued*)

Table 4.2 (*continued*)

Albers' color notes on verso	**Ground**: 3 X Texolite after sanding off former painting **Painting**: Top: Cadmium Green (Shiva) Bottom: Emerald Gren [sic] (Malfa) Center: Reilly's Grey #7 [sic] Monestral Blue Light (Greco) with Permalba (Weber) Cadmium Red Light (Pretested) with Permalba (Weber) All colors pure, without any painting medium Applied with palet knife [sic]	**Ground:** 3 coats of Texolite on front, sand papered […] + 1 coat of semigloss enamel painting: All paints applied with palet knife [sic] No painting medium, no under, no over painting **Painting:** […] Rose: Mixture of Cadmium Red Light (Pretested) with Zinc White (Winsor + Newton) Orange: Mixture of Cadmium Red Light (Pretested) with Cadmium Yellow Deep (Greco) Gray: Schminke Gray IV (Grumbacher) Black: Ivory Black (Pretested) Browns: Venetian Red (Greco) Terra Rosa (Winsor + Newton)	**Ground:** 4 coats of Luminall with linseed oil, varnish, + Turpentine [word] of [word] **Painting:** Colors used (from center) Mixture of Cadmium Red Light (Pretested) + Zinc White (Winsor + Newton) Cadmium Orange (Pretested) Reilly's Gray 7 Reilly's Gray 4 outer margins: Left + right Schminke Gray 3 […] top + bottom Chapin Neutral Gray III (Shiva)	**Ground:** 4 coats Luminall + 1 coat Gesso with little linseed oil, varnish, Turpentine **Painting:** Colors used (from center): Cadmium Orange (Winsor + Newton) Mixture of alizarin Crimson (Pretested) + Zinc White (Winsor + Newton) Cobalt Violet (Winsor + Newton) Thio Violet (Pretested) All applied in one primary coat No under, no overpainting No additional painting medium **Varnished** [19]55 with "M" Varnish
Main Results	**Green** (Upper part): Hydrated chromium oxide (viridian) (Lower part): Cu-based green (Around signature): Green Cu-phthalocyanine **Blue:** Blue Cu-phthalocyanine **Pink:** Cd-based red **Gray**: Bone black Kaolin, barium sulfate, oil	**Brown** and **dark brown:** Iron oxides **Orange:** Cd-based red **Pink:** Cd-based red, zinc oxide Kaolin, barium sulfate, oil	**Gray:** Zinc oxide **Orange:** Cd-based red **Pink:** Cd-based red **White** (borders): Titanium dioxide (rutile) Kaolin, barium sulfate, oil	**Magenta red:** Organic dye, titanium dioxide (rutile) **Purple:** Mn-phosphate (mineral violet), Co-ammonium phosphate, zinc oxide **Pink:** Organic red (similar to carmine), zinc oxide **White** (borders): Titanium dioxide (rutile) Kaolin, barium sulfate, oil

Table 4.3 Josef Albers' paintings examined, belonging to the *Homage to the Square* series, dating 1951–1976, with the main results of the spectroscopic analyses. Studied areas (colors) are named from outside to inside. The last line of results refers to the white ground and to binders or varnishes, if detected. All images courtesy of the Josef and Anni Albers Foundation. Reproduced with permission.

Painting	***Homage to the Square*: *Renewed Hope*** 1951 Oil on masonite 24 × 24 inches Inv. 1976.1.1343	***Homage to the Square*** 1952 Oil on masonite 16 × 16 inches Inv. 1976.1.1312	***Study for Homage to the Square: Aurora*** 1957 24 × 24 inches Oil on masonite 24 × 24 inches Inv. 1976.1.1828	***Study for Homage to the Square, Over the Land*** 1959 Oil on masonite 18 × 18 inches Inv. 1976.1.318	***Homage to the Square*** 1962 Oil on masonite 24 × 24 inches Inv. 1976.1.587
	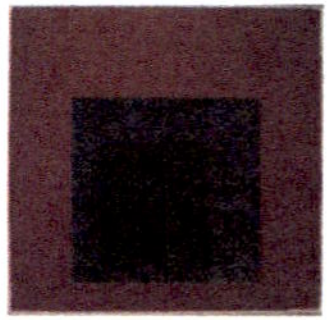	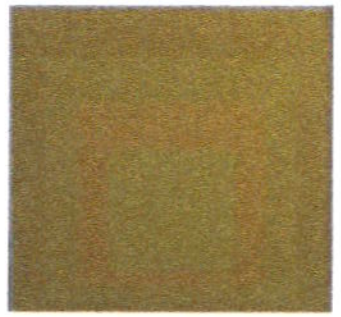	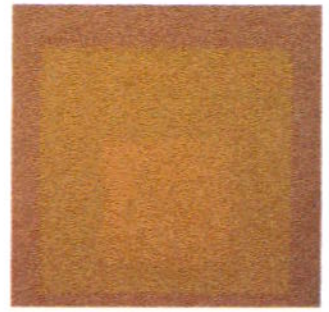	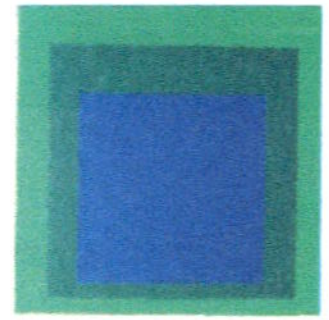	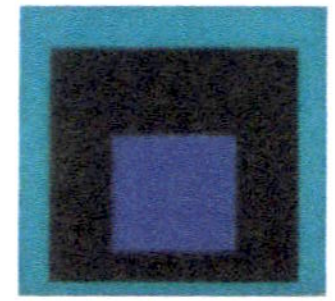
Albers' color notes on verso	**Ground**: 5–6 coats of Texolite with little Linseed oil, varnish, Turpentine **Painting**: Colors used (from center): Cerulean Blue Shade (Pretested) Chrome Oxide Green Light (Shiva) Cobalt Green (Winsor + Newton) all in one coat directly from the tube without any painting medium **Varnish** end of 55 with Eonite	**Ground**: 4 coats of Kemtone with Linseed oil, Varnish, Turpentine **Painting**: Colors used (from center): Cerulean Blue Shade (Pretested) Chrome Oxyde Green Light (Shiva) Cobalt Green (Winsor + Newton) all in one coat, directly from the tube – no painting medium Outer margin overpainted with Cobalt Green Deep (Greco) **Varnish**: Damar Varnish (Grumbacher) brushed on 1955	**Ground**: 2 coats of DuPont Flat White } alkyd + 2 [coats of DuPont Flat] Semigloss "} Enamel **Painting**: Paints used – from center: Cadmium Orange (Rembrandt) Cadmium Yellow Deep (Blockx) Aureolin (Rembrandt) [...] **Varnish**: "M" varnish	**Ground**: 4 coats of Luminall (casein) with Linseed oil, Damar Varnish, Turpentine **Painting**: Paints used – from center: Center top: Cobalt Green (Winsor + Newton) Center bottom: Cadmium Green (Winsor + Newton) Reilly's Gray #4 (Grumbacher) Gold Ochre (Rhenish) **Varnish**: first: Metacrylate resin solved in Xylene Second: Eonite	[**Ground**:] 6 X Liq[uitex?] [**Pigments**:] Paynes Gray Shi[va] Burnt Umber Pret[ested] W + N Reilly's 3 Gru[mbacher]

(continued)

Table 4.3 *(continued)*

Main results	**White**: Titanium dioxide (rutile) **Cyan**: Cerulean blue (mixed with a green pigment?), zinc oxide **Dark green**: Anhydrous chromium oxide green **Blue**: Blue Cu-phthalocyanine, zinc oxide Kaolin, acrylic, barium sulfate	**White**: Titanium dioxide (rutile) **Green**: Cerulean blue (mixed with a green pigment?), zinc oxide **Dark green**: Anhydrous chromium oxide green **Blue**: Blue Cu-phthalocyanine, zinc oxide Barium sulfate, kaolin, calcium carbonate	**White**: Titanium dioxide (rutile) **Yellow**: Cobalt yellow (aureolin) **Orange** (inner frame): Cd-based yellow, zinc and barium (Central square): Cd-based orange Zinc (XRF)	**White**: Titanium dioxide (rutile) **Brown**: Red ochre, zinc **Gray**: Zinc oxide **Blue half square**: Cerulean blue, Co-aluminate, zinc **Green half square**: Hydrated chromium oxide green (viridian), Cd-based yellow pigment, zinc Kaolin, barium sulfate, oil	**White**: Titanium dioxide (rutile) **Gray**: Zinc oxide **Dark brown** (inner frame): Iron oxides, Mn-black **Black Square**: Ultramarine blue Mn-black Kaolin, barium sulfate, oil
Painting	***Homage to the Square: Tap Root*** 1965 Oil on masonite 40 × 40 inches Inv. 1976.1.1809 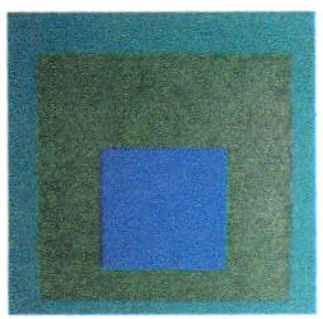	***Homage to the Square*** 1967 Oil on masonite 28 × 28 inches Inv. 1976.1.738 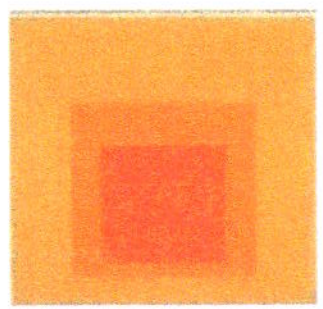	***Homage to the Square: Tropical*** 1968 Oil on masonite 40 × 40 inches Inv. 1976.1.811 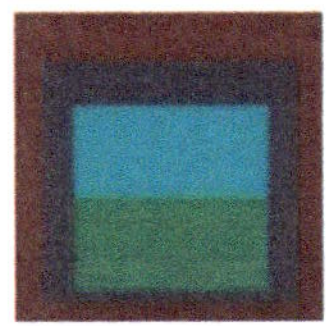	***Homage to the Square*** 1976 Oil on masonite 24 × 24 inches Inv. 1976.1.524 	

Albers' color notes on verso	**Ground**: 7 coats of Liquitex (Permt Pigment) **Painting**: Paints used – from center: Mars Black (Winsor + Newton) Reilly's Gray #3 (Grumbacher) Light Red (Shiva) [...] **Varnish**: Poly vinyl Acetate in Toluene by Charles Tauss	No color notes on *verso*	**Ground**: 6 coats of Gesso: Liquitex (Permt Pigment) **Painting**: Paints used – center first: Cadmium Bar. Yellow Deep (Permt Pigment) large: Cadmium Yellow Medium 3 (Lefebvre) Naples Yellow Reddish (Lukas) **Varnish**: Poly Vinyl Acetate in Toluene sprayed on by Charles Tauss NYC	No color notes on *verso*
Main results	**White**: Titanium dioxide (rutile) **Brown red**: Red-brown iron oxides **Gray**: Zinc oxide, Cd-based pigment **Black**: Iron oxides Kaolin, barium sulfate, oil	**White**: Titanium dioxide (rutile) **Yellow** (outer frame): Lead antimonate yellow, zinc (Second and internal frame): Cd-based yellow, zinc (Inner square): Cd-based yellow, zinc Kaolin, barium sulfate, oil	**White**: Titanium dioxide (rutile) **Orange-pink** (outer frame): Cd-based yellow, lead antimonate yellow **Yellow** (Inner frame and Central Square): Cd–Se-based yellow/orange Kaolin, barium sulfate, oil	**White**: Titanium dioxide (rutile) **Light green**: Green Cu-phthalocyanine, yellow pigment **Dark green**: Cerulean blue (mixed with a green pigment?) **Blue**: Blue Cu-phthalocyanine, zinc oxide, cobalt blue Acrylic, kaolin, barium sulfate, calcium carbonate

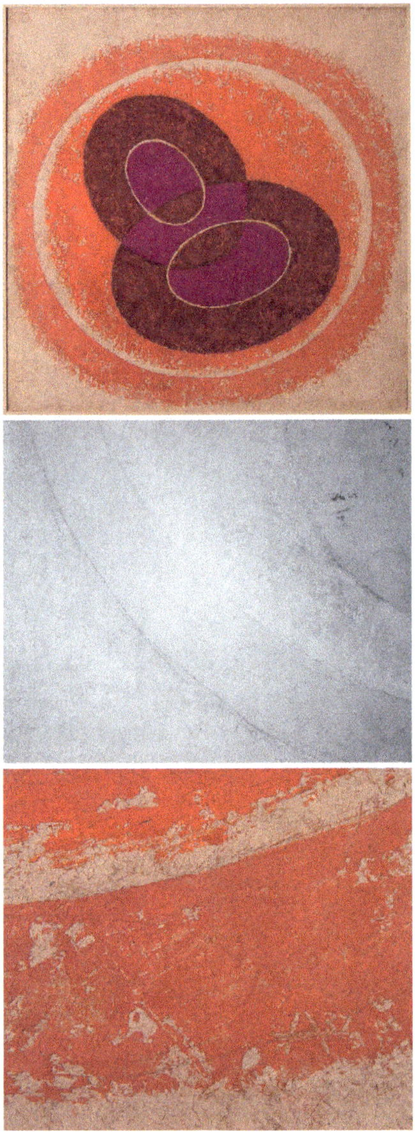

Figure 4.4 Josef Albers, *Proto Form A*, 1937, oil on Masonite, *The Josef and Anni Albers Foundation*, Bethany (inv. 1976-1-1002). The painting (top), an IRR detail (center) and a different visible detail (bottom). Photos: Gianluca Poldi. Courtesy of the Josef and Anni Albers Foundation. Reproduced with permission.

In many cases, IR reflectography allows us to visualize a very thin underdrawing used to organize the space of the composition (Figures 4.4 and 4.7B). This underdrawing is more readable in paintings on paper where small holes due to a pin can be discovered under the corners of the polygonal painted areas, as in the *Adobe* series. Drawing lines are less or not at all visible where the support is rough, like in many works on Masonite: of course its visualization increases under IR-transparent pigments, depending on the paint layer thickness, the smoothness of the support and ground and the contrast between the drawing lines and the ground. The use of a knife usually means quite thick layers and greater thicknesses at the edge of each painted zone, making the underdrawing more difficult to detect, as it lies right along the separating edges between the colored areas. A few changes in the drawing were noted. Sometimes lines were prolonged under uniform areas so as not to lose continuity. In some cases the lines of construction can be seen at the borders of the external colored frame, close to the white painted edges of the painting, over the white ground.

In the less geometric works the underdrawing is free, as we can appreciate under *Proto Form A* (1937) (Figure 4.4) and in some works studied in Bethany.

IRR images confirm the use of a sharp pencil as a drawing tool, probably a hard graphite like the 7H reported by the sources.

In a few paintings, IRR revealed some inhomogeneity of the painted surface, generally not depending on conservation problems but, more typically, on painting process: different thicknesses inside the same area or small stains of other pigments, used first in surrounding zones and then covered when completing the painting. Variations in thickness inside the same shape can be appreciated through IR images.

A few modifications – errors or changes of mind when the painting was in progress – concerning colors used in the same area can be noticed in his earlier paintings, with double layering (*i.e.* blue over red, *etc.*). For example, in the upper right area of *Variant of Related* Albers puts the cadmium red pigment used in the red shapes before hiding it with the pale blue. Nothing can be recovered of the previous composition that was sanded off, as Albers writes on the *verso*.

In *Tautonym B* IRR detects a change of pigment in the lower right corner in the area of the signature 'A 44', and IRC shows that the new green is laid over the original one, as spectroscopic analyses demonstrate (Figure 4.5). He probably didn't put his signature, or wanted to modify it after some time, but he didn't find the exact green pigment used for the original layer.

In the painting *Chalk Green Façade* (1960, private collection) IRR helped to recover an erased pencil writing on the panel *verso*, while UV fluorescence allowed the information about varnish to be read, added on the back by the artist after the color notes with a felt tip pen, now almost completely faded.

4.4.2 Spectroscopic Analyses: Pigments and Organic Compounds

Non-invasive spectroscopies permitted the identification of many compounds, including organic and inorganic pigments, the binder (typically siccative oil) and some varnish like PVA. A synthesis of the results is given in Tables 4.2 and 4.3.

Acrylic signals detected by MIR in the *HTS* of 1976 can be referred to the varnish, as in the *HTS: Renewed Hope* of 1951.

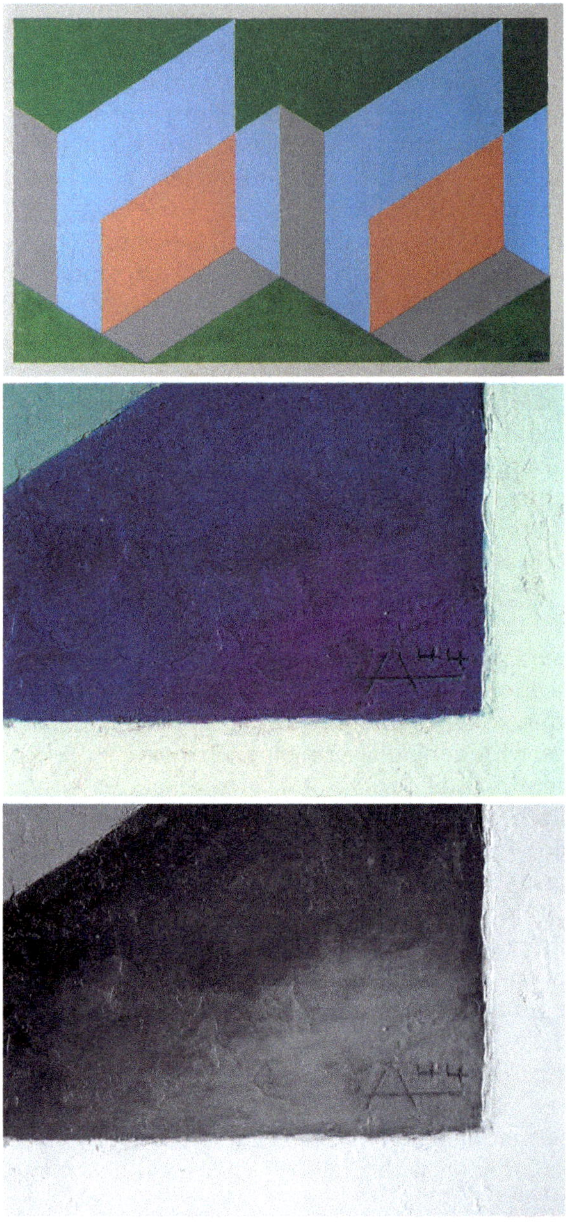

Figure 4.5 Josef Albers, *Tautonym B*, 1944, oil on Masonite, *The Josef and Anni Albers Foundation*, Bethany (inv. 1976-1-1036). The painting (top), a detail of the signature in false color IR (center) and in IRR (bottom). Photos: Gianluca Poldi. Courtesy of the Josef and Anni Albers Foundation. Reproduced with permission.

4.4.2.1 Whites

MIR signals of barium sulfate (Ba, S in XRF spectra) and kaolin can be detected from the 1940s onwards. Kaolin was typically used as an extender, while barium sulfate was used for the white ground. In earlier pieces from 1937 no barium is found in the preparatory layer and calcium is more abundant, as for *White Cross* where calcium sulfate (gypsum) is detected.

The presence of zinc oxide emerges in many paintings, where the Zn signal measured by XRF is confirmed by electronic spectra RS (vis-RS or FORS) because of a 390–400 nm shoulder, 370 nm absorption edge and 485–490 nm inflection point.

Titanium dioxide was found in two forms: anatase and rutile. The polymorph rutile was found first in a painting dated 1947–1952 belonging to the *Adobe* series, named *Pink Orange Surrounded by 4 Grays*, and in many *HTS* (Table 4.2), while anatase was found in the earlier *Tautonym B* of 1944 (Table 4.2). The anatase reflectance spectrum (RS) is similar to the zinc white one, with a shoulder at 390–400 nm but a 370 nm inflection point: the inflection points are different[8] but it's hard to distinguish anatase and zinc oxide by vis-RS when mixed. Where rutile is present in white areas, like the ground, RS spectra show a shoulder around 420–430 nm and a slope increasing around 380–390 nm. Rutile TiO_2 was preferred in the formulation of paints, including house paints,[9] due to its higher refractive index and hence improved ability to scatter light in paint films. It became available after the Second World War, *e.g.* found in paintings by De Kooning,[10] so this finding represents one of the first uses of this pigment.

When the contemporary presence of titanium and barium in some works makes it difficult to distinguish them in XRF spectra at low energies, the joined use of vibrational spectroscopies and RS (FORS and non-FORS) helps.

4.4.2.2 Grays and Blacks

Among the paints frequently used by Albers were tubes of the neutral gray series of pigments produced by many companies from the 1940s through to the mid-1960s that were made to help designers and illustrators to create accurate and repeatable mixtures of colors, from light tints to deep tones. Grumbacher® produced two neutral gray series: the 'Illustrator's Gray' series available in five tones and the 'Reilly's Neutral Gray' set of nine paints in progressive tones of gray, from dark (lower number) to light (higher number), created in the 1940s by illustrator Frank J. Reilly (1906–1967). The gray series produced by Shiva™ was marketed as 'Chapin Neutral Grays'. The Entwistle Company, under the Marabu® label, produced a series of 'Optical Grays' available in both warm and cool tones.[11]

Albers used the neutral grays directly from the tubes, exactly as colors.

In *Tautonym B*, where Reilly's Gray #7 was used, bone black signals were measured by Raman. In the other grays we couldn't easily find signals of the black component, but the multi-technique analytical approach allowed us to understand the differences among some of the gray paints used by Albers, particularly thanks to different chemical elements detected by XRF and the differentiation of RS spectra. *Variant/Adobe, Pink Orange Surrounded by 4 Grays*, painted in 1947–1952, offers a wide gamut of grays:

- the lighter tone at top and bottom is a Chapin Neutral Gray III (Shiva™) containing Zn, Ti, Fe, (Ca), where Fe refers to iron oxide content (coherently with the shape of vis-RS spectra) and Ti derives from rutile (Raman), but also ZnO is present;
- medium gray, Schminke Gray 3, also contains a few iron oxides (here vis-RS shows the 450 nm signal due to d–d Fe^{3+} transitions), while titanium is absent;
- dark gray, Reilly's Gray #7, contains Zn with traces of Ca, Ba, Fe;
- the darker hue is Reilly's Gray #4, with different proportions of the same elements (Cd, Se in a few points derive from the nearby Cd-based orange zone).

As we could verify on the other variants of Reilly's Gray examined, generally RS presents two small absorbance (abs.) bands, at 480–510, 610–660 and 700 nm, and maxima at 550 and 730 nm (weak): this pattern, although less evident on a dark surface, can be associated with the Cu-phthalocyanine pigment (probably PG7) that was added by the manufacturer to reach the needed tone.

In *Study for HTS* (1965) (not described in tables), the gray central square is Reilly's Gray 3, while the black external frame is Payne's Gray (Shiva™), whose spectrum contains ultramarine blue. The same RS spectrum is obtained for the Payne's Gray used in the black center of *HTS* (1962) (Table 4.3), where the external frame is again Reilly's Gray 3, but with some difference in the spectrum. Burnt Umber (Winsor & Newton™) is used in the intermediate dark brown shape. Mars Black (Winsor & Newton™), highly rich in Fe, was employed in the central black square of *HTS: Tap Root* (1965).

4.4.2.3 Blues and Greens

Looking at the studied paintings but also at many proofs on paper, Albers seems to prefer a small number of blue pigments, in particular cerulean (cobalt-based) and Cu-based phthalocyanines. Of course, they could appear under different paint names, and in different formulations or mixtures inside the color tubes, but invariably we find signals of one of these two pigments, particularly the 570–620 and 710 nm reflectance bands (reflectance maximum at 460–470, lower peak around 660 nm) typical of blue phthalocyanine[12] PB 15:3; or the 580, 610 and 660 nm abs. bands of cerulean Co-based blue. Sometimes he also used artificial ultramarine blue, like in *Variant of Related* (Table 4.2) and in *Untitled (Leaf Study I)* painted on paper (*ca.* 1940, not included in the tables).

Slanting Cross (on green) (1938) is the only example among those we studied of the use of cobalt blue. This was soon substituted in Albers' palette by phthalo-based blue, which was named *Monastral Blue* by many suppliers. Copper phthalocyanine blue was available to painters from 1935–1937,[13] but generally used later. *Tautonym B* (1944) is one of about 20 works made by Albers in 1940s and 1950s in which the pigment Monastral Blue (Greco® brand) appears.

The use of bright intense blue given by this organic pigment lasted more than 30 years, as can be seen in the central squares of some variants of *HTS* (Figure 4.6A).

Cerulean blue was frequently used by Albers for the green-blue tone, like that offered by paints named Cerulean Blue Shade (Pretested) or Cobalt Green (Winsor & Newton™): analyses suggest a mix between cerulean blue and another pigment, probably green but not cobalt green, intended as a mixture of CoO and ZnO to modify and shift the reflectance peak to 510–520 nm without affecting the curve shape as yellow pigments or dyes would do, but we cannot exclude a peculiar formulation of this Co-based pigment.

Apart from the green paints, the 1976 *HTS* (in the inner and intermediate square) is similar to *HTS* (1951) (inner and external squares): after 25 years Albers uses the same cerulean blue mixed with a green pigment and the same blue Cu-phthalocyanine; in fact, the RS spectra almost overlap (Figure 4.6A). In this case, the lack of color notes on the *verso* of *HTS* (1976) is replaced by spectroscopic info.

Regarding green colors, Albers used a wide range of green paints, well represented also in his color tests on paper.

In *Tautonym B* he employed a hydrated chromium oxide (viridian) in the upper area and a copper arsenate (Cu–As) green in the lower part. The above-mentioned change occurred in *Tautonym B* and can be detected in the area of the signature because Albers didn't use the same green pigment when he reworked it: he used a green Cu-phthalocyanine, probably PG7, as vis-RS suggests, which was produced from 1938.

The combined presence of cadmium and chrome in the upper part confirms Albers' reported use of Cadmium Green (Shiva™), a term for a composite pigment based on a Cd-containing yellow with viridian.[14] The same green pigments were used in *Slanting Cross (on green)*.

The title of the painting *Chalk Green Façade* of 1960 (private collection) probably refers to the pigment green earth, a term derived from this English translation of Vitruvius (first century BC): "Green chalk is found in many places, but the best is from Smyrna."[15] In this work, the three mentioned green pigments are used (Figure 4.6B): the dark Permanent Green Light (Pretested) is a viridian; the intermediate green Permanent Bright Green (Grumbacher® F.) is a PG7 green Cu- phthalocyanine – abs.

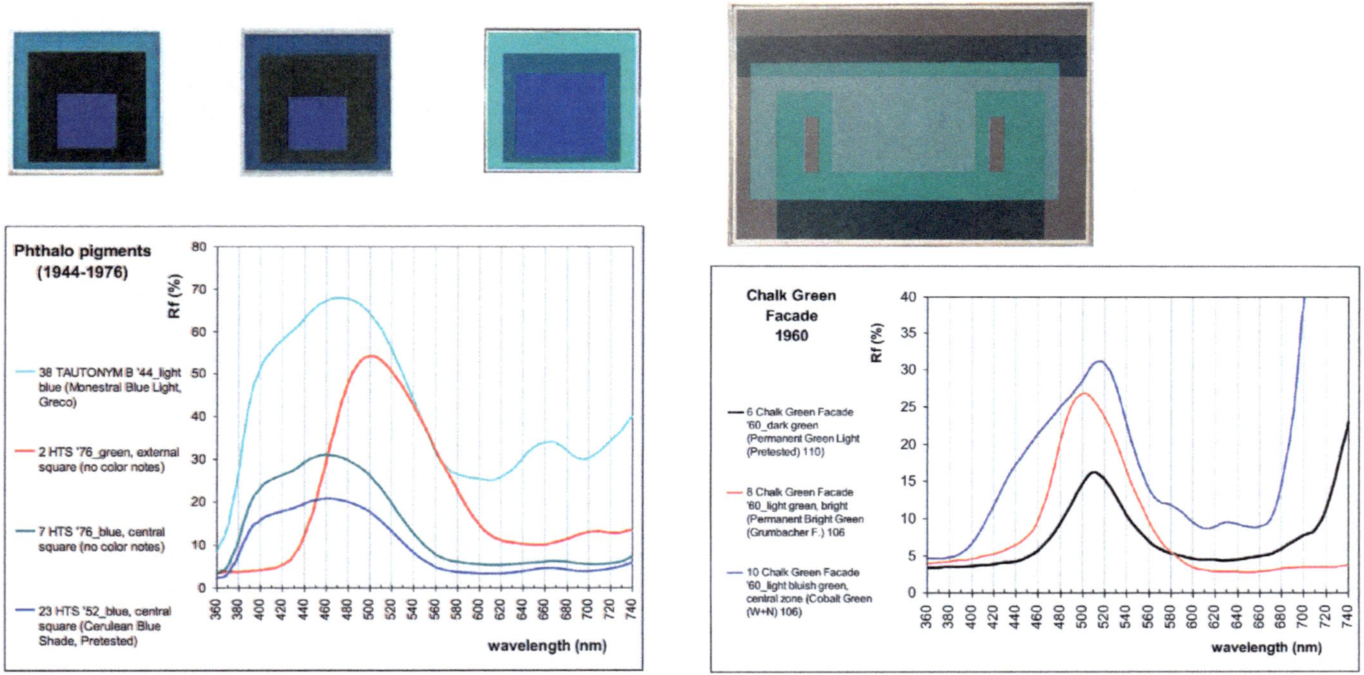

A

B

Figure 4.6 (A–B): (A) Vis–RS spectra of phthalocyanines measured on *Homage to the Square* of 1951, 1952 and 1976 (top left, from left to right) and on *Tautonym B*, 1944, all belonging to *The Josef and Anni Albers Foundation*, Bethany. All central squares are painted with blue Cu-phthalocyanines, the light green outer frame of *HTS* 1976 contains green Cu-phthalocyanine. Courtesy of the Josef and Anni Albers Foundation. Reproduced with permission. (B) Vis–RS spectra of the green areas of *Chalk Green Façade*, 1960 (top right), private collection. Blue line refers to cerulean blue, black line to viridian, red line to a PG7 Cu-phthalocyanine. Photos: Gianluca Poldi. Courtesy of the Josef and Anni Albers Foundation. Reproduced with permission.

bands at 620–650 (650 minimum) nm and 720–730 nm (weak), reflectance peak 500 nm, small peak at 710–720 nm – and the lighter green area is painted with the aforementioned Cobalt Green (Winsor & Newton™) containing cerulean blue.

Both green chromium oxides were easily detected by RS: hydrated chromium oxide reached a deep green color and anhydrous chromium oxide produced a marsh-green hue. We find the latter in two works of the early 1950s: the *HTS: Renewed Hope* (1951) and the *HTS* (1952). Viridian can be characterized by vis-RS and FORS also without its first peak at 420–430 nm – frequently inhibited by the presence of other pigments like yellow ones in the paint – thanks to the second one (510 nm), to the large abs. band and also to the small 710 nm absorption.

4.4.2.4 *Violets*

Cobalt and arsenic detected by XRF in the light violet color at the center of *Protoform A* and in the *White Cross* of the same 1937 date, refer to a Co–Mg arsenate, as RS spectra confirm. Its characteristics include 525–530 and 570 nm abs. bands (590 abs. edge), 450 nm and 550 nm (weak) peaks and a 620 nm inflection point (650 nm shoulder), as reported by literature[16] (see the violet curves in Figure 4.7D). In the lower purple squares of *White Cross* a different RS spectrum can be noticed: 520–600 nm wide abs. band and a 420 nm reflectance peak (black curve in Figure 4.7D), which can be interpreted as a more concentrated pigment in the darker hue, or as the presence of another pigment added to the Co–As-containing paint.

White Cross has no color note on the *verso*, but at least in the upper squares the paint should be the same Cobalt Violet (Winsor & Newton™) written on the back of *Protoform A*.

Despite the same color note, "Cobalt Violet (Winsor + Newton)", a different violet pigment seems to be used in the purple horizontal strips of *Variant/Adobe, Familiar Front* (1948–1952), showing both the features of a Co phosphate (prob. Co-ammonium phosphate) and Mn phosphate (mineral violet): *i.e.* 430+, 490−, 550−, 600 edge, 670+ nm, where the peaks at 430 and 670 nm refer to mineral violet (Figure 4.7C).

4.4.2.5 *Reds, Oranges, Yellows, Pinks and Browns*

Cadmium-based pigments of different compositions were widely used by Albers for obtaining red-orange and yellow hues, at least after *ca.* 1940. Before this date a number of red tones composed by organic dyes were used as such, or overlapped with other colors.

In *Protoform A* (1937) so-called Cadmium Light was used in the pink hues, but since no Cd signals have been detected, it is very probable that the pigment name had nothing to do with its composition, being instead related to the hue it provided, as often happens with commercial pigment names. An

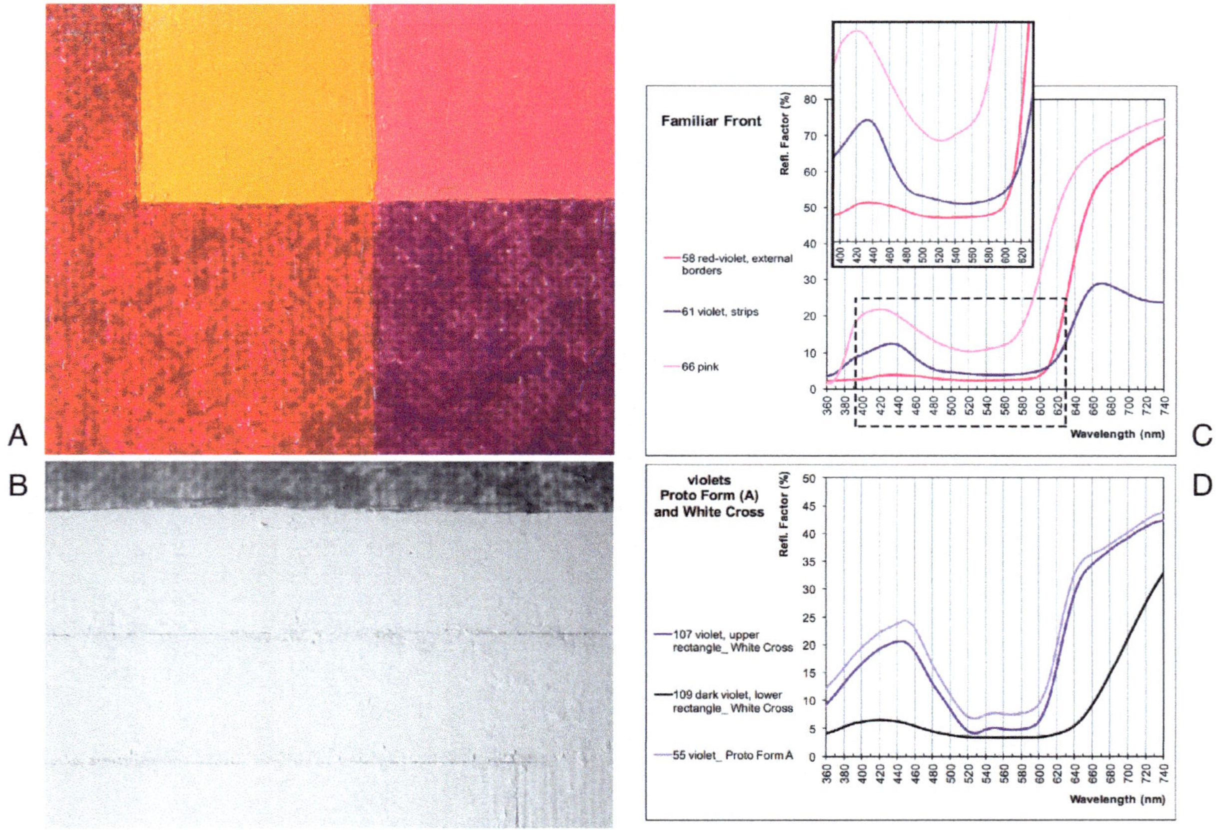

Figure 4.7 (A–D): (A–B) Josef Albers, *Variant/Adobe, Familiar Front*, 1948–1952, oil on Masonite (inv. 1976-1-1383), *The Josef and Anni Albers Foundation*, Bethany. Visible light detail (A) and a different IRR detail (B). Photos: Gianluca Poldi. Courtesy of the Josef and Anni Albers Foundation. Reproduced with permission. (C) Vis–RS spectra of pink, red and violet areas of the previous painting, with a focus on the absorption bands region. (D) Vis–RS spectra of violet areas of *Proto Form A*, 1937 (inv. 1976-1-1002) and *White Cross*, 1937 (inv. 1976-1-1006), both oil on Masonite belonging to *The Josef and Anni Albers Foundation*, Bethany. Spectra are commented in the paper. Courtesy of the Josef and Anni Albers Foundation. Reproduced with permission.

underlayer of yellow strontium chromate constitutes the red circles whose surficial finish is recorded as Harrison Red on the *verso*, a pigment based on organic dye.

Starting from 1938, a wide array of Cd-based reds having different formulations appear within Albers' palette for red, orange, yellow and even pink areas.

In *Related Red I* (1938–1943) two different kinds of red, both Cd-based, are reported on the *verso*: namely Cadmium Red Deep for background and a mixture of Cadmium Red Deep and Medium in the lower area. These areas are characterized by high amounts of cadmium and selenium (XRF) and by a small shoulder at 520 nm in RS spectra. In the light red frame of the central orange areas, where Harrison Red is mentioned, no cadmium signals emerge, with zinc, barium and iron being the main elements instead. Here, an organic red similar to an anthraquinone-based lake is suggested by RS spectra, with abs. band at 540–550 nm; a result that is coherent with Raman, which detected weak skeletal vibrations belonging to poly-fused aromatic rings.

The central dark orange edges in *Variant of Related* (1943) show Raman and RS signals similar to the aforementioned ones. Within this area Harrison Red is once again reported to be used, as for the previous discussed paintings. Under this name at least two kinds of pigments have been known since the early 1890s: Lithol red[17] (PR 49), a very fugitive dye, and Toluidine red (PR3). They both share the structure common to β-naphthol azo pigments.

The Harrison Red sold by Duro® paint company was a Toluidine red.[18] Toluidine was a common bright red pigment in the early 20th century, less expensive than the more reliable cadmiums and vermilion, and more permanent than the various aniline lakes.

It is worth noting that the Harrison Red used by Albers was provided by Winsor & Newton™, which sold also Toluidine red but with the commercial name of Winsor Red.[19]

Furthermore, some peculiar Raman evidences belonging to skeletal vibrations of PR49[20–22] let us hypothesize that the Harrison Red used by Albers was a type of Lithol red, a pigment which deserves a little more attention.

Lithol red is an azo pigment discovered in the late nineteenth century by an Austrian chemist and used also by Mark Rothko in his Seagram and Harvard murals of the 1950s and 1960s.[23,24] Lithol red was widely available from the 1920s onwards and comprises a family of sodium (PR49), barium (PR49:1), calcium (PR49:2) and strontium (PR49:3) salts of diazotised Tobias (2-naphthylamine-1-sulfonic) acid coupled with 2-naphthol. The metal dictates the shade of the pigment, which ranges from yellow to purplish-blue: the red one has Barium as the coordinating metal whose presence agrees with our XRF data, being indeed the Ba and Ca types that are most commonly found. The primary drawback of Lithol red is its poor lightfastness, which could explain the difficulty in finding strong signals by Raman spectroscopy. Lithol red in its Harrison Red version may be well responsible for the dark red edges of *Related Red I* (1938–1943) and *Variant of Related*, as well as for the red areas of *Protoform A* (1937). The red palette of *Variant of Related* (1943) includes another red paint, Carmin, used for the central rectangle which is also a Cd-based one.

A mixture of Cadmium Red Light and Cadmium Yellow Deep is reported for the orange hues of *New Mexico Black Pink* (1947). They both appear to be composed by a Cd-sulfide-selenide, as revealed in their emission spectra, which prove to be a useful and crucial integration when light elements such as sulfur are not appreciable due to matrix effects. A Cd-sulfide-selenide, recorded on *verso* as Cadmium Orange and different from that of *New Mexico Black Pink* (1947), detected by its emission, is found in the orange areas of *Pink Orange Surrounded by 4 Grays* (1947–1952).

The widespread use of Cd-based compounds expands also within the *HTS* series, where yellow hues are the main characters. They are composed of a wide range of cadmium sulfides which differ from each other, to reproduce an ample range of yellow shades, as in *Homage to the Square* (1967) and *Homage to the Square: Tropical* (1968); they are sometimes mixed with Cd-sulfide-selenides to obtain orange tones, as in *Study for Homage to the Square: Aurora* (1957).

Few yellow areas are shown to be painted with a non-cadmium based pigment, as the outer frame of *Homage to the Square* (1967) whose main elements – lead (Pb) and antimony (Sb) – as well as its RS spectrum remind us of the old-fashioned lead antimonate, the so-called 'Naples yellow' which is mentioned as 'Naples yellow reddish' on the *verso* of *Homage to the Square: Tropical* (1968), the reddish hue being due to a Cd-based red.

The other cadmium-free yellow part involves the noteworthy presence of a very sparingly used Co-based yellow pigment, *i.e.* aureolin, strikingly evident both from its elemental and molecular features.

A very limited pink palette emerges with the red-orange one. *Protoform A* (1937) and *Familiar Front* (1948–1952) are characterized by an organic dye resembling a madder lake structure, while a Cd-based pigment reported as Cadmium Red Light was chosen for the pink of *New Mexico Black Pink* (1947) and *Pink Orange Surrounded by 4 Grays* (1947–1952), as well as *Tautonym B*, being in each case mixed with ZnO to obtain lighter hues.

The magenta hue in *Familiar Front* (1948–1952) is indicated as Thio Violet by Pretested®. Interestingly, it is still currently on the market by the same company, and it is composed by quinacridone magenta, a dye patented in 1955, later than the painting creation. Evidently a formulation change occurred which favored the quinacridone component over an older composition based upon an organic dye whose features are coherent and resemble very closely those of thioindigo, an indigo-derived molecule in which two NH groups are replaced with two sulfur atoms ('thio'), and by which the commercial name could have been inspired.

Mostly iron oxides were found in brown areas: in *New Mexico Black Pink* (1947) and *Homage to the Square: Tap Root* (1965) Venetian Red and Light Red

(Shiva™) are indicated on *verso* to be used respectively, whose signature is clearly visible from NIR and from XRF analyses revealing iron as the main component, similar to ochre and brown earth.

Finally, in the *Adobe* painting *Browns, Ochre, Yellow* (1948, inv. 1976.1.1384), studied only by means of vis-RS and not included in the tables, all the pigments have the typical shape of iron oxides, despite the red-brown of the central rectangles which is named Cadmium Red Purple (Shiva™).

4.5 Conclusion

This non-invasive survey of Albers' palette allowed a large part of it to be characterized, addressing at the same time new research items for those pigments that are still unidentified and allowing a database to be built which could serve as a compass with which to navigate the manifold pigment and dye formulations blooming at the beginning of the 20th century.

Going from earlier Albers' artworks – more experimental in the use of pigments, both in their choice and application, including layering – to his later works, his rigorous approach to artistic materials and material characterization emerged, marked out by his need to define a method, through systematic practice. The huge number of proofs concerning colors, materials and tools testify to his rigor and provide a natural foundation upon which an impressive level of knowledge could be achieved in identifying a large selection of 20th-century art materials.

Albers' methodical, we can say taxonomical, approach to rigorous inscriptions is well rooted in Germany's tradition and in that of the Bauhaus, and comprehensible if we look at him also as a great teacher that invited his students to open their minds and experiment, but with a deep asset of knowledge. His rigorous approach emerges also from his major writing, *Interaction of Color* (1963), where he reminds us of the need to experiment with rigor before building a theory and systematic practice. As he wrote in the first page of that book:

"In visual perception a color is almost never seen as it really is – as it physically is. This fact makes color the most relative medium in art. In order to use color effectively it is necessary to recognize that color deceives continually. To this end, the beginning is not a study of color systems."[25]

But Albers' methodical archiving work tells us also the basic role of knowledge of the color as material, the knowledge and proof of some of its characteristics, and his archives constitute an impressive heritage for conservation scientists too, today.

Appendix: Materials and Methods

Non-invasive analyses on paintings during the exhibition were performed without removing them from the wall, by means of portable instruments.[26]

Visible photographs and IR imaging (IR reflectography 0.8–1.1 micron and False Colour IR) were carried out with a modified Sony® camera and proper filters, using a halogen light (1000 W). For colorimetric measures and some Reflectance Spectroscopy measures in the 360–740 nm range (vis-RS) a Konica Minolta® CM2600d handheld spectrophotometer was used (10 nm acquisition step and 3 mm diameter spot). These preliminary exams were made both in Italy and the United States by Gianluca Poldi, Visual Art Center (CAV), University of Bergamo.

The SMAArt Centre, University of Perugia, together with CNR-ISTM, Perugia carried out a wide range of spectroscopic analyses, listed below.

Reflectance MID-FTIR spectra were recorded using a compact portable Bruker® Optics ALPHA-R spectrophotometer. Reflectance FT-NIR spectra were recorded using a compact portable JASCO™ VIR 9600 spectrophotometer equipped with a near-infrared fiber optic sampling probe. UV-Vis reflectance and fluorescence system is a prototype that collects spectra on the same point of investigation within different excitation wavelength ranges, by means of several sources/detectors and a multi-furcated fiber optic cables system. The XRF equipment is composed of a miniaturized X-ray generator EIS P/N 9910, equipped with a tungsten anode and a SDD detector (resolution of about 150 eV at 5.9 keV) cooled with a Peltier element. The Raman prototype is assembled from various components and has two lines of excitation: Laser Nd:YAG with excitation @ 532 nm and diode laser (TOPTICA) @ 785 nm.

Acknowledgements

This research, coordinated by Gianluca Poldi (CAV, University of Bergamo) and by Chiara Anselmi (SMAArt Centre, University of Perugia), and supervised by Jeannette Redensek for The Josef and Anni Albers Foundation, wouldn't have started without the precise interest of the Foundation Director Nicholas Fox Weber, Laura Mattioli and Antonio Sgamellotti, to whom we are indebted. The authors would like to thank the other colleagues who took part to the Perugia spectroscopic campaign: David Buti, Brenda Doherty, Francesca Gabrieli, Chiara Grazia, Costanza Miliani, Patrizia Moretti and Aldo Romani.

We also want to thank Fabio De Chirico, former director of the Soprintendenza di Perugia, the staff of the Galleria Nazionale dell'Umbria and the staff of The Josef and Anni Albers Foundation, Jeannette Redensek above all.

References

1. *Painting on Paper. Josef Albers in America*, ed. H. Liesbrock and M. Semff, catalogue of the exhibition (Basel, Kunstmuseum, 5 November 2011–29 January 2012), Hatje Cantz Verlag, Ostfildern 2011.
2. J. Redensek, On Josef Albers' Painting Materials and Techniques, in *Josef Albers, Minimal Means, Maximum Effect*, catalogue of the exhibition (Madrid, Fundación Juan March, 28 March–6 July 2014), Fundación Juan March, Madrid 2014, pp. 21–40.

3. E. de Kooning, Albers Paints a Picture, *Art News*, 1950, **49**(7), , 40–43 and 57–58.
4. J. Redensek, On Josef Albers' Painting Materials and Techniques, in *Josef Albers, Minimal Means, Maximum Effect*, catalogue of the exhibition (Madrid, Fundación Juan March, 28 March–6 July 2014), Fundación Juan March, Madrid 2014, pp. 36–38.
5. P. Sherwin Garland, Josef Albers: His Paintings, Their Materials, Technique and Treatment, *J. Am. Inst. Conserv.*, 1983, **22**(2), 62–67.
6. P. Sherwin Garland, I could paint that! Forgery of an Homage to the Square, *AIC Paintings Specialty Group Postprints*, 2008, **20**, 82–88.
7. *Josef Albers: Spiritualità e rigore/Spirituality and Rigor*, ed. N. Fox Weber, catalogue of the exhibition (Perugia, Galleria Nazionale dell'Umbria, 20 March–19 June 2013), Silvana Ed., Milano, 2013.
8. M. Bacci, M. Picollo, G. Trumpy, M. Tsukada and D. Kunzelman, Non-invasive identification of white pigments on 20th-century oil paintings by using fiber optic reflectance spectroscopy, *J. Am. Inst. Conserv.*, 2007, **46**(1), 27–37.
9. G. Gautier, A. Bezur, K. Muir, F. Casadio and I. Fiedler, Chemical Fingerprinting of Ready-Mixed House Paints of Relevance to Artistic Production in the First Half of the Twentieth Century. Part I: Inorganic and Organic Pigments, *Appl. Spectrosc.*, 2009, **63**(6), 597–603.
10. M. de Keijzer, The history of modern synthetic inorganic and organic artists' pigments, in *Contributions to Conservation: Research in Conservation at the Netherlands Institute for Cultural Heritage*, ed. J. A. Mosk and N. H. Tennent, James & James, London 2002, pp. 42–54.
11. J. Redensek, On Josef Albers' Painting Materials and Techniques, in *Josef Albers, Minimal Means, Maximum Effect*, catalogue of the exhibition (Madrid, Fundación Juan March, 28 March–6 July 2014), Fundación Juan March, Madrid 2014, p. 34.
12. G. Poldi and S. Caglio, Phthalocyanine identification in paintings by reflectance spectroscopy. A laboratory and *in situ* study, *Opt. Spectrosc.*, 2013, **114**(6), 929–935.
13. M. de Keijzer, The Delight of Modern Organic Pigment Creations, in *Issues in Contemporary Oil Paint*, ed. A. Burnstock, M. de Keijzer, J. Krueger, T. Learner, A. de Tagle, G. Heydenreich and K. J. van den Berg, Springer, Berlin, 2014, pp. 45–73.
14. N. Eastaugh, V. Walsh, T. Chaplin and R. Siddall, *The Pigment Compendium. A Dictionary of Historical Pigments*, Elsevier, Oxford 2004, p. 68.
15. Vitruvius, *On Architecture*, translated by F. Granger, Harvard University Press, Cambridge 1999, VII.vii.4.
16. M.-C. Corbeil, J.-P. Charland and E. A. Moffatt, The Characterization of Cobalt Violet Pigments, *Stud. Conserv.*, 2002, **47**(4), 237–249.
17. R. Mayer, *The Artist's Handbook of Materials and Techniques*, London 1951, p. 51.
18. I. Hebblewhite, *Artists' Materials*, Phaidon, Oxford 1986, p. 124.
19. N. Eastaugh, V. Walsh, T. Chaplin and R. Siddall, *The Pigment Compendium: A Dictionary of Historical Pigments*, Elsevier, Oxford 2004, p. 367.

20. P. Vandenabeele, L. Moens, H. G. M. Edwards and R. Dams, Raman spectroscopic database of azo pigments and application to modern art studies, *J. Raman Spectrosc.*, 2000, **31**, 509–517.
21. M. V. Cañamares, J. V. Garcia-Ramos, C. Domingo and S. Sanchez-Cortes, Surface-enhanced Raman scattering study of the adsorption of the anthraquinone pigment alizarin on Ag nanoparticles, *J. Raman Spectr.*, 2004, **35**, 921–927.
22. S. A. Centeno, V. Llad´ o Buisan and P. Ropret, Raman study of synthetic organic pigments and dyes in early lithographic inks (1890–1920), *J. Raman Spectr.*, 2006, **37**, 1111–1118.
23. *Mark Rothko's Harvard Murals*, ed. M.B. Cohn, Center for Conservation and Technical Studies, Harvard University Museums, Cambridge, 1988.
24. L. Carlyle, J. J. Boon, M. Bustin and P. Smithen, The Substance of Things, in *Rothko*, exhibition catalogue, ed. Achim Borchardt-Hume, Tate, London, 2008.
25. J. Albers, *Interaction of Color*, Yale University Press, New Haven 1963, p. 1.
26. B. Brunetti, C. Miliani, F. Rosi, B. Doherty, L. Monico, A. Romani and A. Sgamellotti, Non-invasive investigations of paintings by portable instrumentation: the MOLAB experience, *Top. Curr. Chem.*, 2016, **374**, 10.

CHAPTER 5

From Identification to a New Insight of Preservation Theory for Contemporary Art: Innovative Approaches to Complex Care in Alina Szapocznikow Case Studies

IWONA SZMELTER*[a,b] AND JOANNA KURKOWSKA[a,c]

[a]Faculty of Conservation-Restoration of Artworks, AFA, Wybrzeze Kosciuszkowskie 37, 00-379 Warsaw, Poland; [b]Interdepartmental Division "Novum" for Care and Conservation of Modern and Contemporary Art, Warsaw, Poland; [c]Department for Research and Documentation Techniques, AFA, Wybrzeze Kosciuszkowskie 37, 00-379 Warsaw, Poland
*E-mail: iwona.szmelter@asp.waw.pl

5.1 Introduction

Ensuring heritage identification and protection in its contemporary meaning challenges the continued distinction between tangible and intangible heritage. Analysing the notion of cultural heritage within heritage studies and conservation/restoration/reconstruction work, objects recognised as belonging to heritage were identified as material, but at the same time as a

Science and Art: The Contemporary Painted Surface
Edited by Antonio Sgamellotti, Brunetto Giovanni Brunetti and Costanza Miliani

Published by the Royal Society of Chemistry, www.rsc.org

"tangible manifestation of intangible cultural heritage". In fact, Laurajane Smith and Natsuko Akagawa suggest that "heritage becomes 'heritage' only when it is recognisable within a given cultural set or social values which in themselves are 'intangible'".

This chapter is devoted to understanding the structural changes that occurred in the last years in conservation-restoration of modern and contemporary art. This change especially concerned artworks which cross the limits of the traditional artistic production—from atypical synthetic sculptures embedded in other materials, to environments, assemblages, installations with light, conceptual art and many other examples.

Thea van Oosten, in her erudite essay titled "Here Today, Gone Tomorrow? Problems with Plastics in Contemporary Art", has widely identified the problems that threaten the durability of heritage in contemporary visual art. This chapter is specifically aimed at answering the following question: what should the optimal attitude of a conservator-curator towards endangered contemporary works of art be?

The introduction of new elements into the modern theory of heritage conservation and care is certainly a complex task. Here we will contribute to the definition of a new approach to safeguard contemporary art objects. It is based on a theory and methodology that results from long experience in the conservation of modern works by internationally recognised Polish artists, such as Szapocznikow, Tadeusz Kantor, Mirosław Bałka, Krzysztof Bednarski and others, as well as on an adequate understanding of conservation–restoration.

After three decades of interdisciplinary work on contemporary heritage, we are of the opinion that the definition of a theoretical paradigm and a conceptual framework for the safeguarding of contemporary art, although difficult, is a task that urgently needs to be undertaken.

Heritage science, intended as the set of studies based on interdisciplinary research on history, interpretation, diagnostics, conservation and management of cultural heritage, undoubtedly has a pre-eminent role in the knowledge, contextualization and transmission to the future of cultural identities. Within this framework, the examination of contemporary works of art involves the holistic investigation of each individual object (*via* scientific and cultural studies, comprehensive of science and humanities) to an extent that goes beyond the boundaries of tradition, with explorations into areas that in our case are connected with new visual art expressions, such as video, film with ephemeral non-material media, interactive art in which the observer is engaged, performance art and others.

Paradoxically, despite the passage of time, Cesare Brandi's definition (1963) remains current: "Restoration consists of the methodological moment in which the work of art is recognised, in its physical being, and its dual aesthetic and historical nature, in view of its transmission to the future."

Looking for innovative preservation of *avant-garde* cultural heritage, we will consider here, among others:

- problems of identification
- presentation and re-installation
- maintenance of ephemeral art and role of simulacra
- preservation of the utilitarian function of objects
- replacement of destroyed ready-made elements
- ethical considerations in decision-making in conservation
- conservation/restoration/reconstruction/emulation/re-staging.

5.2 The Specifics of Alina Szapocznikow's Visual Art Legacy

Alina Szapocznikow (1926–1973) was an outstanding sculptor and graphic artist. During WWII, as a child of Jewish doctors from Kalisz in central Poland, she survived in the Pabianice and Łódź ghettoes and three German Nazi concentration camps: Auschwitz-Birkenau, Bergen-Belsen and Theresienstadt. After the liberation of the camps, she left for Prague where she began studying at the Faculty of Fine Arts, later continuing her studies in Paris. In 1951 she returned to Poland. In 1963 she moved permanently to France, establishing contacts with the *Nouveau Réalisme* movement around critic Pierre Restany, which also included César, Arman and Niki de Saint Phalle. In 1965, for the sculpture *Goldfinger*, she became a laureate of the prestigious Copley Foundation award, selected by an esteemed jury of artists: Marcel Duchamp, Jean Arp, Max Ernst and Roberto Matta.

Szapocznikow held a unique position in the post-war European *avant-garde*. Her legacy numbers more than five hundred items representing a wide spectrum of artwork, including traditional objects such as sculptures in marble or bronze, as well as hybrid, ephemeral, synthetic sculptures made of polyurethane and polyester and complex assemblages and environments (proto-installations) and installations. Several artworks are representative of kinetic art or include electric light as part of their construction and aesthetic.

In the 1960s and early 1970s, she was one of the first artists to experiment with synthetic resins. Szapocznikow started to experiment with various sorts of plastic substances such as polyester, polyurethane and casts of the human body. She combined plastics in new ways with various materials of different provenance. She embedded in polyester so-called ready-made industrial objects, photographs, grass and other living matter, as for example in *Pollution I* and *Pollution II* (both 1968). She lit up sculptures from the inside with electric light, as with *Buste* (1967), and simultaneously made two series, *Fetishes* and *Desserts* (both after 1969), from polyester casts of breasts, lips and gauze shreds. Ultimately, these two series were very transgressive in the eyes of an old-fashioned audience.

Unfortunately, parallel to the popularity of the slogan "plastic is fantastic", problems arose in curatorial-conservation projects associated with collecting, displaying and preserving contemporary works that include plastic and other embedded materials, such as cigarettes, grass, plants, clothes, bodily casts, *etc.*

In Szapocznikow's sculptures from between 1961 and 1972, synthetic material is used to bind elements of so-called "found–lost" ready-made objects. In complex compositions of plasmatic polyester matter, the artist's personal objects are immersed: photographs, women's underwear, men's shirts, women's skirts and shreds of gauze—especially in Szapocznikow's *Tumours* cycle, in which they are immersed deep in polyester as an abstract artwork. In this dramatic iconology Szapocznikow used casts of her illness-ridden body. She also immersed photographs wrapped in bandages in them, adding grass and fibres as impermanent materials "that do not belong to the classical art world".

Preservation of Szapocznikow's *avant-garde* legacy has been one of the author's (Iwona Szmelter) long-term research and conservation projects since 1994 (including conservation and restoration). Until today one hundred and fifty-two interdisciplinary, mostly synthetic artworks have been identified, conserved, restored and reconstructed for several major retrospective exhibitions which took place in important museums in Poland, as well as at MoMA, Centre Pompidou, *etc.* between 1998 and 2018. The conservation and restoration of these artworks—mostly from the Parisian collection of Alina Szapocznikow's heir, Piotr Stanislawski—concerned unconventional synthetic and natural material, problems with identification and difficulties with presentation of the artist's intent and ideas.

5.3 Identification of Materials/Analytical Methods

The unique character and great artistic value of these works of art made it necessary to approach the analysis with a great deal of professionalism but often in an unconventional way. Visual examination of the objects combined with tactile inspection was used first as the simplest non-invasive approach. It gave important information on: colour, surface texture, transparency, degradation signs, production methods and a material's modifications due to the addition of extenders/fillers and colourants. Magnifying glasses and stereomicroscopes (with minimal magnification of 10*x* or 20*x*) were often used, but unaided eye observation was the first method employed.

Describing the physical properties of the materials, such as density, hardness and smell, was the aim of the next step. Materials used by the author have changed in these properties throughout the years of their existence in her works of art—they have become more fragile, brittle and have changed their optical appearance, becoming matte. The surface of the objects has become flaky and started exfoliating. Details attached to each other with the use of different glues have increased in their tendency to detach and delaminate. Parts of objects consisting of foam materials have deteriorated into a powdery substance. Colours have changed into darker and yellower tones. Plasticisers that have migrated towards the surfaces of materials have bound dust particles and formed an incrustation, totally changing the original appearance of the objects. Materials that did not contain fillers or extenders have kept their gloss and have not shown the tendency to crack. On the contrary, resins in which these substances were used have lost their glossy appearance and their surfaces have now become matte and cracked.

Numerous invasive tests were conducted on original materials in order to formulate a conservation treatment program that could stop the polymers' degradation process. Minute samples were taken for the purpose of micro-chemical tests and instrumental analysis. Their sensitivity to polar and non-polar solvents was determined, as well as their reaction to aqueous solutions of hydrochloric acid (3M HCl), sodium hydroxide (4M NaOH) and concentrated nitric acid (HNO_3). The Aburn test was involved to assess the flame colour, the speed of burning, self-extinguishing time, smell and the pH value of the fumes and residues after burning the whole sample.

Instrumental analysis was used for unambiguous identification of material constituents. Fourier-transform Infrared Spectroscopy (FTIR) was useful to determine the structure of samples by showing functional groups and different chemical bonds, which in turn allowed identification of the main components of the analysed materials.

FTIR spectra of organic macromolecules are typically very complex. Thus, only characteristic polymer functional groups were interpreted while reading the spectra. Thus, characteristic peaks for the –N=C=O functional group were identified in polyurethane resins used by the artist. Also, the ester-related carbonyl groups' vibration peaks (–COOR), together with strong absorption bands related to C–H bond stretching vibrations, allowed for the identification of polyester resins. The bands corresponding to polypropylene were interpreted on the basis of a direct comparison with the standard.

In summary, a comparison made between spectra obtained from samples and reference spectra,[1] combined with other aforementioned analyses, led to the conclusion that Szapocznikowa's examined works mostly consisted of polyester, polyurethane resin or polypropylene.

The identification of fillers/pigments/extenders, which the artist added to resins to change their visual appearance, was aided with microscopy, micro-chemical tests and elemental analysis (XRF and SEM-EDX), as well as ultraviolet-induced luminescence observed *via* microscopic techniques. Some of these were micro-invasive techniques, as they demanded minute samples to be taken from the object. Transmitted light observations with the use of an optical microscope gave information on pigment grain morphology. In some instances simple micro-chemical tests were employed, which solved issues unresolvable by elemental analysis alone (*e.g.* phosphate or sulphide ion detection). UV-induced luminescence was helpful with certain pigments, which show a characteristic fluorescence when irradiated with ultraviolet light (*e.g.* zinc white, some organic reds).

5.4 Alina Szapocznikow—If Plastic is Fantastic!? Condition Reports: Searching for Ideas Before the Conservation/Restoration of Szapocznikow's Works

Condition reports revealed that Szapocznikow's artworks were inherently impermanent, as over one hundred and eighty objects were found in a state of decomposition, being akin to pieces of jigsaw puzzles. The state of the works varied according to the execution technique, the means of "renewal" up to this point and the place of storage of each work.[2] The darkened layers on certain works and the dark tone in the case of polyesters were due to the repeated preparation of these objects for expositions by periodically rubbing them with oil. Destruction, cracks and holes, which in some cases constitute missing pieces, are caused as much by mechanical damage as by the actual technique of the works.

Szapocznikow's works from 1956–1972, found in the *avant-garde* collections of museums and in Piotr Stanislawski's private collection, range from traditional pieces, shaped under the influence of the Parisian and Prague schools, to those deep-dyed in mass resins, illuminated pieces, kinetic pieces, assemblages, environments and ephemeral objects.

The techniques the artist used are numerous and diverse: from plaster, bronze and marble to dyed polyesters, polyurethane and compositions of artificial materials linked with non-permanent organic ones, such as grass and newspaper. The famous critic Hermansdorfer commented:

> "the artist devotes the most attention to the castings made in synthetic resins. (...) Plastic allows precise mapping of every detail, the smallest detail. Mapping and stopping in time. For themselves, their relatives, for the contemporary time and posterity. In spite of the fragility, it guarantees survival, the preservation of a trace of presence."[3]

Preliminary examination and preparation consisted of studies of documentation of the artist's work and of her public speeches. Apart from Szapocznikow's own accounts, we also critically analysed information obtained from art critics and the mass media (unfortunately based on "narration" and second-hand knowledge).

The acceptance of any preservation procedures depended on its probable agreement with the artist's intentions and motives, with her credo. Her son was consulted on essential and technical issues. On the basis of Szapocznikow's extant personal archives in Paris, spanning from 1957 to 1972, especially some receipts for art supplies, a list of resins and other materials was drawn up. These receipts bear the artist's signature, address and personal data. From these receipts and from her co-workers' comments, the types of artificial resins used by the sculptor could be determined. The types of synthetic materials used were confirmed by identification with chemists,

Dr Zadrozna and Dr Kurkowska. This data created the basis for the verification of a description of the technique and technology used in Szapocznikow's work.

5.5 Case Studies: Searching for a Strategy for Decision-making in Preservation

Case studies in contemporary art have indicated difficulties in sustaining complex ideas of art while preserving authentic material, including the application of ready-made parts, sculpture made from polyesters, polyurethanes, plants, green grass, chewing gum, *etc.* Szapocznikow's legacy was identified and preserved in the long-term protection project where due to the extremely individual character of her works it is necessary to first of all preserve the intentions and ideas of the artist (immaterial), and then matter (tangible). This requires the creation of a new conceptual framework for the complete care of the recently created visual artworks, both tangible and intangible, with the importance of interdisciplinarity in the preservation/conservation of cultural heritage.

In every case study the conservation diagnosis model and the proposed conservation concept were discussed and accepted by a committee consisting of *trans*-disciplinary specialists. Technical problems with conserving Szapocznikow's legacy also result from the characteristics of the materials and technologies she used, *i.e.* the possibility of halting the deterioration of synthetic materials by natural causes.

The principles and the stages of conservation activity led to the halting of external damaging processes, cleaning the objects, removing overpainting and layers added later, consolidating the layers of work, reinforcing portable constructions, filling in missing pieces, retouching, eventual care of function (*e.g.* kinetic or utilitarian), protecting the surface of works from further degradation and using, among other things, photo-stabilisation coating, which isolated works from UV rays.

The conservation and preservation was entrusted to a team led by the first author of this chapter, as Head of the Interdepartmental Division "NOVUM" for Care of Modern and Contemporary Art at the Faculty of Conservation and Restoration of Works of Art at the Academy of Fine Arts in Warsaw. Here below, examples of innovative approaches for complex care of various groups of objects are presented.

State of Weightlessness—Homage Komarow (Figure 5.1) resembles a mummy, held in a vertical position, slightly at an angle, presented in a so-called "anti-gravity" effect. It is constructed of a metal base wrapped in gauze, bandages and fibreglass, which was then dipped in polyester resin several times. The membranes of the synthetic material separate and it crumbles in raised blisters around the neck, breast and feet of the figure. Firstly, in order to preserve the fragile synthetic sculpture, all the cracks and member separations were filled. Following tests, a hardening substance

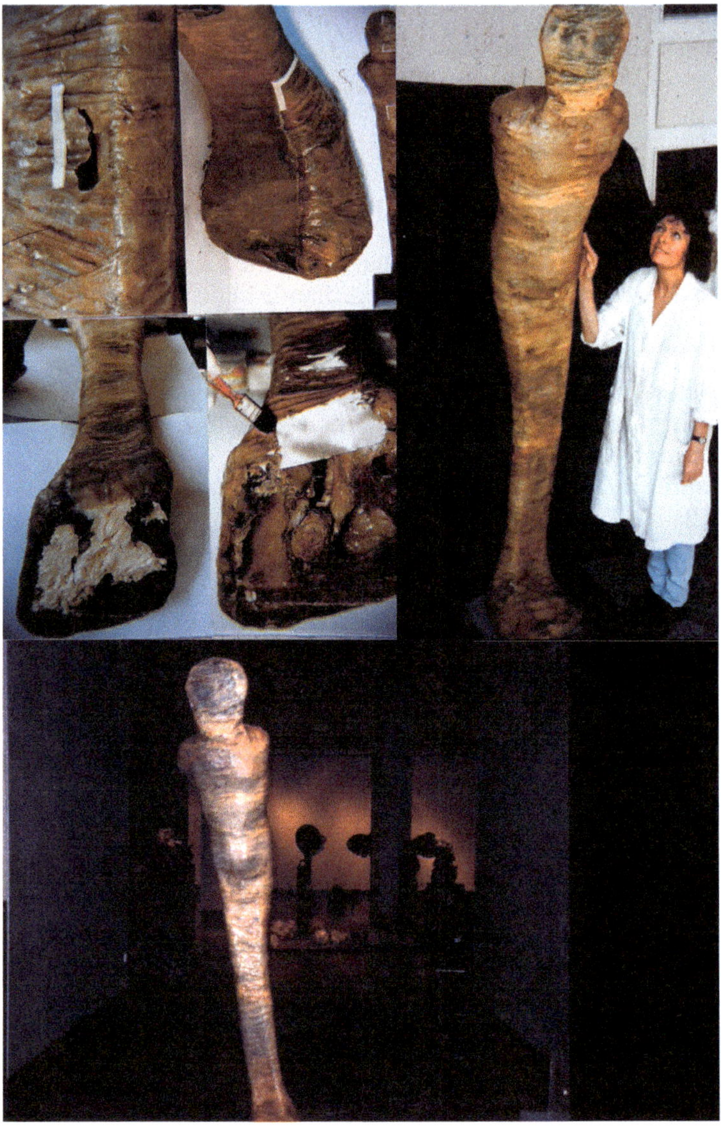

Figure 5.1 *State of Weightlessness—Homage to Komarow, Alina Szapocznikow*, 1967. Polyester, gauze, photographs, metal base. Dimensions: 235 cm × 70 cm × 40 cm. © ADAGP, Paris, courtesy The Estate of Alina Szapocznikow/ Piotr Stanislawski, Galerie Loevenbruck, Paris/Hauser & Wirth, photo: I. Szmelter. Reproduced with permission.

with a catalyst was applied to the polyester resin, Polimal® 140, that had not been soaked. Next, the missing pieces in the fibreglass enforcement were filled with original fragments that had been found. The holes that could not be replaced with original material were filled by casting patches into which gauze and bandage were melted. These patches imitated the original.

Bandage with a similar weave to that available in the 1960s was found in an old pharmacy, and the colour of the filled pieces was matched up with the colour of aged resins by adding pigmentation (burnt ochre, green umber). Unfortunately, in terms of construction, it proved impossible to limit the reconstruction exclusively to the missing places. This operation, though conflicting with conservational theory, was absolutely necessary in order to preserve the object. The sculpture was still liable to crack in three places and was as fragile as an eggshell. The risk factor increased with the mounting of the work on a metal pedestal which was not capable of supporting the unsteady sculpture. The most fragile places were laminated with clear polyester resin, Polimal® 140, with a catalytic hardening solution. This drastic action was necessary because it was associated with the replacement of the internal structure, the function of which was to carry the sculpture. We created a new construction for the object. The hollow space within the sculpture was filled with polyurethane mounting foam (prepared in laboratory), which improved the work's statics. The metal stand with a rod that went into the portable construction in the sculpture was sanded, corrosion pits were removed and it was covered with Paraloid® B-44 isolation. Finally, the entire polyester composition was covered with a protective layer of Paraloid® B-67 with the UV blocker Tinuvin® 770 and with a tarnishing addition of colloidal silica, Silice.

In conclusion, the main problems of the *State of Weightlessness—Homage to Komarow* case study were related to the unconventional intention of the artist. To preserve Szapocznikow's idea it was necessary to reconstruct the object's inner structure and recover its effect of suggesting independence from the law of gravity.

Preservation of the idea and the author's intention was based on the artist's play on zero gravity in her *State of Weightlessness—Homage to Komarow*, its namesake being an astronaut who died during a space flight. It was a strange sculpture, which occupied a separate place. It was a long silhouette, unstable on its feet. This was Szapocznikow's personal tribute to the epic of space exploration. In this case the artist's intention is very important to the arrangement and exposition of the piece. The preservation of the combination of the idea (being against gravity) and intention with material may become a great technical problem.

Before conservation, *Headless Torso* (Figure 5.2) was in extremely bad condition. The polyurethane element of the cushion does not powder, does not peel; due to the colour used it is difficult to tell if there have been any changes in colour. However, in many places larger and smaller cracks and holes are visible. The sculpture is in relatively good condition. Previously, as a result of transport, the lack of support caused numerous cracks in the material, so it was decided to introduce an element supporting the structure from behind (a kind of metal rack welded from two parts), transferring the weight-bearing function of the object onto it. The stand is connected with a wooden element suited to the shape of the sculpture on which the original stands. The surface is covered with a thin layer of surface dirt (as of 2013).

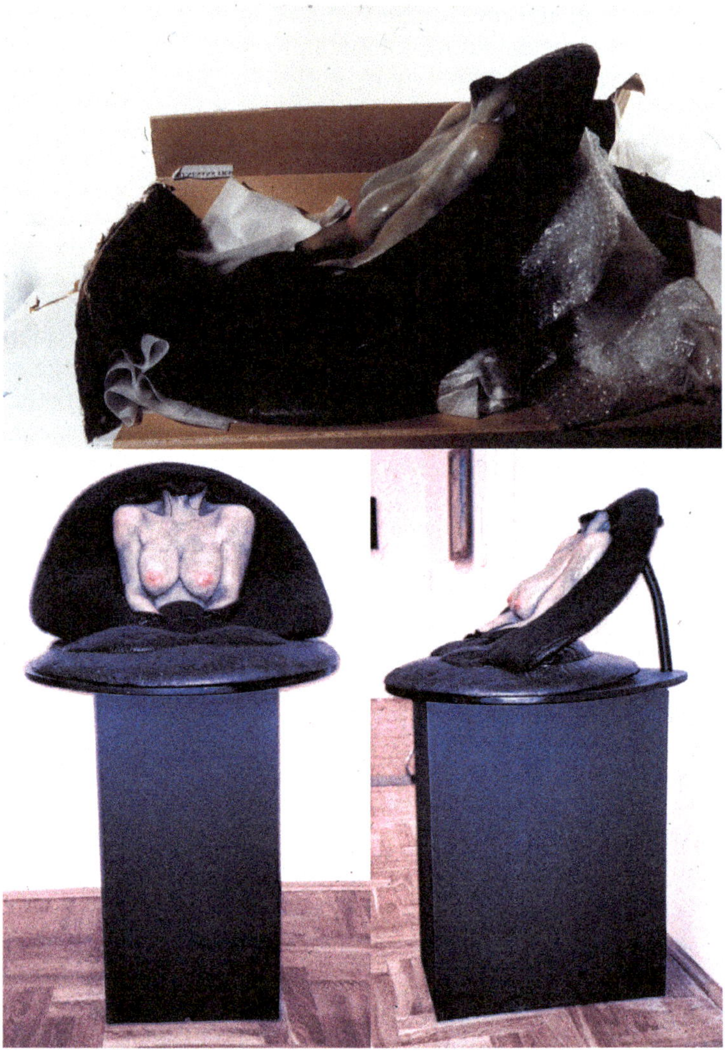

Figure 5.2 *Headless Torso*, Alina Szapocznikow, 1968. Polyester resin and polyurethane foam. © ADAGP, Paris, courtesy The Estate of Alina Szapocznikow/Piotr Stanislawski, Galerie Loevenbruck, Paris/Hauser & Wirth, photo: I. Szmelter. Reproduced with permission.

The polyester part of the woman's body has been preserved in good condition. The whole surface is visible in the air bubbles embedded in the mass, which results from a not very careful technological process. The bubbles located close to the surface are small holes that accumulate dust. Unfortunately, the lack of support during transportation has caused numerous deep cracks (twenty-eight!) in the soft polyurethane material. After cleaning, polyurethane was impregnated by Impranil® DL. The polyester part of the woman's body has been preserved too. Finally, the surface of the sculpture was covered with a thin layer of varnish for protection.

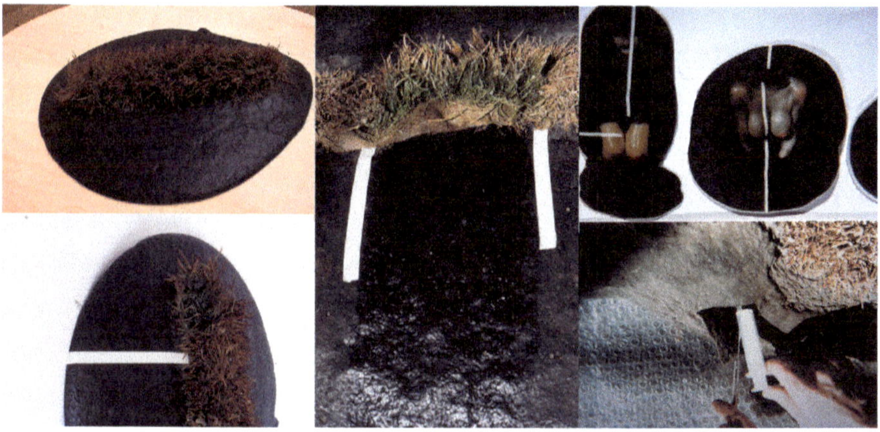

Figure 5.3 *Pollution I*, Alina Szapocznikow, 1968. Artwork made from polyurethane, natural grass on a jute belt. Dimensions: 11 cm × 87 cm × 170 cm. © ADAGP, Paris, courtesy The Estate of Alina Szapocznikow/Piotr Stanislawski, Galerie Loevenbruck, Paris/Hauser & Wirth, photo: I. Szmelter. Reproduced with permission.

Pollution I's support was made of highly degraded hard polyurethane (Figure 5.3); it was cracked, stained, dull and had numerous corrosion pits. The grass was brown, dirty, its leaves broken and crushed. With each exhibition we consider the possibility of replacing the elements. The object's surface has been cleansed both mechanically and chemically, and the cracks in the polyurethane, in danger of further degradation, have been glued closed. The grass has been washed and impregnated, which partly resulted in the visibility of its rusty green colour and natural character. The entire work has been secured with Art-matt aerosol varnish. The conservation of this piece has been limited to superficial cleansing and protection with varnish. Our inability to halt the destruction of the sculpture's base and the obvious neglect of its liability to further degradation is analogous to our accepting cracks in paintings, which are not retouched as evidence of the natural ageing of the material.

Szapocznikow's *Dessert* series was inspired by Roland Topor's work in its use of pure nonsense elements and play with conventions; they were shockingly grotesque and full of black humour (Figure 5.4). The artist's technique of moulding her own breasts, or those of Topor's fiancée Ariadne, who was at that time a willing model for the artist, was reconstructed on the basis of negatives brought from the Paris studio. Their existence made duplicating the moulds possible. These cannot, however, replace the primary moulds made by Szapocznikow at different stages of her work, because the machine-made duplicates would not have the artist's original deformities, such as fingerprints, marks of her formal choices, hesitation or marks left by tools. Bearing this in mind, the conservation program encompassed mechanical and chemical cleansing of the object, gluing and impregnation by injection, as well as filling the missing parts in the structure with dyed polyester resin. The

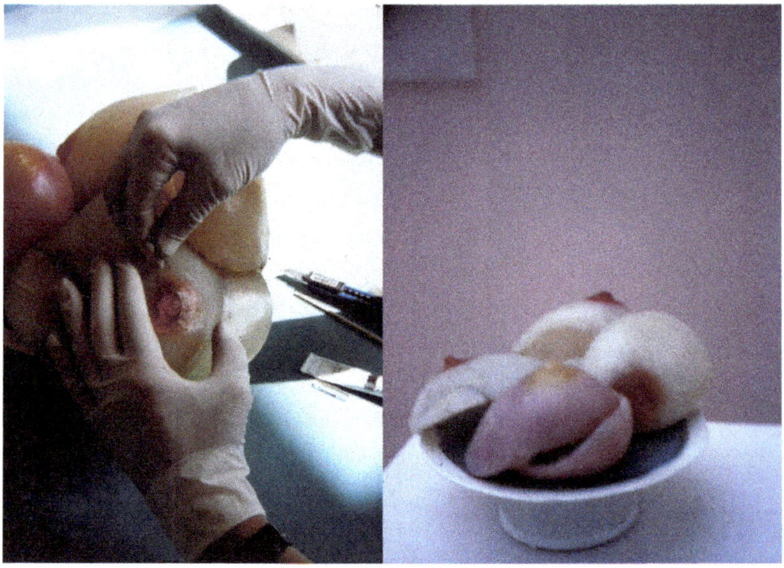

Figure 5.4 *Dessert III*, Alina Szapocznikow, 1971. Assemblage, a composition created out of ready-made objects, *i.e.* a porcelain cake stand which holds a composition of three polyester-styrene mass-coloured, deep-dyed moulds of female breasts. Dimensions: 18 cm (height) × 24 cm (width) × 25 cm (length). © ADAGP, Paris, courtesy The Estate of Alina Szapocznikow/Piotr Stanislawski, Galerie Loevenbruck, Paris/Hauser & Wirth, photo: I. Szmelter. Reproduced with permission.

cake stand was kept along with its historical repairs in order to maintain its authenticity. After cleaning and structural conservation, the assemblage was coated with a protective layer consisting of UV absorbent and some colloidal silica to tarnish the surface.

The objects in *Personified Tumours* are filled with glass wool, gauze and newspaper (Figure 5.5). Most of the works have polyester walls of a thickness varying between 0.5 mm and 80 mm, and different degrees of saturation with pigment. The conservation began with a chemical cleansing of the objects. The interior of the objects, accessible through existing openings, was de-acidified, which should prevent further degradation of the paper. The losses of the walls were then refilled and glued with unsaturated polyester resins, Polimal® 102 and Polimal® 109. With two basic chemical setting systems at our disposal, *i.e.* cobalt and amine, the former was selected for health and environmental reasons. The initial violet colouration of the cobalt accelerating agent made colour selection more difficult. Therefore, some pigments were added. In those fragments which had an additional construction carrying function fibreglass was added to the resin in order to improve its mechanical strength. After replenishment, they were locally polished. Finally, in order to prevent divergent rates of

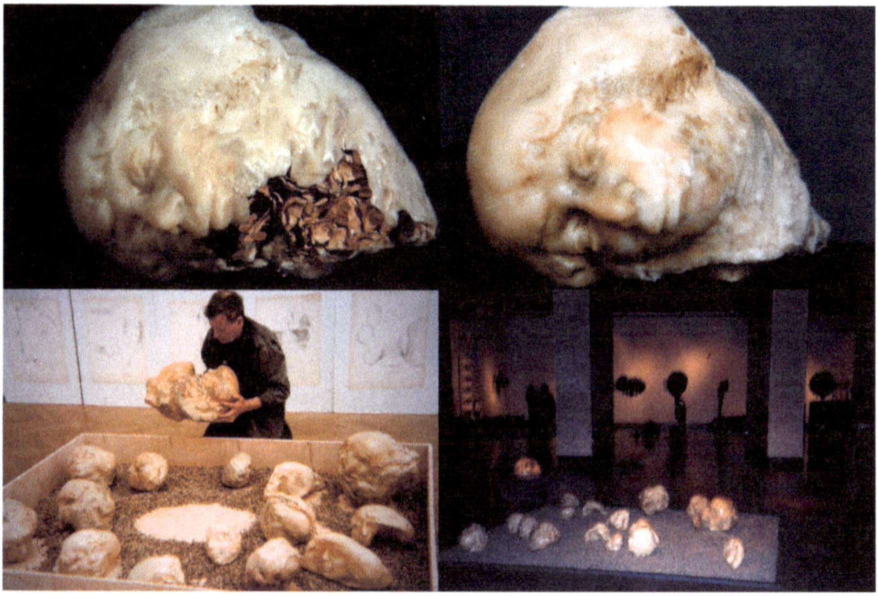

Figure 5.5 *Personified Tumours*, Alina Szapocznikow, 1971. Environment, a set of fourteen self-portraits, cast and modelled in polyester resin on the basis of a gypsum cast of her own head. Dimensions close to real-life. © ADAGP, Paris, courtesy The Estate of Alina Szapocznikow/Piotr Stanislawski, Galerie Loevenbruck, Paris/Hauser & Wirth, photo: I. Szmelter. Reproduced with permission.

polymer photo-degradation, the objects were coated with a protective layer, based on Paraloid® B-67 with an addition of Tinuvin® 770 (an ultraviolet blockade) and silica. In practice, the evident degradation of polymers manufactured in the 1950s, 60s and 70s and the anticipated differences between these materials and those resins that had been used in the conservation process suggested the application of protective coatings to the objects. Based on published results of ageing studies, poly-acryl resins, stabilised with the addition of photo-stabilisers from a group of amines with spatial hindrance, were selected.

Many other polyester-coloured sculptures, such as Szapocznikow's *Journey* (1967), were badly restored in the 1970s. As a result, their original ideas and intentions became lost and it was necessary to reconstruct these. Fibreglass fabric had been covered with polyester resin and cloth and mounted onto inappropriate constructions, destroying the original effect by applying an excessive amount of glue and many patches. The result was highly unprofessional (with the piece resembling a badly restored polyester kayak!). Complex research and preservation were conducted by Aga Wielocha in her MA at Warsaw's Faculty of Conservation under the supervision of the primary author of this paper. The preservation started with time-consuming removal of old materials from quasi restoration, the cleaning and consolidation of original parts, followed by re-integration of original layers and the restoration of missing parts. Finally, some varnish with UV isolation layers was applied to the whole piece, consolidating it.

In summation of the Alina Szapocznikow case studies, no assumptions were made before our conservation/restoration treatment, because in each case research of the individual art history and context of the work was undertaken. There was no fixed technological configuration, theoretical framework or know-how that was universal to the entire diversified (cultural) legacy of Szapocznikow.

5.6 Innovative Conservation/Preservation Issues

The great lesson in understanding the artist's creative process came from analyses of Szapocznikow's motivations, ideas and intentions, alongside her methods and technology. As Hanna Hoelling wrote in her essay in 2017:

> "Today, conservation no longer aims simply to prolong its objects' material lives into the future. It is seen as an engagement with materiality, rather than material—that is, with the many specific factors determining how objects' identity and meaning are entangled with the aspects of time, the environment, ruling values, politics, economy, conventions and culture."[4]

The broad range of conservation issues in the preservation of Szapocznikow's artwork are based on care for the integrity of artwork. In this long-term project we experienced many problems with unconventional approaches and practical know-how, some of which were presented by the author in the 2012 monograph *Innovative Approaches to the Complex Care of Contemporary Art*.[5] Let us, therefore, take a closer look at these and attempt to categorise them according to some issues.

5.6.1 The Problems of Presentation and Re-installation

The cycles *Tumours* and *Herbarium* are Szapocznikow's last works made in polyester and polystyrene. Apart from their value as human testimony, they provide a gripping illustration of the naked truth of the impermanence of modern artwork made from plastics.

Personified Tumours (1971), an environment, is a set of fourteen self-portraits, cast in polyester resin on the basis of a gypsum cast of her own head. The objects are filled with glass wool, gauze and newspapers. The treatments given were: de-acidification, consolidation and reconstruction.

The discrepancy between the original integrity of the set of the artist's fourteen self-portraits and the piece's re-installation indicates the loss of the piece's integrity as an environment (proto-installation).

5.6.2 Ephemeral Art and Simulacra

A very special problem arises in connection with the care of a series of sculptures made in 1971 out of chewing gum—"discovered" spatial forms in tiny "spit-outs", made by Szapocznikow during her stay in hospital—as a mini symbol of monumental sculpture. The question is whether we should attempt to prolong the life of ephemeral pieces. This series of sculptures was photographed. Since the objects have not been preserved, the only items left today are the photographs Szapocznikow asked her husband Roman Cieślewicz (himself a famous graphic artist) to take at the hospital, and which were later distributed in a limited number. This constitutes a typical example of "preservation through documentation" as simulacrum, a prototype of *Simulacra*, since the objects are not extant, the chewing gum has not survived. It has recently been decided that the photographs should be multiplied (in limited editions) in order to follow the artist's intention, as she was amused by both the monumentalising and multiplying of these sculptures, made of chewing gum just taken out of her mouth.

5.6.3 Replacement of a Destroyed Ready-made Element

In the *Pollution I* and *II* works, Szapocznikow experimented with materials in order to achieve an aesthetic contrast between the effect of the black polyurethane volcanic resins and the living matter of fresh green grass. The support of *Pollutions* was made of extremely degraded hard and soft polyurethane sponges which gradually cracked, became stained and dull, developing numerous corrosion pits. The grass wilted, and was brown, dirty, its leaves broken and crushed. After cleaning, it became more colourful, but far from the artist's intention to contrast fresh grass with the harsh post-industrial surface of, as it were, volcanic black polyurethane. The conservation program, as one of the elements of conservation design, was presented with complete objectivity and without preference for specific solutions. The first conservation work only included preservation of the substance, cleaning and reinforcing.[6] We nevertheless think that for subsequent exhibitions of the piece one should consider replacing the old grass (which is hardly recognisable any more) with a strip of fresh lawn grass, putting it in the crack in the polyurethane base, thus reclaiming the object so that it becomes able to communicate the artist's intention, namely drawing the viewer's attention to the contrast between the green grass and the black, dangerous "pollution". While making her assemblages and experimenting with living matter and plastics, Szapocznikow also made casts as body impressions in polyester, vinyl and polyurethane, mostly of herself. By duplicating particular themes she approached Pop Art and New Realism. The complex care included: recognition, contextualisation, treatments (cleaning, consolidation and re-integration). Singular challenges should aim at preserving specific vitality effects, which Pierre Restany picturesquely rendered as: "the decadent splendour of a new Jugendstil, incarnate in a festival of sumptuous bodies, put into relief by the flickering of their internal lighting".[7]

5.6.4 How to Preserve the Utilitarian Function of Objects

A different strategy was employed in Szapocznikow's *Lamps* cycle (1968), in which she used only a few elements: female breasts and lips and a phallus, cast in polyester/polyurethane. In this way the female breasts served as a prototype for a lighting device. Witty plastic artworks served as a prototype for a lighting device. In the "bellies" and *Lamps* cycles, Szapocznikow's pieces made from body elements have the character of "exhumed" matter.

In the late 1960s, Szapocznikow developed works with a strong undertone of vital self-irony. These objects were quasi "utilitarian", or design prototypes, as if decorative, and downplayed both the individual and the popular, mass-produced. We are on the threshold of a decision regarding whether it is legal and ethical to make an emulation, or even replicas, of these utilitarian artworks. As Carol Mangusi-Ungaro has pointed out, the legal context is probably changing because museums have become factories for works of art—producing exhibition copies for installations and even producing replicas made by artists.[8]

From another point of view, the particular "code" given to works by the artist's (mother-like) use of her own materials, techniques and technologies plays a role like DNA in creating a human genome. As a consequence, an artwork is born which incorporates the artist's rapid pace of work and individuality and this tension and strain is implied in the sculpture's layers of plastic. Polyesters and polyurethanes with appropriate activators, fillers or accidental materials form the so-called readiness of the creator (ready objects with a symbolic meaning, enlarged by the context of deposition with synthetic material) and constitute mixed media that are unique in the field of conservation of museum collections.

5.6.5 Arrangement and Exhibiting

Alina Szapocznikow's works of art are still widely discussed and exhibited in ways that provoke controversy. They need long-term curatorial conservation/preservation related to re-installation guidelines. Interpretation of the artist's intention and empathy for this was basic to the preservation of Szapocznikow's works. Preservation was a kind of advocacy of the artist's intention in the course of conservation and reconstruction, as with *Les Cabouchons* (1968). This is strictly related to respect for the law, the work's integrity and its copyright. In restoring and conserving presented artworks and legacy, the main goal was to exhibit the artworks in the form that would be most representative of the artist's intentions, without hiding the materials' natural ageing process. The aim was to "do no harm" and consequently to return the works to their aesthetic and exposition value as much as possible, within the limits set by the technical potential and the natural degradation process of the artificial resins which constituted the material of most of her works. On the other hand, this is clearly significant in terms of conservation ethics—based on selection of aesthetic solutions in restoration, reconstruction and arrangement and exhibiting. A very important factor in the exhibiting of objects is 3D space, the respect for space, the distance between the elements of the installations, as well as light and colour. In many situations the senses were an important part of the artist's means of expression.

5.6.6 Towards a New Philosophy of Preservation: Ethical Considerations

It seems obvious that we have to avoid a situation where mental inertia keeps "old ideas" alive after they are obsolete, as in the care and conservation of Szapocznikow's visual art legacy. First of all, "do no harm", as in the medical motto "*primum non nocere*". Principles, which originate in a traditional way of thinking, develop and evolve in every individual conservation decision. Based on interdisciplinary art, science, technology and research, diagnostics and identification, we tried to achieve maximum respect for the authentic substance of an object. In light of this proposition and with regard to the highest degree of singularity found in contemporary artwork, theory must be flexible and the concept of conservation tailored to the needs of every object.

In order to maintain a flexible approach and balance in decision-making (between the traditional and modern), towards methods of preservation of subject matter and the intentions of the artist that are no longer readable in the work, we needed to do curatorial-conservation re-orientation in traditional methodology. A forgotten idea is in fact "preservable", because in the process of conservation/restoration/reconstruction it is possible to return to the artist's intention and ideas in order to preserve the integrity of the artwork and its potential unity.

5.6.7 Final Thoughts towards a New Philosophy on the Care of Contemporary Art

We have observed a turning point in cross-disciplinary research devoted to the conservation of modern and contemporary cultural heritage in the field of visual art. This is a symptom of a changed philosophy of conservation for the diverse cultural legacy based on new ideas and atypical and impermanent materials "that do not belong to the art world". As a consequence, a new understanding of the processes of structural changes in systems of knowledge in conservation/restoration of modern and contemporary art has arisen.

Ethical aspects of the care of works of art go hand in hand with legal standards, including copyright and preservation of the integrity of the artwork. Many problems arise during preservation of atypical objects of visual art legacy. Firstly, we consider a question related to the replacement of ready-mades. Should the preservation of complex artwork focus on the original materials, or lay more emphasis on the original appearance? This problem was connected with the decision: to preserve the idea of the object or not to preserve it? This was the case for several of the *Lamps* works and the old cake stand in *Dessert III*, with its historical quality thus allowing for consistency in conservation in the spirit of material fetishism. Conservation-restoration of *Dessert III* (1971); an assemblage, a composition created out of ready-made objects, *i.e.* a porcelain cake stand which holds a composition of polyester-styrene mass-coloured moulds of female breasts, lips, cigarettes, trash. Finally, that part of the sculpture which was the cast of the artist's breasts came to be fully restored. After discussions, we decided that the plate should be kept along with its historical restorations in order to maintain its authenticity, as an original element (with its patina). The artist included intimate fragments of underwear in polyester resin, thus breaking a taboo. The critics termed this a cultural transgression but conservators had to respect the artist's intention and preserve this assemblage with underwear.[9] The intent of the pioneering artist is only now being fully recognised.

5.7 Extending Preservation Theory for Modern and Contemporary Art

Now, a hundred years after Duchamp's revolution in *avant-garde* art and the shock related to the introduction of new materials including plastics, the art legacy rapidly begins to deteriorate.

Alina Szapocznikow's legacy remains a specific phenomenon of the *art world*. She was a pioneer in plastic sculptures and kinetic and light effects in art, and her many works are saturated with existential issues and large doses of self-irony. Therefore, complications arose in the preservation of her legacy—theoretical-conceptual and ethical, as well as practical—in identification and conservation methods.

In each case a curatorial-conservation project requires cultural identity and deep understanding of the ideas behind the work for its successful preservation. Specifications for the care and collection of contemporary art and the sharing of conservation decisions are as follows:

- Identification: instrumental analyses, humanistic recognitions, co-operation with the artist (interview if possible).

- Registration: taking into account the specific nature of artworks. Research during pre-acquisition and acquisition procedures.
- Identification, preservation and restoration of objects created with atypical techniques, new materials and their specific use (kinetic, *etc.*).
- Records and archives of objects including ephemeral elements such as noise and smell.
- Care of objects including multimedia elements (video, audio, demonstrations of slides, introductions and computer programs, *etc.*).
- Supervision during the assembly and the disassembly of objects (*e.g.* re-installation of installations). Care of the integrity of artwork and the integral role of elements in complex artworks, as well as of the artist's copyright.

The time is ripe for the definition of a new theoretical paradigm for the care of the cultural legacy of the most recent artistic heritage, enabling the continuity of preservation of these witnesses to contemporary culture. Although they will be based on existing theories, they may extend beyond their limitations.

The beginning of this modern approach to conservation was in the spirit of the historical relativism of Alois Riegl (1903)[10] and the aestheticising theory of Cesare Brandi (1963)[11] to arrive at the modern theories of sustainable conservation for the "common good" in the early twenty-first century. The heritage of visual art is often an open art process—not only the pieces of art, but the contemporary legacy, including elements of conceptual art, total art or hybridised or synthesised art, so-called *Gesamtkunstwerk*, occasionally immaterial artworks and not those oriented around a "final object". This becomes an issue leading to complex tangible and intangible legacy and a departure from the criterion of the necessity to preserve the authenticity of the substance in destructed fragments of objects whose expression is entirely different from the message and code created by the artist in total art.[12]As Hoelling states in 2017:

> "conservation is a complex techno-cultural practice with a strong, retroactive impact on its objects and subjects. Conservation offers an invaluably rich context to study a man-made world. Simultaneously, it allows us to pursue fundamental epistemic questions related to what, when and how artworks exist in the world and how our engagement with them is contingent upon the prevailing cultural–historical conditions."[13]

The elements of the new theory include proposals for extended principles of research on the diverse heritage of visual art, the manner of conducting safe preservation and management, the dissemination of the results, as well as the social issues involved and the contribution of art to sustainable development.[14] The complex approach to preserving cultural heritage in its tangible and intangible, as well as digital aspects, presented in the study is becoming an optimal perspective for preserving cultural heritage, including contemporary art. Bearing in mind that art is changing along with other changes in our lives, and takes on increasingly new forms, it stands to reason that any new paradigms of preservation of the most modern art should have an open and flexible character. Taking care of these aspects reduces the risk of "blank spots" in history as well as maintaining the continuity of civilisation. The area of visual art of the twentieth and twenty-first centuries requires particular protection, even though it might sound paradoxical considering the young age of this ephemeral legacy.

5.8 Final Thoughts: Conclusion

Identification and preservation of contemporary visual art heritage is based on collaboration across boundaries. Starting with identification of the artwork's materials, technologies and ideas, caring for visual art heritage is faced with an entirely new set of problems. There are difficulties with decision-making processes in conservation and more, in preservation strategy including restoration, reconstruction, emulation, and so on.

The new understanding of heritage realises its cultural and social role for contemporary and future generations. This purpose is served by the contemporary evaluation of the heritage of visual arts of the twentieth and twenty-first centuries, which is based on an open and research-based approach to the mixed character of culture, diffusion and hybridisation of disciplines of visual art. Only the passage of time skilfully used to preserve the values of cultural heritage will show us what, after many years, can be added to the cultural legacy.

The phenomenon of modern and contemporary art that lies outside traditional disciplines was presented with case studies of Szapocznikow. Her legacy is atypical: in the period between 1956 and 1972 her works are made from plastic coloured on the inside, with so-called "ready-made" elements, and include several examples of conceptualism, and, moreover, some of the objects have a function of use. This justifies controversial questions about possibilities of deviation from classical doctrines, conservation theories and tenets into reconstruction, uses of artworks, light, *etc.* The complex preservation process transformed Szapocznikow's *avant-garde* legacy into a new cultural phenomenon, widely exhibited in the world (MoMA 2012, Centre Pompidou 2013).

The trans-disciplinary research with many variants may be useful in many curatorial-conservation investigations. This concerns the holistic character of research and many new aspects of preservation based on curatorial-conservation procedure. Beginning with the complex identification, pre-acquisition process–this will also include preventive conservation. For conservation standards for specifically defined groups of materials and complex works of differing microclimatic requirements plus standards of the preservation of authenticity (material, image, idea), conservation "by documentation". The new challenges are connected with the recreation of ephemeral objects through the method of artist's replicas–respecting the opinions on the permissibility (or otherwise) of the replacement of ready-made elements–the creation of replicas for display purposes, or emulations.

These considerations are part of the historical processes of structural changes in systems of knowledge in the preservation of cultural heritage, including modern and contemporary art. Cultural heritage has provided a new strand of understanding based on the artistic and social importance of "man-made" heritage. The review of ethical and technical issues provoked the need for the "*primum non nocere*" (first, do no harm) concept, and analyses for each of the major options and variants, *etc.* The cultural identity and permanence of the witnesses of contemporary visual art legacy in general depend on the creation and application of new innovative concepts of taking care of them—for their value as human testimony. We are now participating in a dynamic process of transformations in the twenty-first century, when crucial fields of research within the scope of cultural heritage will come from separate cross-disciplinary domains that require special attention.

References

1. O. Dieter, Hummel and Friedrich Scholl, *Atlas of Polymer and Plastics Analysis*, Wiley-VCH, 2nd edn, 1999.
2. Y. Shashoua, *Conservation of Plastics. Materials Science, Degradation and Preservation*, Elsevier, Amsterdam, 2008, p. 9.
3. M. Hermansdorfer, Source of joy, pain, truth, in *Alina Szapocznikow. 1926–1973*, ed. Z. Gołubiew, [cat. published], Warsaw, 1998, pp. 24–26.
4. H. Hoelling, The Technique of Conservation: On Realms of Theory and Cultures of Practice, *J. Inst. Conserv.*, 2017.
5. I. Szmelter, Innovative Approaches to the Complex Care of Contemporary Art, in *The Knowledge Tree*, ed. I. Szmelter, Academy of Fine Arts, Archetype Publications, Warsaw, London, 2012, pp. 10–33, Abstract.
6. F. Waentig, *Plastics in Art: A Study from the Conservation Point of View*, Petersburg, 2008, p. 48.
7. P. Restany, *Alina Szapocznikow – Age-Old Body Language*[cat. published], Zachęta, Warsaw, 1998, p. 33.
8. C. Mancusi-Ungaro, Authority and Ethics, in *The Replica and its Implications in Modern Sculpture Workshop*, Tate Modern, 18–19 October 2007, https://www.tate.org.uk/.../carol-mancusi-ungaro-authority/download/file/fid/7372, (accessed 12.10.2018).
9. http://contemporarylynx.co.uk/alina-szapocznikow-human-landscapes-review-experience-translated-through-body, (accessed 20.10.2018).

10. A. Riegl, The Modern Cult of Monuments: Its Essence and its Development, in *Historical and Philosophical Issues on the Conservation of Cultural Heritage*, ed. N. S. Price, M. K. Talley Jr and A. M. Vaccaro, Getty Conservation Institute, Los Angeles, 1996, pp. 69–83.
11. C. Brandi, *Teoria Restauracji*, ed. G. Basile and I. Szmelter, trans. M. Kijanko, MIK, Warsaw, 2006, passim.
12. I. Szmelter, 'Paradigm of Theory and Conservation Practice in Relation to Modern Art' (Paradygmat teorii i praktyki konserwatorskiej w odniesieniu do sztuki nowoczesnej), in *Contemporary Problem of Conservation Theory (Wspolczesne Problemy Teorii Konserwatorskiej)*, ed. B. Szmygin, ICOMOS, Warsaw-Lublin, 2008, pp. 133–145.
13. H. Hoelling, op.cit. (accessed 1.10.2018).
14. I. Szmelter, Elements of the New Theory of the Conservation of the Heritage of Visual Art. Necessity for the Protection of Atypical Old and Contemporary Art, *Art and Documentation*, 2017, pp. 155–180, http://yadda.icm.edu.pl/yadda/element/bwmeta1.element.desklight-d05c006d-f84a-4286-980d-2914cfe79833 (accessed 14.11.2018).

CHAPTER 6

The Colors of Lina Bo Bardi: Analytical Investigations of Lina's Felt-tip Pens

ANTONIO MIRABILE*[a], GIULIA GERMINARIO[b], INEZ D. VAN DER WERF[b, c], LUIGIA SABBATINI[b, c], BRENDA DOHERTY[d] AND PATRIZIA MORETTI[d]

[a]Paper Conservator, 11 rue de Bellefond, 75009, Paris, France; [b]Dipartimento di Chimica, Università degli Studi di Bari "Aldo Moro", Via Orabona, 4, 70125 Bari, Italy; [c]Centro Interdipartimentale "Laboratorio di Ricerca per la Diagnostica dei Beni Culturali", Università degli Studi di Bari "Aldo Moro", 70125 Bari, Italy; [d]Istituto CNR di Scienze e Tecnologie Molecolari (CNR-ISTM), via Elce di Sotto 8, I-06123 Perugia, Italy
*E-mail: am@antoniomirabile.com

6.1 Introduction

The identification of modern and contemporary felt-tip pen[†] (FTP) ink formulations is a current analytical challenge, which may be considered best assessed by brand and model variability through time. Undoubtedly complicated by the global mass production and vast range of pen typologies (*i.e.* ball-point pen, felt-tip pen, markers, highlighter, white-out, *etc.*) and associated ink formulations coinciding with their individualist use by recent artists, the most part of which have been selected and adapted for

[†]The Writing Instrument Manufacturers Association (WIMA) uses the term *porous pointed pen* in ref. 1. WIMA has not undertaken any revision of this nomenclature since 1982.

Science and Art: The Contemporary Painted Surface
Edited by Antonio Sgamellotti, Brunetto Giovanni Brunetti and Costanza Miliani

Published by the Royal Society of Chemistry, www.rsc.org

pure effect and are not necessarily of a professional artist's quality, reflecting the freer and faster acquisition we see in the social and cultural aspects of modern society. Most of the FTP inks that first became available in the mid-20th century degrade unpredictably, suffer from light fastness and color change and other migration or color transfer into adjacent substrates. Furthermore, they have not yet been fully studied with regard to their constituents, elucidation of degradation pathways and response to treatment methods. Such FTP-containing artworks are undergoing conservation treatments for the very first time and require, in addition to conservation ethics and educated caution, considerations on moral rights and ongoing dialogue with living artists or estates.

In this contribution we present the combined results and interpretation of a multi-analytical investigation of 20 Johann Faber FTP (Figure 6.1) commercialized in the 1960s belonging to the Italian-Brazilian architect and designer Lina Bo Bardi (Rome 1914–São Paulo 1992).

Nowadays, there are an increasing number of scientific publications on characterizations of FTP inks,[2–10] yet only very few of them[4,7,9,10] are directly related to real artworks or artists' toolboxes. For conservators and heritage scientists alike, the opportunity to work on the original palette of the artist enhances and contextualizes the discussion of their creative approach. With regard to Lina, as an architect, political activist, furniture and stage set designer, editor, writer and exhibition curator, she largely employed FTP

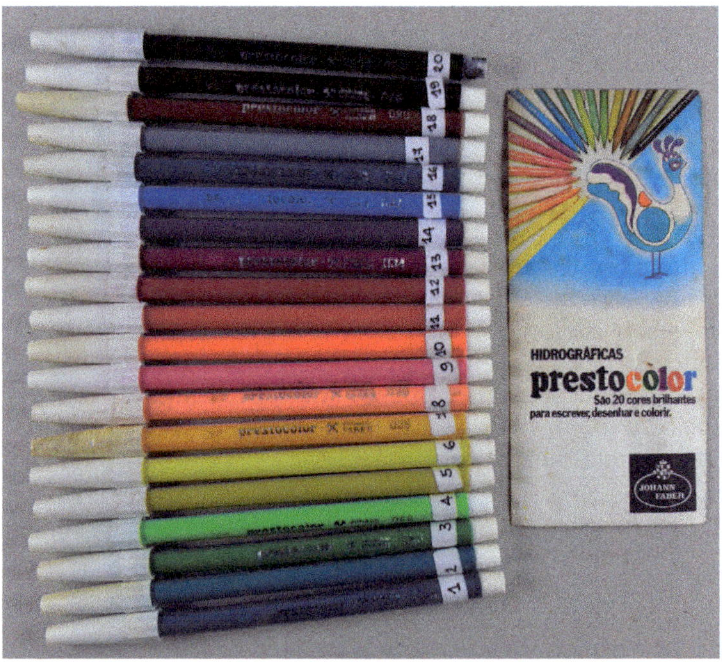

Figure 6.1 Johann Faber™ felt-tip pens (hidrográficas prestocolor).

for her activities, from making fine art, design and architectural drawings to doodling margins, jotting down notes and filling out forms. This also permits an investigation into the history of FTP inks, their features, rapid development/commercialization, sensitivity and response to conservation treatments.

Identifying dyes and colorants used in modern and contemporary inks complements the analysis and the evolution of the art style as well as the development of the artist's material. It provides a precise and transversal knowledge, encompassing the multidisciplinary fields of history of art, conservation science, conservation methods and material history. Understanding the physical and chemical properties and characterizing the inks' composition is necessary to anticipate the range of behaviour and impact of conservation treatments and to create the ideal housing, storing and exhibiting conditions.

6.2 Lina Bo Bardi

Arriving in Brazil in 1946, Lina Bo Bardi embraced the traditional cultures of her adopted country and added mud and straw to her palette. The Brazil in which she worked was never rich and was often politically turbulent. She was someone who could work with whatever was at hand, including off-cuts from the building of stage sets. She published a thesis as a manifesto titled "Introductory contribution to the teaching of architectural theory" where she disagrees with the exterior forms and virtuosities that she felt were devaluing the spirit of modern architecture.

In 1951 Bo Bardi built a Glass House (*Casa de Vidro*) for herself and her husband, raised on columns above an inclined site on the edge of the city. Over the years, vegetation has surrounded it, as she knew it would, such that it has become a more opulent and beautiful place than when first built. The Glass House became the Istituto Lina Bo and Pietro Maria Bardi (ILBPMB) and now hosts Bo Bardi's collection[‡] of artworks, drawings and documentation.

The Glass House was a precursor to her most important mid-century project: a permanent home for the Museu de Arte de São Paulo (MASP), which her husband Pietro Maria Bardi had been directing. The MASP, built between 1957 and 1969, was meant to express Brazil's ascent on the international stage – as well as the ascent of wealthy, cosmopolitan São Paulo against its traditional rival, Rio de Janeiro – and Bo Bardi proposed an audacious, iconic building worthy of the job. On a prime lot on Avenida Paulista, the museum comprises a hulking concrete and glass box suspended eight meters off the ground, held in place by bright red pillars and creating beneath it a public gathering space. She described it as her "fixed tropical greenhouse".[11]

[‡]Personal communication with Anna Carboncini, director of the ILBPMB: The collection encompasses 2600 books, 2300 magazines, 200 movable artworks, 500 popular artefacts, 300 archaeological objects, 6700 Lina Bo Bardi's drawings, 1200 artworks on canvas and paper, 18 000 photos and 11 000 archival items.

In the early 1980s, though, as the dictatorship's rigidity began to moderate, Bo Bardi completed one of the most extraordinary projects of recent Brazilian architecture: the SESC Pompéia (Sketches of the flow diagram are shown in Figures 6.3a and 6.4a), a social and cultural centre adapted from an obsolete oil drum factory in downtown São Paulo. It now includes football fields and a swimming pool and a theatre. Rather than tear down the pre-existing concrete structure, Bo Bardi looked at how locals were already using it, and then refined her design to favour the functions that the community actually needed. In a city and a country with strong class divisions, Bo Bardi's SESC Pompéia is the rare urban space appealing to all; from international gallery-goers to neighbourhood old-timers looking for a chess match.

Among the 6700 drawings of the collection, over 1200 are realized in or contain FTP[§].

6.3 Felt-tip Pens

The first FTP was patented by Lee Newman as a felt-tipped marker in 1910,[12] quickly accompanied by another early version of the pen in 1911 under the name of the European Kaweco® Signier.[13,14] These marker pens were not commercially viable, and soon fell into oblivion. It was not until technical developments in dye chemistry as well as in nib and reservoir design that a more popularized felt-tip pen could be produced and commercialized in 1946, followed by the highly popular Magic Marker created by Sidney Rosenthal in 1953. The year 1962 brought the development of the modern fiber-tipped pen, in contrast to the marker which generally has a thicker point, by Yukio Horie of the Tokyo Stationery Company, which later became Pentel®. Highlighters and fine-line markers appeared in the 1970s. Binney & Smith®, who made Crayola®, bought the name Magic Marker in 1989 and started selling improved permanent markers. In 1996, they also started selling Magic Marker II and dry-erase markers for writing on white-boards and glass.

FTP is a pen that has a tip made of porous, pressed fibers such as felt and a reservoir of colored ink. This reservoir has a core of an absorbent material which serves as a carrier for the ink. One more necessary part of a marker is a cap that prevents drying of ink in a nib. Regardless of form and design of the felt-tip pen as an object, which is portrayed by a chain of brands, nibs and reservoir evolutions, one element of great importance to be further understood is the behaviour upon aging of this media correlated to ink formulation containing dye, solvent and additives.

The ink of a FTP also contains a solvent to maintain its liquid form and falls into two classes: water-based or organic solvent-based. Among the solvents, ethanol and isopropanol could be often found.[15] Regardless of what

---

[§]Personal communication with Anna Carboncini, director of the ILBPMB.

liquid carrier is used, the solvent must be compatible with the nib and the reservoir, as well as with the type of dye used. In addition to the main solvent, other components are added. Among them may be mentioned viscosity modifiers added to improve flow and to prevent drying out in the nib, together with a variably small percentage of pH modifiers, humectants, antifoaming agents, surfactants, biocides, fillers and extenders.[16,17]

The range of dyes used in the composition of these inks is extremely wide; hundreds of dyes are commercially available[18] and many ink colors are feasibly made from dye mixtures. Artworks realized with felt-tip pens present conservation problems; by far the most observed is their notorious lightfastness[19] and color change – some of these inks will fade even in the absence of ultraviolet light.[20] Another phenomenon attributable to the instability of dye-based inks is color transfer, the migration of certain colors into adjacent drawings or wrapping paper. In addition to the typical problems associated with art on paper, they also suffer from the irregular behaviour of the inks in water- and/or solvent-based conservation treatments.

The artists' palette examined herein is a set of 20 markers carrying the logo of the Johann Faber™ brand. Produced and acquired locally from the largest Johann Faber™'s subsidiary stationed in Brazil, which was subsequently and historically acquired by Faber-Castell® in 1967.[21] Thus the study of these commercial felt-tips from this time period could provide an important source of data to the probable alignment of ink formulations and production across the Americas.

6.4 Scientific Analyses

Twenty historical Johann Faber™ felt-tip pens, made available by the Lina Bo and Pietro Maria Bardi Institute (São Paulo, Brazil), were investigated through a multi-analytical approach (results summarized in Table 6.1). The complexity of the formulations and the detection of dyes studied has required a continuous implementation of the method. Specifically, spectroscopic methods (*i.e.* X-ray fluorescence, conventional and coupled to thin-layer chromatography micro-Raman, SERS and reflection infrared spectroscopy) combined with the use of chromatography techniques (GC-MS and Py-GC-MS)[¶] were exploited in order to provide a thorough chemical characterization of the ink formulations.

The elemental composition of the inks was determined by non-invasive XRF analyses performed directly on the tip of each marker, while micro-destructive spectroscopic and chromatographic investigations (felt-tip ink marked on aluminum foil or glass slide) provided data concerning the molecular features of the dyes, binders and additives. In most of the analysed inks Raman spectroscopy prevalently permitted main dye component/s to

¶The instrumental details of each technique are provided in the appendix.

Table 6.1 Summary of the results obtained from 20 early Johann Faber felt tip-pens (hidrográficas presto color).

Felt-tip pen	Raman spectroscopy	Py-GC/MS
FTP1	**Triarylmethanes**	
	Yellowish light green SF	Yellowish light green SF
	Brilliant Blue[a]	Brilliant Blue
	Patent blue V[a]	Patent blue V
	Diamond green G[a]	Diamond green G
	Diethylene glycol, diisooctyl phthalate (plasticizer), phenols (antioxidant), plant gum (binder)	
FTP2	**Triarylmethanes**	
	Yellowish light green SF	Yellowish light green SF
FTP3	**Triarylmethanes + Azo dyes**	
	Yellowish light green SF[a]	Yellowish light green SF
	Tartrazine[a]	Tartrazine
FTP4	**Triarylmethanes + Azo dyes**	
	Yellowish light green SF[a]	Yellowish light green SF
	Tartrazine[a]	Tartrazine
FTP5	**Azines + Azo dyes**	
	Nigrosin	Nigrosin
	Possible mono azo Fast red B	Possible mono azo Fast red B
FTP6	**Azo dyes**	
	Tartrazine	Tartrazine
FTP7	**Xanthenes**	
	Fluorescein	Fluorescein
FTP8	**Xanthenes**	
	Eosin Y[b]	Eosin Y
FTP9	**Xanthenes**	
	Eosin Y[b]	Eosin Y
	Possible mix of other xanthenes[a]	In combination with further unidentified dyes
FTP10	**Azo dyes**	
	Naphthol red[a]	Tartrazine
	(Un-interpreted conventional Raman spectrum)	Possible mono-azo naphthol
FTP11	**Azo**	
	Naphthol red[a]	Possible mono-azo naphthol
	(Un-interpreted conventional Raman spectrum)	
FTP12	**Xanthenes**	
	Eosin Y[b]	Eosin Y
	Possible mix of other xanthenes[a]	
FTP13	**Indigoids + Xanthenes**	
	Indigo Carmine	Indigo Carmine
	Possible Eosin Y[b]	Eosin Y
FTP14	**Triarylmethanes**	
	Crystal Violet	Crystal Violet
FTP15	**Triarylmethanes**	

Felt-tip pen	Raman spectroscopy	Py-GC/MS
	Yellowish light green SF	Yellowish light green SF
	Brilliant Blue[a]	Brilliant Blue
	Patent blue V[a]	Patent blue V
	Diamond green G[a]	Diamond green G
FTP16	**Triarylmethanes**	
	Water blue IN	Water blue IN
FTP17	**Azines + Azo + Oxides**	
	Nigrosin	Nigrosin
	Iron oxide[c]	Possible mono azo Fast red B
	Possible mono azo Fast red B	
FTP18	**Azines + Azo dyes**	
	Nigrosin	Nigrosin
	Possible mono azo Fast red B	Possible mono azo Fast red B
FTP19	**Azines + Azo dyes**	
	Nigrosin	Nigrosin
	Possible mono azo Fast red B	Possible mono azo Fast red B
FTP20	**Azines + Azo + Oxides**	
	Nigrosin	Nigrosin
	Iron oxide[c]	Possible mono azo Fast red B
	Possible mono azo Fast red B	

[a]Identified by TLC-Raman marker band of azo compound by FTIR.
[b]Br from XRF
[c]Fe from XRF.

be detected. However, in order to exhaustively distinguish the different dyes present in the mixture, this spectroscopic technique was coupled to thin-layer chromatography (TLC). Surface-enhanced Raman spectroscopy (SERS) was employed only for those inks which produced too much fluorescence in conventional Raman spectroscopy and to enhance sensitivity. It should be noted that this latter approach can only be performed when felt-tip pens are available and currently not routinely on felt-tip pen inks in drawings due to the micro-destructive nature of such investigations.

The identification of binders, solvents and plasticizer was reached by using FTIR spectroscopy implemented with the chromatographic analyses. The Py-GC-MS has established exhaustive information on the type of solvent and plasticizers and provided indications on the presence of a polysaccharide binder. Subsequently, this latter item was investigated more thoroughly by GC-MS.[22]

In order to establish the most appropriate conditions for the conservation of the Lina Bo Bardi sketches and drawings some of the investigated felt-tip pen inks were subjected to aging. The fading evaluation was performed on the inks stored in the dark for 21 months and on inks exposed to natural light conditions for two months.

Below, the analytical data obtained for the different ink components and for the aging test are shown. Each felt-tip pen is individually indicated with the acronym FTP followed by a specific number.

6.5 Dyes

Micro-Raman analyses supported by TLC and corroborated by Py-GC-MS allowed complex mixtures of dyes belonging to different classes such as xanthenes, indigoids, triarylmethanes, azo and azine dyes to be detected.

6.5.1 Xanthenes

Xanthene dyes characteristically contain a xanthylium or dibenzo-*y*-pyran nucleus as the chromophore, with amino or hydroxy groups meta to the oxygen as the usual auxochromes. They have long provided synthetic coloring agents that were widely employed in the dyeing of paper and the preparation of lakes for use as pigments.[23] They were also used in the dyeing of silk and wool where brilliant shades with fluorescent effects were required, but where lightfastness was not so important. Within the class of xanthenes, fluorenes and fluorones have found important modern applications as fluorescent biological marker probes and in photovoltaics.[24]

A xanthene dye often detected in both early and current commercial pink/red/orange felt-tip inks[4,5,8–10] is Eosin Y (acid red 87, CI 45380), the renowned brominated derivative produced by Caro (1887–92).[23] In Lina's felt-tip pens the presence of this dye was suggested by the specific XRF detection of bromine in four pink-purplish inks (FTP8, FTP9, FTP12 (Figure 6.2a) and FTP13 (Figure 6.2b)) and corroborated by both the spectroscopic and chromatographic examinations. The marker bands of the C–Br vibrational modes at *ca.* 1015 cm^{-1} and in the spectral range from 280 to 198 cm^{-1} are diagnostic in the Raman spectra collected for these inks (Figure 6.2c).

Notably, in a yellow-orange felt-tip pen (FTP7), differently from the most common use of azo dyes for these ink hues, a further xanthene dye, namely Fluorescein (acid yellow 73, C.I. 45350), was found. It is the first representative dye belonging to the xanthene class, introduced by Baeyer in 1871, closely followed by its disodium salt, Uranine. In the Raman spectrum acquired on FTP7 ink (Figure 6.2c), bands at *ca.* 1497, 1333 and 639 cm^{-1} ascribable to both the xanthene and benzoate moiety,[25,26] as well as the COH *bending* of the phenolic group at *ca.* 1188 cm^{-1}, are recognizable.

The presence of the xanthenes in the pyrograms could be recognizable when the 1,4-dioxane, phenol, biphenyl and 9-H-fluoren-9-one could be detected at 550 °C (FTP7, FTP8, FTP9, FTP12 and FTP13), Fluorescein and Eosin Y at this temperature produce the same products. In order to distinguish these dyes it was necessary to increase the pyrolysis temperature up to 700 °C. In this way, xanthene, together with bromobenzene (Figure 6.2d), can be detected, and therefore this pyrolysis temperature can be considered suitable to provide information not only on the class, but also on the specific dye that can be unequivocally identified as Eosin Y.

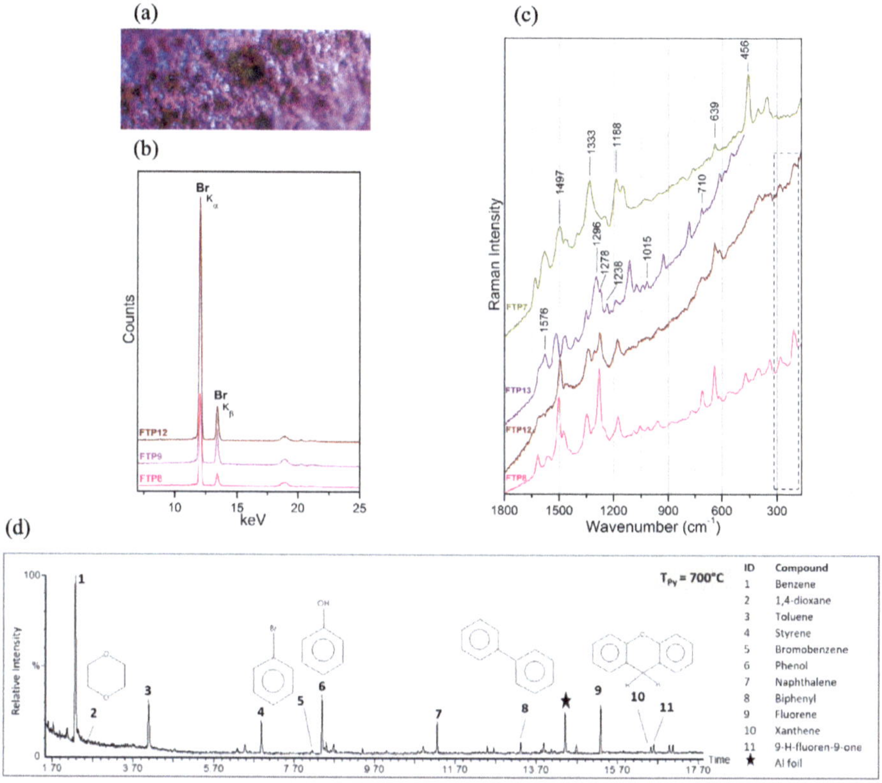

Figure 6.2 (a) OM image of FTP12 ink; (b) XRF analysis of FTP8, FTP9 and FTP 12 containing Eosin Y; (c) Micro-Raman spectra (λ = 785 nm) of FTP7 (Fluorescein), FTP13 (indigo-carmine and Eosin Y), FTP12 and FTP8 (Eosin Y); (d) Pyrogram at T_{Py} = 700 °C of FTP12.

6.5.2 Indigoids

Of the widest and longest used dyes and pigments, the natural plant (indigo and woad) and prestigious animal forms (Tyrian purple) of this class of dyes have been used since antiquity, showing great lightfastness properties. Chemically, 2-(1,3-dihydro-3-oxo-*2H*-indol-2-ylidene)-1,2-dihydro-*3H*-indol-3-one, the sulfonated more soluble synthetic form, as developed by Barth in 1740, has been prepared by treating indigo with sulfuric acid, ultimately replacing the grand use of the plant source.[27] This synthetic form, in mixture with Eosin Y (displaying the characteristic xanthene bands as previously discussed and shown in Figure 6.2b, was found in the ink composition of the red-purple FTP13 by micro-Raman spectroscopy integrated with Py-GC-MS analysis).

The presence of the synthetic indigo carmine (acid blue 74, C.I. 73015) could not be ascertained only by the specific Raman bands at 1576 cm^{-1}, C=C asymmetrical *stretching* vibrations of the two pyrrolidone rings and at 1296 cm^{-1}, due to the possible overlapping combination of the $\nu_{symm}(SO_3^-)$, δ(C–C) and ν(N–C) and the C=O and C–H *bending* modes at *ca.* 1238 cm^{-1} (Figure 6.2c).[28]

The chemical composition of the synthetic indigo carmine–Eosin Y mixture could instead be successfully confirmed by means of Py-GC-MS where some pyrolytic products ascribable to an indigoid dye (*N*-ethyl-*N*-phenylformamide, 3-methoxy-4,7-dimethyl-1-*H*-isoindole,1,2,3-trimethyl-1-*H*-indole and 1-methyl-3-acetylindole) were identified (data not shown). The specific presence of indigo carmine, the sulfonated form of indigo, could be confirmed by the presence of sulfur dioxide at low retention times. The pyrolytic products of xanthene dyes were also identified; however, the presence of Eosin Y could not be confirmed since the pyrogram was acquired at 550 °C.

Interestingly, the synthetic indigo-carmine is not commonly used as dye in the ink formulations of felt-tip pens.[15] Moreover, in a current commercial Stabilo[2] marker with the same color as FTP13 a different mixture of dyes (rhodamine B and erioglaucine) was found.[10] Therefore, despite the limited available data, the detection of an indigo carmine–Eosin Y mixture in Lina Bo Bardi's red-purple pen could be correlated to an early ink formulation.

6.5.3 Triarylmethanes

Triarylmethanes (TAMs) are the oldest synthetic class of dyestuffs derived from aniline, with fuchsine being discovered by the Polish chemist Jakub Natanson in 1856 and subsequently developed as a dyestuff by the notable silk dyers Renard Frères & Franc in Lyon. This dye-class was extended to methyl violet (Lauth in 1860), followed by the blues in 1862 and the greens (methyl green, 1871–1883). The famous Crystal Violet followed in 1883 and Patent blue in 1888. The TAM dyes, belonging to the class of basic dyes, are chemically constituted of a triphenylmethane backbone with an array of substituents (halogen, alkyl, alkoxy, sulfonic acid and carboxyl) on the aryl groups. TAMs have played an important role in the dyeing of textiles, embroideries, wall-paper and furniture and in paintings, and are an object of wide interest in the heritage science and conservation fields, as shown by their wide literature base.[29]

Spectral features ascribable to the presence of three different triarylmethane dyes[30] (Figure 6.3a), namely Yellowish light green SF (acid green 5, C.I. 42095), Crystal Violet (basic violet 3, C.I. 42555) and Water blue IN (acid blue 93, C.I. 42780), were actually detected in seven felt-tip inks (FTP1, FTP2, FTP3, FTP4, FTP14, FTP15 and FTP16), displaying diverse chromatic tonalities from green and blue to violet. Among these dyes, Crystal Violet has been often found in both historical and recent commercial felt-tip pens,[8–10] while Yellowish light green SF has been actually hypothesized only in a Fontana's green ink,[7] suggesting a possible prevalence of this dye in earlier ink formulations. In the modern green-blue inks, indeed, the use of triarylmethanes

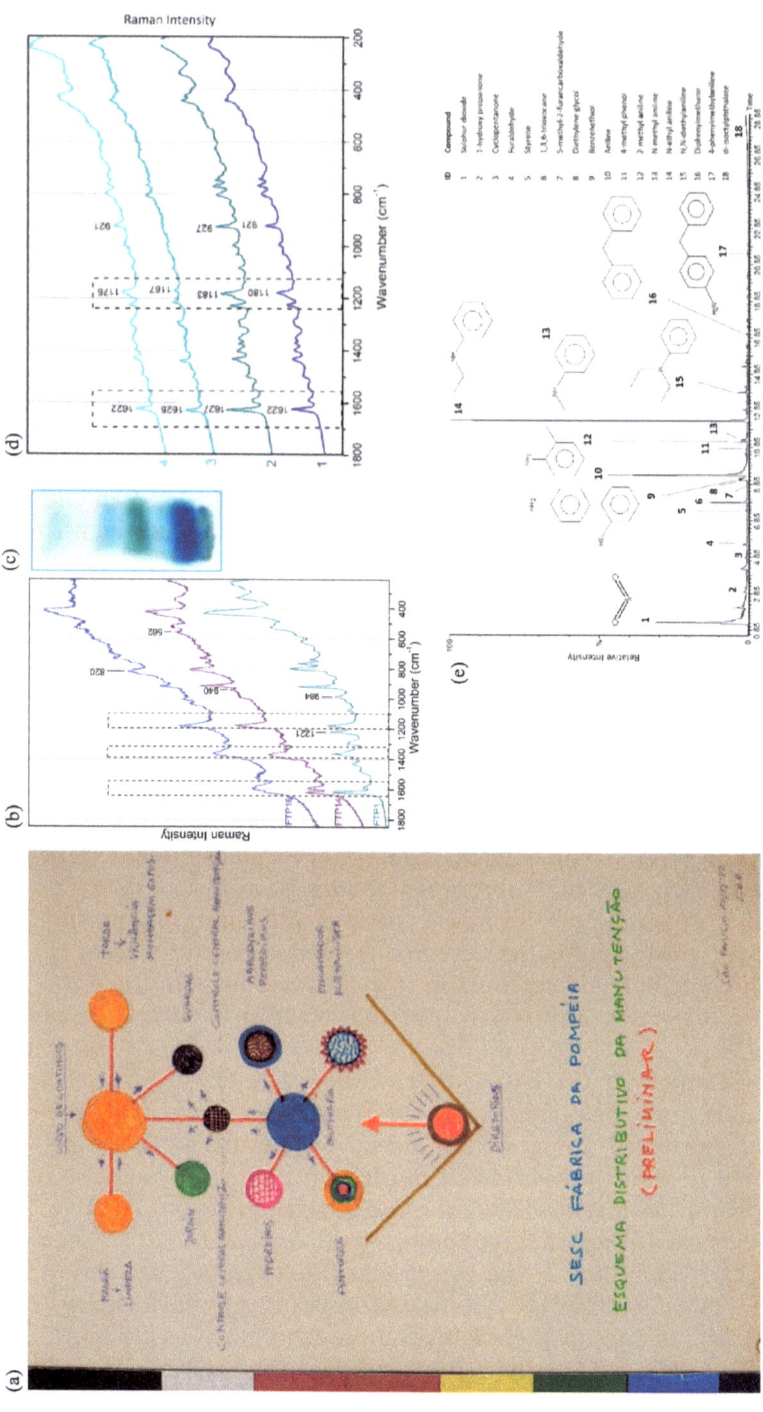

Figure 6.3 (a) Lina Bo Bardi: SESC – Fábrica da Pompéia, Flow diagram, 1982, felt-tip pen and ball-point pen on paper, 21.5 × 31.4 cm, Inv. 094ARQd0654, © Instituto Lina Bo e P. M. Bardi; (b) Micro-Raman spectra of FTP1 (Yellowish light green SF) ($\lambda = 514$ nm), FTP14 (Crystal Violet) and FTP16 (Water blue IN) ($\lambda = 488$ nm); (c) TLC separation of FTP1 using acid eluent; (d) SERS spectra ($\lambda = 785$ nm) of corresponding spots: 1 (Brilliant blue), 2 (Yellowish light green SF), 3 and 4 (possibly Patent blue V and Diamond green G); (e) Pyrogram of FTP1 at $T_{Py} = 500$ °C.

is often replaced by the lightfast phthalocyanine dyes.[2,5,6] However, a triarylmethane dye such as the Brilliant blue or erioglaucine (acid blue 9, C.I. 42090) is still widely used.[5,8–10] Here, the latter emerged in a mixture with ulterior triarylmethanes (Patent blue V and Diamond green G) exclusively in the SERS spectra collected following the TLC separation (Figure 6.3c) on FTP1 and FTP15 inks (Figure 6.3d). The inks with the blue-green shades contain one or more triarylmethane dyes, sometimes with the azo dyes, as also confirmed by the TLC–Raman–SERS (Table 6.1).

All pyrograms show products common to the whole triarylmethane class, such as aniline, *N*-methyl aniline and 2-methyl aniline. The SERS–TLC data show the presence of different triarylmethanes in FTP1 and FTP15 (Table 6.1), so the detection of some pyrolytic products could be due to the fragmentation of more than one of these dyes as for the *N*-ethylaniline and *N*,*N*-diethylaniline products which could be ascribable to Yellowish light green SF and to Brilliant blue (Figure 6.3e). The presence of sulfur-based compounds, *i.e.* sulfur dioxide and benzenethiol, could be justified by the scission of the sulfonic acid groups in Yellowish light green SF, Brilliant blue and Patent blue V, or by the bisulphate counterion of Diamond green G, already identified with SERS analyses.[31]

Water blue IN could be confirmed by the detection of diphenylamine and *N*-methyl-*N*-phenylaniline (FTP16), together with sulfur-based compounds; whereas *N*,*N*-dimethylaniline could be instead considered characteristic of the Crystal Violet dye (FTP14).

6.5.4 Azo and Azine Dyes

The azo pigments are significant in the timeline of modern pigment development, providing a vast color range (yellow, orange, red, bordeaux, carmine and brown, although blue and green shades have no commercial significance) and properties for use in the coatings and paint, ink, plastic and fiber industries.[32] Azo pigments were first created using the process of diazotization discovered by Peter Griess in 1858, characterized by the presence of one or more azo groups (–N=N–) bound to the sp^2 hybrid carbon atoms.[33]

Tartrazine (acid yellow 23, C.I. 19140), first synthesized by Ziegler in 1884, is a yellow, soluble mono-azo pigment among the most widespread in the ink formulations of the felt-tip pens.[4,5,8,10] Indeed, also in this study Tartrazine was detected in the composition of five felt-tip inks, pure (yellowish FTP6) or in admixture with Yellowish light green SF (green FT3 and FTP4) or a further azo dye such as Naphthol red (orange-red FTP10 and FTP11). This dye could be clearly identified in the pyrogram of these inks, since sulfur dioxide, benzenethiol, aniline, benzylnitrile, 1-isocyano-3-methylbenzene, 4-cyanothiophenol and specific recombination products, such as diphenylsulfide, diphenylamine, and 5-methyl-1,10-phenanthroline, can be attributed to the pyrolysis of Tartrazine (data not shown).

In the pyrograms of these inks beta-naphthol can be considered the main fragment produced by this sub-class of azo dyes. In addition, *p*-methylbenzene-isocyanate, 1-isocyanonaphthalene, naphthalene, naphthalenamine and naphthalenthiol seem characteristic products even if, to the best of our knowledge, no reference data are available to identify the specific mono-azo naphthol dye.[34–36]

Evidence of possible earlier ink formulations could be correlated to the unusual employment[15] of the mono-azo Fast red B (acid red 17, C.I. 16180) in mixture with Nigrosin (acid black 2, C.I. 50420) found in five of Lina's felt-tip pens. Nigrosin is an aqueous blue-black dye produced in 1867 by Coupier belongs to the azine series of the early synthetic dyes. Chemically derived from phenazine, azines observe the characteristic connectivity RR'C=N–N=CRR and have been chiefly used to dye wood, leather and textiles and more recently have found great utility as biological stains.[37]

Specifically, in the Raman spectra collected for FTP5, FTP17, FTP18, FTP19 and FTP20 (see representative spectra in Figure 6.4b) the detection of the

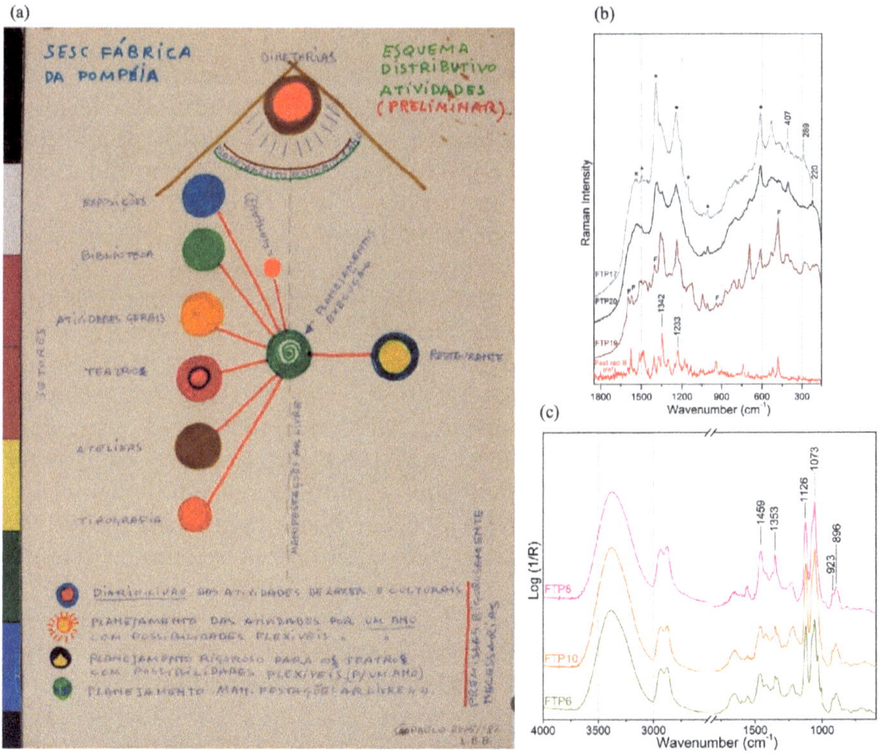

Figure 6.4 (a) Lina Bo Bardi: SESC – Fábrica da Pompéia, Flow diagram, 1982, felt-tip pen and ball-point pen on paper, 21.5 × 31.4 cm, Inv. 094ARQd0655, © Instituto Lina Bo e P.M. Bardi; (b) Micro-Raman spectra of FTP19, FTP20 (λ = 488 nm) and FTP17 (λ = 514 nm) (*Nigrosin bands; F = monoazo fast red B bands); (c) Transflection mode FTIR spectra of FTP6, FTP8 and FTP10.

typical mono-azo bands at ≈1233 and 1342 cm^{-1} (>C–NH–N=C<)[38] ascribable to Fast red B is affected by the overlapping phenomena with the diagnostic signals of Nigrosin.[39] Therefore, the presence of Fast red B could be confirmed by the identification of sulfur dioxide, naphthalene and naphthalenamine in all the pyrograms obtained.

Nigrosin could be recognized in the pyrograms of these inks by the presence of sulfur dioxide, benzene isocyanate, benzenethiol, aniline, biphenyl, diphenylamine and phenazine. Most of these pyrolytic products are in common with other dyes. However, the presence of phenazine could be considered specific to this dye.

Moreover, additional signals at *ca.* 407, 289 and 220 cm^{-1} identified in FTP17 and FTP20 Raman spectra (Figure 6.4b) suggested, instead, the presence of an iron oxide-based compound.[40] The latter was corroborated by the XRF detection of iron in the same inks and it could well be associated with the use of an iron catalyst during the Nigrosin production.[41,42]

6.6 Binders and Additives

Besides confirming the varied dye compositions of most of Lina's examined pens, reflection FTIR spectroscopy permitted the presence of ulterior constituents to be revealed. Specifically, the presence of an additive such as diethylene glycol (DEG), commonly added to the ink formulations in order to avoid them drying out too fast,[15] was revealed. The characteristic bands of this component at 1459 and 1353 cm^{-1}, CH_2 *bending* and *wagging*, respectively, the doublet at *ca.* 1126 and 1073 cm^{-1} associated with asymmetric and symmetric C–O–C or C–O *stretchings*, as well as the absorptions due to primary alcohols at 923 (COH) and 896 cm^{-1} ($-OC_2H_4$), were noticed in the infrared spectra collected for each felt-tip ink (some representative FTIR spectra are shown in Figure 6.4c). Pyrolysis products such as the intact diethylene glycol molecule, 1,4-dioxane and 1,3,6-trioxocane products identified in all the pyrograms have confirmed the presence of DEG. In addition, diisooctyl phthalate, used as plasticizer, together with antioxidants were revealed, corroborating the same additive formulation also found in modern inks of different commercial brands (Figure 6.3e).[5,6]

As to the binders, instead, the acrylic-based compounds (*e.g.* styrene–acrylic copolymers, polyacrylonitrile, *etc.*) often used in the current markers,[6] were not identified. Indeed, in all pyrograms obtained for Lina's pens (Figures 6.2d and 6.3e), 1-hydroxy-2-propanone, 2-furancarboxaldehyde, and 5-methyl-2-furancarboxaldehyde could be related to a polysaccharidic binder. A specific GC-MS procedure for the analysis of plant gums (see Appendix) allowed us to identify the presence of xylose, arabinose, fructose, galactose and glucose which are due to the presence of a plant gum found in both earlier[7] and more recent (Stabilo® felt-tip)[6] ink formulations. Typically in the earlier inks the arabic gum was the most exploited, but in this case the xylose:arabinose ratio of >1 and the lack of some characteristic monosaccharides did not allow for an unambiguous identification of the type of plant gum.[43]

6.7 Aging Behaviour

Preliminary aging data were recorded for a few of Lina Bo Bardi's felt-tip pen inks. The inks were first aged by keeping the colored areas on paper in the dark (for 22 months); the same specimens were subsequently exposed to natural light aging (for two months). Although these conditions could not be considered optimal for an exact aging test, they can give us some useful indications on the behaviour of the inks.

The histogram in Figure 6.5a depicts the ΔE values calculated after dark aging (red bar) and light aging (blue bar) and which were considered as a sum of the two contributions since the data were acquired on the same specimens. These results clearly show that all samples suffer a color change even if stored in the dark, which tends to get worse after only two months of light exposure. Considering the ΔL^*, Δa^* and Δb^* components (data not shown), it was noticed that the contribution of ΔL to the ΔE value is significantly reduced upon light aging, while the influence of Δa^* and Δb^* is more significant, indicating a color change rather than a fading effect. The FTP6 ink, containing only Tartrazine, shows the highest value of ΔE, which is mainly due to dark aging, evidencing a very high instability of the dye. The Yellowish light green SF dye, present in FTP2, tends to degrade very quickly when exposed to light. A similar trend was observed for the FTP3 and FTP4 inks, which contain Tartrazine together with Yellowish light green SF dye, probably depending on the different relative amounts of these dyes in their formulations. In FTP1 a mix of triarylmethane dyes was recognized (see Figure 6.3c and Table 6.1). Its aging trend appears quite similar to FTP2, even if in this ink only Yellowish light green SF was found.

The reflectance spectra of both FTP1 and FTP2 show a decrease of the overall reflectance, but only for FTP1 was a color change observed; the shape of the main band (at *ca.* 480 nm) appears broader after aging (Figure 6.5b).

Inks containing azo dyes (FTP10 and FTP11) tend to change color in the dark, and the effect seems to increase when these dyes are mixed with azines (FTP18 and FTP19). A slight aging difference can be noticed between the FTP9 and FTP12 inks containing the same xanthene dyes (not uniquely identified), but FTP12 exhibits a better light resistance than FTP9.

These preliminary data suggest that many of the inks contain highly evanescent dyes, belonging to the classes of dyes with very low Blue Wool Scale values (between 1 and 2). When an artwork contains fugitive dyes,[44] it must be stored in very strict conditions, since even the 50 lux limit as suggested for these materials can already cause discoloration of these dyes.

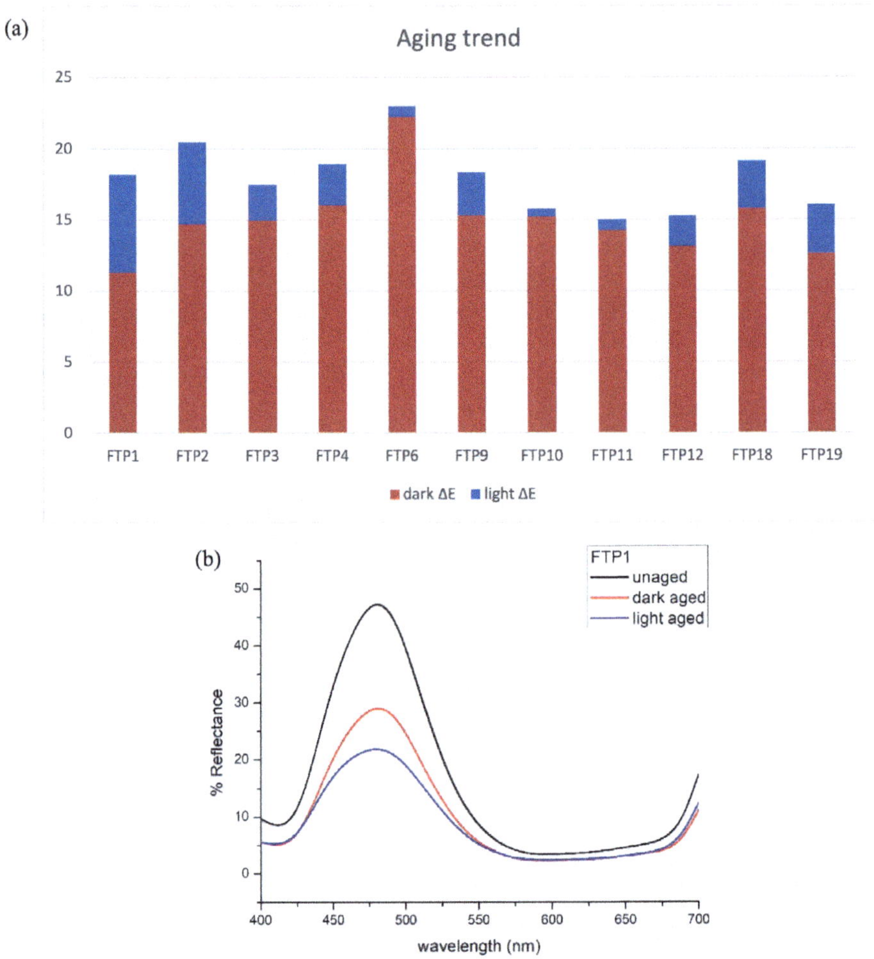

Figure 6.5 (a) Histogram of felt-tip pen inks subjected to the dark and light aging; (b) Reflectance spectra of **FTP1** before (black) and after dark (red) and light (blue) aging.

6.8 Conclusions

The introduction of felt-tip pens was crucial to the innovation of writing instruments. This single invention opened up the marketplace to new and different variations, as well as to different uses from lettering to drawing. Lina Bo Bardi very much appreciated the porous point marker, the smooth flowing ink, the wide range of hues and the comfortable design of FTPs that she largely employed for drawing, colouring, hand lettering and note-taking.

In this work, it has also been possible to sketch history, context and the pervasion of the bright colours provided by the different families of synthetic dyes.

The combined analytical methodology exploited for the study of Lina's felt-tip pens allowed a thorough chemical characterization of these historical ink formulations to be obtained. Disclosing the use of uncommon dyes, pure or in mixture, and that of binders and additives already known in early and more recent markers, respectively, a useful reference to investigate the evolution of felt-tip pen ink formulations has been provided. Moreover, the information obtained, combined with the indications that emerged from preliminary aging tests, offer a valid support to define suitable conservation strategies for felt-tip pen drawings, though aging effects can only be minimized and not avoided, even by storing drawings in the dark.

The scientific examination of Lina's 20 historical Johann Faber™ FTPs has provided valuable information on the constituent and on the aging behaviour of this graphic media. It has contributed pertinent information that is now informing conservation strategies, considering that these inks are directly related to an important part of Lina's drawings and archival documents.

Appendix

X-ray Fluorescence

XRF spectra were recorded using the portable, non-contact spectrometer ELIO by XGLab (Milano, Italy). It is equipped by a Rhodium source with a maximum power of 50 kV and maximum current of 200 μA, a Silicon Drift Detector cooled by a Peltier® system (active area of 25 mm^2 and resolution of <135 eV to the Kα line of Mn), an integrated video camera and a laser-based alignment system. The spatial resolution is about 1 mm. The counts have been acquired for 40 s, keeping the power and the current at 40 kV and 100 μA, respectively.

Micro-Raman Spectroscopy

For the conventional Raman analyses, a laboratory Jasco® NRS-3100 spectrophotometer was utilized in a confocal arrangement equipped with: Ar^+ laser source (emitting at 488 and 514 nm), a diode laser emitting at 785 nm, a triple grating (800 L mm^{-1}, 1200 L mm^{-1} and 1800 L mm^{-1}), an optical microscope provided with four objectives (5x, 20x, 50x and 100x) and a charge-coupled device detector (CCD) cooled by Peltier® effect down to −50 °C. The areas of interest were selected with the 100x objective and the spectra recorded by adopting different values of the exposure time (from 1 to 5 s) and number of accumulations (from 5 to 15). The laser power at the sample reaches a maximum value equal to 5 mW for the Ar + laser and 12 mW for the diode laser. The spectral resolution is from 2 to 6 cm^{-1} and the instrument was calibrated using two standards: polystyrene and silicon.

Analytical TLC was carried out on pre-coated aluminum foils with 0.2 mm thick silica gel (Supelco™, Sigma-Aldrich®, Milan, Italy). The FTP1 and FTP9 inks were directly applied on the foil. The eluent chosen was a basic (butanol:ethanol:ammonia (2.5 M) (5 : 2 : 3 v/v)) system.

Surface-enhanced Raman spectroscopy (SERS) measurements were carried out utilizing silver colloids. Silver nanoparticles were prepared by a modification of the synthesis reported by Lee and Meisel.[45] Nanoparticles were obtained with 500 mL of a 0.018% (m/v) solution of silver nitrate in ultrapure water (18 MΩ cm). The system was heated to the boiling point of water for an hour. Then, 100 mL of a 0.6% (m/v) solution of ascorbic acid was added to reduce Ag+ to Ag0. A solution of sodium chloride 0.4 M was used as activating solution. The colloid showed an absorption maximum at 435 nm and FWHM of 160 nm, as measured with a Shimadzu® UV-1601 UV–Vis spectrometer (following a 1 : 3 dilution with ultrapure water to observe maximum absorbance within the instrumental range). SERS measurements were carried out by adding 3 μL of colloid directly onto the TLC spots where spectra could be obtained between 2 and 10 min with laser excitation at 785 and 532 nm, which remained constant in quality until the evaporation of the liquid. SERS spectra were recorded with an XploRA® (Horiba®) spectrometer equipped with a YAG laser source with excitation wavelengths at 532 and 638 nm, a diode laser emitting at 785 nm, a triple grating (800 L mm^{-1}, 1200 L mm^{-1} and 1800 L mm^{-1}), an optical microscope (Olympus® BX41) with four objectives (5x, 20x, 50x and 100x) and a charge-coupled device detector (CCD) cooled to −65 °C with a Peltier® cooling system. Sample irradiation was accomplished using the 50x and 100x microscope objectives of the microscope.

Reflection Infrared Spectroscopy

FTIR reflection analyses were performed with a portable ALPHA-R spectrometer (Bruker® Optics, Germany/USA–MA) equipped with an external reflectance module with specular optics (22°/22°). The system consists of a SiC Globar source, a "rock solid"-design interferometer (with gold mirrors) and a DLaTGS detector. Data were acquired from areas of *ca.* 4 mm diameter in the spectral range of 7500–350 cm^{-1}, with a resolution of 4 cm^{-1} and using 186 scans. Due to the reflective nature of the aluminum foil used as support and the minor thickness of the ink markings, the infrared spectra of the inks could then be collected in transflection mode.[46,47] The background was acquired from a gold flat mirror.

Aging

Eleven ink samples (FTP1, FTP2, FTP3, FTP4, FTP6, FTP9, FTP10, FTP11, FTP12, FTP18 and FTP19) were subjected to dark storage for 22 months and to photooxidation aging. In the latter case, samples were exposed to the laboratory conditions next to a window.

Since the original inks were not enough to prepare more specimens, only one specimen (3 × 3 mm^2) per ink was prepared on Whatman® Schleicher & Schuell® (⌀ 100 mm) paper for the tests.

Spectrophotometry

Colorimetric measurements and reflectance spectra were acquired in order to estimate the aging trend for the inks subjected to the dark storage and photooxidation aging. Aging tests were performed only on eleven ink samples. The colorimetric measurements were acquired immediately after the ink application, after dark aging for 22 months, and after having subjected these same specimens to light aging for two months.

Spectrophotometric data were acquired with a Konica-Minolta® CM-2600d in CIE $L^*a^*b^*$ color space. The experimental conditions were related to the Normal Recommendation 43/93 ("Colorimetric measurements of opaque surface") and were as follows: spectral range 400–700 nm; lighting geometry and observation d/8; lighting source CIE C; observation angle 10°; reference system CIE $L^*a^*b^*$ color space; UV radiation contribution not considered; acquisition mode SCI; spot diameter 3 mm.

Pyrolysis–Gas Chromatography–Mass Spectrometry

The pyrolysis experiments were carried out with a microfurnace pyrolyser injection system Pyrojector II (SGE, USA). Small amounts (<50 μg) of sample were inserted into a quartz tube (4 cm × 0.53 mm) which was then introduced into the microfurnace operating at a temperature (T_{Py}) of 550 °C and at a pressure of 15.0 psi. Eosin Y-based ink samples were also analysed at T_{Py} = 700 °C. The pyrolysis chamber was directly connected to the injection port (T = 280 °C) of a PerkinElmer® Clarus 680 chromatograph coupled with a PerkinElmer® Clarus SQ8T single-quadrupole mass spectrometer. The MS transfer line temperature was 260 °C, and the MS ion source temperature was kept at 250 °C. The mass spectrometer was operating in the EI positive mode (70 eV) with a scan range of 45–400 m/z. For the gas–chromatographic separation, a 5% diphenyl–95% dimethyl polysiloxane column (PerkinElmer® Elite-5MS; 30 m × 0.25 mm i.d., 0.25 μm film thickness) was used. The injector was operated in split mode, varying between 1:10 and 1:50. The column oven temperature program was 40 °C (4 min hold) to 250 °C at a heating rate of 10 °C min^{-1} (7 min hold). The carrier gas (He) was used in constant pressure mode at 10.0 psi. Data were analysed with TurboMass Ver 6.1.0 (PerkinElmer®) and AMDIS software. Identification of compounds was performed by NIST11 MS library search and comparison with literature data.

Gas Chromatography–Mass Spectrometry

GC-MS was performed in order to characterize the carbohydrates. Therefore, standard solutions ($MONO_1$, $MONO_2$) and some felt-tip marker pen inks were subjected to an analytical procedure, which is a combination of two protocols in use at the Getty Conservation Institute (Los Angeles, USA) and the Department of Chemistry and Industrial Chemistry of the University of Pisa (Italy).

Samples were weighed on the ultramicrobalance in a conical reaction vial and a solution of mannitol as internal standard was added to give a final concentration of 20 ppm in the injection volume. An aliquot of 100 µl of 1.2 M trifluoroacetic acid (TFA) (99%) was added, oxygen was evacuated under a stream of nitrogen for 30 s and hydrolysis was performed in the closed vials at 125 °C for 1 h. After hydrolysis the vial was left to stand until cool. Then 50 µL of a solution of *O*-methoxylamine hydrochloride (98%) in pyridine (99%) (100 mg/mL) were added and the vial was heated at 70 °C for 20 min. It was left to stand until cool and 50 µL of *N,O*-Bis(trimethylsilyl) trifluoroacetamide (BSTFA) were added and the vial was kept at 30 °C for 45 min. An aliquot of 1 µL of the solution was injected (T injection port = 280 °C) and analysed with a PerkinElmer® Clarus 680 chromatograph coupled with a PerkinElmer® Clarus SQ8T single-quadrupole mass spectrometer. The MS transfer line temperature was 290 °C, and the MS ion source temperature was kept at 250 °C. The mass spectrometer was operating in the EI positive mode (70 eV) with a scan range of 40–650 m/z. For gas–chromatographic separation, a 5% diphenyl–95% dimethyl polysiloxane column (PerkinElmer® Elite-5MS; 30 m × 0.25 mm i.d., 0.25 µm film thickness) was used. The injector was operated in splitless mode. The column oven temperature program was 80 °C (1 min hold) to 170 °C at a heating rate of 2.5 °C min^{-1} (15 min hold), from 170 °C to 280 °C at a heating rate of 20 °C min^{-1} (10 min hold). The carrier gas (He) was set in variable flow: 1.2 mL min^{-1} (30 min hold) to 0.8 mL min^{-1} at a rate of 0.4 mL min^{-1} (15 min hold) and to 1.2 mL min^{-1} at a rate of 0.4 mL min^{-1} until the end. The solvent delay was set at 10 min. Data were analysed with TurboMass Ver 6.1.0 (PerkinElmer®) and AMDIS software. Identification of compounds was performed with the NIST11 MS library search and comparison with standard monosaccharides.

References

1. *Nomenclature and Test Parameters for Porous Pointed Pens*, Marlton, NJ, 1982, p. 4.
2. I. D. van der Werf, G. Germinario, F. Palmisano and L. Sabbatini, *Anal. Bioanal. Chem.*, 2011, **399**, 3483–3490.
3. H. Awab, A. D. M. Jar, W. K. Yong and U. K. Ahmad, *Malays. J. Forensic Sci.*, 2011, **2**(1), 1–7.
4. A. Sodo, M. Bicchieri, M. Guiso, M. A. Ricci and G. Ricci, *J. Raman Spectrosc.*, 2012, **43**, 1781–1787.
5. F. C. Izzo, V. Vitale, C. Fabbro and H. Van Keulen, *Microchem. J.*, 2016, **124**, 919–928.
6. G. Germinario, S. Garrappa, V. D'Ambrosio, I. D. van der Werf and L. Sabbatini, *Anal. Bioanal. Chem.*, 2017, **410**(3), 1079–1094.
7. C. Zaffino, A. Passaretti, G. Poldi, M. Fratelli, A. Tibiletti, R. Bestetti, I. Saccani, V. Guglielmi and S. Bruni, *J. Cult. Herit.*, 2017, **23**, 87–97.

8. D. Saviello, M. Trabace, A. Alyami, A. Mirabile, R. Giorgi, P. Baglioni and D. Iacopino, *Talanta*, 2018, **181**, 448–453.
9. D. Saviello, A. Alyami, M. Trabace, R. Giorgi, P. Baglioni, A. Mirabile and D. Iacopino, *RSC Adv.*, 2018, **8**, 8365–8371.
10. D. Saviello, A. Di Gioia, P. I. Turenne, M. Trabace, R. Giorgi, A. Mirabile, P. Baglioni and D. Iacopino, *J. Raman Spectrosc.*, 2018, 1–10.
11. http://www.bbc.com/culture/story/20140611-brazils-best-kept-secret consulted 5 October 2017.
12. L. W. Newman, Marking Pen, *US Pat.*, 946149, 1910.
13. D. Geyer, *Collecting Writing Instruments*, Schiffer Publishing Ltd, West Chester, PA, 1990, p. 168.
14. M. Holbein Ellis, Drawing in fiber tipped pen, new conservation challenges, Conference Papers, *Modern Works Modern Problems*, The Institute of Paper Conservation, Tate Gallery, London, 1994, pp. 114–121.
15. G. Pfingstag, Colorants in inks for writing, drawing and marking, *J. Soc. Dyers Colour.*, 1993, **109**, 188–192.
16. *Surface Phenomena and Additives in Water-Based Coatings and Printing Technology*, ed. M. K. Sharma, Plenum Press, New York, 1991.
17. *Chemistry and Technology of Water-Based Inks*, ed. P. Laden, Chapman and Hall, New York, 1997, ch. 9.
18. K. G. Ponting, *Dictionary of Dyeus and Dyeing*, Mills and Boon, Ltd, London, 1980, pp. 161–171.
19. R. Feller, Felt-tipped Markers and the Need for Standard of Lightfastness for Artist's Colorants, *Bull. Am. Group. Int. Inst. Conserv. Hist. Artistic Works*, 1967, **8**(1), 24–26.
20. R. Feller, Felt-tipped Markers and the Need for Standard of Lightfastness for Artist's Colorants, *Bull. Am. Group. Int. Inst. Conserv. Hist. Artistic Works*, 1967, **8**(1), 25.
21. T. Grant, *International Directory of Company Histories*, vol. 51, St. James Press, 2003.
22. A. Lluveras-Tenorio, J. Mazurek, A. Restivo, M. P. Colombini and I. Bonaduce, *Chem. Cent. J.*, 2012, **6**(115), 1–16.
23. C. Anselmi, D. Capitani, A. Tintaru, B. Doherty, A. Sgamellotti and C. Miliani, *Dyes Pigm.*, 2017, **140**, 297–311.
24. *Van Nostrand's Scientific Encyclopedia*, ed. G. D. Considine and P. H. Kulik, 10th edn, Wiley, London, 2008.
25. L. Wang, A. Roitberg, C. Meuse and A. K. Gaigalas, *Spectrochim. Acta, Part A*, 2001, **57**, 1781–1791.
26. N. G. Greeneltch, A. S. Davis, N. A. Valley, F. Casadio, G. C. Schatz, R. P. Van Duyne and N. C. Shah, *J. Phys. Chem. A*, 2012, **116**, 11863–11869.
27. N. Eastaugh, V. Walsh, T. Chaplin and R. Siddall, *Pigment Compendium – A Dictionary of Historical Pigments*, Elsevier Butterworth-Heinemann, 2004, pp. 194–195.
28. N. Peica and W. Kiefer, *J. Raman Spectrosc.*, 2008, **39**, 47–60.

29. T. Gessner and U. Mayer, Triarylmethane and Diarylmethane Dyes, in *Ullmann's Encyclopedia of Industrial Chemistry*, ed. B. Elvers, Wiley-VCH Verlag GmbH & Co. KGaA, 7th edn, 2002.
30. B. Doherty, M. Vagnini, K. Dufourmantelle, A. Sgamellotti, B. Brunetti and C. Miliani, *Spectrochim. Acta, Part A*, 2014, **121**, 292–305.
31. G. Germinario, E. C. L. Rigante, I. D. van der Werf and L. Sabbatini, *J. Anal. Appl. Pyrolysis*, 2017, **127**, 229–239.
32. W. Freitag and D. Stoye, Chemistry of the textiles industry, *Paints, Coatings and Solvents*, 2nd edn, Wiley-VCH, Chapman and Hall, University Press, Cambridge, 1998.
33. W. Herbst and K. Hunger, *Industrial Organic Pigments, Production, Properties, Applications*, Wiley, 3rd edn, 2004.
34. J. Russell, B. W. Singer, J. J. Perry and A. Bacon, *Anal. Bioanal. Chem.*, 2011, **400**, 1473–1491.
35. E. Ghelardi, I. Degano, M. P. Colombini, J. Mazurek, M. Schilling and T. Learner, *Anal. Bioanal. Chem.*, 2015, **407**, 1415–1431.
36. N. Sonoda, *Stud. Conserv.*, 1999, **44**(3), 195–208.
37. R. E. Smith, Updated by Staff, Azine Dyes, *Kirk-Othmer Encyclopedia of Chemical Technology*, 2013, pp. 1–10.
38. C. H. Munro, W. E. Smith and P. C. White, *Analyst*, 1995, **120**(4), 993–1003.
39. M. M. Abo Aly, *Spectrochim. Acta, Part A*, 1999, **55**, 1711–1714.
40. M. A. Legodi and D. de Waal, *Dyes Pigm.*, 2007, **74**(1), 161–168.
41. A. Bowes, E. Méthé and M. W. Ballard, *Important Early Synthetic Dyes: Chemistry, Constitution, Date, Properties*, M.W. Ballard, Conservation Analytical Laboratory Smithsonian Institution, United States, 1991.
42. H. Panda, *A Concise Guide on Textile Dyes, Pigments and Dye Intermediates with Textile Printing Technology*, NIIR Project Consultancy Services, Delhi, 2013.
43. A. Lluveras-Tenorio, J. Mazurek, A. Restivo, M. P. Colombini and I. Bonaduce, *PLoS One*, 2012, **7**(11), e4938.
44. S. Michalski, The lighting decision, in *Fabric of an Exhibition, Preprints of Textile Symposium 97*, Canadian Conservation Institute, Ottawa, 1997, pp. 97–104.
45. P. C. Lee and D. Meisel, *J. Phys. Chem.*, 1982, **86**(17), 3391–3395.
46. S. C. Boyatzis, A. M. Douvas, V. Argyropoulos, A. Siatou and M. Vlachopoulou, *Appl. Spectrosc.*, 2012, **66**(5), 580–590.
47. F. Rosi, L. Legan, C. Miliani and P. Ropret, *Anal. Bioanal. Chem.*, 2017, **409**(12), 3187–3197.

CHAPTER 7

"Art is Not Science": A Study of Materials and Techniques in Five of Enrico Baj's Nuclear *Paintings*†

LUCA BOCHICCHIO‡*[a], PATRIZIA MORETTI[b], ANNALISA CHIELI[b, c], ALDO ROMANI[b, c], CHIARA RUBERTO[d], LISA CASTELLI[d], BRUNETTO GIOVANNI BRUNETTI[b, c], ANTONIO SGAMELLOTTI[b, c] AND LAURA CARTECHINI‡[b]

[a]University of Genova – MuDA/Asger Jorn House Museum, Albissola, Italy; [b]Istituto di Scienze e Tecnologie Chimiche "G. Natta" - (CNR-SCITEC), Perugia, Italy; [c]SMAArt Centre and Department of Chemistry, Biology and Biotechnology, University of Perugia, Italy; [d]Istituto Nazionale di Fisica Nucleare (INFN), Sezione di Firenze, Via G. Sansone 1, Sesto Fiorentino, 50019 Florence, Italy
*E-mail: luca.bochicchio@yahoo.it

7.1 Introduction: Art is Not Science (in Enrico Baj's Philosophy)

The philosophical and artistic orientation of Enrico Baj (Milano 1924–Vergiate 2003) regarding nuclear energy cannot be exclusively reduced to and traced back to activism in the Nuclear Art Movement, a movement which

†The author of the first two paragraphs, as well as of the five introductions to each paintings' analysis is Luca Bochicchio.
‡Scientist responsible for the project "Baj Nucleare".

Science and Art: The Contemporary Painted Surface
Edited by Antonio Sgamellotti, Brunetto Giovanni Brunetti and Costanza Miliani

Published by the Royal Society of Chemistry, www.rsc.org

he himself founded along with Sergio Dangelo (Milan 1932) in 1951. Baj's interest in science, techniques and technology did not map onto the chronology of this *avant-garde* movement,§ but rather was related to developments in anthropology, philosophy and literature. As the years passed and the artist matured, writing and theoretical re-elaboration took on a greater role in Baj's work, and reflection on the impact of exact science on society and western contemporary spirituality resurfaced, together with the paradigms of chemistry and nuclear physics. These were already used, but with quite the contrary tone, in the *Manifesto della Pittura Nucleare* (Manifesto of Nuclear Painting) and in *Manifesto Bum*, both from 1952. In 1986 we find a short essay titled "*Arte non è scienza*" (Art is not Science), where Baj states:

> Science is our new religion, and with the triumph of positivist rationalism in the name of absolute truth, it has become increasingly dangerous, leading people to blind fideism. Blind faith in science has people arguing that nuclear energy is needed in order to have energy for consumption, energy to make our life systems function: its speakers, its record players and appliances; in other words, that nuclear power plants are necessary for survival; where the opposite is true, survival actually depends on abandoning and neutralizing nuclear energy. Science is grounded in fideist thinking. As a matter of fact, science rests on paradigms that have to be unconditionally accepted by all those who are part of a scientific community. These paradigms are the equivalent of religious dogma.[2]

This excerpt, which can be dated to *circa* halfway between the beginning of Baj's pictorial research in the early 1950s and a later, but very lively period in the early 2000s, lets us hypothesise that the artist would not have embraced and appreciated this diagnostic study of his nuclear paintings *in toto*; however, I do think that he would have actively participated, perhaps by jeering at us "historians", "critics" and "scientists", and disputing or examining our results. Even though a legacy of a certain irony regarding science is covertly present in "*Arte non è scienza*", these more recent texts of Baj dispute the founding manifestos of nuclear painting in their tone as well as in their thesis. In *Manifesto della pittura nucleare* (presented by Baj and Dangelo on February 1st, 1952 at the Apollo Gallery in Brussels) we read:

> The Nuclears want to tear down all the "isms" of a painting that inevitably falls to academism, no matter its origin. They want to reinvent Painting and they can. Shapes disintegrate: the new shapes of man are those of an atomic universe. Forces are electrical charges. Ideal beauty no longer belongs to a caste of stupid heroes or to robots, but rather coincides with the representation of nuclear man and his space. [...] Truth

§For a complete chronology of the Nuclear Art Movement, the reader should refer to Tristan Sauvage's unparalleled book.[1]

> doesn't belong to it: that lies in the atom. Nuclear painting documents the search for this truth.

In *Manifesto Bum*, words and images blend together in perfect futurist synthesis, stating:

> Shapes disintegrate
> The new shapes of man
> are those of the atomic universe.

Baj's militancy in the Nuclear Art Movement coincides with an artistic philosophy and a painting technique that could be defined as integralist or extremist, in so far as its adherence (expressive, poetic, intellectual, linguistic and procedural) to the historical moment and to the definition itself of Nuclearism: "we felt ourselves to be artists of a period in which atomic and nuclear research opened up new and infinite horizons for man; artists were, therefore, rather than nuclear, of a nuclear age"¶. On Baj's part this is a criticism of contemporary society – stained by the double tragedy of Hiroshima and Nagasaki, but at the same time illuminated by the exciting and feverish expectations one had of nuclear power with its potential for clean and inexhaustible energy – which Baj carries out pictorially with the "intention of an epistemological metaphor".[3] This compelling definition can be attributed to Enrico Crispolti, who in an all-important essay written for the exhibit *Alternative Attuali 2* (L'Aquila 1965) pinpoints with precision the expressive poetic passage which we have examined here in the "Baj Nucleare" project from the standpoint of the evolution of materials and executive techniques:

> Towards the middle of the 1950s, a primal instrumental gestural painting gives way – in the nuclear subject, and foremost in Baj's painting – to an inventive gestural painting which leads to more complex formulations. [...] By exploration of these iconological possibilities intrinsic to material primariness, just as its interior and its soul, Baj touches on the dimension of a sort of disinvolvement, away from a vaguely epistemological first intuition, towards something more contrived in which the "*spazialismo fontaniano*" [...] provided lessons of a grand intensity and inventiveness.[3]

Hence, Baj's paintings progressively show an estrangement from Nuclear epistemology. This can be seen not only in his artworks, our primary source, but also in secondary sources, including Baj's writings.

¶Letter from Enrico Baj to Tristan Sauvage (Arturo Schwarz) on August 20th, 1960, published and translated into English in ref. 1, p. 19 and republished in ref. 4, p. 13.

Considering the affirmation of post-war man's new atomic identity, claimed in the 1950s, Baj goes so far as to contest the dogma of nuclear chemistry in the 1980s, a science upon which the present study of analytical diagnostics is based:

> We refuse to entertain the thought that the birth of chemistry in the seventeenth century brought about the wane of alchemy [...].
>
> Among his archetypal visions, man has perennially chased two dreams with physics and chemistry: the transmutation of matter for obtaining infinite riches and the movement of the bodies to discover perpetual motion. [...]
>
> The discovery of the structure of the atom in a certain sense reconsiders the old alchemical theories inasmuch as electrons, protons and neutrons can be considered primary polyvalent particles. As a matter of fact, these particles, structuring themselves in infinite combinations, give rise to various elements and their properties. Modern atomic and nuclear science can produce transmutations, at least in theory, and therefore present itself to our eyes like a definitive philosopher's stone. But in reality [...] modern nuclear science, which appears so close to alchemical ideas, after Hiroshima and Nagasaki, offered us Three Mile Island and Chernobyl: well beyond energy for turning on those lightbulbs which could theoretically indicate to us the way from darkness to the light.[2]

7.2 Inside and Beyond the Nuclear Epistemology of Enrico Baj: Non-invasive Investigations Applied to the Study of Five Paintings From the Archivio Baj Collection (1951–1956)

This study has as its focus the non-invasive *in situ* chemical examination of five of Enrico Baj's paintings which are part of the Archivio Baj collection in Vergiate (Italy). Painting materials were investigated thanks to two distinct work sessions (October and November 2018) of the mobile laboratories of MOLAB-CNR/UNIPG[5] and INFN-CHNet,[||] whose instrumentation and know-how were granted to the "Baj Nucleare" project by access to the E-RIHS.it infrastructure.[6] The study provided us with data that, when cross-checked with previous studies and combined with documents from artistic literature

[||]Overall, the analytical project included point analysis by X-ray fluorescence (XRF) spectroscopy, reflection Fourier Transform infrared (FT-IR) spectroscopy, Raman spectroscopy, ultraviolet-visible-near-infrared (UV-Vis-NIR) reflectance and fluorescence spectroscopies along with hyperspectral Vis-NIR imaging and macro X-ray fluorescence (MA-XRF) scanning techniques.

and other available sources, allowed us to better understand Enrico Baj's work process in terms of choice of materials and their application.

The motivation and start of the "Baj Nucleare" project can be traced to the art-historical research carried out in 2013–2014 for the preparation of the Albisola exhibition dedicated to Asger Jorn (Vejrum 1914–Aarhus 1973), on the occasion of the centennial of his birth.[7] In fact, owing to the important relationship between Baj and Jorn during the years 1953–1956, the exhibition was an opportunity to get to know Baj's work in the "nuclear" period (1951–1957) better. It is important to remember that the Nuclear Art Movement historically represented the Italian gear in a global mechanism of *avant-garde* pictorial research during the years straddling World War II, which included Action Painting in the United States, the Gutai Group in Japan, the neo-expressionist and post-surrealist group CoBrA in Northern Europe and the *informel* and *tachisme*** trends in France. The idea of subjecting some of Baj's nuclear paintings to a non-invasive *in situ* chemical examination, searching for use of materials and their application, originated from a previous collaboration with the Baj Archive, oriented as always to studying Baj's work within the context of the contemporary international pictorial research.[13]

The main aim of the "Baj Nucleare" project was twofold: to attempt to explain experimental techniques in the auroral phase of nuclear art and to verify their evolution in the 1950s, a period in which there was a transition to the use of poly-materials and three-dimensional collages, which hereafter characterises Baj's artistic philosophy for the rest of his life and career.

The planning of the analysis of the works in the Archivio Baj took place first of all recognising previous studies that were important but limited to a specific series of works, which on this occasion were excluded from the diagnostic process. In fact, two different studies were carried out on the *Montagne* (Mountains) series between 2006 and 2015: one at Marina Pugliese and Alessandra Tibiletti's[15] studio on the painting *Montagna con sole*, from 1957 (which was followed by a partial recovery in Corgnati and Tibiletti 2018[16]), followed by the one by Maria Fratelli,[17] which focused on three paintings belonging to the Boschi di Stefano collection in Milan: *Personaggio atomizzato* (1957), *Personaggio nella montagna* (*ca.* 1958), *Agitatevi, pietre e montagne!*

**Regarding the international context of nuclear art, debating his and his *avant-garde* group's refusal to identify with terms and movements such as *tachisme and informel*, Baj states: "fundamental is the consideration that our struggle, the position we had taken up, was above all, anti-abstract in relation to the artistic world and the theories then circulating in that world. *Basically* anti-abstract; that is to say not only was geometric abstraction rejected, but abstraction itself. Because of this we were always opposed to Tapié's *informel* and his *art autre*, which 'other' finished up by proposing nothing more than the substitution of an informal abstract academicism [...]", see ref. 1, p. 19 and, ref. 4 p. 13. On his part Dangelo remembers: "one afternoon, at the Jamaica Bar, I met Peverelli and Dova, who swore only by Pollock and Wols" (ref. 1, p. 20). Other points raised by Baj on Action Painting and *informel* can be found in ref. 8, pp. 24–25. For the relationship between the Nuclear Art Movement and the Cobra Group see: ref. 1, pp. 26–31; ref. 9–11, pp. 113–128. For more general information about *informel* and its international echo, see ref. 12.

(*ca.* 1958). The *Montagne* phase can be considered the one that leads Baj from the nuclear experience to the *Generali* one, and consequently to the most accomplished, politically critical, allegorical neo-representation of the 1960s. Beyond this, however, *Montagne* also holds a specific interest from the viewpoint of scientific studies on materials and techniques, since they allow for the examination of the "material" side of that particular metaphor between pictorial matter and organic terrain, between the heterogeneity of active materials and biological and mineral contamination of a natural landscape. In his *Automitobiografia* from 1983, Baj himself points out that certain strong characters from nuclear painting endure in *Montagne*:

> My paint was teeming, atomized, in movement thanks almost to slippage. This appearance was achieved with oily and bituminous paints, changing colors in a short spectrum from burnt umber to Greek pitch black, paints that were vividly emulsified with water and then let flow on the canvas. I called this emulsion *acqua pesante*,[††] in homage to the previous "nuclear painting" which broadly resulted in the atomization of colors.

Hence, the "Baj Nucleare" project focused on the earlier and, if you will, more enigmatic phase of the experimental activity of the artist, because in it exist only Baj's snippets of memory – or those of other people who shared his life – from 1951 to 1956. One of the aims of our study was to carry out a preliminary examination of the mixed techniques only partially mentioned by the artist – such as, for example, the use of enamel and resin – and in part detected in the studies mentioned above (tar and varnish in Pugliese 2006,[15] vinyl glue and alkyd binding agents present in the enamels, as found by Fratelli 2015 [17]).

Except for the Spatialist movement led by Lucio Fontana, it is hard to find, for that time period, a conceptual coherence between theoretical manifesto and works of art, such as the one we can detect between Baj's paintings, the *Manifesto della pittura nucleare,* and *Manifesto Bum.*

Of the five paintings that were examined (Figure 7.1), two are from the years that launched the Nuclear Movement: *Quamisado II* (1951, 100 × 70 cm) and *Bambino magico* (Magic Child) (1952, 50 × 64 cm). *Due bambini nella notte nucleare* (Two Children in the Nuclear Night) (1956, 115 × 145 cm) also

[††]Ref. 18, p. 149. Among the earliest definitions of "*Acqua Pesante*" (Heavy Water) there is one by Arturo Schwarz: "another technique of the Nuclearists is the one called 'acqua pesante', a prefabricated emulsion (that is, prepared separately, and not directly on the canvas), comprised of water that is violently churned with thick paints, generally of a golden-brown color. This 'acqua pesante', thrown and allowed to ooze on a fresh surface, thick with pigment and oil, produces a crumbled and vibrant look, the effect of 'atomization'. Baj was accustomed to using herbs and flowers on the fresh surface of some of his paintings and then spreading 'acqua pesante' over it, so that only delicate filaments, undefined lines and impressions would emerge from the paste" (in ref. 1, p. 217); this passage was translated by the author, since curiously enough in the dictionary placed at the end of Sauvage's book,[1] "*Acqua Pesante*" is the only term that was not translated into English and French.

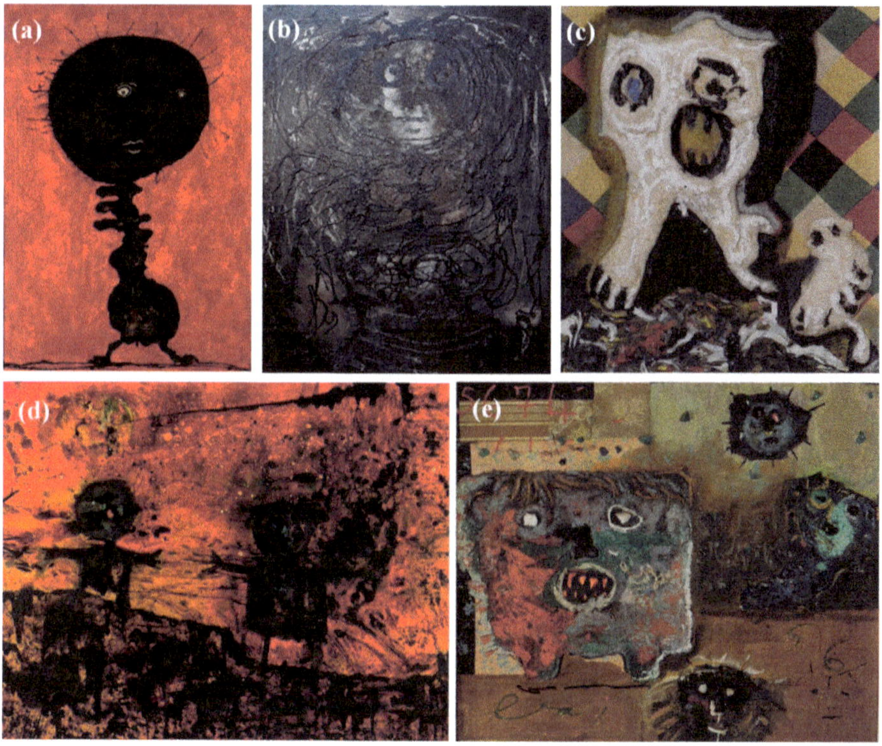

Figure 7.1 Investigated paintings: (a) *Quamisado II* (1951, 100 × 70 cm); (b) *Bambino magico* (1952, 50 × 64 cm); (c) *Personaggio* (1955, 47 × 42 cm); (d) *Due bambini nella notte nucleare* (1956, 115 × 145 cm); (e) *Spettacolo* (1956, 100 × 120 cm). All images: Reproduced with permission from Archivio Baj.

fully reflects nuclear style and artistic philosophy, while belonging to a more mature period, as proven by the enhanced palette and by the intricate technique and combination of materials. The painting *Personaggio* (Character) (1955, 47 × 42 cm) anticipates certain poly-material experimentation which will find full affirmation in subsequent years, whereas *Spettacolo* (Spectacle) (1956, 100 × 120 cm) in our study constitutes the point of achievement and the *summa* of all the techniques observed previously.

The terms by which an art historian can contribute to the analysis of data and scientific results essentially concern the intersection of those findings and the information that is documented in artistic literature, namely secondary sources (such as writings or interviews with the artist), which are usually the only point of reference for a critical, empirical analysis beyond the visual analysis of the primary sources (the works of art themselves).

Hence, let us start to process and cross-check those secondary, literary sources, in light of the results of chemical–physical measurements, which will be illustrated in detail in the following paragraphs and summarised in Table 7.1.

Table 7.1 Summary of the results from the diagnostic campaign on the five investigated nuclear paintings by Enrico Baj.

		Quamisado II (1951)	*Bambino magico* (1952)	*Personaggio* (1955)	*Due bambini nella notte nucleare* (1956)	*Spettacolo* (1956)
Preparation layer		$BaSO_4$ $CaCO_3$ Oil	$BaSO_4$ Oil		$BaSO_4$ Oil	
Paint binders		Styrene component Oil	Oil	Oil Alkyd (glossy black areas) Vinyl (figure)	Oil Alkyd (black areas) Natural resin (black and blue areas)	Oil Alkyd (red and glossy black areas) Vinyl (black area with sand or area with raffia)
Pigments	*Black*	Ivory black	Carbon black Metal powder (Zn, Cu, Al, Ag) Sand	Ivory black (not with alkyd)	Ivory black (not with alkyd)	Ivory black (not with alkyd)
	White	ZnO + TiO_2 (anatase)		ZnO + TiO_2(anatase)		TiO_2 (anatase) $BaSO_4$
	Red/Orange	PR49 Cr-orange		Croceine orange G	$CdS_{1-x}Se_x$	PR3 $CdS_{1-x}Se_x$
	Blue			Phthalocyanine blue Ultramarine	Phthalocya-nine blue	Prussian blue Ultramarine

	Yellow			PY1	CdS	Cr-yellow CdS or $Cd_{1-x}Zn_xS$ (stone)
	Green			Cr-green	Viridian	Phthalocyanine green Cr-pigment + Prussian blue
	Grey			ZnO + TiO_2 (anatase) + Carbon black		Ivory black + Zn-white + Ti-white
Fillers		$BaSO_4$ (red, orange areas)	$CaCO_3$ sulphate/silicate	$BaSO_4$ $MgCO_3$ (blue, yellow, green, red areas) Wax (white areas) Silicate (blue, yellow, black areas)	$BaSO_4$ Wax (red, green, blue areas)	$BaSO_4$ $MgCO_3$ (blue, green, yellow, red areas) Wax (blue, green, red areas)
Degradation compounds		Ca-oxalates Zn-carboxylates	Ca-oxalates	Ca/Zn-oxalates Zn-carboxylates	Ca/Zn-oxalates Zn-carboxylates	Ca/Zn-oxalates Zn-carboxylates

One of the most effective reconstructions of the artist's highly experimental methods in the first nuclear paintings, during a period where the ideological perimeter of the movement had not yet been clearly defined, nor the term "nuclear painting" coined, is due to two interviews with ancestors of the group: Baj himself and Sergio Dangelo. Let us look at how Baj remembers the exciting collaboration in his studio in *Via* Tuilè in Milan, where the painter Gianni Dova (Rome 1925–Pisa 1991) also collaborated from the start:

> Between the end of 1950 and the beginning of 1951, Dova came back to work with me, and Sergio Dangelo arrived at about the same time. This triggered a remarkable atmosphere of creative chaos. There were jars everywhere, jars of enamel which we poured on the canvas.[8]

Dangelo also confirms the use of industrial enamels in that precise historical moment:

> I don't remember the date, but I think it must have been the first week of October, 1950. Dova remarked that he wanted to buy some tins of enamel and quite casually we went with him, first to Crespi the stationer's, and then, tins in our pockets, to the room he occupied in *Via* Solferino, which served as his studio. Here, Dova poured the enamels onto canvases passionately, and we were emotionally shaken by an explosion of colors such as we had never before seen. [...] One afternoon, revisiting Baj, I found that he had again started to paint with enamel: his new pictures were spread out on the floor. As often happens in such cases, several artists, unaware of the experiments of others, and having discovered "something new", were using similar methods to express it.[1]

Beyond the use of enamel in experimental painting in the middle of the twentieth century, which was already extensively documented,[15,19–23] it proves interesting to go back to Baj's testimony regarding his process and the techniques closely related to the use of new materials that became available after the war:

> We started by moving the canvas from its vertical position on the easel to a horizontal position: on the floor. This made all the difference, together with the use of liquid paints and glues, whose movement we managed only minimally. As a matter of fact, lifting the canvas on one side rather than another, we were able to obtain a change in direction both in the filtering and the blending of colors.[8]

Therefore, the canvas gives a support and a surface that holds the image. It assumes an active role for the Nuclear Art Movement and particularly for Baj. We shall see that this procedural aspect is also confirmed by scientific findings, which show how paint was used on the canvas in unconventional ways, letting us speculate about a complimentary movement between the gesture of the artist and the movement of the support.

> The canvas is like a recording device of our excitement, of our movements, of our "influence". It is something that is alive and that lives with us and from us. It is no longer about a coincidental gesture or about gathering power of suggestion from an automatic material, but more so a gesture which "takes a position".[4]

It was 1950 when Baj started the type of experimentation which, as we saw, for some years had the characteristics of a collective process, shared with Dangelo and Dova (and later joined by Cesare Peverelli and Joe Colombo); a time when he was still living the experience of his pictorial studies at the Brera Academy (which he attended irregularly all along).

7.2.1 *Quamisado II*

Undoubtedly one of the most famous works from this period is *Quamisado II* (1951) (Figure 7.1a), a painting that was analysed here first, also from a chemical and physical standpoint. The technique reported for this painting in the general catalogue of Enrico Baj's works ref. 26, n. 44 is oil and enamel on canvas; nevertheless, from the analyses there was no evidence for enamel. This fact paradoxically confirms rather than denies the complexity and anti-conventionality that Baj was testing in his works during this period; in fact, the painter, while resorting to different oil paints, carries out a process that does not conform with academic teachings. This is evident in the two large chromatic areas that make up the painting: the black of the figure and the red of the background. The black areas can be divided into glossy zones and opaque zones. One thing that is certain is that the black was applied in several coats successively and using instruments that included brushes (used first for outlining the figure, but also later on a more surface level) as well as abrasive materials and objects (sponges or rags). As a matter of fact, the black areas have several portions where materials thin out, allowing one to see the light preparation below, producing a grayish color. Baj worked similarly on the red background, but applying further unconventional techniques. The artist probably intervened in the painting layers with a very diluted paint or even solvents that he threw on the canvas with the dripping technique. This can be inferred from the presence of a coalescence of pictorial material such as residues of drops, as if the evaporating or bursting solvents or air bubbles of the paint thinner had left the surface of the painting irregularly spotted.

The unconventional use of traditional materials along with modern paint formulations to achieve the complex optical effects of the black and red areas in *Quamisado II* (1951) was disclosed by vibrational spectroscopy combined with macro X-ray fluorescence (MA-XRF) scanning analysis. A glycerolipid binder was observed by reflection FT-IR in most of the measurements (Figure 7.2a) with evidence of a terpenic component localised in the thick contours of the paint aggregates; probably it can be explained as a residue of the organic diluent applied on the paint surface to obtain variation in the degree of gloss and matteness, especially for the black surface. To enhance the contrast in the surface effects, Baj used different black and red paints. For example, bone black was often observed[24] but not in the abraded areas (Figure 7.2a), suggesting the presence of another organic black. The red background was characterised by Raman spectroscopy as containing both chrome orange ($(1-x)PbCrO_4 \cdot xPbO$)) and the organic synthetic pigment PR49 (Figure 7.2b). A styrene component was also observed in the red colour, hinting towards the possible presence of a styrene-modified oil[25] in the commercial formulation of the red paint. The application of paint layers in sequential steps was evidenced by MA-XRF analysis on a detail of the proper right eye of the central figure (Figure 7.2c). Here the elemental maps of calcium, titanium and zinc in the black background (Figure 7.2d–f) are characterised by a patch distribution not visible to the naked eye, due to a successive application of the organic blacks and the zinc and titanium white paints (ZnO and TiO_2, respectively) over the preparation containing calcium carbonate ($CaCO_3$, see Table 7.1).

7.2.2 *Bambino Magico* (Magic Child)

The second work that was analysed, in chronological order, is *Bambino Magico* (1952, ref. 26, n. 84, Figure 7.1b); less well-known than the previous work, it certainly entailed more of a challenge in its interpretation and at the same time offered more surprises. Thanks to the analyses that were conducted, Baj's method appears clearer in this case, in spite of the title of the work, which without any doubt alludes to a sort of mysterious alchemy in the process of creation. The artist distributed powders (probably metal filings) of silver, aluminum, copper and zinc on the preparation, which was likely still fresh. These powders thicken and spread out almost randomly on the canvas, creating a sort of mineral crust, a seemingly alien surface (magic, or, better to say, alchemical), which, depending on how the light strikes it, takes on shades that vary from dark gray to silver to pink. This effect is reinforced by the use of sand, indicated by the presence of silica that is not homogeneously distributed. The artistic literature is helpful in showing that Baj was knowledgeable about contemporary experimental techniques which implied the use of colored sand. In the interview already quoted, the artist states:

> Among the pioneers of the *informel,* French painters must be mentioned, such as André Masson, who was known to let colored sands fall to the floor, and Hans Hartung, whose gestures anticipated action painting.[‡‡]

[‡‡]Ref. 8, p. 24.

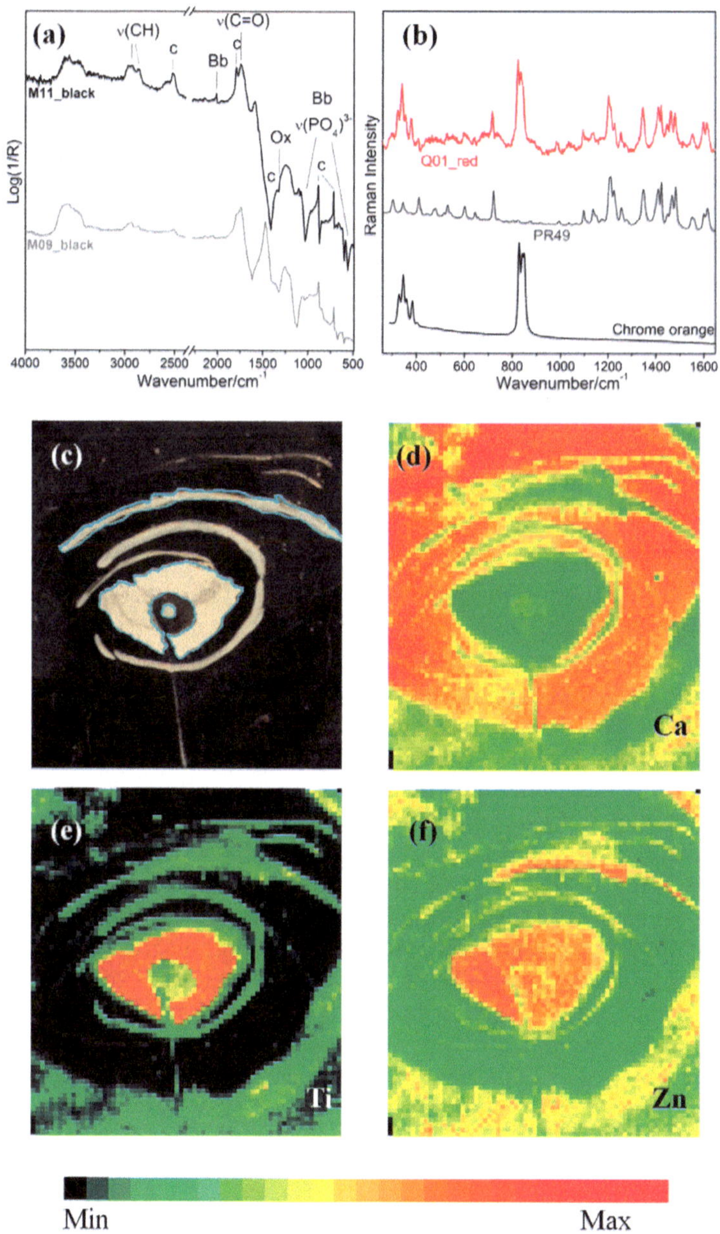

Figure 7.2 Selection of analytical results for the painting *Quamisado II*: (a) Reflection FT-IR spectra of two black areas with (M11_black) and without (M09_black) bone black (Bb, diagnostic bands: stretching vibrational mode of the PO_4^{-3} group and 2013 cm^{-1}). In (a) the bands of CH and C=O stretching modes are marked, suggesting the presence of lipid binder, while the observation of the $CaCO_3$ bands (marked as c) points out the use of this compound as a filler; oxalates (Ox) are also present, possibly as oxidative degradation products of the organic paint components. (b) μRaman spectrum (λ = 785 nm, exposure time = 2 s, number of accumulations = 15) of a red micro-sample (Q1_red, red line) collected at the canvas edge and reference spectra for the pigments chrome orange (grey line) and PR49 (black line). (c) Photograph of a detail of the proper right eye and corresponding MA-XRF elemental distributions of (d) Ca, (e) Ti, (f) Zn. The MA-XRF tube was operated at 28 kV, 40 μA.

In 1935 Vasily Kandinsky was already mixing sand with oil painting (see, for example, his work *Contraste accompagné*, housed at the Guggenheim Museum in New York), and later Lucio Fontana used powders and sands for his *Spatial Concepts* (like the one from 1956, from the Intesa San Paolo collection in Milan).

Baj achieves the figure of the child placing a highly diluted black pigment on a sandy and metallic background. Although we expect it to be enamel, by virtue of its viscosity and fluidity, it turns out to be an organic black with oil as binding medium. This black was evidently spread with the dripping technique, essentially letting it drip by filaments that Baj joined in the typical spiral formation. The presence of oil can also be confirmed by visual analysis, highlighting a homogeneous craquelure, while the typical withdrawals of enamel are not observed (to be found, on the other hand, in the paintings *Personaggio* and *Spettacolo*).

Identification of an oil binder was achieved by reflection FT-IR spectroscopy, as a glycerolipid component (see Table 7.1) is present in most of the measurements, like in *Quamisado II.* MA-XRF was decisive for the characterisation of composition and the distribution of the metallic and siliceous powders. In the thick black dripping paint used to realize, for example, the details of the head, MA-XRF showed an increase in the X-rays of Ca (Figure 7.3b). The presence of this element as calcium carbonate was confirmed by FT-IR. The elemental maps of copper, zinc, silver, silicon and aluminium (Figure 7.3) reveal the presence of these elements in correspondence of the powder aggregates unevenly distributed on the surface. The combined red–blue MA-XRF maps of copper and zinc also evidence that copper is mainly present in association with Zn, possibly in the form of an alloy (brass), while Zn was also used alone (Figure 7.3g). Lastly, MA-XRF helped to understand the composition of the layers underneath the visible paint, as shown by the elemental distributions of Pb and Ba/Ti[§§] (Figure 7.3d and f), probably present as white pigments, which clearly show in negative the features of the overlaid black and metallic paints.

7.2.3 *Personaggio* (Character)

Continuing in chronological order, we find the concurrence of different technical solutions in the painting *Personaggio* (1955, ref. 26, n. 199, Figure 7.1c), solutions which from that moment on would have characterised Baj's work, his use of poly-material and assemblage. Primarily, the supporting canvas is no longer the traditional, canonical one, but is rather a wall covering, a large knit canvas printed with an industrial process, with an ornamental rhomboid pattern. Baj works on this directly without laying down any preparation

[§§]The characteristic energies of the Ti K_α line (4.51 keV) and the Ba L_α line (4.47 keV) are too close to be resolved, giving rise to a single peak in the spectral range of 4.5–5 keV. This is also true for the β lines of these two elements (Ti K_β 4.93 keV – Ba L_β 4.83 keV).

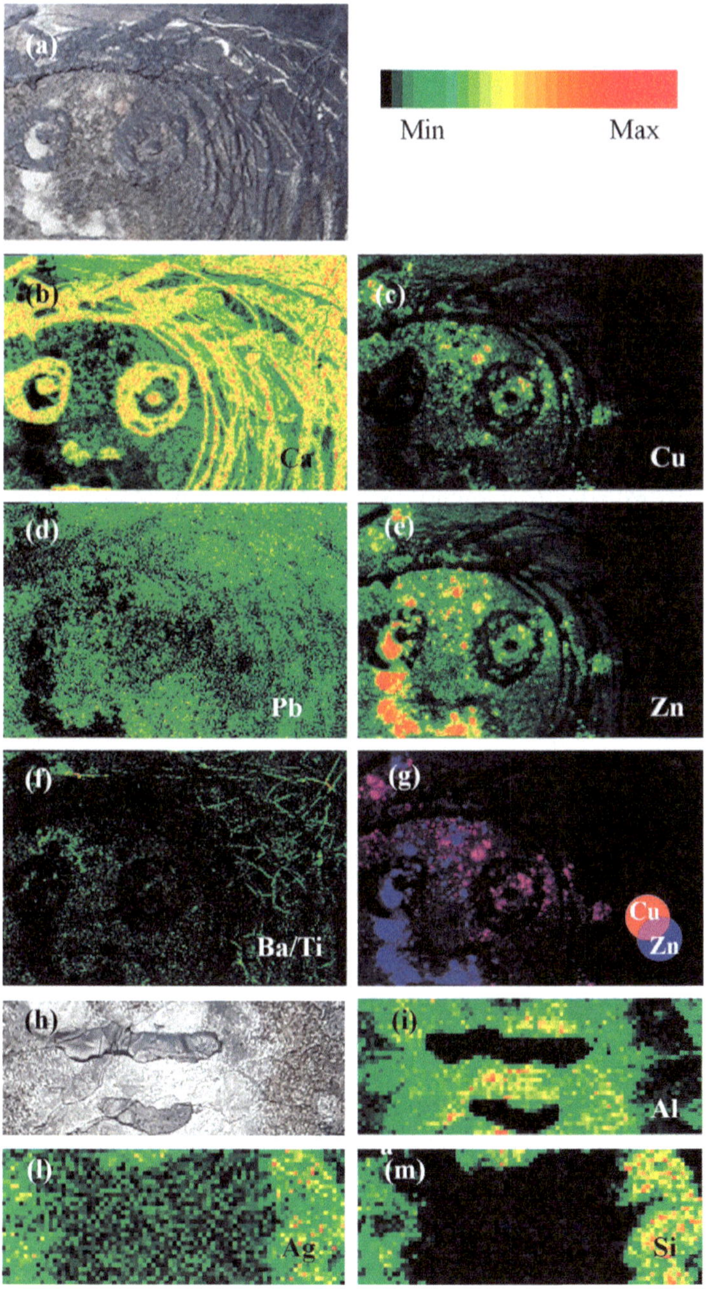

Figure 7.3 Selection of MA-XRF results for the painting *Bambino magico*: (a) Photograph of a detail of the face and corresponding MA-XRF elemental distributions of (b) Ca, (c) Cu, (d) Pb, (e) Zn, (f) Ba/Ti; (g) Red–Blue combined map of Cu (red) and Zn (blue); the MA-XRF tube was operated at 25 kV, 30 μA. (h) Photograph of a detail around the lips and corresponding MA-XRF elemental distributions of (i) Al, (l) Ag, (m) Si; the MA-XRF tube was operated at 25 kV, 30 μA with He flow.

on it[¶¶]. Hence, we find a wide palette of oil paints under the figure, spread with a gestural force and strongly textured either with a brush, a spatula or directly from the tube. The figure that stands out of this colored background is for all intents and purposes an *Ultracorpo* (Ultrabody), a creature that had inhabited Baj's imaginary world since the days of *Trillalì-Trillalà* (1955), effectively described by Breton in 1963.[27] This figure is essentially formed by two elements that are cornerstones in Baj's artistic philosophy, and that appear in his studio right about this time: cotton wool and Vinavil®. Baj composes a true "poem" to vinyl glue in 1983, when he publishes his *Automitobiografia*. The artist's discovery of this material dates back to the same period where he was fascinated by tapestries and fragments of Murano glass that Lucio Fontana gave him:

> I have finally made the discovery of the century: Vinavil, a white glue, thick and stringy like condensed milk and other liquids mentioned before, which allowed me total freedom when combining materials. The collages that can be made this way consist in the voluntary and systematic warping of the elements, diverted as they are from their usual uses. [...] A few days later, having visited Lucio Fontana at his studio in Porta Monforte, I saw bowls and fragments of colored glass. [...] piled up, this sparkling, colorful, opalescent material came from Murano [...]. I was attracted by that firmament of stellar crystals and Lucio gave me a small supply as a gift.[‖‖]

In the piece *Personaggio*, considered here, we can see how Baj goes on to mold his primordial figures, using vinyl glue as a binding agent and giving a sculptural quality to a soft and delicate material such as cotton wool. The artist worked on the modeled figure with black and white brush strokes. Finally, it is important to note here the presence of black enamel, indicated by the presence of alkyd resins in the glossier brushstrokes which mark the left contour of the figure (on our right), and in some areas of the lower part of the painting. The enamel gives a three-dimensional and shiny effect, contributing to the prominence of the figure.

For this painting, FT-IR and Raman spectroscopies were essential to disclose the transition of Baj from the use of traditional materials to the research of new material resources and new ways of working with them. In this logic, *Personaggio* displays a variety of modern synthetic pigments and polymers newly developed for many diverse uses (house paints, industrial paints, synthetic glues, *etc.*) and made available in the market. Oil, vinyl and alkyd paints were unambiguously distinguished by reflection FT-IR spectroscopy[28] (Figure 7.4a) along with the presence of different types of fillers and additives: magnesium carbonate, barium sulphate, silicates and wax

[¶¶]The choice to paint on tapestries and canvases with pre-existing patterns is explained by Baj himself in his *Automitobiografia*,[18] p. 169, and at this point must be considered carefully as documentation of the artist's attention towards materials external and heterogeneous to academic artistic production.

[‖‖]Ref. 18, pp. 169–171.

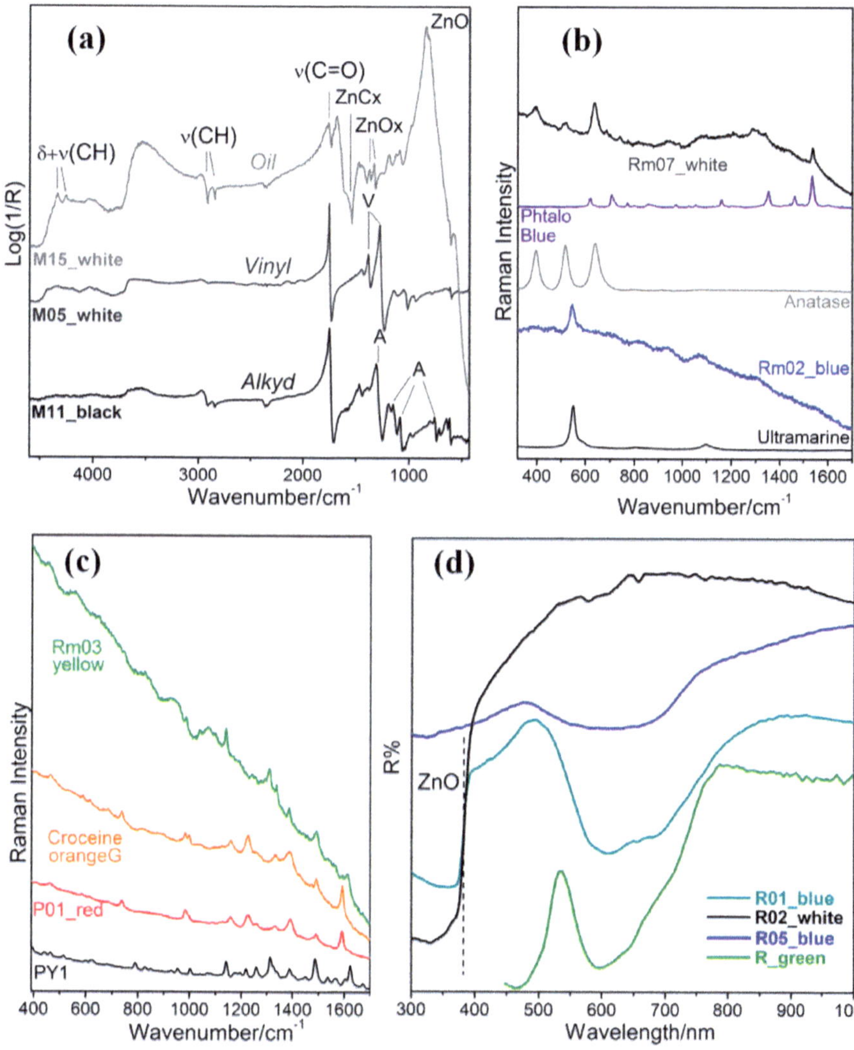

Figure 7.4 Selection of analytical results for the painting *Personaggio*: (a) Reflection FT-IR spectra collected on different paints characterised by oil (diagnostic bands: CH and C=O stretching and $\delta + \nu$ CH modes), vinyl (diagnostic bands marked with V) and alkyd (diagnostic bands marked with A) binders. In the oil paint, diagnostic bands for zinc oxalate (ZnOx, at about 1318 and 1360 cm^{-1}) and carboxylate (ZnCx, at 1540 cm^{-1}) are also observed, owing to the interaction of the binder with the zinc white pigment (ZnO). (b) Raman spectra of white and blue areas where anatase, phthalocyanine and ultramarine blue were identified, respectively; reference Raman spectra (at the same laser excitation line) of these pigments are also reported for comparison. (c) Raman spectra of a yellow area and of a red micro-sample (P01_red – brushstroke on the bottom part); reference spectra for the pigments Croceine orange G (orange line) and PY1 yellow (black line) are also shown. (d) Punctual Vis-NIR reflectance spectra collected on blue, white and green spots of the painting. R01_blue = ZnO mixed with phthalocyanine blue; R02_white = ZnO; R05_blue = ultramarine blue; R_green = chromium oxide (spectrum extracted from the Vis-NIR hyperspectral image-cube on a green area).

(see Section 7.3). Alkyd binder was found specifically in the organic black paint enclosing the central figure, while bone black, probably mixed with oil, was clearly identified only in the eye contours. Raman spectroscopy, supported by Vis-NIR spectroscopy, evidenced the use of more traditional pigments such as ultramarine blue, zinc white and chromium green, but also new synthetic pigments such as Croceine Orange G, PY1 yellow or blue phthalocyanine, along with titanium white (anatase form, already present in *Quamisado II*) (Figure 7.4b–d). The variety of rheological properties of the paints used in *Personaggio* enabled Baj to experiment with new forms of artistic expression and to play with many different optical and physical effects of the three-dimensional surface thus obtained.

7.2.4 *Due bambini nella notte nucleare* (Two Children in the Nuclear Night)

Even if it shows a richer palette and a more complex use of colors compared with paintings from the early 1950s, *Due bambini nella notte nucleare* (1956)[29] (Figure 7.1d) can be grouped under nuclear painting in the closest sense. It has several broad areas where heavily diluted black paint was dripped, probably distributed on the canvas in several rounds with liquid emulsions that included oil (as a binding agent), enamel (alkyd component) and resin. The technique of continued overlapping and grouping of chromatic matter decries a gestural matrix from which we cannot exclude the movement of the canvas, already documented from Baj and Dangelo's statements on the experimental process of the first nuclear paintings. These technical aspects, evident in a visual examination, taken together with the documented presence of enamels and resins, could connect this painting with the "acqua pesante" technique, which alludes to the deuterated form of water used in nuclear physics. In a metaphorical process (that epistemological discourse that Crispolti glimpsed and that was mentioned in the first paragraph), Baj treats his nuclear paintings with a technique that derives from atomic science, but whose practice simply and chaotically comes out from progressive emulsions of oil paints, water and solvents, to which enamels, pitch or industrial resins may have been added. The use of pitch and of resins was documented in Baj's *Montagna con sole* (1957), from Alessandra Tibiletti's examinations (in Pugliese,[15] p. 126).

Until now, between 1951 and 1956, we were able to document Baj's use of oil paints, metallic powders, sand, enamels, vinyl glue and natural fibers, and we tried to hypothesise the voluntary use of resins and waxes (even if it is not possible to rule out that these substances are included in the enamels and pigments themselves). It can't be underestimated how Baj used copper plates, acidic and waxy substances, at least since his first engraving for *De Rerum Natura of* Lucrezio, in 1952:

> Between the swirling spirals of imaginary atomic magma and evocative spots, I continued to etch steel engravings, carrying out the cycle for *De Rerum Natura* [...] The first plate, a solar head, was dramatic. I had bought

> the copper plates from Necchi Metalli on Montenero St. I waxed them myself; after engraving the first one, I placed it in a bowl containing pure nitric acid. The reaction was brutal: the acid started boiling over and attacked the plate violently, eating at it and provoking a dense smoke, which, alas, I breathed in***.

The novelty of the painting *Due bambini nella notte nucleare* with respect to the previous "nuclear" works stays in the choice and use of different pigments to produce the dominant chromatic combination of black and red/orange. The dripping dark areas were found by reflection FT-IR to be composed of an organic black with an alkyd binder in the thick glossy paints, while an oil-resinous component came out in the faded matt black paints. An iron-based pigment (probably an earth) was also found in the dark brown hues. For the red-orange areas the palette of the artist is enriched by cadmium-based yellow and orange pigments, which have been characterised by UV-Vis-NIR reflectance and fluorescence spectroscopy.[30–32] The working technique of the paints was, instead, efficaciously displayed by the complementary use of MA-XRF scanning and hyperspectral Vis-NIR imaging. For example, in the detail of the head of the proper left child (Figure 7.5a), the elemental maps of Ca and Fe (Figure 7.5b and c) highlight the distribution of the dark brown pigment while, by comparison with the photograph of the area, the organic black paint remains undetected. Furthermore, the map of chromium (Figure 7.5d) helps to better observe the distribution of the green paint otherwise not clearly perceivable, being mixed with the brown-black paint. Raman spectroscopy also revealed the combined use of phthalocyanine blue for the blue and green areas.

The peculiar luminescence properties of cadmium-based pigments in the near infrared, supported by their Vis-NIR reflectance profiles, allowed different types of cadmium sulfide and sulfo-selenide pigments (CdS and $CdS_{1-x}Se_x$, respectively) to be identified and their spatial distribution to be visualised by means of the fluorescence and reflectance hyperspectral imaging techniques.[32,33] In particular, the red-orange and yellow areas of the painting background can be described by three endmembers†††[33] whose luminescence, spatial distribution and spectra are shown in Figure 7.5e and f. To better emphasise their correlation with respect to each other, a RGB composite image has been created, where the red, green and blue channels represent the red, orange and yellow compounds, respectively. The analyses identified the presence of two slightly different $CdS_{1-x}Se_x$ pigments employed for the red and orange tones and the use of a CdS pigment for the yellow areas (in Figure 7.5e red, green and blue, respectively). Specifically, the photoluminescence profiles of the two $CdS_{1-x}Se_x$ pigments display a weak signal in the visible (600 nm for the red and 572 nm for the orange) and two strong

***Ref. 18, p. 174.

†††*Endmembers* are a basic set of spectra that best describe the majority of spectra in the image-cube and represent a molecular fingerprint which the elaboration software is able to recognise and spatially map on the analysed surface.

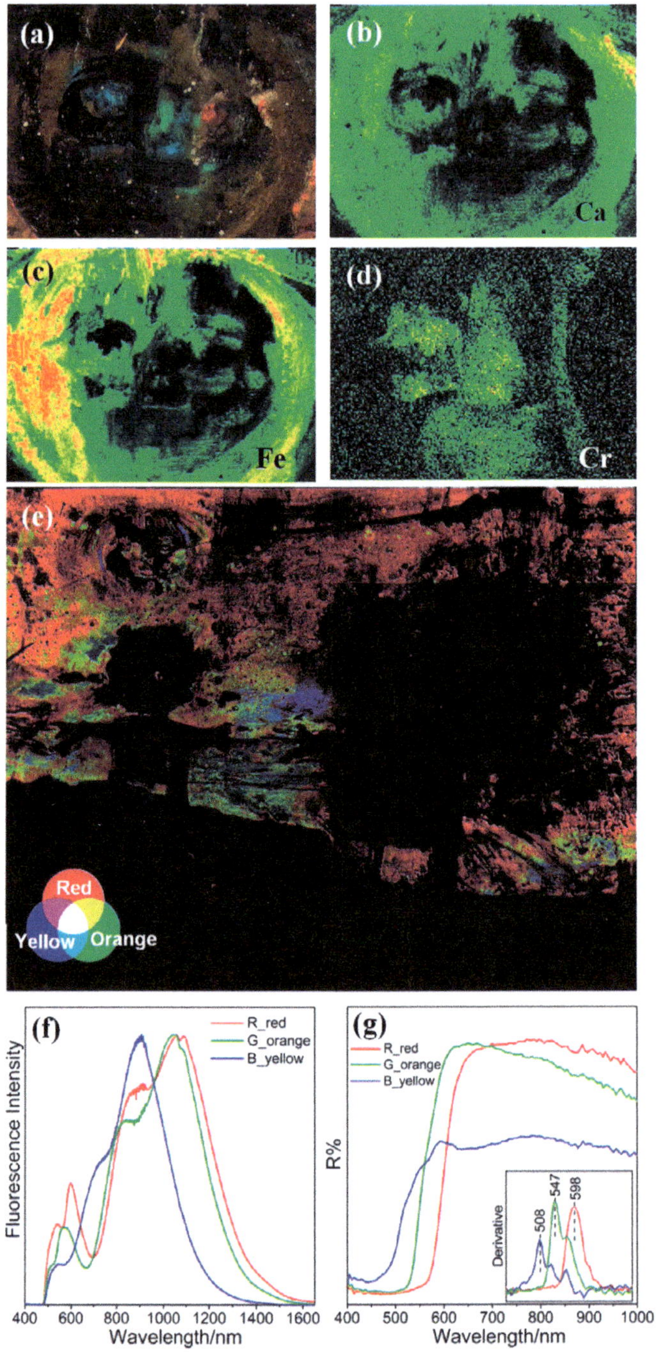

Figure 7.5 Selection of MA-XRF results for the painting *Due bambini nella notte nucleare*: (a) photograph of a detail of the head of the proper left child and corresponding MA-XRF elemental distributions of (b) Ca, (c) Fe, (d) Cr; MA-XRF tube was operated at 35 kV, 30 μA. (e) Results of luminescence imaging spectroscopy using excitation at 405 nm: RGB composite image created using the spatial distribution of the three identified endmembers. R = red pigment (type 1 $CdS_{1-x}Se_x$), G = orange pigment (type 2 $CdS_{1-x}Se_x$) and B = yellow pigment, (CdS or $Cd_{1-x}Zn_xS$). (f–g) Punctual Vis-NIR emission (λ_{exc} = 445 nm) and reflectance spectra corresponding to the R = red, G = orange and B = yellow endmembers identified. In (g) the inset shows the first derivative of the reflectance profiles useful for a precise evaluation of the inflection points.

bands in the NIR (890 and 1070 nm for the red, 831 and 1043 nm for the orange) that can be assigned to near band-edge emission (NBE) and deep-level emission (DLE), respectively.[30] While the NBE emission can be affected by the contribution from the binder emission and self-absorption effects – meaning it is not strictly diagnostic – the differences in the DLE emissions of the two pigments give a clear differentiation in their chemical composition. The RGB composite image highlights a strong correlation between the red and orange pigments, used together in different relative amounts, in most of the reddish-orange paints. The endmember with luminescence spatial distribution reported in blue in Figure 7.5e is, instead, due to the presence of a CdS yellow pigment. Its fluorescence profile exhibits a maximum at 900 nm and a shoulder at 710 nm, suggesting: (i) the coexistence of cubic and hexagonal CdS pigments or (ii) the use of a $Cd_{1-x}Zn_xS$ solid solution.[31] The corresponding Vis-NIR reflectance spectra of the three identified compounds are reported in Figure 7.5g (the spectra displayed are extracted from the image-cube).

7.2.5 *Spettacolo* (Spectacle)

The last work that was analysed for this project was *Spettacolo* (1956, ref. 14, n. 273, Figure 7.1e). It constitutes a safe, parallel harbor in our research and in Enrico Baj's, as it simultaneously presents all the technical elements which we came upon partly or singularly in the previous works. New ones also appear, as, for example, Murano glass fragments and tufts of hemp used like hair in the main figure (molded in gray-brown cotton wool). The background of the painting is made of different pieces of fabric, from preprinted tapestries to sackcloth. Enamels were evidently used in the blacks and reds, and the presence of sand and vinyl glue was also found in the blacks. As far as the palette of oil paints, the options multiply and the colors are either pure or mixed.

All this confirms the thesis according to which, from the first experiments to the more mature ones of the Nuclear Art Movement, Enrico Baj articulates a poly-material executive system in which traditional painting materials, non-conventional ones from industry (enamels and resins) and other heterogeneous materials coming from nature but also from the everyday world of commodities (cotton wool, tapestries, ropes, glass, vinyl glue) are balanced in an extremely complex process of figurative variation (from pre-representation to representation). The techniques and typical processes of Action Painting and the *informel* (from gestural painting to dripping, from the use of industrial enamels to the incorporation of fragments of everyday materials and use of modified and fragmented material, such as cotton wool, sackcloth, rope or glass) support a collage technique which becomes the predominant vocabulary in almost the totality of the expressive research of Enrico Baj[‡‡‡].

[‡‡‡]Ref. 4, pp. 42–44.

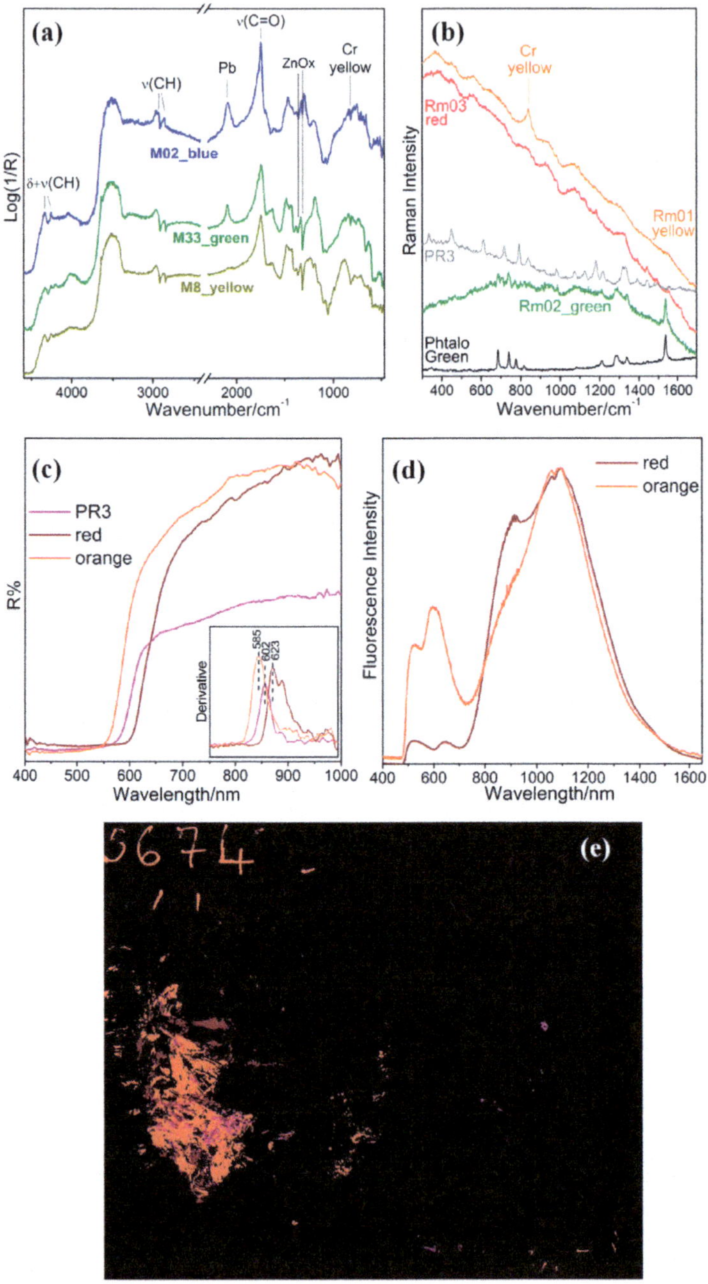

Figure 7.6 Selection of analytical results for the painting *Spettacolo*: (a) Reflection FT-IR spectra collected on a blue and green oil paint (binder diagnostic bands: CH and C=O stretching modes) where Prussian blue (Pb, diagnostic band at about 2097 cm^{-1}) was detected. Zinc oxalate (ZnOx, diagnostic bands at about 1318 and 1360 cm^{-1}) was also identified. (b) Raman spectra of yellow, red and green areas where the pigments chrome yellow, PR3 and phthalocyanine green were detected; reference Raman spectra (at the same laser excitation line) of PR3 and

The matured expressive capacity of Baj manifested in *Spettacolo*, due to the complexity of materials and painting techniques used, is very challenging to approach by analytical investigations. The complementary use of elemental and molecular techniques, in this case, was decisive to disclose all the materials and technical aspects. The integrated use of oil, alkyd and vinyl paints, the richer chromatic palette and the diverse commercial paint formulations were characterised in a totally non-invasive way (Table 7.1). As an example, Prussian blue was found to be used as blue pigment aside from ultramarine blue and applied in mixture with a chrome-based pigment in some green areas as an alternative to phthalocyanine green (Figure 7.6a and b). In the red paints, the synthetic pigment PR3 (Figure 7.6b) is introduced in addition to $CdS_{1-x}Se_x$-based pigments; the spatial distributions of both types of pigment were revealed by Vis-NIR hyperspectral imaging in reflectance, as shown in Figure 7.6e. The imaging analyses, supported by the UV-Vis-NIR reflectance and fluorescence point measurements (Figure 7.6c and d), distinguished the presence of two kinds of $CdS_{1-x}Se_x$ pigments for the red and orange-red tones (mapped in red and orange colors, respectively in Figure 7.6e). In particular, such cadmium-based compounds show different Vis-NIR reflectance profiles (Figure 7.6c), characterised by the typical sigmoidal shape with an inflection point at 585 nm for the orange and at 623 nm for the red pigment, but have similar emission properties, differentiated only by the higher fluorescence intensity of the red compound (Figure 7.6d). The photoluminescence spectra have both two strong bands in the NIR at 1085 and 900 nm, assigned to DLE, and a NBE emission in the visible range between 600 and 640 nm (of less interest because of the reasons reported above). Since XRF analyses showed a larger amount of cadmium in the orange paint – thus excluding a possible dilution of the red pigment with a white compound – these experimental findings might be explained by the addition of an amorphous yellow CdS pigment, whose emission contribution is negligible with respect to the red $CdS_{1-x}Se_x$ one. Such addition would justify the decrease in fluorescence intensity of the orange compound while keeping the emission profile almost the same.

7.3 Analytical Approach and Methodology

The scientific project "Baj Nucleare" is composed of more than 700 points of analysis by complementary elemental and molecular spectroscopic techniques integrated with hyperspectral imaging by Vis-NIR spectroscopy and

---

phthalocyanine green are also reported for comparison. (c–d) Vis-NIR reflectance and emission (λ_{exc} = 445 nm) spectra corresponding to the three endmembers identified for the red painted areas: Magenta = PR3, Red = $CdS_{1-x}Se_x$, Orange = $CdS_{1-x}Se_x$, probably mixed with CdS. In (c) the inset shows the first derivative of the reflectance profiles. (e) Vis-NIR hyperspectral reflectance imaging results displaying the spatial distribution of the three endmembers (colour code same as in (c–d)).

MA-XRF scanning analysis. The wide analytical campaign carried out in two weeks was aimed at providing a representative overview of materials and painting techniques used by Enrico Baj in the five "nuclear" artworks considered in this study, working in a totally non-invasive way and *in situ*, at the place where the paintings are stored, in the old house of Baj in Vergiate (Italy).

The results of the diagnostic campaign are summarised in Table 7.1, integrated with the information provided by micro-FT-IR and micro-Raman analyses of six micro-samples. These micro-samples were collected in marginal areas of the paintings or at the edges of the canvases in order to solve specific questions that otherwise would have remained unanswered.

Elemental analysis by pXRF and MA-XRF scanning supported molecular identification of the pigments achieved by molecular techniques: reflection FT-IR and Raman spectroscopies and UV-Vis-NIR reflectance and fluorescence spectroscopies. Imaging techniques, *i.e.* hyperspectral Vis-NIR and MA-XRF, were then necessary for a full understanding of the painting technique of Baj. Technical details of the analytical methodology and instrumentation are reported in Appendix I.

The diagnostic campaign also provided significant information concerning the different paint formulations of the commercial pigments and paint tubes used by Baj and, more importantly, evidence of the state of conservation of the materials, as well as identification of paint components susceptible to degradation. In this regard, reflection FT-IR spectroscopy was of great relevance for identifying different binders (oil, alkyd, vinyl or resin (Figure 7.4a)) as well as fillers and plasticizers, as already discussed previously in the text.[34] In more detail, calcium carbonate ($CaCO_3$) (Figure 7.2a), barium sulphate ($BaSO_4$), silicates and magnesium carbonate ($MgCO_3$) (Figure 7.7a) were found to be the most common fillers[35] differently distributed in the five paintings; also wax was sporadically detected, probably as paint plasticizer (Figure 7.7a).

Reflection FT-IR significantly contributed also to the identification of alteration products which may affect oil paints.[36] In particular, zinc and calcium oxalates were often observed as a product of the oxidative degradation of the paint organic components.[37] Furthermore, zinc carboxylates were found owing to the reaction of zinc ions from the ZnO pigment (present in three of the five paintings) with the fatty acids of the binder (Figure 7.4a).[38] Very important was also the identification by Raman spectroscopy of lead chromate-based pigments in *Quamisado II* (Figure 7.2b) and in *Spettacolo* (Figure 7.6b). This class of pigments was largely used by Van Gogh and, being renowned for its scarce lightfastness, has been widely studied for years to understand the effects of pigment chemical composition, crystalline structure and environmental conditions on the darkening of yellow paints.[39] In a similar way, the identification of Cd-based pigments was crucial for preventive conservation purposes, due to their high tendency to undergo (photo-) chemical transformations.[40]

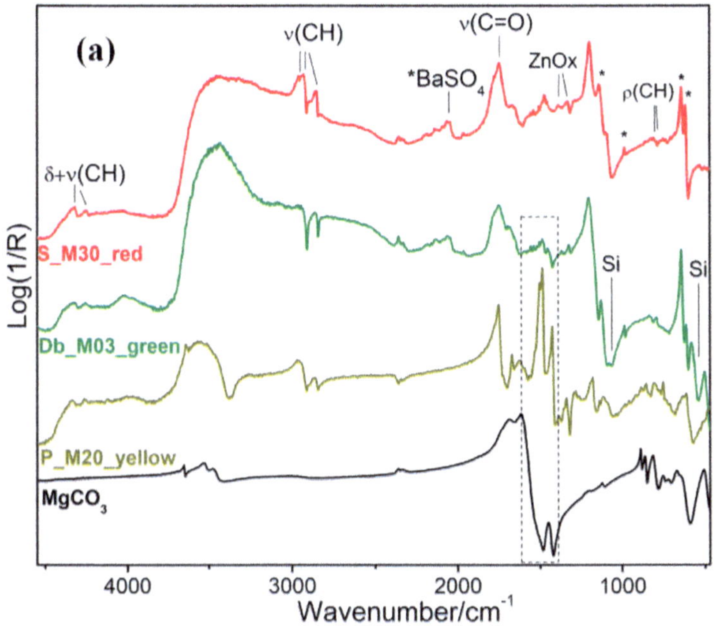

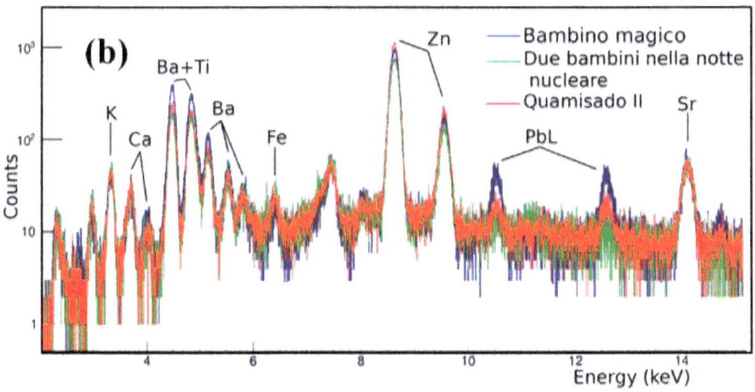

Figure 7.7 (a) Reflection FT-IR spectra from paints of *Spettacolo* (S_M30_red), *Due bambini nella notte nucleare* (Db_M03_green) and *Personaggio* (P_M20_yellow) showing diagnostic bands for different paint fillers and plasticisers: wax (ν(CH) stretching and ρ(CH) rocking modes of the methylene chains) and barium sulphate (*$BaSO_4$) in *Spettacolo*, a silicate component (Si) in *Due bambini nella notte nucleare,* magnesium carbonate in *Personaggio* (a reference spectrum of $MgCO_3$ is also shown for comparison, black line). (b) pXRF spectra of the canvas verso for *Bambino magico*, *Due bambini nella notte nucleare* and *Quamisado II*. XRF tube was operated at 35 kV, 30 μA, acquisition time 120 s.

The preparation layer for three of the paintings executed on canvas (*Bambino magico*, *Due bambini nella notte nucleare* and *Quamisado II*) was also investigated by pXRF (in this case the same instrumental head of MA-XRF was used), by analysing the *verso* of the canvas. The results, reported in Figure 7.7b, show the same behaviour for the three spectra, the only difference being a higher peak of Pb-L for *Bambino magico*, probably due to the contribution of the *recto* of the painting, since a corresponding increase in the Pb-M peak (not shown) is not observed.

7.4 Conclusions

Science is not art, but science can be applied “to art” to understand the side of art “in shadow”, a sort of behind the scenes view that it is necessary to see in order to proceed with greater knowledge in the conservation of artworks. These first diagnostic analyses of Baj's early nuclear works provide useful information for better understanding the artist's technique, while at the same time leaving ample margin of doubt about several aspects of his creative process, a mystery that we like to think remains unfathomable about the meaning of art, despite its total visible manifestation: aesthetic phenomenology as a vehicle for that spiritual dimension that Baj wished for man, in reaction to a growing rationalism and technophilia prevalent on the threshold of the third millennium.

Appendix I: Experimental Methods

Point X-ray Fluorescence Spectroscopy (pXRF)

pXRF measurements were carried out by employing the Elio instrumentation by XGLab (Bruker® Nano Analytics). The instrument has a measurement head mounted on a tripod for easy access to the painting surface. X-rays are generated by a Rh anode tube that was operated at 50 kV and 80 μA with 40 s of acquisition times. The beam was collimated on the surface at 1 mm of diameter.

MA-XRF

MA-XRF measurements were performed with a portable XRF scanner developed by the INFN-CHNet network.[41] A Mo anode tube was used, the beam spot on the surface collimated at ~1 mm of diameter. Tube operating conditions are reported in the captions of the figures.

Reflection FT-IR

Infrared spectra were collected by a portable FT-IR spectrophotometer (ALPHA, Bruker® Optics, Germany/USA-MA) equipped with an external reflectance module (optical layout of 22°/22°) that permits infrared radiation reflected from a surface located about 1 cm away to be collected. The instrument consists of a SiC Globar source, a RockSolid™ design interferometer (system of parabolic gold mirrors) and a DLaTGS detector. The spectra, shown in pseudo-absorption [log(1/R); R = reflectance], were acquired from areas of about 5 mm in diameter, in the spectral range between 7000 and 375 cm^{-1}, with a spectral resolution of 4 cm^{-1} and using an average of 186 acquisitions. Spectra from a flat gold mirror were used as background.

Raman

Raman analysis was performed using the laser excitation at 785 of the portable Xantus™-2 by Rigaku®. It works with a CCD detector (Peltier cooled) and a spectral resolution of 7–10 cm^{-1}, while the spatial resolution is about 4 mm^2. The laser power at the sample can be software-modulated, thus avoiding any sample damage.

UV-Vis-NIR Reflectance and Emission

A compact portable instrument projected and assembled for non-invasive *in situ* diagnostics was used. The excitation sources are: a deuterium–halogen lamp (AvaLight-DHc, Avantes™) for reflection measurements and an ultra-compact diode laser source (Toptica Photonics AG, DE; excitation wavelengths 445 nm and laser power 1 mW) for steady-state fluorescence measurements. A highly sensitive CCD spectrometer (AvaSpec-2048 USB2 Avantes™, 200–1100 nm range, spectral resolution 8 nm) is used for reflection measurements. High-sensitivity calibrated CCD spectrometers, AvaSpec-ULS2048 XL-RS-USB2 (300–1150 nm range, spectral resolution 9.2 nm) and AvaSpec-NIR256-1.7TEC (950–1600 nm range, spectral resolution 24 nm) detect the steady-state photoluminescence (PL). A multi-furcated fiber-optic system (Avantes™) with a standard reflection probe inclined at 21° to the analysed surface directs all the excitation sources to the same area and collects both the reflected and emitted light.

Vis-NIR Reflectance and Fluorescence Hyperspectral Imaging

A SOC710 produced by Surface Optics Corporation, San Diego (USA) has been used for hyperspectral measurements. The system utilises a whisk-broom line scanner producing a 696 × 520-pixel hypercube in the 400–1000 nm spectral range with 128 bands and about 4.5 nm spectral resolution. The spatial resolution can be continuously modulated by adjustable focal length of the mounted objective. Two Elinchrom™ Scanlite 350 W Halogen lamps with diffusing umbrellas have been used for reflectance measurements, while two Hoenle Ledline 500 LED systems emitting at 405 nm have been used for luminescence measurements. ENVI classic software was employed for the data processing and sample angle mapper (SAM) function was used to map the endmembers' distribution.

Acknowledgements

The entire work group gives its sincere thanks to Roberta Cerini Baj and to Andrea Baj for their support during the survey period and for the active collaboration granted to the "Baj Nucleare" project from the beginning. Financial support by the Access to Research Infrastructures activity of E-RIHS.it (www.e-rihs.it) is also gratefully acknowledged. All the authors dedicate this essay to Enrico Crispolti (1933–2018), with whom the methodological introduction of the "Baj Nucleare" project was discussed at the time.

References

1. T. Sauvage, *Arte Nucleare*, Milano, Galleria Schwarz, 1962.
2. E. Baj, *Cose, fatti, persone*, Milano: Elèuthera, 1988.
3. *Alternative attuali: 2. Rassegna internazionale di pittura, scultura, grafica: omaggio a Magritte, opere 1920–1963 omaggio a Mirko, opere 1932–1964; omaggio a Baj, opere 1950–1965*, ed. E. Crispolti, Milano, Lerici, 1965.
4. E. Baj, *Autodamé. Collage e scritture*, ed. L. Cappelli, Nuova Casa Editrice, Bologna, 1980.
5. B. Brunetti, C. Miliani, F. Rosi, B. Doherty, L. Monico, A. Romani and A. Sgamellotti, Non-invasive Investigations of Paintings by Portable Instrumentation: The MOLAB Experience, *Top. Curr. Chem.*, 2016, **374**, 10.
6. www.e-rihs.it.
7. *Asger Jorn. Oltre la Forma/The Form and beyond*, ed. L. Bochicchio and P. Valenti, Genova University Press, Genova, 2014.
8. *Conversazioni con Enrico Baj. Mezzo secolo di avanguardie*, ed. L. Caprile, Elèuthera, Milano, 1997.
9. *Baj Jorn. Lettres 1953–1961*, ed. M. Fréchuret, Saint-Etienne, Musée d'Art Moderne Saint-Etienne, 1989.
10. *CoBrA. Una grande avanguardia europea 1948–1951*, ed. D. Femfert and F. Poli, Skira, Milano, 2015.
11. L. Bochicchio, Build to Destroy. Enrico Baj, CoBrA and the Possible Revolution, in *Baj: Play as Protest*, ed. H. de Bruijn and C. Pilto, Cobra Museum of Modern Art, Amstelveen, 2017.
12. E. Crispolti, *L'informale: antologia di poetica 1943–1961 [1971]*, De Luca Editori d'Arte, Roma, 2017.
13. *Enrico Baj: figure dell'immaginario (1951–2003)*, ed. L. Bochicchio and R. Cerini Baj, Albissola Marina, Vanillaedizioni, 2015.
14. L. Bochicchio, Modifications contre le (mouvais) goût dominant, in *L'invasion des ultracorps*, ed. C. Gatti and R. Cerini Baj, Silvana Editoriale, Cinisello Balsamo, 2016, pp. 25–33.
15. *Tecnica mista. Materiali e procedimenti nell'arte del XX secolo*, ed. M. Pugliese, Bruno Mondadori Editore, Milano, 2006.
16. *Enrico Baj. Dal De rerum natura alle Montagne*, ed. M. Corgnati and A. Tibiletti, Skira, Milano, 2018.

17. M. Fratelli, R. Bestetti, S. Bruni, S. Caravà, V. Guglielmi, S. Minotti, G. Poldi, I. Saccani and A. Tibiletti, Pittura italiana 1950–1960: materiali, metodi esecutivi e questioni conservative. Il progetto diagnostico su alcune opere di Enrico Baj e Lucio Fontana della Collezione Boschi Di Stefano, in *CESMAR 7*, 2015, pp. 81–90.
18. E. Baj, *Automitobiografia. Dai giorni nostri alla nascita*, Rizzoli, Milano, 1983.
19. O. Chiantore and A. Rava, *Conservare l'arte contemporanea: problemi, metodi, materiali, ricerche*, Electa, Milano, 2005.
20. E. Di Martino, *Arte contemporanea: conservazione e restauro*, Umberto Allemandi Editore, Torino, 2005.
21. S. Bordini, *Arte contemporanea e tecniche: materiali, procedimenti, sperimentazioni*, Carocci, Roma, 2007.
22. M. Kokkori, F. Casadio, K. Sutherland and M. Vermeulen, Charting the development of oil-based enamel paints through the correlation of historical paint technology manuals with scientific analysis, in *Issues in Contemporary Oil Paint*, ed. A. Burnstock, M. de Keijzer, J. Krueger, T. Learner, A. de Tagle, G. Heydenreich and K. J. van den Berg, Springer Cham, 2014, pp. 117–125.
23. R. Ploeger, D. Scalarone and O. Chiantore, The characterization of commercial artists' alkyd paints,*J. Cult. Herit.*, 2008, **9**(4), 412–419.
24. A. Daveri, M. Malagodi and M. Vagnini, The bone black pigment identification by non-invasive, *in situ* Infrared Reflection spectroscopy,*J. Anal. Methods Chem.*, 2018, 1–8. Article ID 6595643:8.
25. L. L. Carrick, Vegetable oil paints, *J. Am. Oil Chem. Soc.*, 1950, **27**(11), 513–522.
26. *Catalogo generale Bolaffi dell'opera di Enrico Baj*, ed. E. Crispolti, G. Bolaffi, Torino, 1973.
27. A. Breton, *Enrico Baj*, in *L'Œil*, 103–104, Paris, juillet-août, 1963, pp. 34–39.
28. F. Rosi, A. Daveri, P. Moretti, B. Brunetti and C. Miliani, Interpretation of mid and near-infrared reflection properties of synthetic polymer paints for the non-invasive assessment of binding media in twentieth-century pictorial artworks, *Microchem. J.*, 2016, **124**, 898–908.
29. *Catalogo generale Bolaffi dell'opera di Enrico Baj: dal 1996 al 2003*, ed. R. Cerini Baj and E. Crispolti, Bandecchi & Vivaldi, Pontedera, 2004, n. 259a.
30. C. Grazia, F. Rosi, F. Gabrieli, A. Romani, M. Paolantoni, R. Vivani, B. Brunetti, P. Colomban and C. Miliani, UV–Vis-NIR and microRaman spectroscopies for investigating the composition of ternary CdS1− xSex solid solutions employed as artists' pigments, *Microchem. J.*, 2016, **125**, 279–289.
31. F. Rosi, C. Grazia, F. Gabrieli, A. Romani, M. Paolantoni, R. Vivani, B. Brunetti, P. Colomban and C. Miliani, UV–Vis-NIR and micro Raman spectroscopies for the non-destructive identification of $Cd_{1-x}Zn_xS$

solid solutions in cadmium yellow pigments, *Microchem. J.*, 2016, **124**, 856–867.

32. M. Thoury, J. K. Delaney, E. R. de la Rie, M. Palmer, K. Morales and J. Krueger, Near-Infrared Luminescence of Cadmium Pigments: *In Situ* Identification and Mapping in Paintings, *Appl. Spectrosc.*, 2011, **65**(8), 939–951.
33. C. Cucci, J. K. Delaney and M. Picollo, Reflectance Hyperspectral Imaging for Investigation of Works of Art: Old Master Paintings and Illuminated Manuscripts, *Acc. Chem. Res.*, 2016, **49**, 2070–2079.
34. C. Miliani, F. Rosi, A. Daveri and B. Brunetti, Reflection infrared spectroscopy for the non-invasive *in situ* study of artists' pigments, *Appl. Phys. A*, 2012, **106**, 295–307.
35. T. Learner, *Analysis of Modern Paints*. Getty Publications, 2004.
36. F. Rosi, L. Cartechini, L. Monico, F. Gabrieli, M. Vagnini, D. Buti, B. Doherty, C. Anselmi, B. Brunetti and C. Miliani, Tracking metal oxalates and carboxylates on painting surfaces by non-invasive reflection mid-FTIR spectroscopy, in *Metal Soaps in Art – Conservation and Research*, ed. F. Casadio, K. Keune, P. Noble, A. van Loon, H. Hendriks, S. Centeno and G. Osmond, Springer Nature, 1st edn 2019, pp. 173–193.
37. L. Monico, F. Rosi, C. Miliani, A. Daveri and B. Brunetti, Non-invasive identification of metal-oxalate complexes on polychrome artwork surfaces by reflection mid-infrared spectroscopy, *Spectrochim. Acta, Part A*, 2013, **116**, 270–280.
38. L. Baij, J. J. Hermans, K. Keune and P. Iedema, Time-Dependent ATR-FTIR Spectroscopic Studies on Fatty Acid Diffusion and the Formation of Metal Soaps in Oil Paint Model Systems, *Angew. Chem., Int. Ed.*, 2018, **57**, 7351–7354.
39. C. Miliani, L. Monico, M. J. Melo, S. Fantacci, E. M. Angelin, A. Romani and K. Janssens, Photochemistry of Artists' Dyes and Pigments: Towards Better Understanding and Prevention of Colour Change in Works of Art, *Angew. Chem., Int. Ed.*, 2018, **57**, 7324–7334.
40. L. Monico, A. Chieli, S. De Meyer, M. Cotte, W. De Nolf, G. Falkenberg, K. Janssens, A. Romani and C. Miliani, Role of the Relative Humidity and the Cd/Zn Stoichiometry in the Photooxidation Process of Cadmium Yellows ($CdS/Cd_{1-x}Zn_xS$) in Oil Paintings, *Chem. - Eur. J.*, 2018, **24**, 11584–11593.
41. F. Taccetti, L. Castelli, C. Czelusniak, N. Gelli, A. Mazzinghi, L. Palla, C. Ruberto, C. Censori, A. Lo Giudice, A. Re, D. Zafiropoulos, F. Arneodo, V. Conicella, A. Di Giovanni, R. Torres, F. Castella, N. Mastrangelo, D. Gallegos, M. Tascon, F. Marte and L. Giuntini, A multipurpose X ray fluorescence scanner developed for *in situ* analysis, *Rendiconti Lincei. Sci. Fis. Nat.*, 2019, 307.

CHAPTER 8

On Video Art Preservation: A Conversation with Fabrizio Plessi†

COSETTA SABA*[a] AND LISA PAROLO*[a]

[a]Università degli Studi di Udine, Udine, Italy
*E-mail: cosetta.saba@uniud.it, lisa.parolo@gmail.com

8.1 Introduction

The dialogue transcribed here concerns the distinctive features of Fabrizio Plessi's artistic practice in relation to the conservation of his works, starting from the central theme: water. This latter is inexhaustible and it is elaborated by Plessi in a conceptual way, through thick inter-textual weaving, multiple techniques, supports and different formats: drawing, design schemes, film, videotape, video performance, video sculpture, single-channel and multi-channel video installation, often on a large scale, in environmental mode.

As Achille Bonito Oliva points out, these are ramifications of "a path within which different materials compete and relate. The constant presence is that of the video that is incorporated in its value as a container, as an emitter of images and sounds, of a box that marks the space, of a simulacrum and of a temporal stabilizer" always, it must be added, brought back to the

†The conversation took place in Venice, February 14, 2019, at the Venetian studio of Fabrizio Plessi and was prepared and transcribed with the help of Costanza Lo Cicero (who dedicated her master's thesis to the reconstruction of Plessi's video works created in collaboration with the Centro Video Arte of Palazzo dei Diamanti in Ferrara[1]). The text was subsequently integrated and licensed by the artist.

Science and Art: The Contemporary Painted Surface
Edited by Antonio Sgamellotti, Brunetto Giovanni Brunetti and Costanza Miliani

Published by the Royal Society of Chemistry, www.rsc.org

symbolic form of the screen.[2] The realization of the first works in videotape – *Acquabiografico* and *Travel* (1974) – marks an important point of intersection in the artistic trajectory of Plessi with the operative path of the Centro Video Arte (Video Art Center) of Palazzo dei Diamanti in Ferrara that engages, on different fronts, Lola Bonora (art direction), Carlo Ansaloni (technical direction) and Giovanni Grandi (operation). It is from this collaborative experience that the series of questions addressed to Plessi have arisen, bringing us to the crucial question of the conservation of videographic works. This issue stemmed from the research carried out to support the preservation activities performed by La Camera Ottica[‡] Laboratory, University of Udine on analog video works produced in Ferrara between 1973 and 1994.[§]

In particular, the focus on the artist's intentions is fundamental to face the series of problems that have arisen following the digital migration, such as, in single-channel works, the presence of several variants or versions and the gaps in the audiovisual text, and, in complex forms such as video performances, video installations and video sculptures, the absence of some components, as well as the need to reconstruct the "concrete" spatial elements that accompany the audiovisual text and that have often been lost. Added to this, in all cases, is the need to identify re-enactment methods that take into account the technological obsolescence of the devices (monitors, video recorders and media) without betraying the aesthetics and the concepts the artist is talking about. The ultimate conservative purpose is to make possible renewed experiences of these works (through the elaboration of resubmission protocols) and no longer just an archival access to them, with the implicit goal of reconstructing the vast production and exhibition history of which it is impossible here, in short, to explain even the main hubs.

8.2 The Interview

Cosetta Saba: Looking at Italian contemporary art's historiographies of the Seventies, we realize that a lot of information on your artistic production – whose principle is conceptual and whose figure is the liquid dimension – are partial or dispersed. Yet there is no lack of documentary sources and of a transnational context in which – in many ways – artistic research explores media technology and uses video. Not only for what pertains to the cultural transmission, but also to the theme of the conservation of technologically based works. This conversation aims to reconstruct that period, focusing on some issues related to the use of technology and intermedia practices that in Italy, for various reasons that cannot be addressed here, have often been marginal in the context of theoretical-critical reflection. In a context of re-evaluation of the history of relations between arts and technologies in Italy,

‡The Laboratory develops theoretical and applied scientific research in the field of contemporary technological arts with the task of protecting cultural transmission and conservation.
§The works in analog and digital format are preserved by the Gallerie d'Arte Moderna e Contemporanea di Ferrara (GAMC). For more information on the research carried out, see ref. 3.

your artistic research – which proceeds from the film to the single-channel video through video performance and installation – represents an important part in the complex and fruitful panorama of experimentation starting from the Seventies.

Fabrizio Plessi: I agree, the look at the technological arts has often been partial because it is more aimed at investigating other aspects and phenomena of contemporary art. In my opinion, Italy has been excluded from a real international movement. In Germany I did more than sixty major exhibitions and I taught at the Kunsthochschule für Medien in Cologne. There I was recognized.[¶] It cannot be said that I had the same luck in Italy.

And this happened even if Vittorio Fagone[‖] considered that the relation between art and technology was so important in the contemporary Italian scene. In the Seventies, the gaze of the critics was in fact turned essentially to *Arte Povera*, which I myself approached. In a very foolish way it was felt that the technology stood in opposition to *Arte Povera*.

CS: As Omar Calabrese pointed out, probably the intermediary complexity and monumentality of your installations have, so to speak, obscured the fact that *Arte Povera* was a starting point, but then revisited through "the use of the contemporary mediological imaginary".[5] I think, for example, of particular works dedicated to cinema such as *Hollywater* (videotape, 1976) realized in "found footage" and *Casablanca* (video installation, 1984). Apart from a few rare exceptions,[6] in Italy, in academic studies, the analysis of the technological arts has been lacking, and in particular with respect to the uses and experiments on video. As you say, Fagone in the Seventies and Eighties reflected and worked in historical-critical, theoretical and curatorial terms on cinema and artist video. He in fact analyzed/themed/defined video sculpture, video installation, video performance as complex forms. This gives rise to an idea of multimedia and intermediation that involves complex works, with an environmental extension made up of different materials and media that relate to waste and/or interactive reciprocity, often through the presence of film and video devices. This leads to a transformative process that produces meaning by involving spectators/visitors.

Nevertheless, only in the Nineties did the issue of technology come to the fore, given that the spaces of contemporary art – the Biennials, Documenta – were increasingly exponentially invaded by multi-channel video installations, even on a large scale. At that point it was difficult to ignore that video

---

[¶]In the Kunsthochschule für Medien in Cologne, Plessi was professor in Humanization of Technologies (1999–2000) and Electronic Scenography (1994). In Germany since the Seventies, Plessi has exhibited on numerous occasions, including the Folkwang Museum (Essen, 1977–1979), Heidelberger Kunstverein (Heidelberg, 1981, 1983, 1996), Kunstverein Freiburg (Freiburg, 1984), Documenta 8 (Kassel, 1987), Kölnischer Kunstverein (Cologne, 1989), Akademie der Künste (Berlin, 1992), Museum Ludwig (Cologne, 1993), Zentrum für Kunst und Medientechnologie (Karlsruhe, 1993–1997), Martin-Gropius-Bau (Berlin, 2004), Wilhelm-Hack-Museum (Ludwigshafen, 2006).

[‖]To go beyond the numerous contributions for catalogues of video exhibitions and screenings – among which were *Camere Incantate*, Palazzo Reale, Milano, 1980 and *Plessi, Water Video Projects*, Heidelberger Kunstverein, Heidelberg, 1983–1984 – see ref. 4.

– and, with it, the relation between art and technology – had become established in the artistic practices.

FP: Few people saw the potential of video technology right away. Hardly anyone used it in the early Seventies.

Lisa Parolo: Was your goal, even then, to explore the possibilities of video devices?

FP: Yes. Nevertheless, over the years I have not changed a lot in my artistic career, I have not adapted to anything and I have made sure that the technologies were modified to meet my work, not the opposite. By critics I was associated with video art because it was more convenient, because I used the "television". However, in fact, I'm not just a video artist: my research was soon focused on my own language. I did not believe and I still do not believe that videos are something in front of which one sits down to watch what is happening, as it is in a movie theater. Furthermore, I was sure that the new technologies of the time would have become a very important part of our culture, of our life. For this reason, since then, I have maintained a coherent personal course.

CS: As evidenced by the catalog entitled *Fabrizio Plessi. Azioni, interventi, videotapes, films, performances, 1970/76* (1977) and curated by Gillo Dorfles,[7] the use of video technologies is immediately extended to performative actions, to the environment and, in this way, to time and temporality. The theme of the water then emerged together with a conceptual approach to the work of art that, in this first phase, manifests itself through a direct action on the water. From this point of view the *Acquabiografico* videotape is particularly interesting. It is a work connected to your graphic works collected in the book *Plessi Acquabiografico*[12] and evident also in the short Super8 films (*Taglia/acqua*, *Un buco nell'acqua*, *Improvviso/imprevisto*, *Tosa/acqua pratico* (Cut/water, A hole in the water, Sudden/unexpected, Shear/water practice))[8] realized in the form of a sequence-plan in 1973 with a conceptual and performative system. Furthermore, the *Acquabiografico* videotape seems to be related to a sort of score for six "actions" ("how to saw water in two equal parts"; "how to plant a nail in the water"; "how to cut water in four different ways"; "How to break water into many pieces"; "how to compose liquid geometry"; "Do it yourself") published under the title *Fatevelo da soli* (Do it yourself!) in "Gala" 1974 (May, pp. 46–47).

Reflecting on the historical-cultural context of the early Seventies, with reference to the definitional field in which Fagone then placed – in a complex differential relationship – the "artist's cinema" and the "video art",[8,9] I always thought that, in that moment, you were working with the intent to oppose, alternate, contrast and identify different media – drawings, performances, films, videotapes – in your own language. In this sense, I've always thought you did not have a specific medium. You worked on emulsified canvases taken from film sequences, made photo sequences extracted from videotapes, photographed and filmed photographs, *etc.* However, on this basis, the fact remains that – as Lola Bonora states – with *Acquabiografico* and *Travel*, you intended to experiment with your work and express yourself through the videographic medium.

FB: My research is in the middle; certainly the videographic medium has been fundamental for me, starting from the works carried out with the Centro Video Arte, but the rest of my work is certainly not secondary, it served as a tool for the video, but also *vice versa*.

LP: It is evident that, in addition to the historiographical one, there is also another dimension of investigation to which attention must be paid and which, as anticipated, concerns the cultural transmission and preservation of video art and complex artworks.

FP: I have always had very negative experiences on conservation: unfortunately, many of my original video works have been lost, maybe due to the institutions where I've exhibited them. I am always so busy with my research that I have no time to take care of historicizing my work, *i.e.* to piece together a detailed chronology in order to search my old videos and films and to put a point in my artistic activity.

LP: It is not up to the artist to historicize his own work. The art historian's role in this sense is essential for an adequate reconstruction.

CS: Nevertheless, a way to contribute to historicizing one's work cannot but translate, for the artist and his team, into archival work and commitment to conservation.

LP: Although it is evident that in the contemporary panorama the attention to the conservative aspect of the moving image on magnetic or digital tape has become urgent, there is still little awareness of the preservative strategies to be adopted in the long term. To stay with your artistic research, as we have said, the video is undoubtedly the medium that, since the early Seventies, you have explored the most. The extensions and the possibilities of use in the artistic field of the video have been measured by you using it, from time to time, for the realization of mono- and multi-channel videotapes, actions, performances, installations. Going back over your career, the videotape *Acquabiografico* (1973–1974) seems to mark the starting point of your "technological" research. It is also in line with the research – already started in previous years – on the subject of water; finally, as emerged from an interview with Lola Bonora and Carlo Ansaloni, it seems that two versions of this work were shot, perhaps because the first presented a defect in the image, probably caused by a technical problem.[10] Do you remember the genesis of this work?

FP: Yes, I shot the *Acquabiografico* videotape in Ferrara with the Centro Video Arte. The video equipment had recently been purchased** and with Bonora and Ansaloni we decided to make a very simple video; at the time it worked like this. Often the technology used was not professional and the

**According to Ansaloni, the first video equipment purchased by the Centro Video Arte of Ferrara is a Sony® AV 3400 portable recorder – commonly known as a portapack – a Sony® AV 3650 table recorder and two Sony® CVM 110 control monitors.[11]

image quality was low. Despite the limited possibilities of the media available at the time, I always thought that video would be a very important element and that later on it would become a very important element in art. So, even though I was also a designer, I abandoned the more pictorial aspect of my work to dedicate myself to video installations. For me this was, in fact, the future of art. From that moment on, I have worked with video for almost thirty years.

LP: The long collaboration with the center in Ferrara led to the creation of numerous works between the Seventies and the Nineties – for example, the video installations *Arco Liquido* (1981), *Liquid Gravity Center* (1982–1987), *Roma I* (1987) and *Roma II* (1988, Figure 8.1), *Materia Prima* (Raw Material, 1989). What was the role of the Centro Video Arte in actually carrying out the work? In the period of your collaboration it seems that the technical realization of the video and video installation was entrusted mainly to Ansaloni. What was Ansaloni's role in the process of creating the work?

FP: Ansaloni took care of everything, but in particular he was an operator and followed the post-production. Giovanni Grandi, on the other hand, devoted himself to setting up the technical problems. In short, we had a balance, just like the fifteenth century. Ansaloni oversaw the audiovisual sector – naturally under my direction – and Grandi, on the other hand, as I said, took care of the technology necessary for the installations. The monitors have always had operational problems, they have never been equal to each other in terms of color and often a specialized technician was needed to

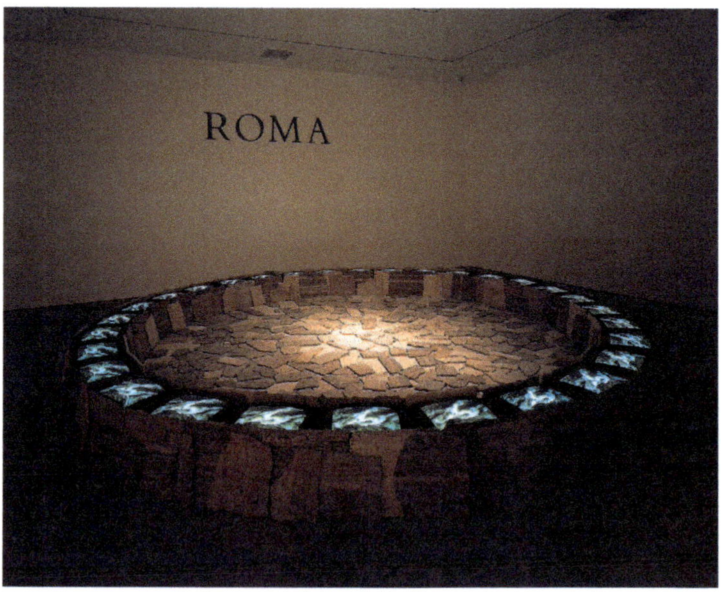

Figure 8.1 *Roma II* (1988), video installation, Prato, Museo Luigi Pecci – Travertine marble slabs and thirty CRT monitors. Plessi Private Archive. Reproduced with permission.

make them work properly, especially in video installations. In short, Grandi was brilliant. As for my role, as always, everything came from a project in a drawing. The Centro Video Arte was, in fact, a performer of my projects.

CS: There was, therefore, a transition from drawing to video, and your book *Acquabiografico*[12] testifies to this. It is a particular autobiography – a collection of projects with a conceptual matrix that investigate the water – which bears the same name as the videotape filmed in collaboration with the Centro Video Arte. In this collection the performative dimension is very present.[12]

Concerning the performative actions, the period of your collaboration with the musician and performer Christina Kubisch is particularly interesting. From this collaboration, during the Seventies, were born projects such as *Liquid Piece* (1975), *Performance for Alto Flute, Voice, Violoncello and Electric Vibrator* (1977) and the performance installations *Two and Two* (1976–1977), *Liquid Time* (1978) and *Tam-tam* (1979–1980). The video seems to be the catalyst for research that combines performance, concert and installation at the same time.

FP: Yes, ours was a research on the relation between music and visual studies, in which we integrated a sonorous aspect – Kubisch is an experimental musician – and a visual one. The videos and performances that were created during this project were very particular and very contemporary.

LP: Concert/video/performance installations made with Kubisch show how "video" is intended as a structure articulated in space, with a sculptural sense. This concrete dimension of the video then became essential in your installation and sometimes a starting point to which other things are added (*Bronx*, 1985, Figure 8.2). In this new format the image is no longer transmitted live, but is often pre-recorded and conveyed by cathode ray tube (CRT)

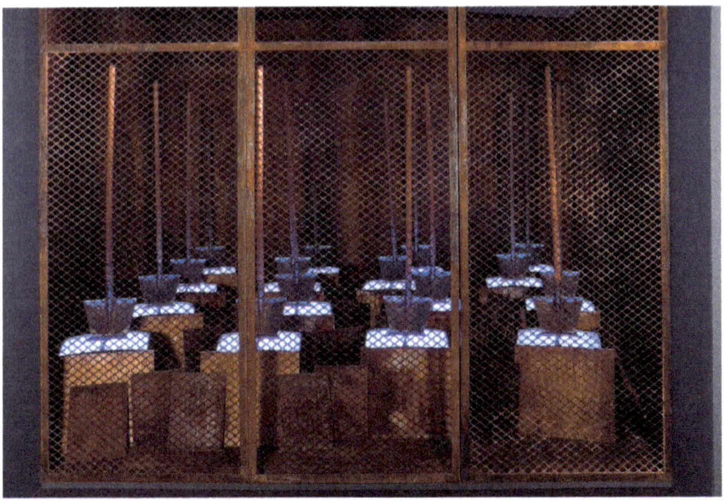

Figure 8.2 *Bronx* (1985), video installation, Milan, Rotonda della Besana – Travertine marble slabs, wind machine and 100 CRT monitors. Plessi Private Archive. Reproduced with permission.

monitors organized in a precise way inside the exhibition space. Can you define your conception of the electronic medium?

FP: My artistic position was very precise already in the Seventies: Plessi, an artist who deals with new technologies. I have always seen it as a means to face my work, my language. My language has not changed, it is made by video, but also film, photographs and drawing. I believe that video is a material, a medium like any other and so it is for video sculptures, video installations and video performance. Depending on the "message" I wish to give to the spectator I use any medium I need. In all cases it all starts from a drawing in detail of what I want to show and then giving this project to the team working for me. In this sense, my creative process can be compared to the architect's one. Video installation where a natural development of the researches I carried out during the Seventies. In order to express my language, I had to consider the sculptural essence of the audiovisual image.

LP: Your artistic journey has been based on experimentation, on heterogeneous exploration of artistic means and languages. To witness this trend there is undoubtedly the already mentioned book *Acquabiografico*.[12] The publication preceded the creation of the homonym videotape and among the numerous "aquabiographic" projects there are multiple directions of investigation based on the research on performance, installation and the moving image.

FP: The book came first: sponges, taps, everything I did is collected there. Afterwards, I adapted it from time to time. But I've never done "film" because to me – I must say – the narration is not interesting. I believe that the "video" becomes very boring when it is telling a story. Well, I don't want stories to be told. My intent is to give emotions that have no narrative development. In the last forty years I have made five hundred projects/works on "water" and they are all different, all mobile, all fluid. Someone asks me: "But how can you invent new things every time?" I answer that it is my life, I am in it. My exhibition *Plessi's Splash*,[13] which is now in Paris, is on a stone that falls into the water. There is nothing more. I make the whole show around this falling stone.[13]

CS: For you, therefore, the meaning of "film" concerns the presence of a narrative component. The difference between "film" and "video" does not include the type of media – film or magnetic tape – but the presence or absence of narration.[††]

FP: Exactly. For this reason, the works *Travel* (1974) and *Back Water* (1984), in which there is a narrative dimension, are indicated in the opening titles as "film", although they are shot entirely on analog videotape.[‡‡]

[††]We refer to the artwork descriptions in ref. 3. In the opening titles of the video the latter is indicated as a "film by Fabrizio Plessi". The same happens in *Back Water*.

[‡‡]In *Back Water* (U-Matic, color, 1984) Plessi focuses on the journey into the past, building an analogy between the flow of water and that of memory. In the work some fragments of *Travel* – the second videotape created by the artist with the Centro Video Arte – are reused (½" open reel, b/n, 1974).

LP: About *Back Water* it is interesting to note that the copies conserved at the Galleria d'Arte Moderna e Contemporanea of Ferrara present substantial color differences. Are these differences desired or are they caused by the transcription of the original work on several supports (from U-Matic to U-Matic, but also from U-Matic to VHS)? Between one and the other copy is it possible or does it make sense for you to define which, for a conservative question, is the "original"?

FP: As far as I am concerned I can say that it is indifferent, because slight chromatic differences do not affect the meaning of the work. One copy has warmer colors, another a bit colder. I do not see this difference as decisive for my work.

LP: We have already addressed some crucial questions concerning your relationship with the electronic medium. There would be a preliminary question to face: you are known for your work with video and installations, but there is also a moment in your career, that of the late Sixties, little known and less historicized than your subsequent research. These are more pictorial works, but also linked to sculptures and installations that did not yet include the use of monitors. For example, in 1967, in a display at the Galleria del Cavallino, the word-sculptures were arranged in space as if they were sinking.[14]

FP: The work of that period was very specific, focused on non-sense. At the end of the Sixties I had already started to think of moving from a two-dimensional to a three-dimensional language. After all, my work has always been linked to clichés and in that period my style was at the border between conceptual and pop. I was representing, by sculpting or painting, different words (*i.e.* salire – get in; affondare – sinking) colored with very bright shades but tautologically the letters composing the words were represented like they were doing the same action – to get in, to sink – described. In this sense my work was conceptual.[15]

The same was in the Seventies with different *media*. In my action *Camminare sull'acqua* (1975),[§§] I walked on water. It was not an arbitrary, individualistic, crazy gesture. It was to make it clear that you can make gestures without having a result. We are used to the fact that every gesture of ours has a response or a result. If we don't see the result, it seems like the gesture is useless. This is the meaning. I wanted to do the opposite, do things to make people think, to activate their perception. Then, when I walk on water, I don't really walk on water. Because, if I walked on water, I would be God. I focus on trying to walk on water and not succeeding. If I cut the water, it is the same. Like I did with the performance *Operation Antwerpen* (1975), cutting the Schelde river into two equal parts. On that occasion I also proposed a questionnaire addressed to the population, in which they were asked to express a preference between the "first" or "after" river cut. Everyone said they preferred "before", because people don't understand. Too bourgeois, tied to habits, to take on such a

---

[§§]The action is filmed in 16 mm and then migrated on magnetic support (the two productions are by Deutsch-Französische Jugendwerk). See also ref. 7.

crazy gesture. Instead it made one think, because the result was not visible: the water had returned as before, but I had really cut it.

LP: So the role of the spectators, their reception of the works and their reaction to them, as often happens with the conceptual turn of art between the Sixties and Seventies, is fundamental for the success of the works.

FP: Of course. When people saw – for example, in *Operation Antwerpen* (1975) – that water had returned as before, perhaps they thought it was as if the gesture had never been done. We are used to seeing a result from an action. If we do not see the result, it is as if the action had not been done. But all these works – watering the Seine, planting nails in the Seine – were not the result of mad exaggeration. It was a work on perception. So the meaning is deeper. With water I did everything in a thousand ways, but the goal was always to try to give real meaning to our actions. I am also thinking of *Schiaccia Acqua* (Crush the water, 1972):[16] I put a sponge in a vice and crushed it, compressed it. It wasn't madness, nor was there any sense. My research of that period is often read as if there had been a Dada sense, a playful sense. Instead it was a reflection on perception: all my work, even now, is about perception. If I now think of a stone falling into water, it is just a stone falling into the water. People say: "But how? It's just a stone." But no, it's everything else. Either the water changes color or becomes another moving element and then it changes. There is always something profound, not ironic even if this latter is the easiest, most superficial reading one can have. All my work has always been linked to the relationship between man and nature.

CS: To remain in the Seventies, while not ignoring that, as Gillo Dorfles said, "Cutting, stretching, shearing, stroking, watering [you] objectify the metaphor and give it a force of impact that it would otherwise not have",[16] I ask: why has the aquatic element, compared to others, assumed such central value in your work?

FP: I always say: "I am a solitary navigator." I travel into this sea of art, which is always a storm, because the sea of art always has gigantic waves to face. But I am tied to my helm and I know exactly where to go. Therefore, all that is apparently a game, an irony, a little unrealistic thing, instead it has, underneath, a conceptual depth. I see water not as a naturalistic element; I really see it as a metaphor of our living. That's why I've gone fifty years to manipulate water and I never get tired. Because, after all, I never see it in an illustrative, narrative, anecdotal sense, but I always see it as a metaphor for us. So water is the fluidity of thought, of the mind, of the soul. It's a much deeper thing.

CS: Water is a playful, paradoxical, critical and generative dimension. Water – as Fagone claims – "is not the sea, but the infinity of the mind".[17]

LP: Of course, water is the *fil rouge* that runs through your research over the decades. With a retrospective look, what can you say about the other already mentioned work, *Travel* (1974), the second produced in collaboration with the Centro Video Arte? In this case the theme is very different from the aquatic one and yet this work is almost always mentioned in your biographies and, sometimes it is given more importance than *Acquabiografico*.

FP: Reviewing my old videos, I think the depth of thought is evident. In *Travel*, but also in *Up Down* (1982), a work of almost ten years later, seeing a female back and the small movements of the muscles means bringing the viewer's attention and his gaze to details that often are not considered. In my videotape *Travel* the woman takes an absurd journey in her bathroom, as if she were searching for herself. The intent of the work is not to narrate an impossible journey, but to tell an experience. When [Eleonora Mialich], with the light bulb – a "light shower" – illuminates her body, in fact, she is in search of her own essence. Reviewing my works of that time, even if apparently they are distant things, I always find the same fluidity, the same mobility. On the other hand, we always have a kind of liquidity of the soul. Goethe said that the soul is liquid, it is fluid. Living in Venice, I always say that I have become tolerant in everything, because it fits me like communicating vessels. I adapt to situations, to thoughts. I, a square Emilian, coming to Venice saw this kind of liquid, fluid light that became part of my life. I am infinitely more tolerant than I used to be. And I can also change a lot of arguments and defenses. Why? Because basically I have nothing solid to defend. The fluidity of thought is what leads us to support even the opposite of what we had argued before. And this means freedom. On the other hand, I have always loved the sea and, on the contrary, deeply hate the mountains, because they prevent me from seeing the horizon, what is beyond. And this is an impediment that I can't stand, which makes me feel imprisoned.

CS: But your museum, the Plessi Museum, is in the mountains. In this regard, I would like to investigate the reasons that led you to create a museum there.

FP: Of course. The Brenner highway had bought a large installation of mine, *Vertical Sea*, a thirty-meter giant sculpture I had made for the Hanover Expo in 2000. A large habitable sculpture – there are sixty to seventy people inside – which is an iron mountain, in which there are waterfalls, rivers, lakes, all in digital. The Brenner highway has decided to put it at the border, where the road begins. Then, what was supposed to be a small museum became bigger and bigger: a very beautiful building. Inside there are a lot of works of mine and many other sculptures and large drawings. Also, there is a very special library that I built myself, a library without books. It is made of iron, twelve meters high and inside there are embedded trees. Since the tree is the beginning of the paper, I went back to the origin, I took back what is basically the archetype of the paper.

LP: Now, about the works present at the Plessi Museum, you have spoken about "sculptures". In other cases, instead, you had mentioned the fact that "video installation" is a term that was born with your works, specifically related to them. Is there a difference in the way you use the two terms? How would you define your work?

FP: In fact, when I talk about sculptures, I refer to works that have the video inside them. When I started working, the monitor was a horrible object to see. It weighed a lot and was almost a meter wide: in short, a bulky box. My job was to make this physicality of the video disappear. So, I had to "carve it" – this is the right word – with sculptures that had "the horrible artifact" inside. Today, since it is also possible to attach LCD monitors to the wall as if they were paintings, there is no problem. But the philosophy of the work does not change. It changes why I did these things. The cathode ray tube has always been my enemy. As it still is for me the light. LCDs have solved many problems, but the fact of having to obscure a room in order to be able to see the video images is an incredible limitation. Today it is infinitely easier to do art or video art, if we want to call it in this way. It's all simple, all easy. You get everything.

LP: So technological progress has not taken away from, but has added to your work, giving new and wider aesthetic and "immersive" possibilities. Your intention is therefore to hide the medium, particularly in video sculptures, rather than show it. As if to say: technology, artifice is not everything, we also need ideas and nature.

FP: Yes, when I taught at the Kunsthochschule für Medien in Cologne for ten years I understood something essential: the more sophisticated technology is, the more creativity diminishes. When we had few things available, there was a great inventiveness to investigate the medium. As all these technologies have arrived, the artist has become paralyzed and has no longer created as in the Seventies. If I have to answer the question "When was the happiest time for the video?" I think of the Seventies. From then, as technology has evolved, creative thinking has devolved. In fact, the first five or six years I taught were extraordinary. Then less and less, and today when I go to a video art exhibition I get bored to death. It seems to witness a kind of revenge of the art: as if those who cannot tell through the "film" use the "video" because it is easier, simpler, cheaper. On the giant screen, as happens so often at the Biennials, I see banalities. Put on large dimensions, with perfect brightness, with perfect shots, with high definition, but without the substance inside. This is the big problem. I left the Kunsthochschule für Medien when I realized that I couldn't teach anything anymore. In fact, the students did not want to learn a method, that is, that technology is at the service of thought and not *vice versa*.

C.S.: Is it the question of the "medium as a message" that continues to show itself up from the time of Marshall McLuhan?

FP: McLuhan made a horrible disaster by stating that the medium is the message: some of the biggest nonsense imaginable. The medium is a medium. If I speak into the microphone, the latter is the medium but what is expressed is my thought. But if I have to think that the microphone is instead the final part of the work, then whatever I say is fine. This is a very important thing today. That's why I held the *Humanization of Technologies* course. Because humanizing technologies also means having a historical awareness of the past, not denying anything of non-technological art and not becoming slaves of technology. Most of my students were technology fundamentalists. If there was no video, they didn't go to see the show. It's like one who says: "Do you draw with a pencil or do you draw with the computer?" It's the same for me. There is no difference. Everything is in what you say. They seem to be elementary discourses, but they are the basis, today, of our civil and cultural life. Why, now, at a time when we are witnessing the take-off of these incredible technologies, are there no artists working with technologies in a sensible way? Today there should be thousands of extraordinary artists. But if I compare the current period with what happened in the Seventies, early Eighties, I realize that, now, there is no artist like Nam June Paik. Even Bill Viola did an extraordinary job, I did it, Gary Hill and Wolf Vostell did it. I do not find artists of this caliber dealing with video now. Or maybe they are there, but they do another job. It is no longer the thought of art that is conjugated with technology. Art, nature and technology are very important to me. There is no difference and there is no narrative.

CS: You refer to narrativity in the contemporary. Indeed, from the Nineties onwards, there has been a narrative turning point in art as well as a sort of confluence of the filmic image-movement and the video image-movement, which have generated a sort of "*lingua franca*" used in multiple ways. This is an aspect. Another aspect concerns the use of technologies, and in particular IT technologies, in an illustrative way. That is to say: it is shown how the technological device works in relation to its technique, without any other profound speech.

FP: Yes, this too. I have always been on the crest of the wave, always the same, for forty years. Because I do my job related to my language, I have a very individual way of seeing things. I don't care at all about these *avant-garde* drafts that come and go, but I always keep technology under control. And I never let myself be drunk on technology, even if today I could do it very well, because, if I wanted three hundred televisions, I could use them. Once it was hard to find two of them. That's why, for a moment, I've also become theatrical, even a little over-monumental – then critics point to this ... I use technology like medicine, sparingly. I want to work on my thinking and sometimes the use of technologies in such a discreet way that it becomes difficult to notice.

LP: Of course, but important technological revolutions of video, such as the transition from black and white to colors and from low to high definition, as well as the increased possibilities of post-production have contributed in making the image and sound much more "malleable" compared to the technical possibilities allowed in the Seventies and this, as you said, was fundamental for your journey.

FP: In part. Being an artist is like being on an island, outside the circuits; technology can change but the sense that is given to works must always be consistent. It is not enough to think that using video is equivalent to making art. Sometimes I use the image in black and white, a bit dirty, a little *délabré*, just because I like, after all, the soul of "television" and I'm not at all fascinated by high definition. Everything has been brought back to see better, but no one has thought about what was seen, this is the big problem. I am writing a book called *The Life of Plessi with Low Definition* precisely because my life is low definition. If we looked at this table through an extraordinary television, in high definition, we would see horrible things! We would see the dirt of life that emerges. So I prefer not to be in this situation and see the low-definition image.

CS: With regard to digital technologies, there are two other issues to which I would like to draw attention. The first concerns, once again, conservation: digital technology becomes a game of force, a means of transcription and preservation. Migrating from analog to digital means protecting, but only if it is possible to rely on ethical protocols and preserve the aesthetic and therefore historical dimension of the works. The second concerns the digital means of re-mediation with respect to which hyper-vision is also sought and applied to what is already recorded in another way, with low definition or with another definition. Consequently, we can find ourselves in front of a John Ford film that presents a hyperrealist image surface.

FP: Exactly, it is an exasperation. It's like in painting: in any museum we go to, we see that some great works have been restored. But how does the restoration take place? Peeling off everything on the surface? Doing this, what happens? The artist made the painting red, green, yellow, blue and finished it. When it was finished, he sometimes put six, seven hands of sandarac, just to make it more homogeneous. Now, if we remove all these layers, we do not go to the origin of the picture, but to the pre-origin, that is where the artist absolutely did not want to go. In fact the colors are out of tune now. They all look like Guttuso's paintings. Here, let's apply this to the video: it's the same thing. In my opinion, technology is an instrument of thought, not the other way around. I have always thought that technology should help me live better, understand things better. But also to change a way of seeing. In Venice there is a horrible saying that is "do not make waves". My work, on the contrary, has always been making waves, moving things, altering the static sense of our time. My goal has always been to take people and tell them: "Attention! Look at this. This thing can be seen like this or, if you climb on the table, like that. You can see things from different perspectives." So let us get used to the elasticity of thought, to moving it. That's why Venice is important to me: because this mobility of water has meant that I am a fluid man, just like thought. Apart from the mobile phone, the television device is the most beautiful invention of our time. For this reason I am very sorry to see two or three generations that have really lost themselves. Because I expected these young generations to surprise me and instead I see them as old people. You have to have a very cold relationship with this medium, otherwise you fall into the net and become a living soap opera. You fall into the commonplace of thought. My works – I always say – are like magnesium flashes that illuminate the dark and secret areas of our perception.

LP: Going back to preservation, who now takes care of the video material that is produced for your digital works? I think for example of some of your latest works, such as *Under Water* (2016),[18,19] *Navigare necesse est* (2017),[18,20] *Memoria dell'acqua* (2017).[18,21]

FP: My assistants take care of them, but actually I've never planned a concrete preservation plan. I just keep everything hoping not to lose anything, as happened with analog. In my studio here in Venice, instead, I keep all the catalogs, the press reviews and photos, all drawing and graphic projects of my exhibitions, artworks and my experiences with the theater. I don't have an inventory of all the things I'm storing but I know that I need one and it is my will to remedy this in the near future. I also would need to go in search of film and video material that I've lost and maybe I would need an archivist to create order.

LP: Yes, also because drawings and graphic projects are actually fundamental when speaking about preservation. From this kind of documentation it is easier to understand how some video installations were designed and could be exhibited again. In fact, since the Seventies, this dimension is central, if not prevailing over the actual realization of the work. During the Eighties, when your research moved away from the "concept" (which, however, as we have said, remains the basis of all your works) to focus on video installation, the projects seem perhaps even more important elements, and sometimes they are exhibited together with the "realized" video installation (think of the *Water Video Projects*, 1983).[22] Are the projects themselves artworks?

FP: Yes, they are artworks in all respects. For the making of the videos I send my drawings to my assistants and they make them exactly the same. I know what I want, then I know how to arrange them and this also concerns the sound part.

LP: What is your relation with restoration instead? Are there any works that have been restored?

FP: This is the big problem of today. We are assisting with the changing of technologies, so do television formats. Then we have two possibilities: either we modify the structure based on today's technology or, as is more logical, we keep the structure and change the technology inside. But there is also another question: once the "television" was female, it was soft, it was curved; now it is masculine, orthogonal, straight and much colder. I work on the water flow or on the fire, on the use of continuous images and I liked the first aspect of the object more. Today, now that the TV is flat, the image becomes infinitely colder and the work more anonymous, less emotional. For this reason I still have so many vintage cathode ray tubes televisions.

CS: For the conservation of your work *Tempo liquido* (*Liquid Time*, 1989), located at the Luigi Pecci Museum of Contemporary Art in Prato, was there a problem inherent to technology? The 21-inch Hantarex™ monitors – encapsulated in a wheel of a mechanical-electronic mill that runs on its axis (Figure 8.3A–D) – are now out of production.

FP: The simplest and most immediate conservative action to put in place could result in acquiring other monitors that have the appropriate measures and replacing them; since today it is possible to use various types of screens of all sizes, this process is facilitated: the monitors, which previously had a thickness of 45 centimeters, are now 3 cm thick. It would be enough to put them in place of the old broken monitors.

LP: So you believe that, as a conservative strategy, the replacement of a cathode ray tube with LCD monitors could work, because inside the structure in any case you would see nothing but the screen?

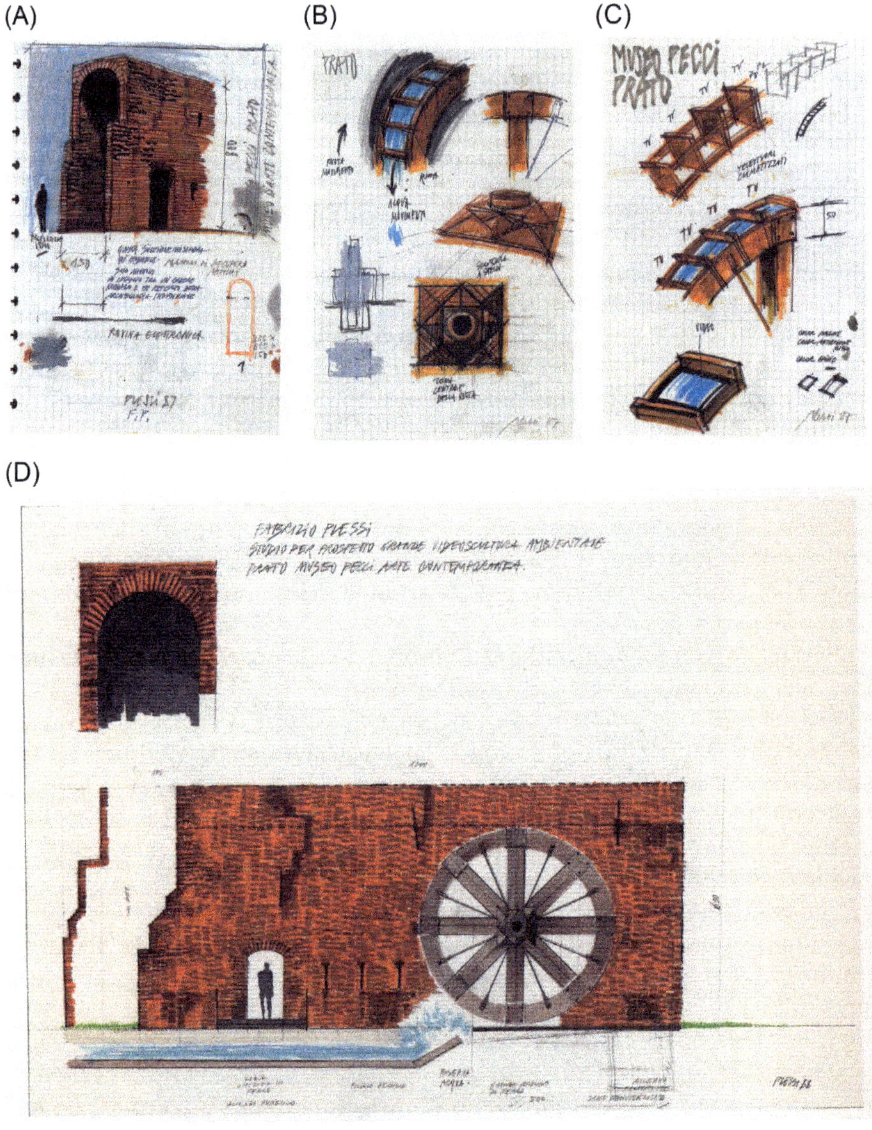

Figure 8.3 *Tempo liquido* (1989), four drawings (A–D) for the video installation, Prato, Museo Luigi Pecci. Plessi Private Archive. Reproduced with permission.

FP: In Germany, at the Zentrum für Kunst und Medientechnologie (ZKM) in Karlsruhe, for the second version of *Tempo liquido* (1993, Figure 8.4) they did it, and that's fine.[23,24]

CS: You therefore believe that the replacement of the "screens" does not change the distinctive features of the work, because the work does not necessarily consist in the material that concretizes it or in the devices through

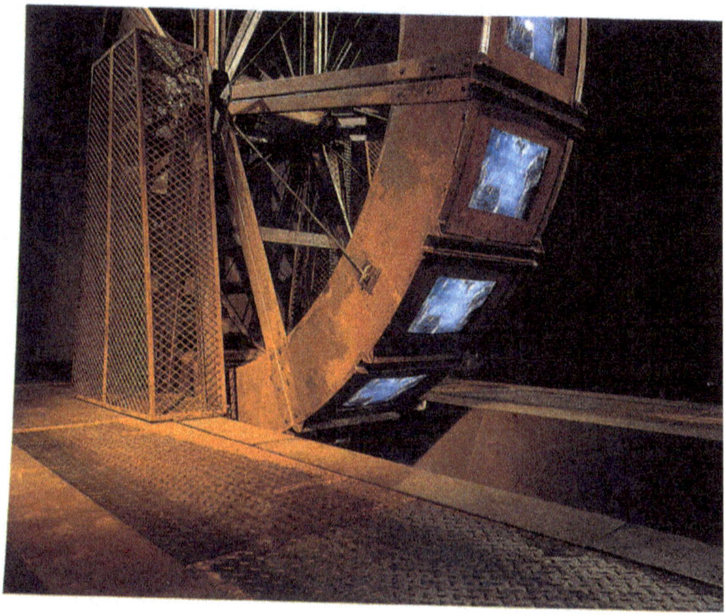

Figure 8.4 *Tempo liquido* (1993), video installation, ZKM – Industrial iron grid structure, hydraulic pump, moving electrical part, water, twenty-one CRT monitors and sound. Plessi Private Archive. Reproduced with permission.

which it is manifested and composed. Nevertheless, changes in the material composition could change the historical meaning of the work and alter the artistic intent.[25]

This is why it is important to know what a work is made of, how it is composed and which elements work in it. For the restoration and re-presentation of a complex work, do you consider it important to proceed with the preservation of the artistic intent when defining the criteria and procedures for replacing, integrating and replicating its compositional components?

Did you share the guidelines and conservative protocol of the *Tempo liquido* (1989 and 1995) and did you take part in the decision-making process? If so, according to what procedures?

FP: I have never been asked what was my point of view in relation to the installation *Tempo liquido*, neither by the ZKM nor the Pecci. I would always like to be involved when an institution is considering the restoration of one of my works, but it depends also on the institution. In the case of ZKM they have substituted the native CRT with new and very similar CRT, and I agree with what they have done. But ZKM is a particular museum which is very acknowledged for media art preservation. With Pecci it is different; when the first CRT monitors broke, they didn't do anything and they put the installation in their storage room. They say it was because they weren't able to find working CRT monitors but in my opinion it is always a question of money and, if they wanted, they could also have used LCDs. It always depends on the will and the economic situation of the museum.

LP: To conclude, remaining on the strategies that you consider most appropriate for the conservation, it seems useful to bring attention to a particularly complex case study, namely the already-mentioned video installation, *Liquid Gravity Center* (1982–1987). Exposed on several occasions, it has taken on many forms, some even radically different from each other.[¶¶] The extreme variability of the installation and some gaps related both to the sources and to the materials have opened numerous questions. Given the "open" nature of the work, the problem of identifying the object to be preserved emerged. Are there one or more versions and variants of *Liquid Gravity Center* that you consider to be the most significant or representative? Moreover, in the absence of a part that seems to contribute in an essential manner to the construction of the meaning of the work, what strategy do you think a museum should adopt in its hypothesis of the same?

FP: If I had to choose which of the different installations could be more representative and would deserve to be re-installed, I'd go for the version realized in 1987 during *Immagina*, a television program of the Italian National Television (RAI). It is the last step of the evolution of *Liquid Gravity Center*. If one or more of the videos that are part of the installation were lacking, I would have no problem in re-making it using today's technologies. In conclusion, I would like to underline that the auditory component is also essential for the interpretation of the artwork and thus it deserves as much attention as the visual part. Sound permits me to create a more immersive space because amplifiers are often put in different places from monitors and the sound is often adding something to the image. For what concerns the relation between the abstract and the concrete dimension, I agree with what Calabrese writes; I've always been at the border between the two, as I was with conceptual and pop. My drawings represent an abstract idea which then develops into a concrete form of existence. And then, again, as in a loop, this concrete part has to do with the abstract concept I want to express, and abstract is also the sound that often gives the final meaning to the work.

8.3 Final Remarks

Looking at the work of Fabrizio Plessi as a whole, it becomes evident that water – in its real and apparent mobility of reflections – is transmuted in a vibrating and fluid way into the electronic image. The latter itself becomes liquid in the digital dimension. In this sense, "video" assumes, like "water", a conceptual scope. For example, referring to *Plessi's Splash* exhibition in

[¶¶]At first the artwork consisted of a single video transmitted by a monitor embedded in the floor and surrounded by earth and stones (Sala Polivalente, Palazzo dei Diamanti, Ferrara, 1982), then by two different videos – in fact, the first one is *Up Down* (1982) – and a growing number of monitors set up in static or moving three-dimensional structures. (Musée d'art moderne di Villeneuve d'Ascq, Lille, 1984, then Rotonda della Besana, Milano, 1985, and then a TV program of the Italian National Television *Immagina*, 1987).

Paris, the artist has repeatedly spoken about the fall of a stone in the water that becomes an event. Although in a different way, even in *Liquid Gravity Center* it is the sound of a stone falling into the water that suddenly moves the element. As Fagone claims, referring to *Liquid Gravity Center* (1985), "in a large dark room, the sound, the concentric opening of the water on a monitor placed horizontally and the presence of some stones arranged all around [damage] clear 'illusion' of a stone that falls into the water";[26] this produces an unusual relationship between the "concrete" and the "abstract".

Significantly Calabrese (1990) has seen in each of Plessi's installations a both concrete and abstract "perpetual motion". The first invests the materials, operates the present "things" placed in the space. The second concerns "the perennial video recording circuit, which repeats the same micro-infinity endlessly (the shovel that digs the water, [...], the pages of the book that are moved by the air, the wind on the water, *etc.*)".[5] Thus, for Calabrese, each installation twice presents a perpetual motion, constituted "through concrete and abstract interactions" and, it is added, starting from a project of which it is a manifestation – as happens, differently, in the versions of *Tempo liquido* (1989–1993) and *Bombay-Bombay* (1993–1997, Figures 8.5 and 8.6), for example.

As Plessi points out, he seeks "to put together electronics and nature", inducing "a metamorphosis of these kinds of celebrated machines of movement;[27] he is in search of a different perception in which true and false often

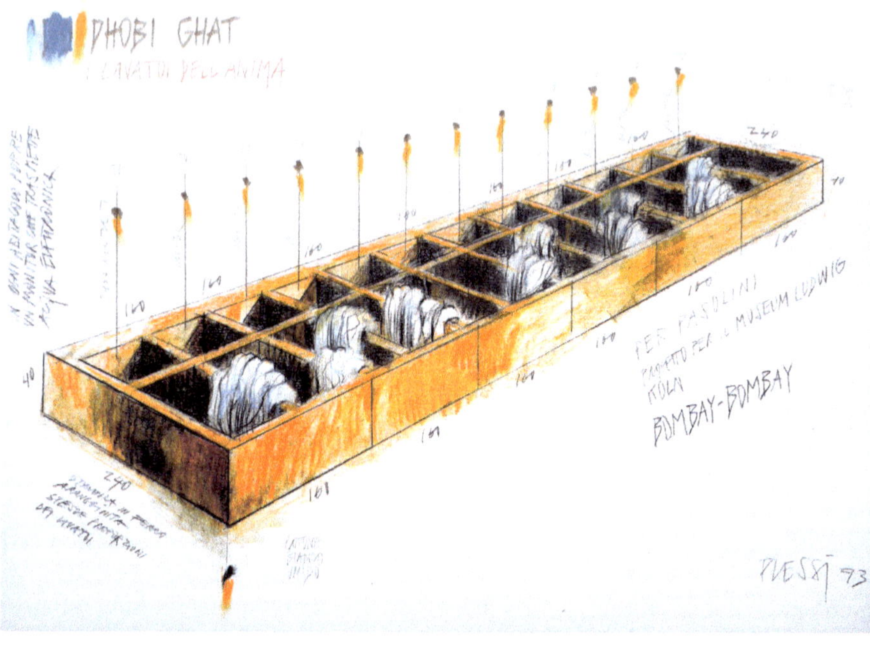

Figure 8.5 *Bombay-Bombay* (1993), drawing for the video installation, Private Collection. Reproduced with permission.

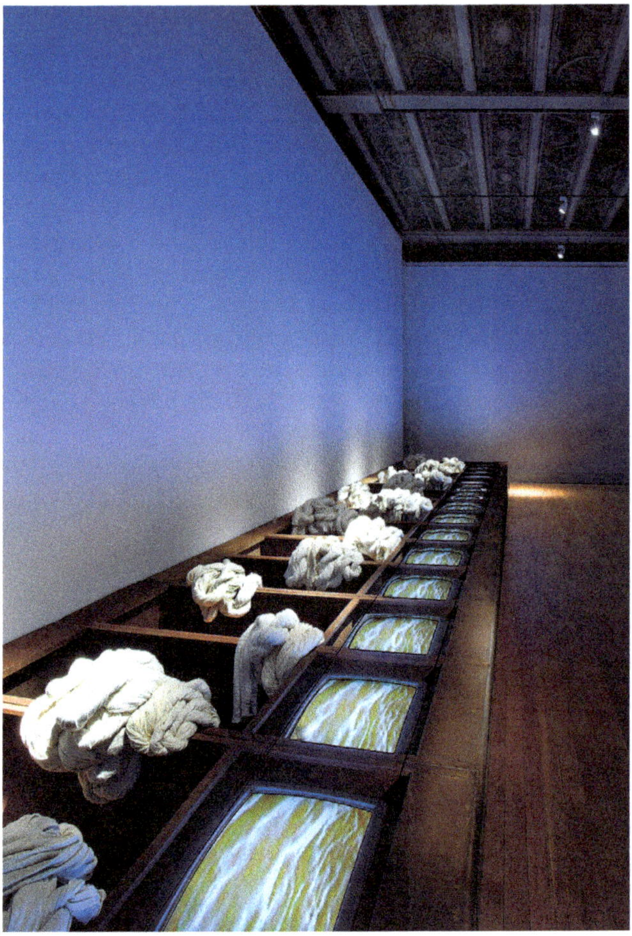

Figure 8.6 *Bombay-Bombay* (1993), video installation, Cologne, Museum Ludwig - Modular structure in iron, mounds of damp white cotton, monitor, sound. Plessi Private Archive. Reproduced with permission.

unite and often the false is truer than truth".[28] Furthermore, in Plessi's work there is an attention to the "places" where the work dialogues with space. The work relates to the place, but it is the place that gives strength to the exhibition project.

Plessi has always thought of himself not as a "video-artist", but as an "artist who also works with video", because he believes that what is crucial is what the artist says and not the media used. The technology must be used, not suffered, and this implies the reinvention of its codes of operation in a context, that of contemporary art, where the artistic specificities "shatter, swallow, are manipulated".[28] With regard to conservation, Plessi thinks that the preservation and restoration of technological art in his work can be dealt with on a case-by-case basis without fetishizing materials, because "these exist because the artist uses them". In this sense, he argues: "If I have to redo one of my works, the material I use does not necessarily have to be that of the first work: it is the artistic creation that I make around the assembly that matters."[28]

The generative process of the work is the plan, it concerns the project, whose realization is delegated to others. The conceptual dimension implies in its own way an intention that concerns the manifestation of the work, implies the principles of cultural transmissibility and, by this way, its preservation on an aesthetic and historical level.

References

1. C. Lo Cicero, *Le videoinstallazioni di Fabrizio Plessi: storia della collaborazione con Centro Video Arte di Ferrara e ipotesi di conservazione*, Masters thesis, University of Udine, 2018.
2. *Plessi*, catalogo della mostra (Reggio Emilia, Civici Musei, 1990), ed. A. Bonito Oliva and O. Calabrese, Reggio Emilia, 1990, p. 13.
3. *Videoarte a Palazzo dei Diamanti, 1973–1979. Reenactment*, catalog of the exhibition (Ferrara, Gallerie d'Arte Moderna e Contemporanea, Palazzo dei Diamanti, 2015), ed. C. G. Saba, L. Parolo and C. Vorrasi, Ferrara, 2015.
4. V. Fagone, *L' immagine video. Arti visuali e nuovi media elettronici*, Feltrinelli, Milano, 1990.
5. O. Calabrese, La tecnologia umanistica di Plessi, in *Plessi*, catalog of the exhibition (Reggio Emilia, Civici Musei, 1990), ed. A. Bonito Oliva and O. Calabrese, Reggio Emilia, 1990, pp. 86–87.
6. S. Bordini, *Videoarte & arte. Tracce per una storia*, Lithos editrice, Roma 1995.
7. G. Dorfles, *Plessi. Azioni, Interventi, Videotapes, Films, Performances, 1970/76*, Mastrogiacomo Editore, Padova, 1977.
8. V. Fagone, *Arte e cinema. Per un catalogo di cinema d'artista in Italia 1965–1977*, Centro Internazionale di Brera, Marsilio, Venezia, 1977, pp. 78–79.
9. V. Fagone, *L'immagine video. Arti visuali e nuovi media elettronici*, Feltrinelli, Milano, 1990, p. 51.
10. C. G. Saba and L. Parolo, Lola Bonora e Carlo Ansaloni in conversazione con Cosetta G. Saba e Lisa Parolo, in *Videoarte a Palazzo dei Diamanti, 1973–1979 Reenactment*, catalog of the exhibition (Ferrara, Gallerie d'Arte Moderna e Contemporanea, Palazzo dei Diamanti, 2015), ed. C. G. Saba, L. Parolo and C. Vorrasi, Ferrara, 2015, pp. 43–57.
11. C. G. Saba and L. Parolo, Lola Bonora e Carlo Ansaloni in conversazione con Cosetta G. Saba e Lisa Parolo, in *Videoarte a Palazzo dei Diamanti, 1973–1979 Reenactment*, catalog of the exhibition (Ferrara, Gallerie d'Arte Moderna e Contemporanea, Palazzo dei Diamanti, 2015), ed. C. G. Saba, L. Parolo and C. Vorrasi, Ferrara, 2015, p. 45.

12. *Plessi. Acquabiografico*, catalog of the exhibition (Milano, Galleria d'arte Vinciana, 1973), ed. P. Restany, Milano, 1973.
13. G. Dyan, *Plessi's Splash*, Opera Gallery Paris, Parigi, 1–14 March 2019, www.operagallery.com.
14. *Fabrizio Plessi*, catalog of the exhibition (Venezia, Galleria del Cavallino), ed. G. Marchiori, Venezia, 1967.
15. *Arte al bivio. Venezia negli anni sessanta*, ed. N. Stringa, Edizioni Canova, Venezia, 2008.
16. G. Dorfles, *Plessi. Azioni, Interventi, Videotapes, Films, Performances, 1970/76*, Mastrogiacomo Editore, Padova, 1977, p. 12.
17. V. Fagone, *L'immagine video. Arti visuali e nuovi media elettronici*, Feltrinelli, Milano, 1990, p. 172.
18. www.fabrizioplessi.net.
19. Under water (2016) was installed in 2016 at Fondaco dei Tedeschi in Venice between October 1, 2016 and April 2, 2017. Fondaco dei Tedeschi, Venezia.
20. The installation Navigare Necesse Est was installed in Palazzo Van Axel/The Pushkin State Museum of Fine Arts, Venice between May 2017 and September 2017. Palazzo Van Axel/The Pushkin State Museum of Fine Arts, Venezia.
21. The installation Memoria dell'acqua was installed at the Beck & Eggeling Gallery in Dusseldorf between February and March 2017. Beck & Eggeling, Dusseldorf.
22. *Water Video Projects*, catalog of the exhibition (Grenoble, Musée de Grenoble; Heidelberg, Heidelberger Kunstverein; Villeneuve D'Ascq, Musée d'art moderne du Nord), ed. V. Fagone, H. Gercke and T. Raspail, Grenoble, 1983.
23. T. Zirlewagen, Fabrizio Plessi's Liquid Time II at ZKM Center for Art and Media: (PART I) The Revages of (Run)time – Maintenance of Electronic Hardware During Long-term Presentation, in *Inside Installations. Theory and Practice in the Care of Artworks*, ed. T. Scholte and G. Wharton, Amsterdam University Press, 2012.
24. F. Y. Tykwer, Fabrizio Plessi's Liquid Time II at ZKM Center for Art and Media: (PART II) The Revages of (Run)time – Maintenance of Electronic Hardware During Long-term Presentation, in *Inside Installations. Theory and Practice in the Care of Artworks*, ed. T. Scholte and G. Wharton, Amsterdam University Press, 2012.
25. A. Rava, Problemi di metodo, in *Conservare l'arte Contemporanea. Problemi, Metodi, Materiali*, ed. O. Chiantore and A. Rava, Electa, Milano, 2005, p. 58.
26. V. Fagone, *L'immagine video. Arti visuali e nuovi media elettronici*, Feltrinelli, Milano, 1990, p. 176.
27. A. Bonito Oliva, Plessi, in *Plessi*, catalog of the exhibition (Reggio Emilia, Civici Musei, 1990), ed. A. Bonito Oliva and O. Calabrese, Reggio Emilia, 1990, p. 205.
28. F. Plessi, L'elettronica e l'arte, in *Arte Contemporanea. Conservazione e Restauro*, ed. E. Di Martino, Umberto Allemandi & Co., Torino, 1996, pp. 203–205.

CHAPTER 9

Arte Concreto Invención: Towards Industrialized Surface Aesthetics

FERNANDO MARTE*[a], PINO MONKES[b], FLORENCIA CASTELLÁ[a] AND MARCOS TASCON[a, c]

[a]Instituto de Investigaciones sobre el Patrimonio Cultural, Universidad Nacional de San Martín, Benito Quinquela Martín 1784, Ciudad Autónoma de Buenos Aires (1296), Argentina; [b]Museo de Arte Moderno de Buenos Aires, Av. San Juan 350, Ciudad Autónoma de Buenos Aires (1147), Argentina; [c]Instituto de Investigación e Ingeniería Ambiental (IIA-CONICET-UNSAM), Av. 25 de mayo y Francia, San Martín, Provincia de Buenos Aires (1650), Argentina
*E-mail: fmarte.iipc@gmail.com

9.1 Introduction

Following the publication of the only issue of the magazine *Arturo* during the summer of 1944, a renewing line was to develop within Argentinian art, characterized by abstract/geometrical forms, known as *Arte Concreto Invención*. The line was intended to address artistic practice, starting from analytical and scientific reflections. In its development of a new ontology for the work of art, from systems of uncompromising objectivity and rationalism so as to avoid all subjective/emotional fact, and in clear opposition to the idea of painting as an opened window to the phenomenal

Science and Art: The Contemporary Painted Surface
Edited by Antonio Sgamellotti, Brunetto Giovanni Brunetti and Costanza Miliani

Published by the Royal Society of Chemistry, www.rsc.org

world, the group produced significant ruptures in the local context, and established, at the same time, contributions of universal significance to this line of work.

The unique issue of *Arturo* had texts and poems by Carmelo Arden Quin, Edgar Bayley, Vicente Huidobro, Gyula Kosice, Rhod Rothfuss, Murilo Mendes and Joaquín Torres García, as well as vignettes and reproductions by Tomás Maldonado, Lidy Prati, Murilo Mendes, Wassily Kandinsky, Piet Mondrian, Viera Da Silva and Joaquín Torres García. The group from Río de la Plata was composed, at the beginning, by Carmelo Arden Quin, Tomás Maldonado, Gyula Kosice, Raúl Lozza, Rhod Rothfuss and Lydi Prati. Then Manuel Espinosa, Alfredo Hlito, Enio Iommi, Martín Blaszko, Claudio Girola, Alberto Molenberg, Juan Melé and Gregorio Vardánega joined the group, with others for whom the term Invention, as opposite to creation, had a capital significance.[1] The rich and prolific original programmatic structure would have later resulted in the subsequent fragmentation: Madi and Perceptism.

After installing a still-pending debate in Buenos Aires, the group directed its efforts towards the *objectification of the concrete ideal*. This new ontology for the work of art starts from the liberation of color and the form of any signification process, expression and symbolism, in order to put in evidence the substantial properties of the material. The figuration, and any process of abstraction based on it, should be left behind to reach a complete autonomy of the visual plane.[2] The work of art would be from then on, although of social transcendence, a daily object. For this, concretism reformulated its work strategy. It resumed the use of rigid supports (at the beginning cardboards and plywood later) instead of the traditional canvas. The purpose behind this was probably of a practical nature, based on the need for more specific surfaces, so that the color will reject any application that affects its planimetric and opaque reading. The idea of the "real materials, real space" slogan from *Art Concret*, led by Théo Van Doesburg, would take them to new supports and to the use of industrial household paints, in some cases as a unique resource and in other cases in mixture with other artistic materials. Such a choice could find an explanation in the search for specific surface finishing, a high degree of opacity, excellent leveling properties of the enamel paints or just faster drying times.[3] The contribution of these new efficient and economic materials was accentuated in the concrete *atelier*.

Among concrete artists, Várdanega and Lozza exhibited a very personal technique, and, through different resources, found an unprecedented topography in the local artistic environment. Such topography had a single antecedent in European concretism, but it was unknown in Río de La Plata in those first years of the group. In particular, Lozza, far from the artistic orthodoxy, employed a fine mixture of oils and synthetic enamel, which he polished between layers with pumice stone and water in order to obtain a uniform finishing surface.[3] Conversely, Arden Quin used an industrial synthetic enamel of a French brand which, according to his declarations, he used as white paint for Madi sculptures from the 1950s onwards, before changing years later to Peugeot's spray paint for cars.[4] This is how new materials came into play in the concrete artists' practice which were more suitable for their demands, either due to their faster drying time, or structural strength, or resistance to polishing or other pressure activities on the surface (Figure 9.1).

According to information revealed by the artists themselves during interviews, unpublished until now, we have been able to confront important information on their execution techniques and aesthetics. The interviews were conducted from 1999 to the beginning of 2001. Artists involved were part of the first vanguard of Río de La Plata, a group that at the threshold

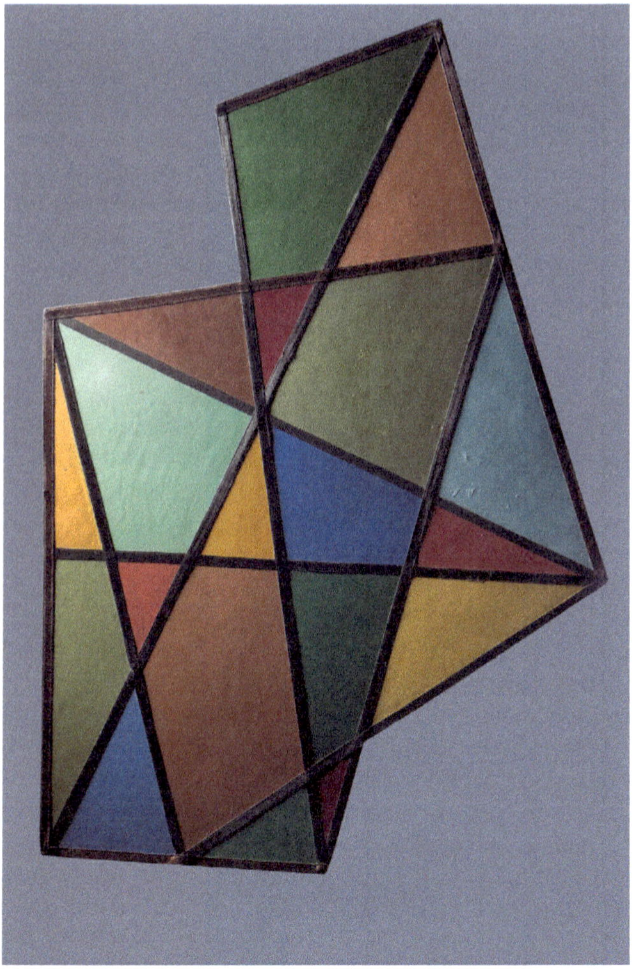

Figure 9.1 Raúl Lozza, "Pintura Nro. 15", 1945. 50 × 37.5 cm. Col. Museo de Arte Moderno de Buenos Aires, Raúl Lozza and Carlos Malvestitti. Reproduced with permission.

of the new millennium still exhibited most of its members working alongside young artists. The themes addressed departed from the inevitable chronology of life and work of each of the artists, their vision of the historical role that the group played, conformation, aesthetic proposal, influences and contributions. Then the interviews delved further into the different conceptual approaches and material crystallization of their thoughts into the artworks; that is to say: materials, execution techniques and aesthetic purposes.

9.1.1 Arte Concreto: Influences and Contributions

A large number of group members made the usual trip from Argentina to Europe. However, unlike their predecessors from the beginning of the century, these artists did it after having consolidated their aesthetic position in their country. This fact significantly distinguishes the Arte Concreto Invención group from the previous generation of Latin American abstract artists who, since the late nineteenth century, responded to theoretical/aesthetic legacies assimilated during years of study in Europe.

As a fundamental background, we remind the reader here of the work of Xul Solar in the 1920s, with his esoteric and astral fantasy. We cite also the contribution from Emilio Pettoruti and his exhibition in 1924, with his works framed in the aesthetics of late cubism. Worthy of consideration is the work of a great pioneer of Latin American abstract art, the Uruguayan Joaquín Torres García, who came from Europe with the intent of emancipating local art from decadent classicist influences. Also relevant was the personality of Juan Del Prete with his sustained affirmation of modernity,[5] who in 1933 made the first exhibition of non-figurative art in Argentina. We further mention the contributions by Antonio Sibellino, creator of the first non-figurative work in the country in 1926, and by Pablo Curatella Manes and Lucio Fontana who, stimulated by their activity in the Florida group, wrote a brilliant page on Argentinian art. On the other hand, the solitary work of *Esteban Lisa*, characterized by a more sensitive abstraction, also deserves a mention, as he knew how to exert a great influence on a good number of young students. All these artists shared a common inspiration from the European *avant-garde*, as well as a great disinterest and incomprehension from the local media and the press.[6]

The members of the Arte Concreto group, once in Europe and with a consolidated position in Argentina, were able to come into direct contact with artworks and artists in Buenos Aires of which they had knowledge only through a few publications, having generally entered the country through European immigration. In their first encounters with these artworks, they felt that those works did not finish taking the great step they had expected, from abstraction to concretism. On the other hand, they could experience a material reality with these artworks which they did not imagine at that time. For some of the theorists from the group, for instance Tomás Maldonado, all forms of abstraction were exhausted, and there was a great need to *concretize* the space.[3] As Maldonado wrote in the manifesto of the *Arte Concreto Invención* magazine, the issue remained unresolved and had not been raised in Russia by Larionov, who reduced the problem to the imitation of light rays (Rayonismo, 1910), while Malevich and Rodchenko noted that, although they work with geometric elements, space and time were still represented on their canvases.[1]

Regarding the term *Concreto*, the first formulations had occurred with the publication in 1930 of the only issue of the magazine *Art Concret*. Van Doesburg wrote a manifesto to establish the theoretical basis of this proposal, describing the way paintings should be constructed and their significance.[7] The words of the Dutch artist set out an aesthetic that encompassed the entire geometric line, but our artists intended to go further towards the absolute objectification of the pictorial event. While in the Netherlands, Mondrian, Vantongerloo and Van Doesburg proclaimed the real abstract as a synonym for concrete, Maldonado perceived the ambiguity of the proposal and worked with his peers to overcome this argument.[1]

In this context, and from what has been exposed by our artists, it was clear that the next step to take was to break the orthogonal characteristic of the plane, to avoid the static and spatial sense that this produces (the painting as a window). In the concrete manifesto, the dialectical problem of abstraction/concretion is clearly posed. For these artists, concrete art does not abstract but invents new realities, so the artwork appears as synonymous of the highest realism: realism of the pure plastic elements and materials in their most direct presentation, as opposite to re-presentation.[1]

As the German theorist Hubertus Gassner said regarding the function of color in Russian Constructivism: "The color of the painting has a double existence; the raw material is at the same time a means of expression. Both aspects go together as the body and the shadow, however, in the course of history of art, the shadow has had more and more transcendence, has fallen on the body and made it invisible".[8] The aforementioned paragraph of Gassner is transcendent in terms of the need for concrete art to free raw material, so it can impose itself explicitly as self-referential and not as a device to elaborate illusory facts.

Among the interviewed artists, Melé and Lozza were those who most emphasized concepts such as concretion, objectification and specificity. The objectification of the concrete ideal would be materialized from the most structural components of the artwork: the support. The latter would have an essential role for the foundation and development of the shaped canvas and the consecutive coplanar. On the other hand, M. Espinosa said: "Regarding the concrete artworks we worked on cardboard, we did not work with other material until it was replaced by wood. (...) It was said in the group, that the practice of established colors, such as oil, should be abandoned (...) But in those days there was a whole change towards can colors, the same ones that are used to paint furniture or objects, industrial synthetic paint."[†] In addition, *A. Molenberg* expressed: "All this was at the beginning, and we agreed that the artwork should be polished, clean, impeccable (...) synthetic enamel on cardboard."[‡]

The proposals developed by the concretism of Río de La Plata would make the material structure that constituted the work of art, support and color indissoluble. The resource of new materials and execution techniques was given from new aesthetic/conceptual speculations, which appeared as inaugural for the pictorial practice of the different vanguards of the twentieth century, especially the idea of functionality for the plane/color (Figure 9.2).

9.1.2 Polygonality: The First Approach

Undoubtedly, the choice of a rigid support is closely linked to the materialization of a rigorous aesthetic/conceptual approach, the modification of the orthogonal format: the shaped canvas and, fundamentally, what was

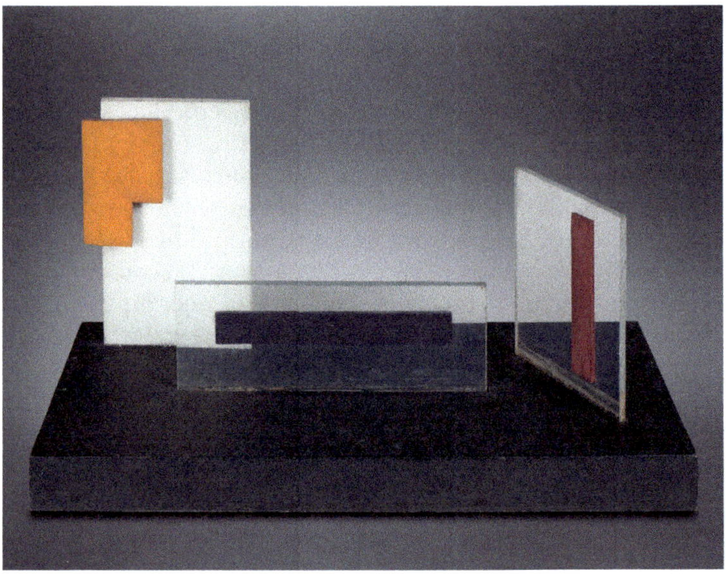

Figure 9.2 Juan Melé, "Colores en el espacio real", 1947. Col. Museo de Artes Plásticas Eduardo Sívori, Juan Melé and Carlos Brasero. Reproduced with permission.

[†]Manuel Espinosa interviewed by Pino Monkes, Buenos Aires, 2000: "En cuanto a las obras concretas se trabajó sobre cartón, no se trabajó con otro material hasta que se reemplazó por madera. (...) decía que se debía abandonar la práctica de colores establecidos como el óleo; (...) Pero en aquellos días se dio todo un cambio por los colores en lata, los mismos que se usan para pintar muebles u objetos, pintura sintética industrial." Translation made by the authors.

[‡]Alberto Molenberg interviewed by Pino Monkes, Buenos Aires, 2001: "Todo esto era en un principio, y conveníamos en que la obra debía ser pulida, limpia, impecable (...) en esmalte sintético sobre cartón." Translation made by the authors.

one of the major contributions from the group, the coplanar function. The latter meaning the separation of the different planes of the artwork by the penetration of space (Figure 9.3). Evidently, such rigid materials made the operability easier and offered greater neutrality from the visual point of view. These qualities were essential for achieving the objectification of painting, the true purpose of the concrete ideal. M. Espinosa and A. Molenberg, who joined the group shortly after its formation, mention in their interviews that they were entering into Arte Concreto Invención as they would enter a scenario with a strong programmatic structure, which had somehow induced them to use industrial paint for coloring earlier works in cardboard and plywood.[1]

Illustrations from *Arte Concreto Invención* magazine, 1946,[1] exhibit a series of pieces that are mostly inscribed within the line of the shaped canvas. However, this format would be left aside by many of the members of the group, particularly Maldonado, Hlito and Melé, to return to the regular format. The precariousness of the cardboard used in the first years, as well as the fact

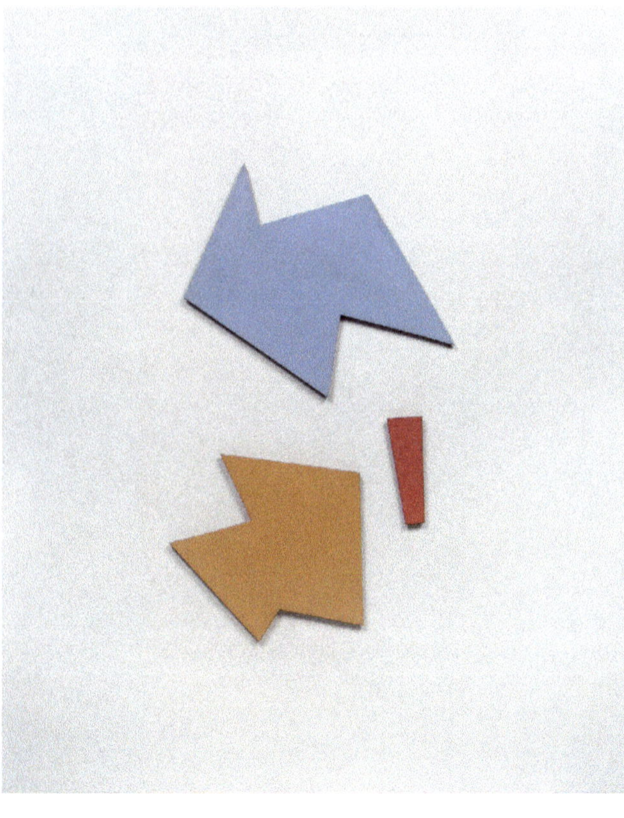

Figure 9.3 Raúl Lozza, "Síntesis Mural Nro.310", 1953. 136 × 110 cm. Col. Museo de Arte Moderno de Buenos Aires, Raúl Lozza and Carlos Malvestitti. Reproduced with permission.

that artists were moving their artworks by public transportation, as declared by Molenberg,[9] would lead to the loss of many pieces: almost all of them. These artworks had to wait for a later reconstruction when the market had requested it. When Melé and Vardánega joined the concrete group the first experiences of shaped canvas and coplanar had already been developed. This may be the reason why their first concrete artworks did not include cardboard as a support. In his interview, Melé does not mention this material, giving the idea that he did not use it in those days. For instance, both paintings by Melé from 1946, *Marco recortado N° 3* (Figure 9.4), belonging to Sívori

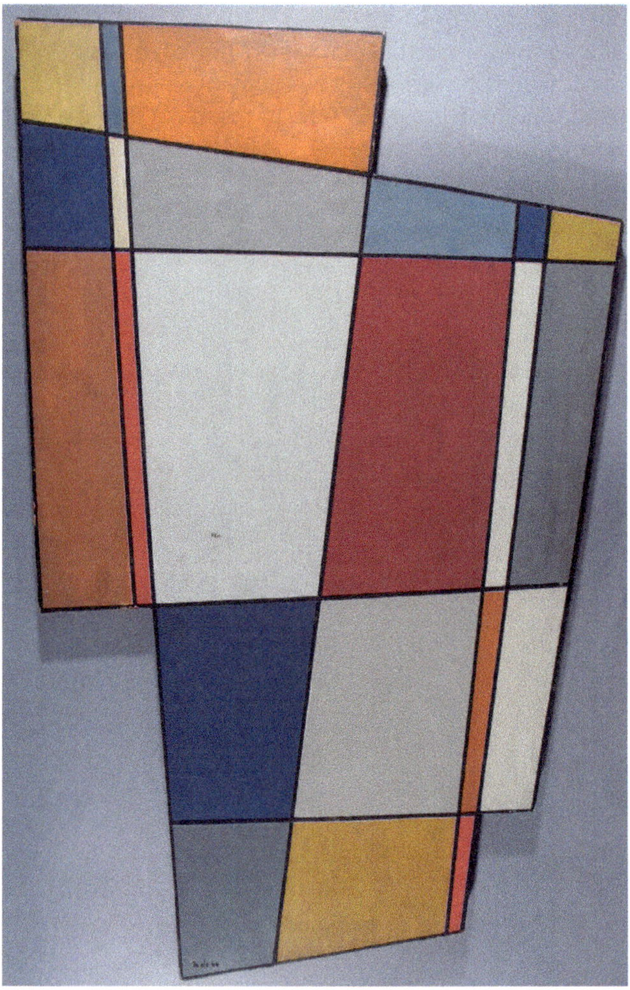

Figure 9.4 Juan Melé, "Marco recortado Nro. 3", 1946. 85 × 55 cm. Col. Museo de Arte Moderno de Buenos Aires, Juan Melé and Carlos Brasero. Reproduced with permission.

Museum and *Marco recortado N° 2* from the Cisneros collection, are made on hardboard. Due to the fragility of the cardboard used until 1945, another more solid material, plywood, would be used. Lozza, for his shaped canvas, as well as for the coplanar modules, would invariably use plywood. According to his interview, he used a hand saw to cut the plywood, and then he covered it with several layers of caulk, sanding in between, in order to hide the woodgrain.[2] The final aesthetics of his works would require this previous treatment.

When A. Molenberg presented *Función Blanca* to the group, the separation of the constitutive planes of the artwork would then be a resource that most of the artists would adopt, although not for a long time. Molenberg comments in his interview that due to the fragility of the cardboard, he had to redo it "many years later" using plywood for it, and made it somewhat larger than the original in cardboard.[9] However, cardboard would continue to be the support of many Carmelo Arden Quin artworks, who prepared it in those days with a recipe by Torres García: ground chalk, animal glue and milk, and protecting the back of the cardboard with naphtha.[4]

9.1.3 Towards Another Aesthetic of the Surface

"We exalt optics" *Maldonado* said on one page of *Arte Concreto Invención* magazine.[1] For these artists, the artwork was understood as a set of purely perceptive properties, posing a color without limitations of nuances or ranges, unlike the neoplasticism that gave symbolic/mystical content to its chromatic alternative.

Lozza, in search of an ideal surface for his aesthetic proposal, adopted an unprecedented technique in the local scene that he used in both his "*muro portátil*" (portable walls) and his planes by polishing the entire piece. However, towards the end of his career, he would replace this procedure for industrial satin paint. Lozza declares in the interviews how exhausting this technique was on the largest portable walls. This was a slow and very careful process, which in the case of minimal imperfection or accident led him to redo the entire layer of color.[2] As a precedent of such procedure appeared the solitary figure of the Belgian artist Georges Vantongerloo, who shares with Lozza a scientific conception of art; both artists worked to link science and aesthetics with the consequent dilemma: rigor/freedom. After his departure from the concrete group, Lozza continued with his investigation, arriving at a mathematical foundation that regulates the form/color/background relationship. This so-called "*cualimetría*" of the flat form was a rather hermetic operation.[11]

Gregorio Vardánega made use of a similar technique to Vantongerloo for two of his artworks from 1950. The design of one of these paintings, *Aries o la línea melódica del radar*, is made with the same technique of incision on the polished surface, which will be retouched in black to give a virtual modulation of the groove. It also shows the use of molds; a possible explanation for this particular resource is that any error by freehand, would force the artist to completely remake the background. The use of a mold would ensure a more controlled process. Vardánega and Melé visited Vantongerloo for the first time in 1948 in his Parisian studio,[12] where they could observe the artworks and learn about this singular technique. When asked for references on Vantongerloo's execution technique, Melé responds: "His surfaces were enameled, like porcelain, he worked loading with a large amount of color applied by spatula; oil for artists; white, and then he sanded it with water. They were very polished surfaces, and then on that surface he traced lines of pure colors."[§] Melé seems not to have observed those grooves, on which the colored lines are arranged, a resource that surely interested Vardánega more.

9.2 Material Characterization

Materials play a specific role on concrete art aesthetics. To understand the absolute relevance of materials in this movement, it is important to focus the attention on how the works of art were created. Were the artists able to obtain the type of paint required for their purposes or did they adapt their artistic technique to the materials available on the market at that time?

Significant contributions to answer this question may arise from direct analytical studies on techniques and materials employed, together with research in industrial archives and interviews with the artists themselves. However, regarding analytical investigations, if one considers the broad availability of new materials just after World War II, classical analytical approaches could be not always adequate and, in some cases, *ad hoc* strategies should be employed.

In particular, new materials adopted by concrete artists were mostly of an organic nature and for the identification of these compounds non-invasive analytical techniques can only lead to the characterization of the compounds' family (proteins, lipids, synthetic resins, *etc.*) without any possibility of achieving specific molecular identifications. This leads to the need for using micro-invasive investigations, such as, for example, those based on gas-chromatographic techniques, which imply the collection of small samples from the painted surfaces.

Having this in mind, one of the most challenging issues in this matter is the sampling, where the smooth and polished surfaces that characterize the

---

§Juan Melé interviewed by Pino Monkes, Buenos Aires, 1999: "Sus superficies eran esmaltadas, como porcelana, él trabajaba cargando con gran cantidad de color aplicado a espátula; óleo para artistas; blanco, y luego lo lijaba con lija al agua. Eran superficies muy pulidas, y luego sobre esa superficie trazaba líneas de colores puros." Translation made by the authors.[10]

works of concrete art limit the possibility of a sample removal that does not leave noticeable traces. Thus, there is the need to reduce samples to the minimum size (a few micrograms) and to increase the efforts in developing a comprehensive analytical procedure that could allow significant conclusions to be achieved.

The materials to be examined on painted surfaces consist generally of binding media, pigments and additives, all related to the industrial development of the historical moment of the artist's life. Following this line and taking into consideration the variety of compounds that could be encountered, a multi-analytical approach is generally indispensable, covering the whole spectrum of organic, inorganic and hybrid phases.[13] On one side, the organic phase consists of the binding medium and other organic compounds, added to tailor viscosity, gloss and drying speed of the paint. On the other side, the inorganic phase is mainly composed of pigments and fillers, mixed to modulate the consistence and color of the pigments. Further, hybrid components can generally include special additives, as occurs for example in the case of metal stearates, which are of common use in modern painting technology. Regarding organic analysis, as already mentioned, chromatographic techniques, such as gas-chromatography coupled with mass spectrometry (GC-MS), have been widely exploited for the identification of binding media and organic additives, as well as for searching for markers that could possibly help to identify brands or provenances. Regarding inorganic components, a combination of vibrational spectroscopic techniques, such as FTIR and/or Raman spectroscopy, and elemental techniques, such as XRF or scanning electron microscopy coupled with an energy dispersive X-ray (SEM-EDX) detector, have been frequently used for the successful characterization of pigments and fillers.

9.2.1 Organic Components: Binding, Additives and Markers

In our case, the organic phase was analyzed by GC-MS. Figure 9.6 shows a summary of the results of the analyses performed on a variable number of minute samples taken off from the examined paintings. These samples include four works by Lozza (*Pintura N.15*, *Pintura N.171*, *Pintura N.117*, *Pintura N. 72*), three by Melé (*Colores en el espacio real*, *Marco recortado*, *Volumen policromatico*), and two by Verdánega (*Aries* and *Relieve*). As shown, in most samples, independently of the examined painting, an alkyd resin or an oil/resin mixture has been found *via* GC-MS, while in a very minor number of samples a pure drying oil has been encountered.

As is known, drying oils are chiefly composed by glycerols esterified with fatty acids forming mono-, di- or triglycerides. The classification of drying, semi-drying and non-drying oils depends on the presence and the number of unsaturations present in the fatty acids, which are responsible for the process of autoxidation and polymerization of the drying film. On the other hand, alkyd resins are constituted by a backbone formed by polyhydric alcohols (generally glycerol) and polybasic acids (generally phthalic acid) which are polyesterified.[14] The presence of fatty acids or fatty oils, as dangling side chains, introduces modifications of the final properties of the synthetic resin. In particular, the drying becomes slower as the fatty acid proportion increases, due to the different drying processes to which the polyester backbone is subjected during solvent evaporation and the autoxidation and polymerization of fatty acid side chains.

The discrimination between alkyd resin and pure drying oil was achieved in our case on the basis of the presence (or not) of phthalic acid, since glycerol and fatty acids are present in both. As already mentioned, the samples characterized by pure drying oil were the absolute minority (3 out of a total of 31). However, it should be noted that the samples classified as "alkyd resin/mixture" should not be considered in absolute as characterized by a specific enamel paint, as they could also be the result of a global sampling including layers of drying oil and alkyd resin within the paint stratigraphy, or even the result of an intentional mixture of the two types of media made by the artist to accelerate the drying process.

Regarding the execution technique by Raúl Lozza, cross sections of his paintings have evidenced the superposition of paint layers and the aforementioned polishing process, applied layer by layer. During the artists' interviews, the intentional mixture of oil/resin paints was mentioned by *Lozza*, when he explained: "From the beginning I painted with Rembrandt colors, but in order to polish the paint layers I had to add a 'strong' varnish, a synthetic one. I mean, I had to mix it to make it resistant. Otherwise, it would pass at least one year until the oil was completely dried".[¶] In practice, the mixture of oil and resin allowed the artist to achieve faster drying, along with the execution of his specific surface polishing. In this context, the term "varnish" could be imprecise, but as it was verified through industrial contemporary magazines, the unpigmented synthetic enamel constituted the so-called varnish.[15]

The results from *Lozza*'s artworks (Figure 9.5) indicate that they are all "alkyd/oil mixtures"; for the works of Melé and Vardánega just 3 of the 14 samples were found to be composed of pure drying oil (*i.e.* phthalic acid was not detected). Again, this finding *per se* cannot be interpreted as an indication of a specific decision of the artist, since a representative number of samples

---

[¶]Raúl Lozza interviewed by Pino Monkes, Buenos Aires, 1999: "Yo desde un principio utilizaba colores Rembrandt pero para poder pulirlo le tenía que agregar un barniz fuerte, sintético, es decir mezclarlo para que tenga resistencia, porque sino hay que dejar pasar por lo menos un año de tiempo para el secado del óleo." Translation made by the authors.

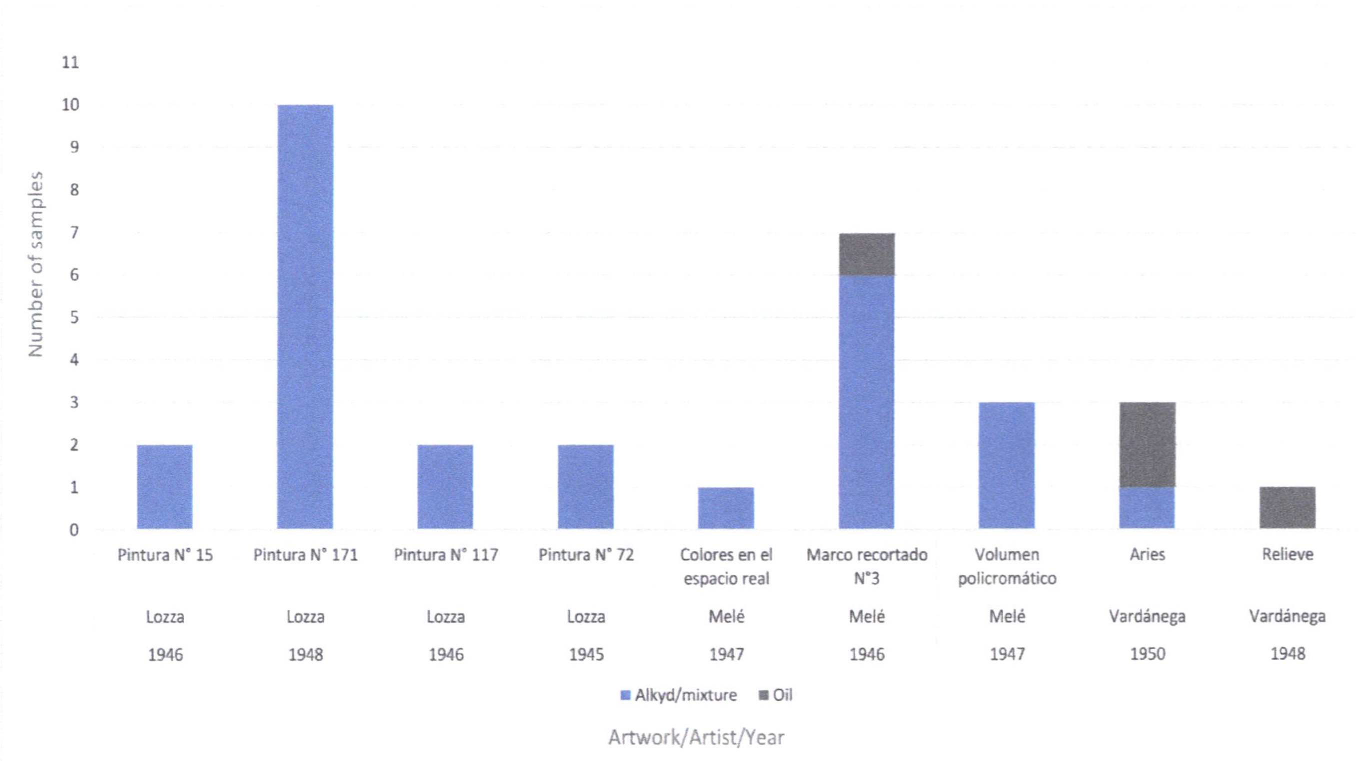

Figure 9.5 Summary of the GC-MS results showing the type of paint employed for a group of artwork authored by Raúl Lozza, Juan Melé and Gregorio Vardánega.

should be analyzed facing, in this manner, the aforementioned complications related to sampling. The challenge that turns out to be essential in order to get to further conclusions, could be the quantification of the three components of concern: polyols, polybasic acids and fatty acids.[14] However, also in this case limitations can be introduced by the reactions needed to derivatize alcohols previous to GC-MS analysis, since multiple derivative products could be formed.

An important issue to be taken into account is the possible presence of organic additives in the paint. Typical examples of additives are metal stearates, normally of aluminum and zinc, which have been frequently included in paint formulations, starting from the 1920s. These compounds were suitable ingredients to avoid phase separations between the pigment and the medium and for optimizing the medium/pigment ratio in the final product.[16] The presence of metal stearates alters the palmitic to stearic acid ratio, calling for specific attention in oil type identification in the case of drying oil paints.[17]

Other organic compounds were used in the manufacture of alkyd paints or enamels, to improve the final paint properties. An example is found in pentaerythritol, a substance that was introduced into the alkyd resins' manufacture, as a substitute of glycerol, during the late 1950s. The use of this polyol, largely employed during World War II as a raw material for the production of explosives,[18] led to significant improvements on the alkyd paint surfaces, such as water resistance, flexibility, better gloss and gloss retention. In the case of the samples analyzed, three of Lozza's paintings (*N° 15*, *N° 117* and *N° 171*) showed the presence of this polyol.

Trimethylolpropane was another compound generally used as a substitute for glycerol during alkyd resins manufacture. This polyol was used for highly branched resins, as its amount and distribution of hydroxyl groups allows for achievement of a particular esterification with the polybasic acid.[14] In our cases, the artwork *Aries o la línea melódica del radar* by Vardánega presented this compound in the paint film.

9.2.2 Inorganic Compounds: Pigments and Fillers

Raman spectroscopy and SEM-EDX were applied for the study of pigments and fillers. On one side, Raman spectroscopy allows for the identification not only of the chemical nature of materials, but also of crystalline structure. This aspect is relevant when considering some technological improvements which appeared in the paint manufacturing industry during the 1940s and 1950s.[13,19] One example of this type is offered by the possibility to distinguish rutile, a crystalline variant of titanium dioxide which was introduced during the 1940s due to its better refractive index and reduced chalking effect in comparison to the other crystalline form of titanium white called anatase, employed since the 1920s.[18] Instead, SEM-EDX allows for an elemental analysis of the samples. These techniques, in particular, contribute to the understanding of the paint chemical composition of the stratigraphic sequence when more than one layer is present. As an example, the identification of zinc and aluminum is of interest, possibly related to the presence of metal stearates. In other cases, the identification of aluminum and silicon could be due to special coatings of the rutile pigment, with the purpose of reducing its chalking effect.

In summary, regarding the white pigments, titanium white and lithopone ($ZnS + BaSO_4$) were found in the studied artworks. Titanium white was present in its anatase form in Vidal paintings, as well in the case of Melé but mixed with some small amounts of Lazurite. On the other hand, works by Vardánega presented a more complex mixture of pigments. For instance, in *Aries o la línea melódica del radar* there is a mix of lithopone and titanium oxide as rutile, while in *Desintegarción cromática* by the same author, two different strata of the painting were analyzed. The superficial stratum showed the presence of titanium oxide, as rutile. However, the underlying layer was mainly composed of zinc white (ZnO) and titanium oxide, but as anatase. This would imply that the artwork was painted by applying different layers with diverse paint compositions, aiming to achieve a particular aesthetical goal.

Finally, it is worth mentioning that several elements found by SEM-EDX did not show the corresponding Raman spectra. In other words, there are elements that do not form a crystalline structure. For instance, aluminum and zinc were detected in this manner in some samples, possibly due to the presence of metal stearates, commonly used to accelerate the drying process of the paint.

9.3 Conclusion

The artistic intentions of the group Arte Concreto Invención were to find a broad spectrum of materials to experiment with. The conjunction of exhausting polishing processes and synthetic enamel combined with drying oils seemed to have found a way of achieving smooth surfaces and the absence of a personal footprint. The rigid supports offered them the proper base to obtain the desired aesthetics: flat finishes and non-orthogonal shaped pieces. The pursuit of particular markers that could lead us to the specific characterization of the paints employed by the artists must be carried on with a combination of analytical techniques, together with archive research for a contextual interpretation.

Acknowledgements

The financial support of the Jean Paul Getty Foundation is gratefully acknowledged. The authors are grateful to Tom Learner and Luiz Sousa for valuable discussions and support, to Joy Mazurek, Michael Schilling and Herant Khanjian for help in the study and data interpretation, and to Alan Phenix for the cross section preparation and analysis. Access to the works of art was possible thanks to the Museo de Arte Latinoamericano de Buenos Aires, Museo de Artes Plásticas Eduardo Sívori and the Museo de Arte Moderno in Buenos Aires.

References

1. Revista *Arte Concreto Invención*, 1, Buenos Aires, 1946, p. 8.
2. Raúl Lozza Interviewed by Pino Monkes, Buenos Aires, 1999.
3. P. Monkes, *Arte Concreto en el Río de La Plata – de los materiales al ideal ontológico Concreto*, Degree thesis, Universidad Nacional de las Artes, 2008.
4. Carmelo Arden Quin Interviewed by Pino Monkes, Buenos Aires, 2000.
5. J. A. García Martinez, *Pintores argentinos del siglo XX*, Centro editor de América Latina, 1981.
6. N. Perazzo, *El arte concreto en la Argentina en la década del 40*, Ediciones de Arte Gaglianone, Buenos Aires, 1983.
7. T. Van Doesburg. *Concrete art*, 1, Bruxelles, 1930.
8. H. Gassner. *Rodchenko: Construcción 1920 o el arte de organizar la vida*, Siglo XXI Editores, México, 1995.
9. Alberto Molenberg interviewed by Pino Monkes, Buenos Aires, 2001.
10. Juan Melé interviewed by Pino Monkes, Buenos Aires, 1999.
11. R. Lozza, *Teoría estructural del color*, Talleres gráficos Córdoba, Buenos Aires, 2004.
12. J. Melé. *Vanguardias del '40. Memorias de un artista concreto*, Ediciones Cinco, Buenos Aires, 1999.
13. F. Castellá, M. Pérez-Estebanez, J. Mazurek, P. Monkes, T. Learner, J. F. Niello, M. Tascon and F. Marte, A multi-analytical approach for the characterization of modern white paints used for Argentine concrete art paintings during 1940–1960, *Talanta*, 2020, 208.
14. M. R. Schilling, J. Keeney and T. Leamer, *Stud. Conserv.*, 2004, **49**, 197.
15. Industria y química: revista de la Asociación Química Argentina, a. 6, n. 6, Noviembre, 1944, pp. 184–188.
16. C. S. Tumosa, A Brief History of Aluminum Stearate as a Component of Paint, *WAAC Newsletter*, 2001, **23**, 10.
17. *Issues in Contemporary Oil Paint*, ed. A. Burnstock, K. J. van den Berg, M. de Keijzer, J. Krueger, T. Learner, A. de Tagle, *et al.*, Springer International Publishing AG, 2014.

18. *Modern Paints Uncovered: Proceedings From the Modern Paints Uncovered Symposium*, ed. T. Learner, P. Smithen, J. Krueger and M. Schilling, Getty Conservation Institute, Los Angeles, 2007.
19. M. Pérez Estebanez, P. Monkes, M. F. Castellá, N. Mastrangelo and F. Marte, Estudio de los pigmentos blancos utilizados en la pintura concreta en Argentina, *18a Jornada de conservación de arte contemporáneo*, 2017, 243–254.

CHAPTER 10

Judith Lauand: Art and Technology of a Brazilian Concrete Painter

LUIZ ANTONIO CRUZ SOUZA*[a], ALESSANDRA ROSADO[a], YACY-ARA FRONER GONÇALVES[a], RITA LAGES RODRIGUES[a], HUMBERTO FARIAS DE CARVALHO[b], MARIA ALICE SANNA CASTELO BRANCO[a], GIULIA GIOVANI[a] AND VÍTOR PAIXÃO AMARAL[a]

[a]Federal University of Minas Gerais, LACICOR – Conservation Science Laboratory, School of Fine Arts – Federal University of Minas Gerais – UFMG, Av. Pres. Antônio Carlos, 6627 – Pampulha, Belo Horizonte, MG, 31270-90, Brazil; [b]Federal University of Rio de Janeiro, Brazil
*E-mail: luiz-souza@ufmg.br

10.1 Introduction

In the period between the end of the Second World War and the beginning of the 1960s, Brazil had a leap in economic and industrial growth, as well as fast urbanization. The development programs carried out in the democratic governments of Gaspar Dutra (1946–1950), Getúlio Vargas (1950–1954) and Juscelino Kubitschek (1956–1961) aimed at the insertion of Brazil in the framework of Western capitalism. In that period São Paulo and Rio de Janeiro quickly became cosmopolitan metropolises. In them, the surplus of private capital was invested in the modernization of urban cultural equipment. In

Science and Art: The Contemporary Painted Surface
Edited by Antonio Sgamellotti, Brunetto Giovanni Brunetti and Costanza Miliani

Published by the Royal Society of Chemistry, www.rsc.org

this regard, the foundation of Museu de Arte de São Paulo (MASP) in 1947, and of Museu de Arte Moderna do Rio de Janeiro (MAM/RJ) and Museu de Arte Moderna de São Paulo (MAM/SP), both in 1948, have been paradigmatic initiatives, together with the organization, in 1951, of the I São Paulo Biennial.

The adoption of government policies to encourage industrial progress involved the substitution of imported goods with those nationally produced. However, by the mid-1950s, the Brazilian industrial park, mainly concentrated in the state of São Paulo, was still incipient compared to those of European hegemonic countries and the United States. Due to this, the process of Brazilian industrialization underwent development during the 1950s, in the postmodern molds of the international system, whose logic in this temporal cut presupposed mass consumption. Alongside this, the Brazilian social structure remained fundamentally archaic, a condition stemming from the conservative and patriarchal cultural legacy of the agrarian export model, aggravated by extreme social inequality.

In the field of plastic arts, the assimilation of constructive theories occurred in this diachronic context, marked by the significant gap between the utopian ideal of modernity, sustained by the new generation of artists of the *concrete* movement and the geopolitical and cultural relations that permeated the process of modernization of the country. A fact that leads, according to the art historian Aleca Le Blanc, to an "ostentatious anachronism: the development of modern visual languages before the large-scale modernization of the region".[1]

In the early 1950s, intellectual adherence and the resulting conceptual reinterpretation of the language of *concrete art* compelled the artists of the *Grupo Ruptura* (São Paulo), under the leadership of Italo-Brazilian Waldemar Cordeiro (1925–1973), to experiment with techniques and materials that met the contents and methods proposed by the Swiss Max Bill (1908–1994). In 1954, the artist Judith Lauand (Pontal/São Paolo, 1922), who had already produced paintings in abstract geometric language, was invited by Cordeiro to participate in the group. Judith Lauand was the only female representative in the Grupo Ruptura, mirroring a society at the beginning of an effective transition between the countryside and the city, where the role of women was severely curtailed by the prevailing patriarchal regime.

Historical sources, found in archives of artists, in interviews or in specialized publications, indicate that the vast majority of technical solutions of concrete artists have been developed collectively, in an empirical manner, that is, on the basis of trial and error. In the artisanal process, the attempt of using new industrialized materials – belonging to industrial activities or civil construction – made it easier to achieve the aesthetic objectivity required in concrete art. Such use of materials contributed to the depersonalized finishing of the works produced, supposedly removing any traces of subjectivity from them, generating a similarity to the visual of industrially manufactured functional products, which concurrently proliferated in the markets of Brazilian cities.

In approaching Brazilian concrete art production in the 1950s through the bias of technical and material solutions, it is possible to say that the articulation between art and modernity was achieved through the incorporation of materials that were fully identified with the industrial world of modern societies and, at the same time, were easily accessible in the regional market. As metaphors, the use of industrialized materials gave the aesthetics of concrete art works, incisively, the tropological sense of modernity. Not only the paints, but also the new types of rigid supports that were derived from different forms of processed wood, which have been widely used in civil construction and furniture industries, especially hardboard panels manufactured in Brazil since 1954 by the companies Eucatex® and Duratex®.

Regarding this perspective, way beyond the reaction to the figurativism or the denial of academicism's manifestations, the new forms of creation of art objects reflected the artistic-ideological attitude of this generation in the search and desire for new forms and means of artistic expression, in harmony with the country's emerging modernization. On the other hand, in creative processes, concrete artists did not exclude the application of previous knowledge of academic codes acquired by previous artistic training or experiences, nor did they abandon classical materials and resources. After all, as will be seen, Judith Lauand painted the work *4 grupos de elementos* in 1959 using egg tempera, an ancestral technique known in all periods of European art referenced, including *Il librodell'arte*, written at the beginning of the fifteenth century by Cennino Cennini.

In contrast, throughout the 1950s, the emphasis of art criticism with regard to the use of new industrialized materials in concrete works was one of disapproval. The article "Os Pintores e a falta de tintas" (The Painters and the Lack of Paints), written by the renowned critic Antônio Bento (1902–1988) in 1953, exemplifies that rejection. In dealing with the effects of the crisis of importing artistic paints, which had been dragging on since the previous year without a satisfactory solution for the category of artists – due to the policy of import substitution – Antônio Bento devoted most of his article to reproving the material experimentalism practiced by concretists. From his point of view, in Brazil, painters zealous about the tradition of their work refused to use national products, except the concrete painters who, besides rejecting oil paints, preferred experimenting with new Brazilian synthetic resins. In his words, "concrete painters replaced the *Ripolin* with *Quentone*[2] and with Ypiranga paints. [...] They use national paints of inferior quality, counting on a smooth and uniform painting similar to that done by spraying automobiles and refrigerators." After emphasizing the inferior quality of the national commercial products, Bento launched the question: "Will these paints last a few years? And this is a question that will only be objectively answered through time." In a scenario marked by the crisis of importing of artistic paints, the art critic concluded that the concrete artists were the only group of Brazilian artists not affected by that.[3]

In retrospect, Antônio Bento's critique of experimentalism, which was spelled out in this antique document, fosters some considerations. The first is about the dualism between the visual and conceptual references that guided the manifestations of concrete art in Brazil with the freedom of material experimentation assumed by artists of that movement and the materials used in these works. Those materials were manufactured by national industries and were considered of inferior quality and therefore unstable, according to the technical parameters shared in the 1950s. The second consideration concerns the misleading technical and material standardization of concrete practices, not only repeated in this article, but also verified in other texts of the art critics of the time.

In addition to criticism of technical experimentation, another relevant issue concerns the rejection of works in a concrete trend by the Brazilian art market in the 1950s. On this subject, Judith Lauand stated that at that time "there was no buyer (of artworks) at all. Once Fiaminghi found me a job to draw prints for men's shirts. I also taught. It was the way to get some money to cover the expenses. No one wanted to know about our work."[4] (Interview of the artist by the curator of the exhibition *Judith Lauand: the 1950s and the Construction of Geometry*, held at the Instituto de Arte Contemporânea (IAC) in São Paulo in 2007).

These are the conceptual questions and contextual conditions that, articulated, permeated both the interdisciplinary research on materiality and the evaluation of the state of preservation of the works of Judith Lauand approximately sixty years after the execution, presented here.

10.2 Methodology

The methodology adopted in the present study followed the principles of Technical Art History and the parameters established and published by the International Network of Conservation of Contemporary Art (INCCA). In addition, publications by the Getty Conservation Institute and recent Brazilian texts about modern and contemporary works of art have also been consulted.[5,6]

Technical Art History is a relatively new methodological approach to understanding artistic production and objects, through a combination of material physico-chemical analysis and traditional art historical methods. In this context, methodological analysis of an art object systematically approaches subjective judgment criteria (discourses uttered by critics, art historians, experts in the attribution of art status to an object, which pass through purely technical judgment) and of judgments that excel by the scientific rigor, desire for objectivity (formal and stylistic classification categories, documentary, historical and physico-chemical analyses). This syllogism, carried out by an interdisciplinary reading of the object of study, allows, therefore, a more in-depth analysis of the work that, although complex, leads to an understanding of the specificities of the form of artistry – considering a whole set of elements such as technique, stylistic rupture, sensibility, society and commerce, among others.

There is a saying in the science of conservation and restoration that to conserve and/or restore a work of art, one must know it. In this direction, INCCA promotes the sharing of information and knowledge among various fields of science and the arts that is necessary for the preservation and conservation of modern and contemporary art. One of its premises is the accomplishment of interviews with the artists, collaborators or contemporaries with the purpose of obtaining information on the technique, projects and conceptual and stylistic intentions of the artist.

Thus, the methodology used for the knowledge of techniques and materials of the works selected for this study was based on eight distinct but complementary and fundamental fronts: organoleptic examinations, verbal description, survey of the state of conservation, scientific imaging documentation, historical and documentary research, interviews with the artists and their contemporaries, as well as simulating the making of the works under study, in the search for, through the use of the identified materials and techniques, aesthetic and material effects as close as possible to those obtained by the artist, allowing for a practical and deepened understanding of the material and aesthetic aspects of the work.

It is important to point out that some of the research activities were carried out *in situ*, at the museums where the objects were stored and exposed, *e.g.* the Museu de Arte Moderna do Rio de Janeiro and at the Pinacoteca do Estado de São Paulo. They are: organoleptic examinations, descriptive analysis, state of conservation assessment, scientific documentation by image, some methodologies of material analysis with EDXRF (Energy Dispersive X-ray Fluorescence), FTIR (Fourier Transform Infrared Spectroscopy), in addition to the removal of micro-samples for further analysis in the laboratory, such as Raman spectroscopy, pigment dispersions and others.

10.3 The Contribution of Women Artists in the Brazilian Concrete Art

As already mentioned, Judith Lauand was the only female artist of Grupo Ruptura, invited by Waldemar Cordeiro in 1954. To understand the importance of studies on female artists, it is fundamental to look at those who, even without calling themselves feminists, acted in the society in a way that would break with male predominance in the visual arts.

Since the 1960s, parallel to the feminist movement, studies related to the presence of female artists in the History of Art have sought to shed light on women's participation in the world of art. The classic text by Linda Nochlin,[7] "Why have there been no great women artists?", published in 1971, highlights the search for insertion of the female presence in the narrative of the History of Art. A few years later, in 1981, in the United States, a museum dedicated to this subject, the National Museum of Women in the Arts was created. In Brazil, studies on this subject arose in academia during the 1980s and 1990s in a timely manner, but they gained breath over time and today there are studies that seek to reflect on the presence of women in Brazilian art, such as those made by Ana Paula Cavalcanti Simioni[8] in 2008.

Before Judith Lauand's presence, other female artists participated notably in *avant-garde* movements in Brazil, such as Anita Malfatti (1889–1964), responsible for one of the first solo exhibitions of modern art in Brazil in 1917 and a member of the Modern Art Week of 1922, alongside Zina Aita (1900–1967). Another striking name is that of Tarsila do Amaral, an essential painter for the understanding of the modern movement in the arts in Brazil. Nevertheless, the mark of these artists was still the figurative painting and, in this sense, Judith Lauand was one of the first female artists to dedicate herself to non-figurative art in Brazil.

Still in Brazil, in the same decade of the 1950s, but this time in Rio de Janeiro, another group of artists of constructive origin gave rise to the Grupo Frente, this one counting on the female appearances of Lygia Clark (1920–1988) and Lygia Pape (1927–2004), who were already exponents on the international scene.

In addition to these, it is relevant to mention Mira Schendel (1919–1988), Maria Leontina (1917–1984) and Mary Vieira (1927–2001 – the latter acting mainly as a sculptor), like other women to be remembered in the field of non-figurative art. Each of these artists developed their own very particular artistic expressions, while sharing concrete art as a conceptual basis.

Judith Lauand was born in inner São Paulo and studied at the School of Fine Arts in the city of Araraquara. On the occasion of her move to São Paulo in 1952, she attended the woodcut engraving course of Lívio Abramoc (1903–1992) and, at the end of 1953, while working at the II São Paulo Biennial as an educator, she contacted concrete artists and also the critic Mario Pedrosa (1900–1981). According to the artist, "here [referring to São Paulo], in contact with Geraldo de Barros and Alexandre Wolner I became concrete."[9] In the wake of her artistic, traditional and academic training, the new environments and concepts discussed and lived by the artist, in addition to the innovations launched by the São Paulo Biennial, allowed her exposure to new abstract and geometric artistic forms, bringing new questions to her artistic production.

In 1954, she held a solo exhibition at Galeria Ambiente in São Paulo, already with concrete works of art. Following this exhibition came the invitation of Waldemar Cordeiro to participate in the Grupo Ruptura. "He liked my painting. He thought it was similar to what they did."[10] When joining the group, she participated in the main concrete art exhibitions of the following years: the *I Exposição Nacional de Arte Concreta* which took place at the MAM/SP in 1956 and at the Ministério da Educação e Cultura in Rio de Janeiro in 1957, the *Konkretekunst* in Zurich in 1960 and in 1977 the *Projeto Construtivo Na Arte Brasileira* at the Pinacoteca.

Judith Lauand, in her interview, reveals the place for women, the restriction on the activities they could carry on, saying that her father had prevented her from traveling to Italy to study: "[...] I attended the free classes at the School of Fine Arts in Araraquara. I was drawing and Helio Morganti came and examined the drawing, he thought it was very good and asked to speak with my father if he could send me to Italy and that he would pay for my studies. I talked to my dad and my dad refused. He said that I could not, I could not study in Italy, that his daughter had to study at home."[11] In modern times, *bourgeois* values intended for women the universe of the private; studying at home meant that the woman would be under the tutelage of the family, guarded from the dangers of life and the world. The public sphere was destined for men.[12]

Women artists such as Judith Lauand, who entered into predominantly male groups, occupying a space as professional artists, contributed to women leaving the private universe and occupying public space and the artistic world.

Nowadays, due to the valorization of the concrete art produced in Latin America and the studies carried out on the presence of women in the History of Art, individual exhibitions have been carried out on the artist: *Judith Lauand: experiências*, at the MAM/SP in 2011, *Judith Lauand: 1950 e construções da geometria*, at the IAC in São Paulo in 2015, and *Judith Lauand: Abstractions of Brazilian Concretism*, held in New York at the Driscoll Babcock Galleries in 2017. However, more detailed research is still lacking from the perspective of studies on the work and biographies of the artist. When we approach the work of the artist reflecting on its materiality, we seek to elucidate one aspect of her art.

10.3.1 Materials and Techniques

The artist, in a testimony given in 1960, seeks to objectify the plasticity problem in the production of her works: "A picture cannot be explained. A picture is seen. Words do not replace the direct view of the formal structure, the relations of colours, of spaces, of plasticity."[13]

The two works analysed – *4 grupos de elementos*, belonging to the collection of the Pinacoteca/SP, and *À base de hexágonos*, from the collection of the MAM/RJ (Figures 10.1 and 10.2) – both from 1959 and of equal size, are examples of Judith's praxis. In these works it is possible to observe, besides the synthetic precision of its forms, the poetic license and the concerns of the artist with the cataloging and preservation of the artworks. The signatures, in both paintings, are preceded by the letters "C" – short for "*concrete*" – and chronological sequential numbering of their productions. Some of the inscriptions are from the artist herself, located in the backside of the panels, painted directly by brush and including handwritten labels. The data contained in these notes refer to the time of production of the work, together with information on the materials and techniques employed.

The paintings have the same type of structural assembly: the hardboard is fitted in front of a wooden frame and, on the backside, which has a corrugated surface, there are two wooden slats fixed in parallel in the central part of the support. These slats, besides fulfilling the structural function of support of the hardboard (avoiding possible warping), additionally allow a distance of the plane of the painting from the wall, creating with this a differentiated

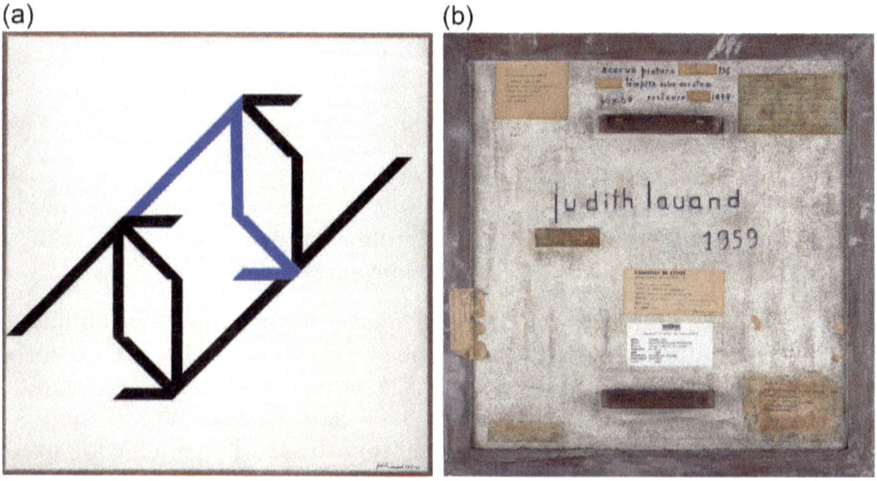

Figure 10.1 Judith Lauand; *4 Grupo de Elementos*, têmpera s/Eucatex®, 1959 (60 × 60 cm). Pinacoteca-SP. Photos front (a) and back (b). Photo courtesy of iLAB-CECOR-UFMG.

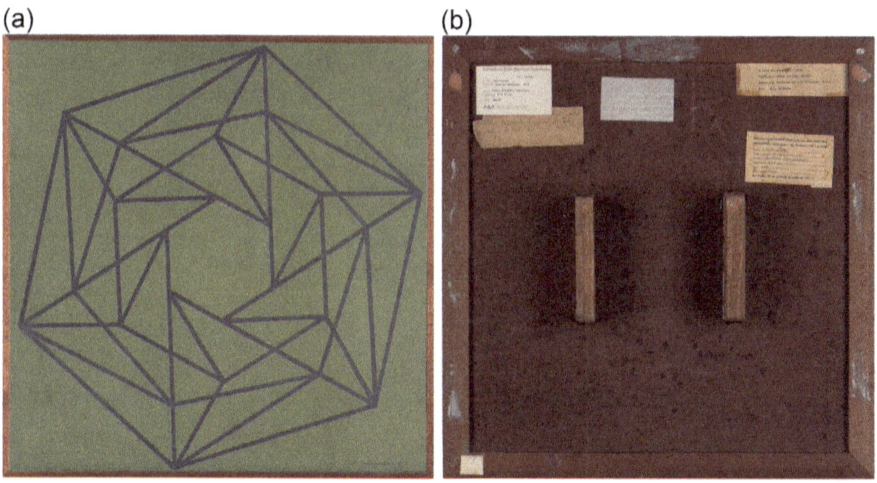

Figure 10.2 Judith Lauand; *À base de hexágonos*, Alquídica s/Eucatex®, 1959 (60 × 60 cm). MAM Rio de Janeiro. Photos front (a) and back (b). Photo courtesy of iLAB-CECOR-UFMG.

form of visualization, contributing significantly to a conceptual change in the appreciation of the work, which would become a mark of concretism.

In relation to the pictorial technique, we highlight the following singularities: in the painting *4 grupos de elementos*, Judith Lauand explores the power of line and color. The work consists of the insertion of black and blue lines on a white square background. As described previously, on the back of the work

(Figure 10.3) the artist records on a label the technique and pigments used in painting and how to clean the surface of the same:

> It is a thin temper: egg yolk emulsion.
> titanium white + leadwhite = s.
> blue = Ultramarine + 3 cobalts (with wax)
> black = ivory black
> clean with bread crumb or ultra-soft rubber and in some cases with a squeeze of alcohol, if this is the case. God protect it.

The results of the physical-chemical analyses performed corroborated with the data indicated above. The ground layer is composed predominantly of zinc white ($ZnO + CaCO_3$). It is emphasized that the white lead referred to by the artist is zinc oxide. The entry word *alvaiade* (lead-white) was adopted by the paint industry in the twentieth century to identify paints composed of zinc oxide. The white background consists of white pigments of titanium oxide (anatase) and zinc oxide, with carbonate and calcium sulfate inerts in an egg tempera medium, according to the results of FTIR and Raman. This paint was applied predominantly in the horizontal direction, with short, fluid, non-homogeneous brushstrokes, leaving the ground layer foreseen and, with the flat brush, as shown in the UV-Visible fluorescence image of Figure 10.4.

The paint layers of the blue and black stripes were applied with a flat tip brush. The edges of these stripes present a slightly thicker surface than the rest of the layer, together with some smearing, which characterizes the use of adhesive tape. It was used to precisely delimit the boundary zones of these areas with the white background, as can be seen in the detail of

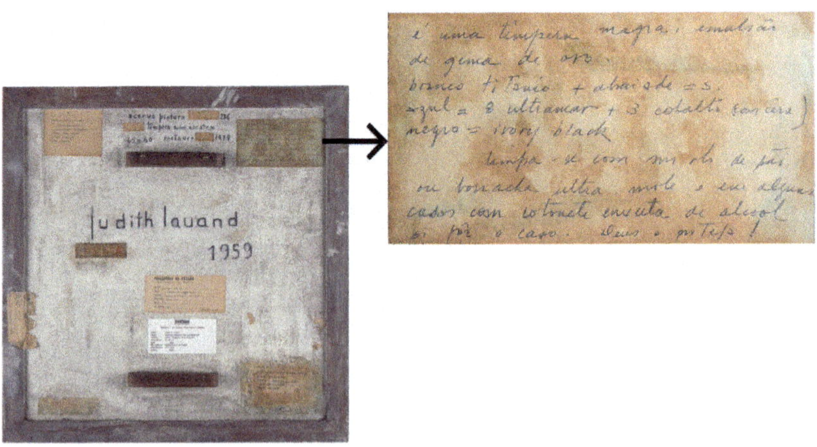

Figure 10.3 Back of the work *4 Grupo de Elementos*, têmpera s/Eucatex®, 1959 (60 × 60 cm). Pinacoteca-SP. Highlighted section is the sticker with the artist's inscription. Photo courtesy of iLAB-CECOR-UFMG.

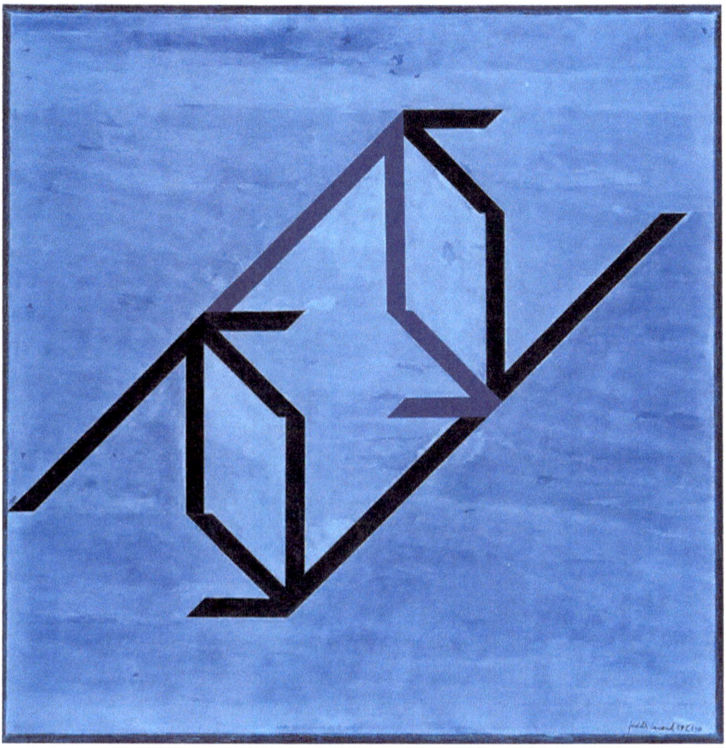

Figure 10.4 Photography Visible UV Fluorescence of the work *4 Grupo de Elementos*. The non-fluorescent white background in purple/bluish color is characteristic of titanium white. There is also the yellowish-green fluorescence of the zinc white that makes up the apparent ground layer in some areas. Photo courtesy of iLAB-CECOR-UFMG.

Figure 10.5. The performance of simulations, reproducing a detail of one of the blue stripes under a white background, using egg tempera, round brushes and adhesive tape (Durex®), produced a similar result to the one verified in the work under study. The adhesive tape fixed on the totally dry quench, when removed, does not cause the detachment and delamination of the paint layer. Holes, present in intersecting areas or at the beginning or end of bands, indicate the use of a compass or pins to mark the width and length of the bands.

In the black and blue bands, respectively, ivory black ($Ca_5(OH)(PO_4)_3$) and cobalt blue ($CoO{\cdot}Al_2O_3$) were identified, according to the recorded FTIR and EDXRF spectra (see Figure 10.6).

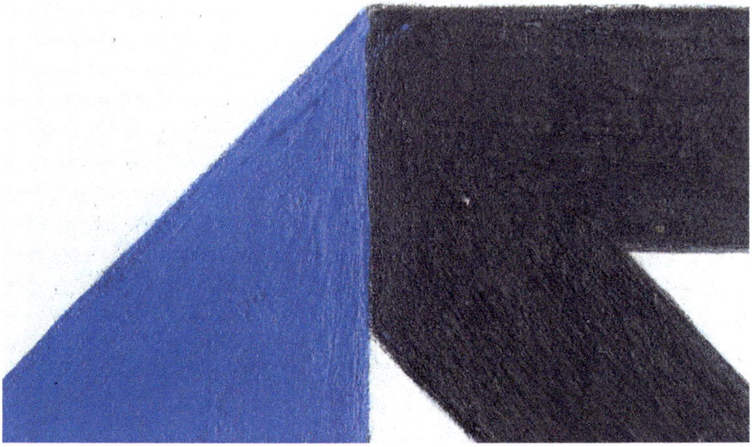

Figure 10.5 Photomacrography with tangential light of the work *4 Grupo de Elementos*. Photo courtesy of iLAB-CECOR-UFMG.

The work *À base de hexágonos* draws attention to the dynamism of the graphic image above all, built only by means of the articulation of black lines, of the same thickness, on a background of neutral color. The impression of movement counterclockwise occurs by the inclination of the hexagons, in relation to the square plane in which they are inserted, and especially by the force of the six more oblique lines that depart from the vertices of the hexagon of the region near the edges of the painting.

The grayish-green background was painted with a paint roller brush. It is a matt granulated alkyd paint consisting of chromium green resulting from the mixture of yellow lead chromate ($PbCrO_4$) and Prussian blue ($Fe_4[Fe(CN)_6]_3$), together with talc, the later added with the purpose of providing the matt finishing. The white ground is a thin layer and was applied on the smooth surface of the hardboard, being composed of white titanium oxide, zinc oxide and calcium carbonate (Figure 10.7).

The detailed verification of the black stripes by optical microscopy and photomacrography allows us to conclude that the black bands would have been painted over the green layer, with the borders delimited by adhesive tape (according to the smearing). The presence of holes near the intersecting areas indicates the use of compasses or pins, marking the edges of the geometric elements that determine the black stripes (Figure 10.8). On the back of the work there is a label, partially covered by another, with the following inscription done by the artist: "[...] base of hexagons, synthetic paint on *Eucatex*, national exhibition of concrete art MAM, Rio 1960". The words "synthetic paint" as well as "enamel" or "synthetic enamel" were used in the paint market in the 1950s and 1960s and are still used nowadays to indicate, among others, alkyd paints.

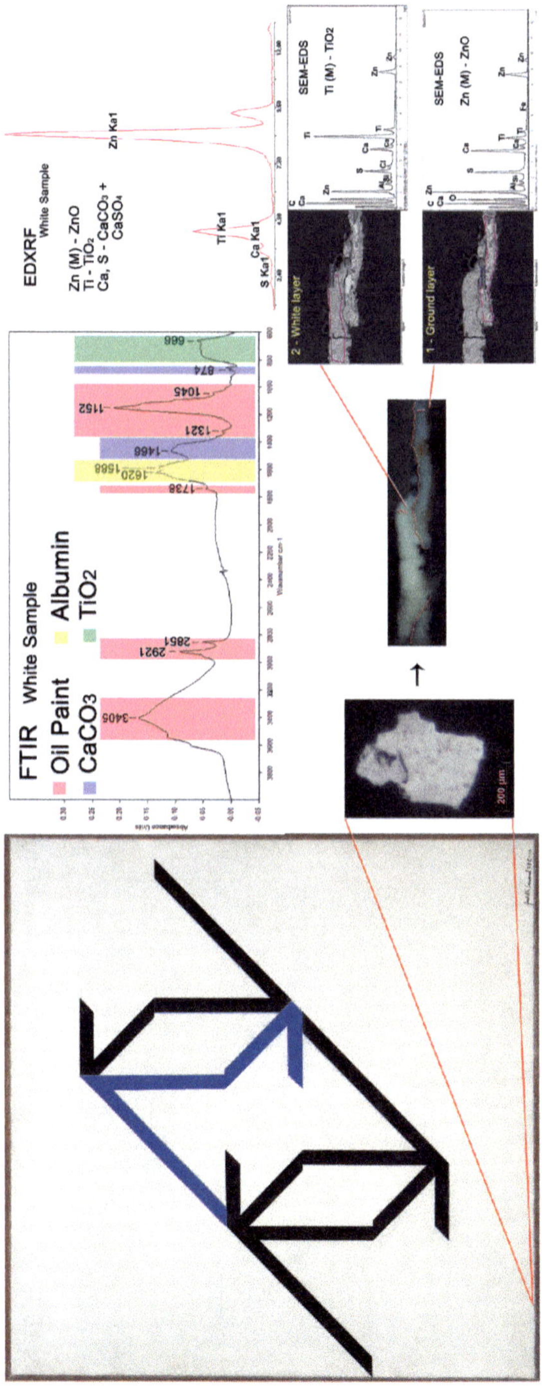

Figure 10.6 *4 Grupos de elementos*, identification of the materials used from FTIR, EDXRF, Stratigraphy and SEM-EDS layers evaluation.

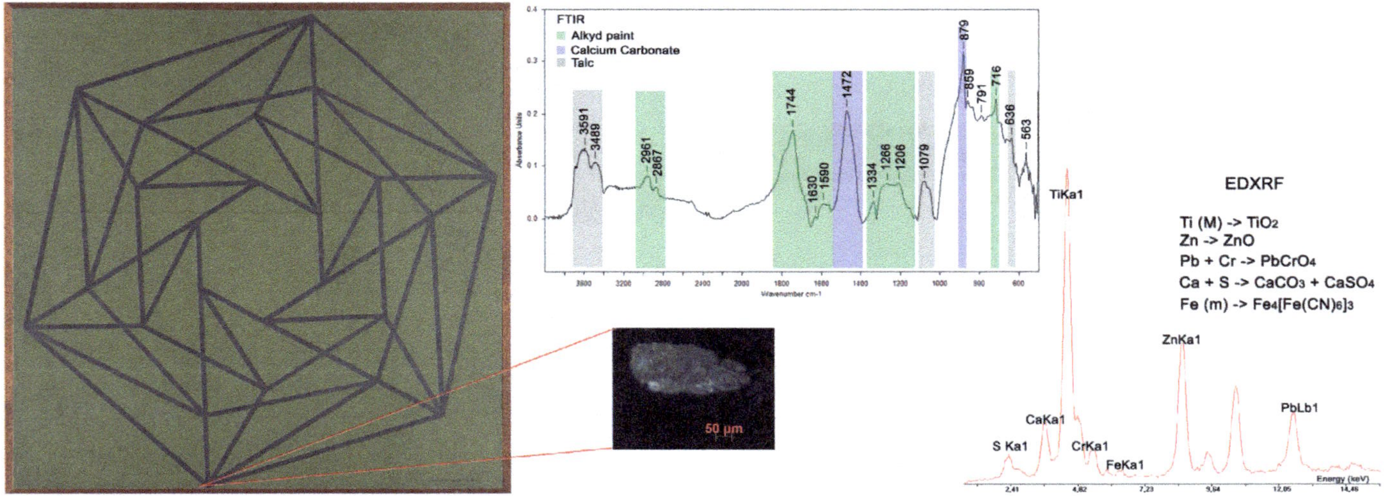

Figure 10.7 *Á base de hexágonos*, identification of the materials used from FTIR and EDXRF analytical techniques.

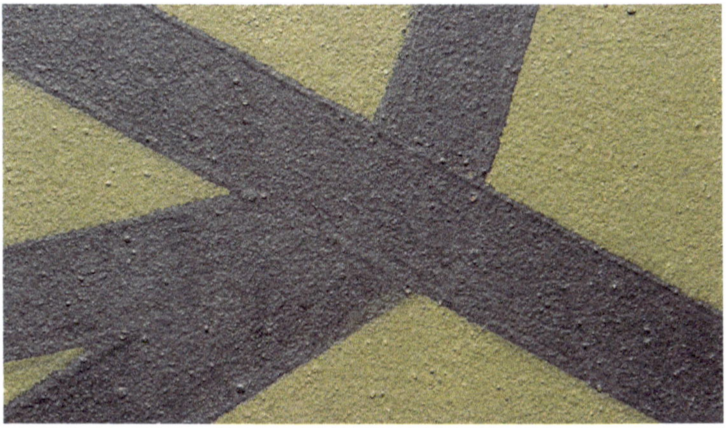

Figure 10.8 Photomacrography with tangential light of the work *Á base de hexágonos*. Photo courtesy of iLAB-CECOR-UFMG.

10.3.2 Aesthetics and State of Conservation

The works of Judith Lauand start from a principle of activation of the movement through arrangement of the line in the monochromatic space background of the painting. The visible geometry is the analytic stereometry constructed by means of coordinates and vectors, the result of a Cartesian perception of the organization of the space from mathematical principles. However, displacements, dissonances and dissent activate the movement from a living "line" that is detached from the surface of the frame.

In the work *4 grupos de elementos*, the influence of the constructive relations of *concretism* is potentialized. The exteriorization of the structural logic of the cube as a geometric solid occurs through the linear fracture, accessible from the perception of timeless displacement of the vertices by the external segments that look like rails. The structure elaborated graphically and synthetically exposes the nature of form, whose mirroring results from its own stereometric reality for the visible exterior. In the artist's own words, form was primordial and preceded the color, and the symbolic economy of the palette was derived from this conceptual choice.[14]

The work *À base de hexágonos*, from 1959, is a geometric construction of rotational symmetry whose efficiency is reinforced by the integration between the square format of the support and the organizational structure of the language elements – lines or straight segments – that the artist sorts in the execution of a graphic design based on hexagons. Moreover, the relation between the rational idea and its construction in a quadrangular space runs through the specification of a rigid support, such as that chosen by the artist. After all, this construction requires the use of instruments specific to the technical drawing projects, in which precision and quality of execution are fundamental for a "good form", whose transposition of the idea into the support, by extension, presupposes a work similar to that of the technical designer who, to better execute the tracing of his object, places the support on a board or a table.

Regarding the state of conservation, the two works are in good condition, with some punctual degradations that do not damage them aesthetically or structurally modify the objects.

The painting *4 grupos de elementos* shows some yellow stains distributed along the white pictorial surface, mainly on the edges of the work. It is possible to observe some damages such as: small cracks, scratches, small areas of losses and dark spots located on the white surface, but these are only observable with a more detailed analysis. The black and blue surfaces do not show visible changes. The back of the work has generalized soiling, traces of white paint on the wooden chassis, traces of paper and labels adhered to the back, which in turn seems to be acidified and stained.

The work *À base de hexágonos* presents subtle dark stains distributed throughout the pictorial surface. There are chromatic retouching and two bulging areas on the top, one on the left side and one on the right side, supposedly caused by nails placed on the back (already removed). The back shows a number of rounded-shaped spots distributed throughout.

In general, the degradation patterns observed do not cause any aesthetic inconvenience or impair the enjoyment of the artworks. In addition, the irregular surface of the objects contributes to the visual perception of the degradations being diluted in the vibration of the background color and the colors of the geometric lines. It is emphasized that the geometric elements and the colors are preserved and the supports, in spite of visible degradations in the back, remain stable.

It should be noted that there are evidences of previous interventions, which can be observed by both UV-Vis fluorescence examination and conservation reports. A thorough consultation of the institutional archives allows us to affirm that restoration interventions were commonly performed by either the artists themselves or professionals close to them.

10.4 Final Considerations

It is interesting to point out that, contrary to what was usually discussed and assumed by the art critics of the time, the well-disseminated idea and concept that the concrete artists should not leave evidence of the manufacturing process on the final object, such as paint brush marks or any other particular signs (meaning the production of a impersonalized artwork) is not effectively what we have found. In principle, the use of industrial materials (hardboard, synthetic materials, spray painting, *etc.*) would allow something like a industrial production line.

The works of Judith Lauand, however, indicate to us that the artist's experimentation, inserted in the concrete language, is beyond this consensus. The artist mixes knowledge of ancient artistic techniques with technical design and the use of industrial materials to transform her creative thinking into matter. In 1954, the artist's freedom to express her declared love to synthesis, precision and exact thought[15] (C.) is clearly visible in her works. This liberty is perceived in her decision to use egg tempera in the *4 Grupos de elementos* and in the use of a granular alkyd matt paint in the *À base de hexágonos*, as well as in the paint delivery method, with the help of the adhesive tapes and the diverse types of paint brushes, as demonstrated.

The voice of the only female artist in the Grupo Ruptura does not appear in the debates or in contemporary articles in the 1950s and 1960s. However, the artist expresses herself through her art, letting her work speak for itself through its visual enjoyment and through subtle details which were incorporated into her works, such as the letter "C" of concrete after the signature and the labels fixed on the back of the works indicating materials and/or techniques. The results of the historical context and physical-chemical analyses indicate that there is a close relationship between the conception of the work and the concerns regarding its conservation through time. There is in this sense the intention of painting to continue expressing its visual force across the ages, as close as possible to the creative moment.

Acknowledgements

The contents of this paper are a component part of the results of the project "Concrete Art in Brazil – The Material of Form: Industrialism and the Latin American Avant-Garde" sponsored by the Getty Foundation, within the Getty Trust Project "Pacific Standard Times – Los Angeles/Latin America – Getty PST LA/LA". The authors are grateful to the partners at the Universidad San Martin in Buenos Aires, and at the Getty Conservation Institute, in Los Angeles. A special thanks to the Pinacoteca do Estado de São Paulo and the Museu de Arte Moderna in Rio de Janeiro, for their technical support, as well as their authorization to perform the field research and access to the paintings.

References

1. A. Le Blanc and P. Gottschaller, *Making Art Concrete: works from Argentina and Brazil in the Colecíón Patricia Phelps de Cisneros*, The Getty Conservation and The Getty Research Institute, Los Angeles, 2017, p. 3.
2. Quentone refers to the original Sherwin Williams brand name Kem-tone®.
3. A. Bento, *Os pintores e a falta de tintas*, Jornal Diário Carioca, 29 November 1953, Ed. 07793, p. 6.
4. C. Fioravante, *Judith Lauand: os anos 50 e a construção da geometria*, IAC, São Paulo, 2015, p. 73.
5. C. Freire. Preservar o Futuro?, in *Arte contemporânea: preservar o quê?*, Ed. Museu de Arte Contemporânea da Universidade de São Paulo, São Paulo 2015, pp. 9–14.

6. O. Chiantore and A. Rava, *Conserving Contemporary Art. Issues, Materials and Research*, Los Angeles Getty Conservation Institute, Los Angeles, 2012, p. 327.
7. L. Nochlin, *Why have there been no great women artists? The Linda Nochlin Reader*, Thames and Hudson, London, 2015, p. 42.
8. A. P. C. Simioni, *Profissão Artista: Pintoras e escultoras acadêmicas brasileiras*, Editora da Universidade de São Paulo, FAPESP, São Paulo, 2008.
9. J. Lauand, *Judith Lauand. Enciclopédia Itaú Cultural*, s/d. https://www.youtube.com/watch?v=ZiIqmHZENWY&t=4m06s.
10. J. Lauand, *Judith Lauand. Enciclopédia Itaú Cultural*, s/d. https://www.youtube.com/watch?v=ZiIqmHZENWY&t=4m25s.
11. J. Lauand, *Judith Lauand. Enciclopédia Itaú Cultural*, s/d. https://www.youtube.com/watch?v=ZiIqmHZENWY&t=2m04s.
12. M. Perrot, *Minha história das mulheres*, Contexto, São Paulo, 2015.
13. J. Lauand, Judith Lauand, 1960, in *Projeto Construtivo Brasileiro na Arte: 1950–1962*, Aracy Amaral, 1977, vol. 214, p. 357.
14. R. T. Barros, *Arte Construtiva na Pinacoteca de São Paulo*, São Paulo, Pinacoteca, 2014.
15. C. Fioravante, *Judith Lauand: Experiências*, Museu de Arte Moderna de São Paulo, São Paulo, 2011, vol. 7, p. 72.

CHAPTER 11

A New Substance Under the Sun: How Synthetic Polymers Were Selected and Transformed Into Works of Art by Ângelo de Sousa, Julião Sarmento and Lourdes Castro

MARIA JOAO MELO*[a], JOANA LIA FERREIRA[a], SARA BABO[a], ANA ISABEL PEREIRA[a], MARIA ELVIRA CALLAPEZ[b], MARÍA JESÚS ÀVILA[c] AND JULIÃO SARMENTO[d]

[a]Faculty of Sciences and Technology of NOVA University Lisbon, Department of Conservation and Restoration and LAQV-REQUIMTE, 2829-516 Caparica, Portugal; [b]University of Lisbon, Centro Interuniversitário de História das Ciências e da Tecnologia, 1749-016 Lisboa, Portugal; [c]Centro de Artes Visuales, Fundación Helga de Alvear, Pizzaro 8, Cáceres, Spain; [d]Studio Julião Sarmento, Capa Rota, Sintra, Portugal
*E-mail: mjm@fct.unl.pt

11.1 Introduction

Plastics have proliferated since the middle of the twentieth century, and we are now forced to accept the prediction made in 1988 by a retired DuPont® chemist, who said that "humanity would eventually perish by

Science and Art: The Contemporary Painted Surface
Edited by Antonio Sgamellotti, Brunetto Giovanni Brunetti and Costanza Miliani

Published by the Royal Society of Chemistry, www.rsc.org

being smothered in plastic".[1] However when Ângelo de Sousa, Lourdes Castro and Julião Sarmento were making polyesters, their materials of choice, plastics, were seen as "a new substance under the sun, (...) to improve upon nature, or even transcend it".[2] Plastics were fascinating materials which looked different from any other human invention, with their use in "everyday life", according to Meikle, offering "a sense of greater material possibility".[1]

In this chapter, the "miracle" qualities of poly(methyl methacrylate), or "PMMA", are explored with great ingenuity and creativity by Lourdes Castro and Ângelo de Sousa. Their fascination with this novel material is revealed through several interviews.

For Lourdes Castro, PMMA substituted older materials like canvas, but also offered previously unattainable qualities. Lourdes' attraction to PMMA was sparked by its transparent qualities that would offer a perfect "immaterial" body to her shadows. Ângelo de Sousa worked and shaped PMMA to its extremes, exploring properties that earlier generations could have never imagined. Ângelo was fascinated with the possibility of creating complex three-dimensional shapes from a sheet, exploring the thermomechanical properties of PMMA. Both artists revealed the very essence of PMMA as if they could indicate, at a molecular level, the very substance of which it is made. Julião Sarmento discovered poly(vinyl acetate), or "PVAc", during the second half of the twentieth century, as plastics became an increasingly visible presence in the material world. As was the case with Lourdes Castro and Ângelo de Sousa, this synthetic material allowed Sarmento to explore possibilities in his art that would have been impossible with traditional media. In the interviews, we also discuss conservation issues with him, as well as how this binding medium functions, and may or may not be durable.

Short histories of PMMA and PVAc, as well as biographies for the artists, are presented in Box 11.1 through 11.5. These introduce the interviews with the artists, where we discussed the making of their works of art. Ângelo de Sousa's interviews cover his work with PMMA and PVAc, and were conducted in his home, in Porto, in October of 2005 and June of 2008. Lourdes shared many details of her material selections and how she would work with them in our laboratories, from the Department of Conservation and Restoration, 2005–2009. These meetings were follow-ups to previous contacts, which included letters, *e.g.* Figure 11.1. Finally, for Julião Sarmento we summarized the conversation that took place in a master class he organized in the conservation and restoration programme at FCT NOVA, May 2010. In this workshop Sarmento explained how he made one of his paintings in detail (from the *White series* created in the 1990s). He is a brilliant professor and the students enjoyed this opportunity and the detailed information he shared. The master class was filmed, and a video selection is available.†

These artists are inviting us to step into their workshop and sharing with us their fascination with these novel plastic materials, and their unique properties which they explored with ingenuity and creativity.

†Electronic supplementary information (ESI) available: Julião Sarmento master class video. See DOI 10.1039/9781788016384.

11.2 The Artists

Box 11.1 Ângelo de Sousa

Ângelo de Sousa (1938–2011), or Ângelo, as he chose to sign his works, is one of the major names of Portuguese contemporary art. Ângelo worked extensively on different artistic media – painting, drawing, sculpture, photography and film – however, as Fernando Pernes observed, he was a painter, deeply and essentially a painter.[3]

Ângelo described himself as a technology enthusiast. From the time that Ângelo started painting, in the 1950s, he experimented with many materials and felt that research was part of his nature. Ângelo stated, "from the time I was little, whenever I would do something, I needed to understand how it worked". Ângelo once mentioned that, in part, his work derives from the materials he uses.[4]

In his early years as a painter, not long after he arrived from Mozambique in 1955, he bought Dörner's book, *The Materials of the Artist and their Use in Painting*, which was a major investment at that time (a fifth of his monthly income). He also purchased *The Artist's Handbook of Materials and Techniques* by Ralph Mayer in 1959 and collected books about plastics. Unlike other artists, Ângelo named precisely the materials he used in order to supply the correct information to anyone who may need to restore his works.[4] Each painting has a label describing the support, preparation, paint and protection layers (Figure 11.2).

Box 11.2 Lourdes Castro

Lourdes Castro was born in Funchal in 1930. She studied painting at the Lisbon School of Fine Arts but could not adapt to the prevailing academicism. She left Portugal in 1958 with René Bertholo and settled in Paris after a brief stay in Munich. The following years, during which she founded the international group and journal *KWY*, were rich with artistic experimentation. Lourdes Castro abandoned traditional painting supports and began to create assemblages with everyday objects, which she then uniformly covered with paint. It was by working in these assemblages and serigraphy that the artist discovered the "shadow", and started to explore the line she has followed to the present day: the exploration of contour and shadow.

In an excerpt taken from 1963, the artist explains how, "Serigraphy is what brought my first shadows to light. I made collages with objects and I wanted to make prints, so I placed those same objects on the pre-treated silk. That is how I got actual projected shadows. My fascination has endured in the drawing, the simplicity of the form, the outline of a shadow, by its invisible presence. For me, a shadow has more meaning than simply the object described. It is a way of contemplating things and people around me."[5]

Box 11.3 Julião Sarmento

Julião Sarmento was born in 1948 in Lisbon, Portugal, and lives and works in Estoril, Portugal. He studied painting and architecture at the Lisbon School of Fine Arts. Throughout his career, Sarmento has worked in a wide range of media: painting, drawing, sculpture, photography, film, video, performance, sound and installation. He also developed several significant site-specific projects. He has had numerous solo and group exhibitions throughout the world over the past five decades.

Julião Sarmento represented Portugal at the 46th Venice Biennale (1997). His work was also shown at Documenta 7 (1982) and Documenta 8 (1987), the Venice Biennale (1980 and 2001) and the São Paulo Biennale in 2002. His work is represented in many public and private collections in North and South America, Europe and Japan.[6]

11.3 The Polymers

Box 11.4 An organic glass, poly(methyl methacrylate) or PMMA

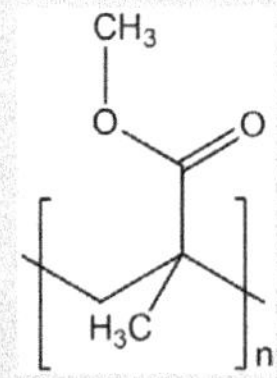

PMMA is one of the most important acrylic polymers, and is also referred to as organic glass. Its transparency is due, in part, to the fact that poly(methyl methacrylate) is soluble in its own monomer. This allows it to be synthesized by chain (radical) polymerization very slowly in bulk with very few impurities, resulting in a polymer with outstanding optical properties.[7]

Research into these materials started in the second half of the nineteenth century. In 1901 Otto Röhm's PhD thesis was on acrylates, but it was only in the early 1930s that both Röhm and Haas®, in Germany, and Imperial Chemical Industries® (ICI), in England, independently discovered the rigid, transparent and thermoformable plastic, PMMA. However, it was not until 1932 that John Crawford discovered a commercial process to produce the monomer, thereby making it possible to produce acrylic sheets on an industrial scale.[8]

Depending on the company that produced them, these sheets were commercialised under different trade names, such as Plexiglas®, Perspex®, Lucite® or Altuglas®.

Its transparency, light-weight nature, strength and ability to be moulded, made it highly suitable for aircraft canopies, and so PMMA gained huge importance during the Second World War.

These characteristics were also attractive to artists, such as Naum Gabo, Antoine Pevsner and Moholy-Nagy, who had been using plastics like cellulose acetate and nitrate in their artworks since the 1910s.[9] However, PMMA, like plastics in general, gained importance as an artistic material during the 1960s, when these synthetic materials grew more common in everyday life.[9,10] Portuguese artists Lourdes Castro and Ângelo de Sousa also began using PMMA sheets in the 1960s. They were among the first Portuguese artists to have used PMMA sheets in their artistic production.

Box 11.5 The advent of poly(vinyl acetate) or PVAc

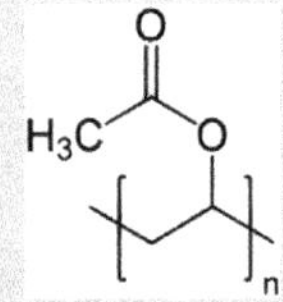

Poly(vinyl acetate) was introduced in 1928 and for the first formulations an organic solvent was used.[11,12] Safer, non-toxic aqueous emulsions followed very quickly and were developed in the late 1940s.[11] After the Second World War, the paints and coating industry underwent a massive shift, due to the lack of natural raw materials following wartime consumption. Those natural resources were soon replaced by synthetic polymers.[4]

The novelty of vinyl and acrylic emulsions as paint attracted artists, and these emulsions were first used in works of art sometime in the 1940s and 1950s, respectively.[12] During that same period of time, artists made their own formulations based on these synthetic emulsions, which were introduced to the market and usurped the primacy of oil paints. Artists' quality paints are similar to household paints in their general formulation, although they generally have higher quality pigments and higher application viscosity. These paints are thus formulated with less extender, more pigment and thickener. According to Crook and Learner, the first Polymer Tempera, an artists' PVAc-based emulsion paint, was developed in 1946 by the paint manufacturer Borden Co., working in collaboration with the American artist, Alfred Duca.[13] The first artists' quality PVAc binding medium in the UK was marketed by Rowney® as an unpigmented binder, and the same company introduced a series of PVAc colours in 1966. Other colourmen, such as Spectrum, Reeves® and New Masters®, also produced artists' quality vinyl and vinyl-acrylic paints, and Flashe® by Lefranc & Bourgeois®, a PVAc/VeoVa emulsion-based artists' paint, is still sold today. In Portugal, alongside household commercial paints like Robbialac™ REP paints, which have been produced since the 1950s, there was also an early production of artists' quality PVAc-based products by A Favrel Lisbonense – Sabu, Geo and Vulcano V7. Vinyl paints have been used by Portuguese artists since at least the early 1960s.[4,10–12]

11.4 Ângelo de Sousa on Colour: Exploring Space and Shape With Vinyl Emulsion and Acrylic Sheet

After an initial period working with oil, casein and wax (encaustic), Ângelo started using a vinyl medium for his paintings (Figure 11.3). Later he used offset inks, acrylic paints and numerous drawing materials. He also used acrylic sheets (PMMA) and metals like iron and steel for sculpture, in addition to exploring photography and film.[4]

Ângelo was curious and enjoyed experimenting. He was able to take advantage of all possibilities of materials both in formal and aesthetic aspects. When asked about how he would find and select new materials for his work, Ângelo replied, "That is a matter of luck." He has an idea in his head and sooner or later something will come up. In his own words: "I am not looking for something, but it happens, so I use it. Most people do not notice circumstances. I feel very strongly about circumstantial situations. I have this idea that circumstances determine and influence your choices. I have felt this for a long time, not only in my artistic life but in my life in general. The practical circumstances mean a lot."[14]

11.4.1 PVAc Like an Enamel

In 1961 Ângelo started working with PVAc.[4,12] In an English magazine called *The Studio*, he had glimpsed a painting by Sidney Nolan, an Australian artist, which was listed as having been made with PVAc. One day he bought some white wood glue, which was described as PVAc, and then he thought it might have been what Nolan was using and tried it for himself.

For Ângelo it was important to have a material that was inexpensive and would allow him to work quickly. He explains, "If I have a lot [...], it works."[4] The idea was to feel comfortable and not worry about wasting paint. He was first able to accomplish this with the vinyls and later with offset inks. When

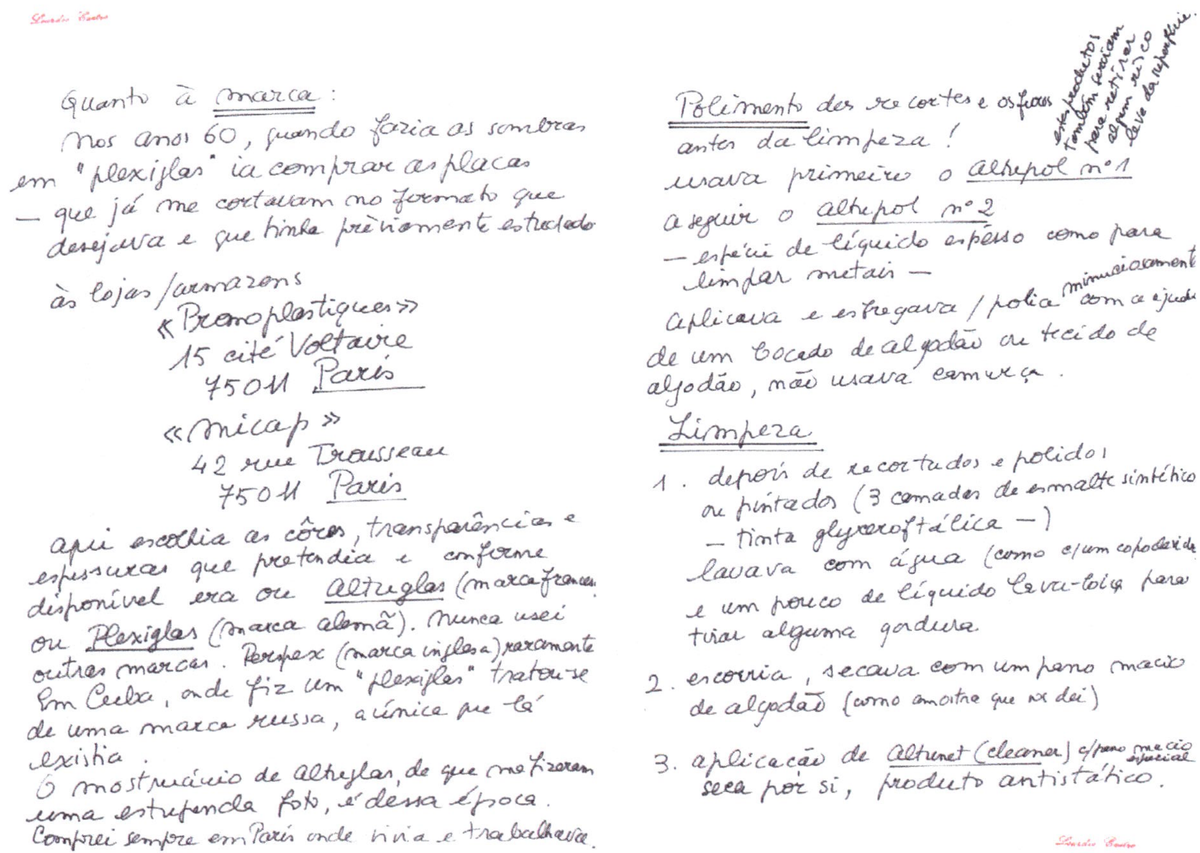

Lourdes Castro

Quanto à marca:
Nos anos 60, quando fazia as sombras
em "plexiglas" ia comprar as placas
— que já me cortavam no formato que
desejava e que tinha prèviamente estudado
às lojas/armazens
«Bromoplastiques»
15 cité Voltaire
75011 Paris
«Micap»
42 rue Trousseau
75011 Paris
aqui escolhia as côres, transparências e
espessuras que pretendia e conforme
disponível era ou Altuglas (marca francesa)
ou Plexiglas (marca alemã). Nunca usei
outras marcas. Perspex (marca inglesa) raramente
Em Ceuta, onde fiz um "plexiglas" tratou-se
de uma marca russa, a única que lá
existia.
O mostruário de Altuglas, de que me fizeram
uma estupenda foto, é dessa época.
Comprei sempre em Paris onde vivia e trabalhava.

Polimento dos recortes e os furos
antes da limpeza!
usava primeiro o Altupol nº1
a seguir o Altupol nº2
— espécie de líquido espesso como para
limpar metais —
aplicava e esfregava / polia minuciosamente com a ajuda
de um bocado de algodão ou tecido de
algodão, não usava camurça.
(estes produtos também serviam para retirar algum risco leve da superfície.)

Limpeza
1. depois de recortados e polidos
ou pintados (3 camadas de esmalte sintético
— tinta glyceroftálica —)
lavava com água (como c/um copo de detergente)
e um pouco de líquido lava-loiça para
tirar alguma gordura
2. escorria, secava com um pano macio
de algodão (como amostra que vos dei)
3. aplicação de Altunet (cleaner) c/pano macio especial
seca por si, producto antistático.

Lourdes Castro

Figure 11.1 Letter sent by Lourdes Castro with information on the PMMA brands she used and where she acquired them, *left*; and, on how she cleaned and polished her PMMA artworks, *right*. Reproduced with permission from Lourdes Castro.

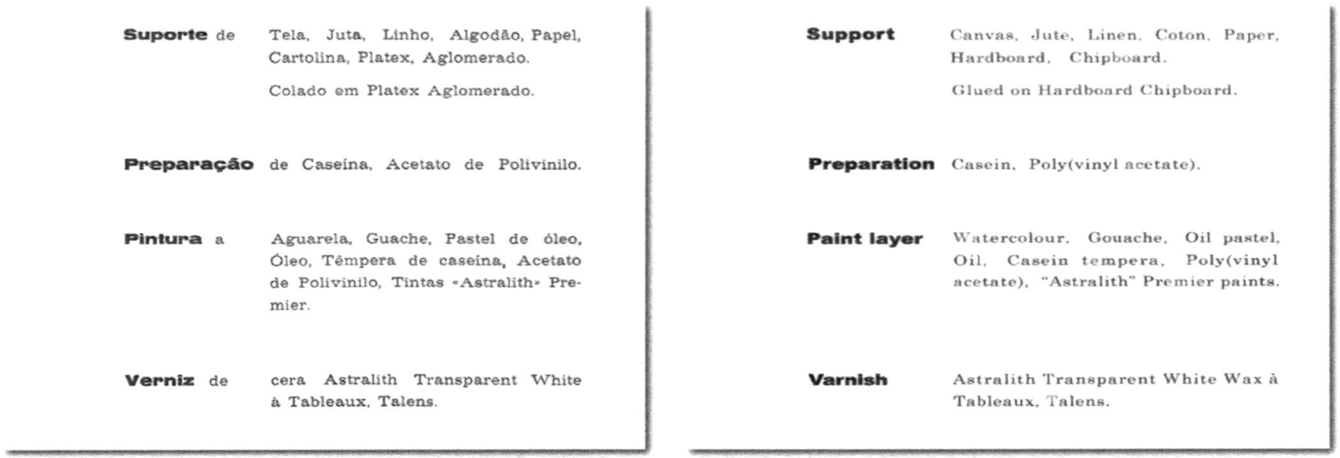

Figure 11.2 Label used by Ângelo de Sousa to identify the materials used in the support, preparation, paint and protection layers. The labels are placed on the back of his paintings, in Portuguese (English translation is ours).

he saw the reproduction of a PVAc painting by Nolan in *The Studio* magazine, Ângelo was excited to discover that artists were using something other than oil. The PVAc had several advantages. "With polyvinyl, there was total impunity. [...] there were no bad results."[4] He appreciated the advantages of fast drying, low cost and availability, and he produced a series of abstract paintings (Figure 11.3). They were made with PVAc mixed with watercolours or gouache. The result was similar to enamel; he was no longer creating textures. Although these works were very rational and geometric, there was no preliminary design. They would just happen. Nevertheless, Ângelo created some rules that he would adapt throughout the process, to make it faster and more efficient. By his own description, "it was almost industrial painting".[4] These pieces were made of collages of painted pieces of paper on a white support, by painting random areas with a certain geometric figure or using adhesive tape to create areas that could be further painted. Ângelo exclaimed, "There is a great arbitrariness in the process. The idea is to have rules of the game, rules for the game, but without being inflexible."[4]

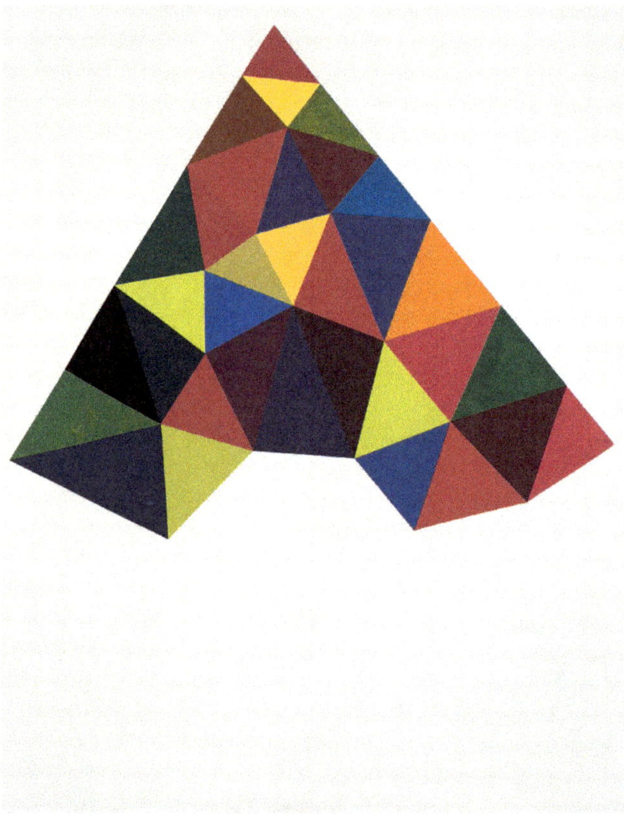

Figure 11.3 Ângelo de Sousa, *Geométrico Grande (Large Geometric)*, 1967. PVAc on hardboard, 170 × 138 cm. Caixa Geral de Depósitos collection; Photo: José Fabião. Reproduced with permission from Miguel de Sousa, Caixa Geral de Depósitos collection and José Fabião.

11.4.2 PMMA in Motion

There is a clear relationship between Ângelo's sculpture and painting. Like the abstract paintings, his sculpture is both geometric and organic at the same time. The acrylic sheets were heated and then folded, but because this process was difficult to control, it was impossible to plan a shape in advance. These works are not sculpture in its traditional sense; Ângelo discovers three dimensions through a pre-existing volume or even the creation of that volume. He researches the ways in which a flat form can be transformed into a three-dimensional object, and any possibilities for that to happen, with the goal of providing mobility to that sculpture.[4,10]

11.4.2.1 A Question of Curiosity and a Yellow Dinosaur

We have transcribed a question and answer between Ulrich Loock and Ângelo, published as an interview in 2006 in the Gulbenkian exhibition catalogue,[14] to better introduce the "Yellow dinosaur" that was not included in the publication.

Ulrich: You have these amazing constructions where the interior shapes of a sculpture were dependent on its exterior shape or *vice versa*. This economical sense reappears in Carl Andre, for instance, or in Stella's *Black Paintings*. They are very close to Rodchenko's work. It is quite important in your work that you use the same kind of economy. You make your cuts but you don't throw any material away. You cut it, you bend it, you fold it up but you don't lose anything. Everything stays with the piece and you can always trace back the process. We can even say that there is a performative aspect in your sculpture.

Ângelo: I think that this is a very important factor. My idea is that everybody starts from the drawing. Even the Minimalists started with drawings. Contrary to that, I start directly from the materials. I don't know what is going to happen, it is a kind of action sculpture. [...] Sometimes I have a very clear idea of what I want to do, sometimes I spend a big amount of money and time, but then I think, what am I going to do with that? It was so overdetermined that I just lost interest. In other cases I had a clear idea as well but when I decided to make it I knew that something unexpected was going to happen in the middle. It is curiosity.

There was one piece called "the dinosaur", which I sold or gave to someone, but before I did, I made a copy for myself in aluminium. Yesterday I tried to remake it in plastic, but it did not work. It is a dinosaur with five legs [...] At the beginning it was something else, I do not remember what. At the time, I bought this very kitschy yellow, this cheap acrylic that was used to make buttons.[†] Unfortunately, they are not for sale anymore. I thought they were so outrageous that they would go very well with those apparently geometric sculptures (Figures 11.4 and 11.5).

[†]Ângelo de Sousa purchased acrylic sheets in Porto, and they were likely from a small Portuguese company, *Plásticos do Sado*, which produced pearlescent acrylic sheets for the button and cutlery industries. Their production process was different from that used by Plexiglas® and Altuglas®.[10] Whether this will significantly influence the long-term ageing of this material is something our department is currently investigating.

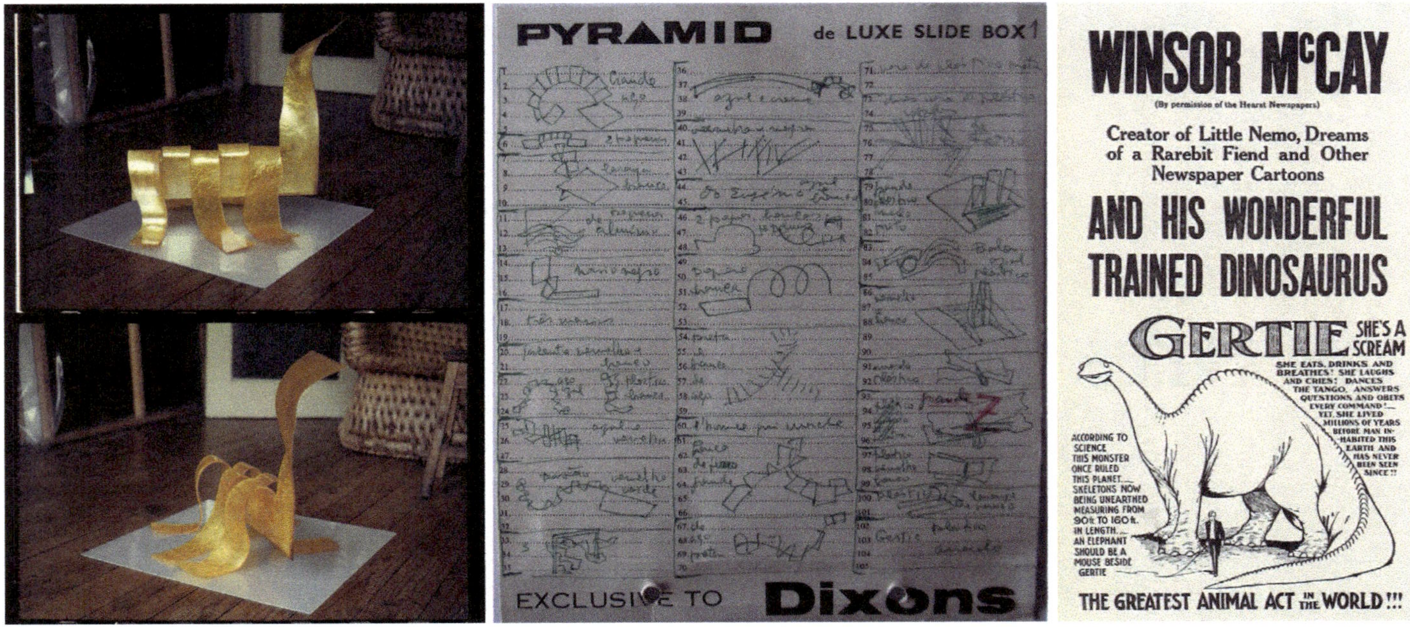

Figure 11.4 *From left to right*: Colour slides by Ângelo de Sousa showing one of his sculptures in acrylic sheet produced in 1964–65. The original sculpture is a lost piece that was described as *Gertie Plástico Amarelo (Gertie Yellow Plastic)* in a list of slides of his sculptures that was found in Ângelo's studio. All reproduced with permission from Miguel de Sousa. Newspaper advertisement for Winsor McCay's film “Gertie the Dinosaur”, published in *Variety* and other magazines in December 1914.

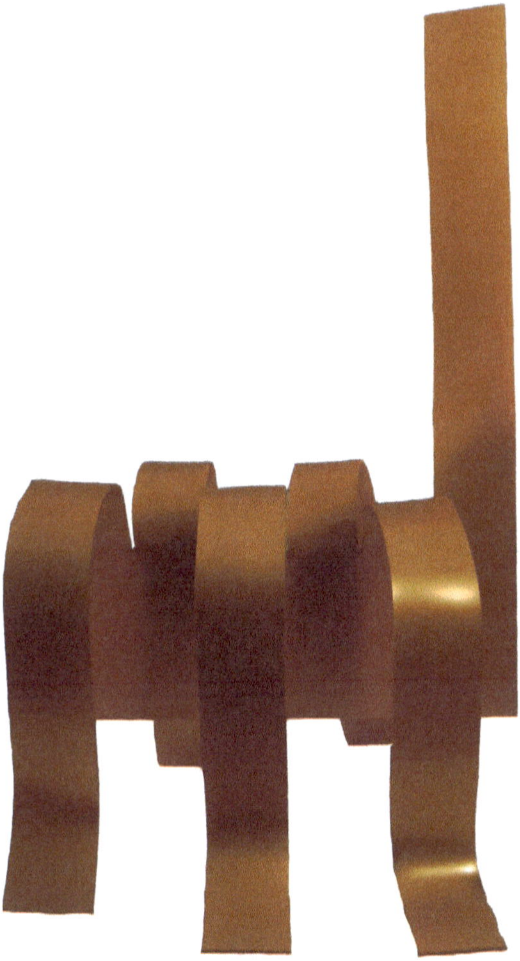

Figure 11.5 Sculpture in acrylic sheet by Ângelo de Sousa, 2006. This piece is likely a new version of the previous "yellow dinosaur" sculpture, for more details please see text. Reproduced with permission from Miguel de Sousa.

11.5 PMMA "Shadows" by Lourdes Castro: "I Made the Shadows Come Out of the Shadow, I Gave Them Colours, an Independent Life"

From 1964 until 1968, Lourdes Castro worked with PMMA to create "projected shadows". In these artworks she cut, painted or serigraphed silhouettes of friends in everyday actions on the acrylic sheets (Figures 11.6–11.8). By overlaying different transparent sheets, often coloured, the artist has materialized the shadows in an almost immaterial way, since the light can travel through the artwork and reproject the shadows on the walls, fusing them within the environment. The way that Lourdes Castro "made the

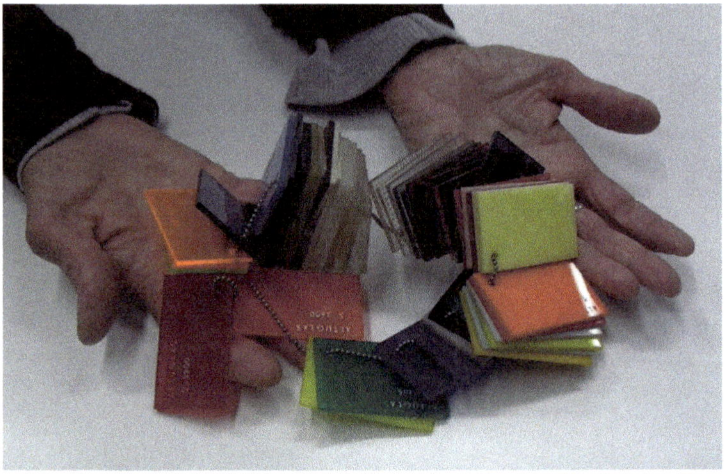

Figure 11.6 Lourdes Castro colourful PMMA palette that she is showing us in the laboratory and that we studied.

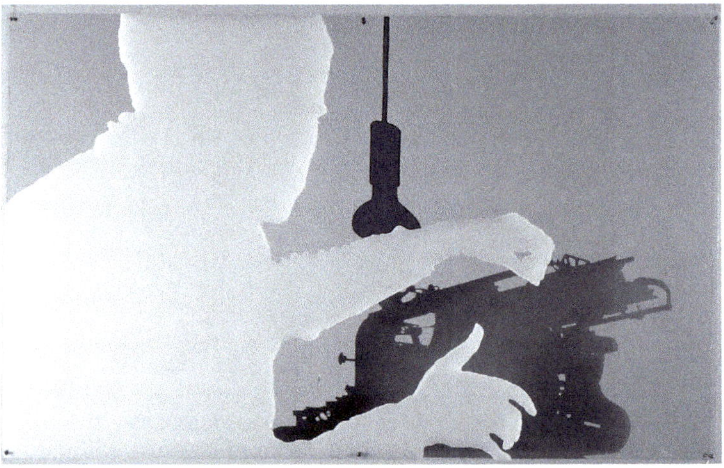

Figure 11.7 Lourdes Castro, *Sombra Projectada de René Bertholo (Projected Shadow of René Bertholo)*, 1965. Glycerophthalic paint (alkyd resin) on acrylic sheet. Calouste Gulbenkian Museum – Modern Collection. Photo: Reinaldo Viegas. Fundação Calouste Gulbenkian. Reproduced with permission from Lourdes Castro.

shadows come out of the shadow, gave them colours, an independent life" is next described, based upon several interviews given to the authors that were supplemented using other published information.[15–18]

11.5.1 Why Was PMMA Selected?

"Searching for a textureless material that had more to do with the result I wanted to obtain, in 1964 I created my first piece in plexiglas. At last, an immaterial material like the shadows.

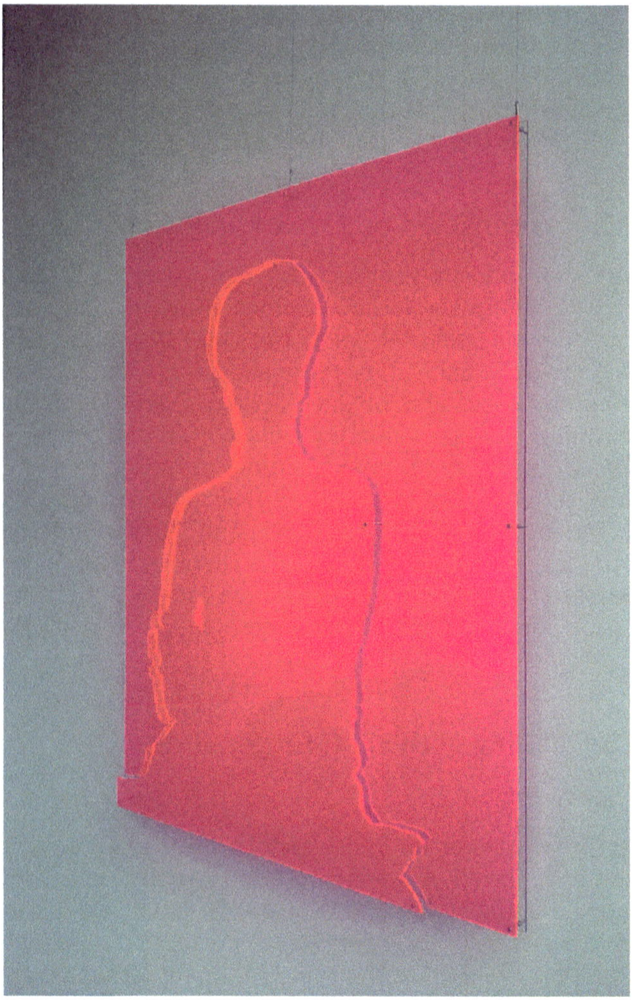

Figure 11.8 Lourdes Castro, *Sombra Projectada de Christa Maar (Projected Shadow of Christa Maar)*, 1968. Cut acrylic sheet and glycerophthalic paint (alkyd resin). Calouste Gulbenkian Museum – Modern Collection. Photo: Sara Babo. Fundação Calouste Gulbenkian. Reproduced with permission from Lourdes Castro.

Working on canvas lost its meaning. There was a constant dialogue between the plexiglas and me. Because of the plexiglas' transparency and translucency, the shadows grew fainter and (re)projected onto the wall.

I painted underneath, cut it out and made serigraphy over it, I even engraved (books). I never wanted to mould, volume was not part of my projects.

I made the shadows come out of the shadow, I gave them colours, an independent life."[15]

Please note that the artist always uses the trade name of the German company Röhm and Haas®, for this material, Plexiglas®. However, "Plexiglas" is used as a generic term for a sheet of acrylic. In fact, the samples loaned by the artist for analysis were manufactured by the French brand Altuglas® (Figure 11.6).

Lourdes states in writing that she acquired sheets from both brands, "based upon availability, it was either Altuglas® (French brand) or Plexiglas® (German brand) (Figure 11.1). I have never used other brands. Perspex® (English brand) rarely. In Cuba, where I made once a plexiglas, I used a Russian brand, which was the only kind they had."

11.5.2 What Is Depicted by These Shadows?

The PMMA shadows are always of friends "who lent her their shadow".[5] They were people with whom she felt at ease. Only once did she accept a commission from someone she did not know, and afterwards she stated that she "would never do it again".[17] In these shadows, friends are usually seen performing common gestures or daily activities: reading a book, talking, smoking, combing hair, *etc.* Sometimes more than one person is depicted, sometimes people and objects and sometimes only objects.

As Lourdes Castro explains, "I never used the person's name because there is a natural tendency to seek a resemblance, and that was not what I wanted. I knew that the essential characteristics were there, but I would like people to just see "someone". However, I once wrote in a catalogue that after ten years I would identify participants. After those ten years, I started using peoples' names."[5]

11.5.3 The Making of the PMMA "Shadows"

11.5.3.1 Drawing

"It would start with the person first, which is why they were just friends and people I knew well. Usually, friends would come to the house for dinner and then I would put the paper on the wall and draw the shadow; from the paper I would then transfer it to canvas or later to plexiglas. I remember one time I did an exhibition in Basel, in the same gallery as [Jean] Tinguely, and he asked me, "How do you do this, is it from photography?" And I replied, "No, it is just drawing." He hugged me and said, "That is the way it should be done" (laughs)."[5]

11.5.3.2 Cutting the PMMA Sheets

According to Lourdes Castro, the acrylic sheets were protected with sticker paper, upon which the shadows outlines would first be drawn. In her first work on PMMA, *Projected shadow of my mother*, 1964, the sheet was cut out with a hand saw. After this initial experiment, she started to use a Stanley electric saw to cut the acrylic sheets. Due to the heat produced by the saw, the sheets would end up glued back together and were then separated again with a manual metal saw. In an attempt to circumvent this laborious procedure, Lourdes Castro was told to drip oil over the drawing outlines and found that it worked, since it minimized the heat, thereby avoiding the second cut. After this step, the holes necessary to attach more than one sheet together or to hang the work on the wall were made. For finishing of the cutting edges, Lourdes Castro used iron files and then a polishing kit from Altuglas® – first *Altupol 1* followed by *Altupol 2* (currently Altuglas Polish 1 and 2), which were applied with a cotton pad or cotton cloth (Figure 11.1). Light surface scratches were also removed with these polishing agents. As a final treatment, the artist washed the acrylic shadows with water and soap, and after drying applied *Altunet* (currently *Altuglas Cleaner*) as an antistatic agent with a special smooth cloth. The works were signed using a pyrography tool (Figure 11.1).

11.5.3.3 Painting and Serigraphy on PMMA Sheets

When the sheets are painted, Lourdes Castro would use a glossy paint she called "glycerophthalic". The designation indicates an alkyd-based paint, the so-called enamel paints. Sheets were painted from the back, and the artist would usually paint three layers to make it very opaque. The outlines were drawn with a fine brush and the surface painted with a wide brush or roller.

No specific documentation has been found about the process of creating the serigraphs on the plexiglas sheets, but according to information supplied by the artist, the work was performed by hand, with the help of her partner, the artist René Bertholo.

11.5.4 How the Work Should be Displayed in an Exhibition

The transparency of the acrylic allows objects to "merge" with the context. Thus, the display of acrylic artworks requires special attention. "Normally I always show up at least a week prior to the exhibitions. When the cut-out shadow is projected onto the wall, that shadow is really real, it seems that the person is there. [...] It is like seeing the very shadow of a person who cannot be there." In some catalogues it is possible to see the reflection of people or objects that are in front of the artwork. However, the artist states that it is not her intention to create this reflection. The work should only project its shadow backwards.[18]

11.6 Julião Sarmento's Making of the *Frozen Leopard* (1991): "Everything Is a Pigment"

In May of 2010 Julião Sarmento organized a master class for the Conservation and Restoration course[‡] to show the making of the *Frozen Leopard*, a 275 cm × 315 cm vinyl painting that is included in his *White series*, created in the

[‡]At the Department of Conservation and Restoration, Faculty of Sciences and Technology of NOVA University Lisbon, Campus de Caparica. Participants: Julião Sarmento, Romeu Gonçalves, Maria João Melo, Ana Isabel Pereira, Joana Lia Ferreira, Leslie Carlyle, Master Students of Conservation and Restoration attending the History and Techniques of Artistic Production Lectures (academic year 2010–2011); Jorge Imaginário (video production).

1980s and 1990s (Figures 11.9–11.11). The text below is a summary of the event, which can also be found in video form in the supplementary files at https://doi.org/10.1039/9781788016384.

11.6.1 Fundamentals of an Artist at Work

Julião Sarmento: In this workshop, we will first reproduce the *Frozen Leopard*. I think this is very important, because at a national and international level, nobody really understands how I work. My work is often described as "oil on paper"; I should point out that I have nothing against oil, I like it very much; I worked with oil while at school but, during my professional career, I have only painted two paintings in oil. Two!

Maria João Melo: What do you mean when you say that you like oil very much?

JS: I also like many other things that I do not use, because they are not appropriate for my work. Oil is not appropriate for what I want. I am much faster than oil.

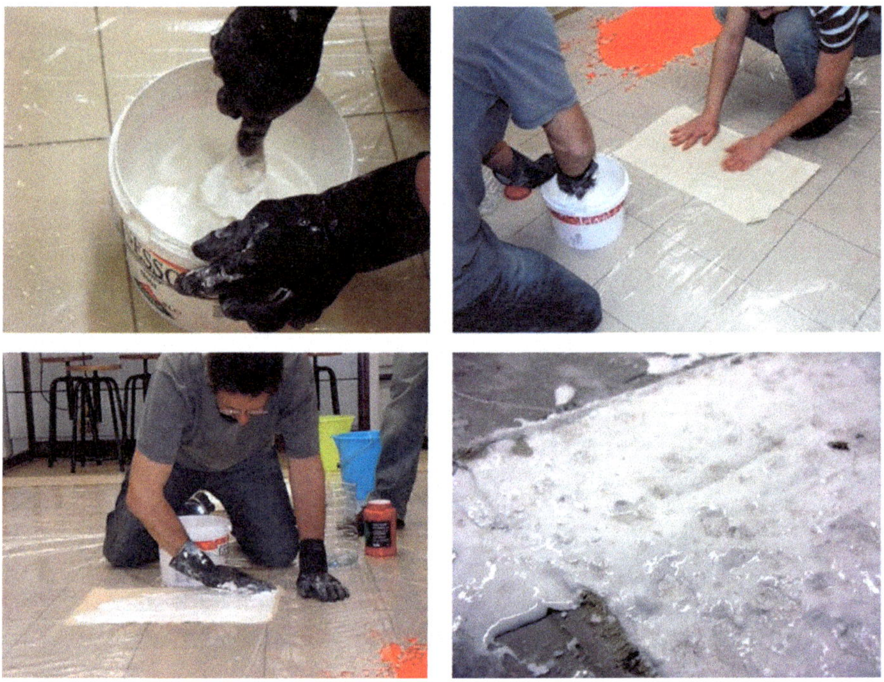

Figure 11.9 Julião Sarmento preparing a white background with the materials and techniques used to create the *Frozen Leopard*, 1991. *From left to right*: first a vinyl glue and *Cenógrafa* white pigment were mixed in a bucket until the desired consistency was like "chantilly", in the artist's own words. Next, the cotton fabric was wetted and stretched over plastic, and the white paint was applied on it. The last photo shows the final appearance of the paint before drying.

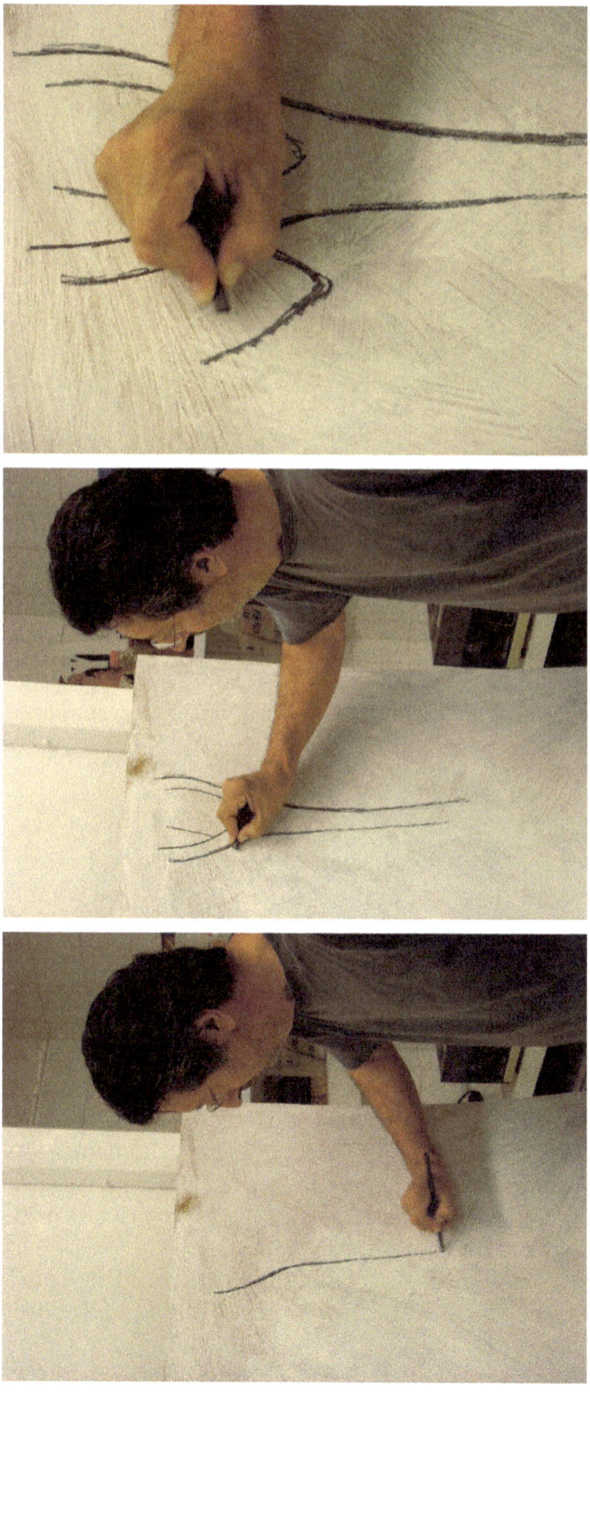

Figure 11.10 Julião Sarmento exemplifying how the drawings were made in the *Frozen Leopard*, using soft graphite sticks; when the artist places his hand on top of the outline, the loose graphite leaves marks on the white surface ("ghosts", as he named them).

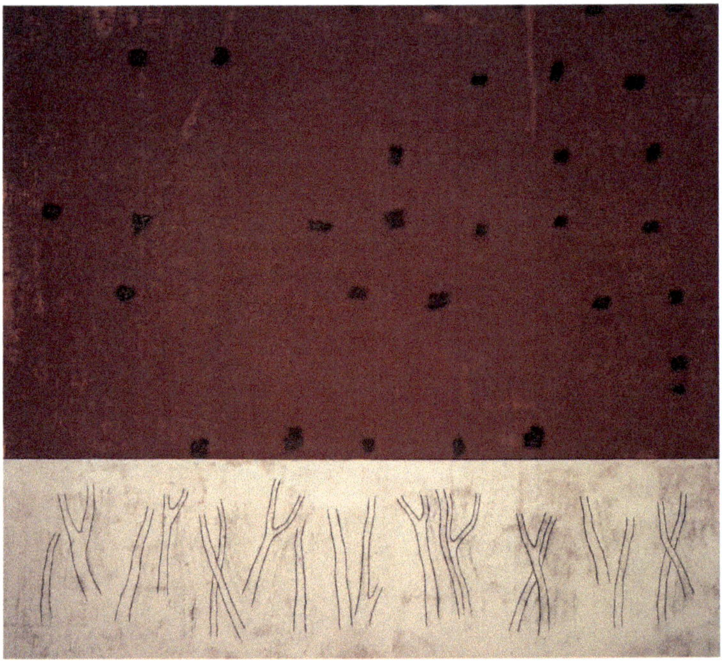

Figure 11.11 Julião Sarmento, *The Frozen Leopard*, 1991. Calouste Gulbenkian Museum – Modern Collection. Mixed media on canvas (basically PVAc on canvas; for more details on the materials and techniques please see text, Section 11.6). Photo: Mário Gastinger. Fundação Calouste Gulbenkian. Reproduced with permission from Julião Sarmento.

JS: There are some basics that you should know about my work. I select materials based on what I want to do. But at other times, I can adjust what I want to do based upon the materials available. I have always been very inventive regarding materials. Although I do not have a Jewish or Scottish bone in my body, it may look like I have one because I am such a cheapskate. At the start of my career, it was borne of necessity: I did not have much money. Now that I can spend more money, I still do not think it is worth it. I have also seen experts lost in endless discussions as to why I used this or that kind of paper, this pigment or that glue. It is very simple: I started using dry pigment and glue (PVAc) because it was much cheaper than buying acrylic paints. At the time, I thought that it would produce effects that I could not achieve with acrylic paints, and, I could get a 10 m × 2 m surface for the same price as a 70 cm × 70 cm in acrylic. This is straight economics.

MJM: We are finding that these older PVAc formulations are probably more stable than the acrylics. So, they would be cheaper, more stable and more suited to the technique.

JS: Back at the start I was painting on grocery bags, a greyish-brown paper. Why did I start working with this paper? Because about two hundred of those sheets cost me the same as a single sheet of Fabriano. It's economics.

As restorers, there are three fundamental periods you need to know. Until 1974, all my paintings were acrylic on canvas (except for the one or two that were on oil).

MJM: Which brand?

JS: Acrylics from Talens®. We dreamed of painting with what the Americans painted with, but here in Portugal we did not have access to those sorts of materials. Except for a few works from around 1969, until 1974 I applied one or two coats of acrylic gypsum, also made by Talens®, on top of the industrial ground layer. And then I painted with acrylic paint. Between 1962 and 1969, I made some works in which the support was hardboard; I would do a frame in pine that would be covered with hardboard. And then I would paint on top of the hardboard.

Joana Lia Ferreira: Just out of curiosity on which side of the hardboard would you paint?

JS: On the smooth side. And normally I would use household paints...

JLF: Robbialac?

JS: No! That was very expensive. I discovered one in a drugstore in Oporto and I used it for everything that I wanted to be a dull matte. For bright colours, I would paint with Robbialac™ because it was better quality.

So, 1962–1969 was my "rigid canvas" period, and then I went back to regular canvas, to linen canvas. Unfortunately, you will not find many of my works from that period because there were a series of accidents in my life that made me lose a large part of my work.

In 1974 I abandoned painting, in a desperate act; between the end of 1974 and 1981 I did not paint, I did not do a single drawing. I worked with photography, with video, sound, with Super8 film, installations – ultimately, the work I was doing related to mechanical reproduction.

In 1981, with the triumph of David Hockney the painting fever returned, and I painted the series you now want to learn how to reproduce. At that time, people went back to this sort of ground zero, and objects needed to assert themselves through their size. People would paint very big things. I once again started working on canvas in 1983;[§] I used raw cloth and bed sheets. I still have some of those raw cloth paintings. I would take the cloth, stretch it out and paint on it with Sabu paints. I concluded that the Sabu paint lasted a long time, it was kind of an acrylic.[¶] I would buy tons of bottles of Sabu.

In 1985, my work shifted again. That was when I started working with Vulcano V7 [a vinyl emulsion[4,11]]. I started thinking that everything is a pigment and I started using different pigments. If I were to grab all of you and cover you in Vulcano glue and paint, you are pigments. Pigment is everything that can cling to the binder.

MJM: The colour wouldn't be great, right?

JS: Yes, but you can see my point. Forty ruined chairs are pigment. In the end, that's what it is, they are particles bound together through a binding

---

[§]Between 1981 and 1983 I only worked on paper (if anyone shows you a canvas from that time, it is a fake).

[¶]The binding medium is a vinyl emulsion.[4]

medium. I also wanted to do big paintings, and in Portugal the widest width of raw cloth I could find was 1.50 m. So, I ordered it from Spain. I realised that in addition to being higher quality, it was cheaper. Nowadays I still order canvas from Spain.

JLF: Do you wash it?

JS: I do not. I wet it. You will see in a minute how it is done.

JLF: Earlier you said that everything is a pigment.

JS: Exactly. During the 1980s I used the Cenógrafa pigment formulations from Casa Varela.

At that time, I would do everything with my hands, I mixed the paints with my hands. But then my skin started to crack, and I started to wear gloves and after the gloves, I started to use assistants.

I used a restricted colour palette first because the Cenógrafa brand did not offer a wide selection. Then, I started using colours I liked which narrowed even more the choice and, from 1990, I would use white and black paint or, mix the white with the black. Usually, I would mix two packages of dry pigment (2 kg) with one bottle of glue (5 l).

The fact that I work with Vulcano or Bizonte glues, like other poly(vinyl acetate) can be a problem, because it yellows (as you know). Many of the *White* paintings which once were white, nowadays are yellow. Some areas are whiter or yellower than others. The more poly(vinyl acetate), the more yellow it is. That said, it does not bother me, I know that it yellows.

MJM: The PVAc-based whites made by Joaquim Rodrigo, which are thirty to forty years old, are almost as white as Joana's reproductions [made now, for her PhD thesis[4]]. But his white was made with titanium white. On the other hand, some new PVAc-based formulations are much more prone to yellowing.

JS: And the *Cenógrafa* white is made of what? Is it zinc?

AIP: It is lithopone. It has barium and zinc.

JS: Well when you look at art this technically [...] It is a good thing you exist, but then I start to wonder that if you think about this too much, you end up doing nothing.

11.6.2 Reproducing the White Background for the *Frozen Leopard*

This was the method used in the *White series* paintings from the 1990s. Julião Sarmento put some Bizonte glue into a plastic container and was ready to add the powdered pigment as he began his explanation. Meanwhile, his assistant prepared the canvas, first by dunking it in a bucket of water and stretching it over the plastic that covered the floor (Figure 11.9).

JS: Here is where the artist's soul resides. The proportions in the mixture are very important. It depends on what I want. The proportion between the pigment and the glue is not always the same. To make it just like I used to do, I would dilute the glue with water and would mix it up slowly. It is a good consistency when it looks like whipped cream. The water is added slowly, just like cooking.

Now, I must ask you all to step back. Romeu, can you go outside and bring me some soil?

MJM: Why did you start using soil in these white paintings?

JS: Sometimes I would look at the painting and it seemed too white. It bothered me.

MJM: But, why soil?

JS: Because it was right there to be picked up.

At this point, his assistant had gathered soil from outside and put it in a bucket with water. Sarmento speckled this mixture over the paint's surface with his hand.

JS: There you have the *Frozen Leopard*. It is done. Now you just leave it to dry. There are things that I would never do in the real world that I ended up doing here in this demonstration. For example, here you can see the marks left by my hands and fingers. In real life, I never let that happen. I could spend half an hour concealing it all.

MJM: So, the painting will dry and then you put it right on the stretcher?

JS: No, the stretcher is the last thing done.

Student: So how do you draw it straight if the last thing to be done is putting it on the stretcher?

JS: Because I take the canvas and staple it to the wall. But sometimes I would do it on the ground, it depends. It is also important that you know that in the 1980s, I only stretched canvas until I could afford someone that would stretch them for me. My first assistant only had one job, and that was to stretch the canvas. Because after stretching around 500 canvases, I was tired of it. These days, Romeu stretches the canvas.

MJM: So, besides soil, is there any other material you use to create these effects?

JS: Usually I prefer what is readily on hand, but it depends on the circumstances, on the occasion. I want you all to understand one thing, and this is essential. I am highly permissive regarding the materials I use. I mean there are no taboos. Therefore, if at some point, I figure out that I want to put sticky tape in it, I will put sticky tape in it.

11.6.3 Reproducing the Drawing of the *Frozen Leopard*

As there was no mock-up of a highly textured paint on which Sarmento could draw, a piece of hardboard with a titanium white paint layer on top was used for the demonstration (Figure 11.10).

JS: I used soft graphite sticks; the ones that are soft and smudge a lot. [He draws] If you now approach this drawing you will see that here on the side the shapes of the "ghosts" are starting to appear. They usually appear because your hands get dirty from the graphite powder.

AIP: And afterwards, you apply something over the drawing like a fixative?

JS: Nowadays I apply fixative (only over the drawing). Back in those days I used the hairspray *Elnett Satin®*.

AIP: But the paintings would smell like a hair salon...

JS: Yes, they would smell like a hair salon. That was a problem. But it was cheaper.

JS: When you are drawing over a highly textured surface, over canvas, there is much more powder and the line is not so smooth. It is completely different from what is happening here.

11.7 Final Thoughts

The dialogues and interviews with Ângelo de Sousa, Julião Sarmento and Lourdes Castro show us how these artists selected and used novel materials, which they then transformed into works of art.

The novel materials, PMMA and PVAc, are polyesters that were appropriated by these and other artists as soon as they became available. PMMA was used as "organic glass" and PVAc as emulsion paint. Our systematic studies in polymer photodegradation mechanisms have shown that these two polymers are among the most durable synthetic materials.[11] However, formulations play a crucial role in the overall polymer matrix stability.[11,19] Although the aesthetic play with these materials and experimentation with them as physical matter are more tied to social and artistic values,[20] to know how artists used these materials will also help with more accurate predictions of the materials' lifetime and inform better conservation and restoration procedures.

The case study approach, using specific examples, illustrates the relationship between the molecular properties of PMMA and PVAc and factors related to social and artistic choices. The outstanding optical properties of PMMA served Lourdes Castro's shadows superbly. Both the optical properties and the ability to transform sheets of PMMA into complex three-dimensional shapes upon heating and shaping are explored by Ângelo to break with a traditional definition of sculpture. In his own words, "The relationship between the parts that could be seen in different positions was more important than a fixed point of view (just one), as we can see even in Michelangelo's work and in sculpture in general. I found that it was time to get rid of that privileged point of view."[14]

PVAc in Ângelo's hands shines like a colourful enamel, whereas in the paintings by Julião Sarmento, it is used in large formats, as a binder that may hold potentially everything, to create white and black, very thick and textured, paint layers that yellow over time. The way both artists played aesthetically with vinyl emulsions will have an immediate effect on their ageing. Julião Sarmento accepts that the pristine white backgrounds of his *White series* paintings will yellow over time. Although he is interested in all aspects of material stability, he believes that thinking too much about these issues would affect his creativity and, as he confessed, "you end up doing nothing". Whereas Ângelo made every effort to learn about material durability and to protect his works on PVAc from ageing, adding layers of PVAc varnish on top of the painting and selecting more durable colourants.[12]

Acknowledgements

The authors would like to thank the financial support of Fundação para a Ciência e Tecnologia, Ministério da Educação e da Ciência (FCT/MCTES), Portugal, through doctoral programme CORES_PD/00253/2012 and doctoral grant SFRH/BD/52318/2013; FCT Research contract IF/00653/2015; Associated Laboratory for Green Chemistry – LAQV (UID/QUI/50006/2019). Jorge Imaginário is deeply thanked for the video production of Sarmento's Master class. We are very grateful to Lourdes Castro and Ângelo de Sousa, for all that they have shared with us most generously. To Miguel de Sousa, Ângelo's son and President of the Directory Board of NEÂdS – Núcleo de Estudos Ângelo de Sousa, for his trust in our work and friendship.

References

1. J. F. Meikle, in *Molecules that Matter*, ed. R. J. Giguere, The Frances Young Tang Teaching Museum and Art Gallery at Skidmore College and the Chemical Heritage Foundation, Saratoga Springs and Phildadelphia, 2008, pp. 65–79.
2. What man has joined together, *Fortune*, 1936, **13**, 69.
3. F. Pernes, in *Ângelo: 1993 uma Antológica*, ed. M. H. Pérez and M. Ramos, Fundação de Serralves, Porto, 1993. pp. 8–10.
4. J. L. Ferreira, PhD thesis, Universidade NOVA de Lisboa, 2011, http://hdl.handle.net/10362/5342, accessed January 2019.
5. C. da Ponte, *Efeméride Boletim Cultural*, 2010, vol. **6**, p. 4.
6. www.juliaosarmento.com/biography/short-bio, accessed February 2018.
7. P. C. Painter and M. M. Coleman, *Essentials of Polymer Science and Engineering*, DEStech Publications, Lancaster, Pennsylvania, 2009.
8. J. A. Brydson, *Plastic Materials*, Butterworth-Heinemann, Oxford, 7th edn, 1999, pp. 398–400.
9. F. Waentig, *Plastics in Art – A Study from the Conservation Point of View*, Michael ImhofVerlag, Petersberg, 2008, pp. 74–83.
10. S. Babo, J. L. Ferreira, A. M. Ramos and M. J. Melo, in *Future Talks 015 – Processes. The Making of Design and Modern Art. Materials, Technologies and Conservation Strategies*, ed. T. Bechthold, Die Neue Sammlung – The Design Museum, Munich, 2017, pp. 160–170.
11. J. L. Ferreira, M. J. Melo and A. M. Ramos, *Polym. Degrad. Stab.*, 2010, **95**, 453.
12. J. L. Ferreira, M. J. Ávila, M. J. Melo and A. M. Ramos, *Stud. Conserv.*, 2013, **58**, 211.

13. T. J. S. Learner, *Analysis of Modern Paints*, The Getty Conservation Institute, Los Angeles, 2004.
14. *Ângelo de Sousa Escultura* (exhibition catalogue), Fundação Calouste Gulbenkian, Lisboa, 2006.
15. *Além da Sombra* (exhibition catalogue), Fundação Calouste Gulbenkian, Lisboa, 1992, p. 53.
16. S. Babo, E. S. Fragoso, R. J. C. Silva and M. J. Melo, in *Preprints from the 16th ICOM-CC Triennial Conference*, ed. J. Bridgland, et al., International Council of Museums (ICOM), Lisbon, 2011, paper 1001.
17. *Lourdes Castro e Manuel Zimbro: À Luz da Sombra* (exhibition catalogue), Assírio & Alvim, Lisboa, 2010, pp. 21–25.
18. R. Macedo, PhD thesis, Universidade NOVA de Lisboa, 2008, http://hdl.handle.net/10362/13743, accessed January 2019.
19. J. Lemaire, J. L. Gardette, J. Lacoste, P. Delprat and D. Vaillant, in *Polymer Durability: Degradation, Stabilization and Lifetime Predictions*, ed. R. L. Clough, N. C. Billingham and K. T. Gillen, American Chemical Society, Boston, 1996, pp. 577–598.
20. H. Lechtman Heather, Cloth and Metal: The Culture of Technology, in *Andean Art at Dumbarton Oaks*, Dumbarton Oaks, 1996, p. 34.

CHAPTER 12

Miquel Barceló: The Rotten and the Waste as a Fundamental Part of the Process

IRENE BIOLCHINI*

Guest Curator MIC, International Museum of Ceramics in Faenza, Faenza, Italy
*E-mail: irene.biolchini@gmail.com

12.1 An Introduction to Barceló's Practice and to his Aesthetic of the Waste

Miquel Barceló (b. 1959) started to work at a very early age in his native island, Majorca. The cultural context surrounding the artist was populated by some Conceptual movements that were acting as disruptors with the aim of subverting the dominant Franco's regime, operating mostly *via* events and performances.[†] That was the case of the occupation of Sa Dragonera,

[†]The dominant groups in those years were the Taller Lunatic and the *Jove Plastica* and they were fighting against the traditional way of painting landscapes. Describing the activities of the *Jove Plastica*, Damià Ferrà-Ponç declared: "The Francoist regime absorbed this way of painting as a way of crating order thanks to several prizes, such as the *Ciudad de Palma*. For this reason, the change towards plastic arts had a political and cultural connotation. It was not merely a change in style, passing from cubism to expressionism. No. It was a way to confront the system, and for this reason it also developed a radical approach, something that had never been seen before."[1] Originally in the text: "El régimen franquista había asimilado esta pintura como una pintura de orden en sus certámenes, en los premios Ciudad de Palma. Por ello, el cambio en el mundo de las artes plásticas tuvo una connotación política, cultural. Y por eso el cambio fui muy radical. No fue solamente un cambio d estilo, pasar de un cubismo a un expresionismo. No. Fue un enfrentamiento contra todo el sistema y por eso adquirió una radicalidad que nunca se había visto."

Science and Art: The Contemporary Painted Surface
Edited by Antonio Sgamellotti, Brunetto Giovanni Brunetti and Costanza Miliani

Published by the Royal Society of Chemistry, www.rsc.org

when the group of the Taller Lunatic guided the occupation of the island, in order to stop the construction of a series of hotels on the natural soil facing Majorca.[2] The result of that occupation was the cancelling of the building speculation, leading to one of the first conquests of the island by the green activist movement.

The involvement of Miquel Barceló in such occupation was mostly liminal: the artist went to Sa Dragonera and lived there for several days, but the trip was mostly a personal and human experience, connecting the artist with the natural landscape, reading and working in contact with the wild surroundings, rather than having an active part in the political fight.[‡] The marginal role that the artist had in this occupation is a good example of the controversial participation and connection that the artist felt with the group: whilst taking part in their performances, the artist was mostly interested in another kind of aesthetic, something that he felt was closer to the Surrealist, Dadaist and *Informel* practices. In this regard, it is necessary to keep in mind that the island of Majorca was still a remote province of Spain under Franco's age, but at the very same time it was also the place in which one of the most influential Spanish artists, Joan Miró, decided to live. Miró was a key and central figure[§] and certainly influenced Miquel Barceló's way of thinking and his practice. It is very well known that the young artist, after winning a prize, was invited to the house of the Spanish master and once there he was particularly impressed by Miró's collection of small objects found on the beach, both natural and artificial. The practice that was typical of the master remained in the imagination of the artist, who – till today – has a variety of animal parts and objects in his atelier (mixed with plastic masks and other curiosities). But not only was the collection of strange items important for the young painter: knowing Miró meant for the young Barceló to deal with the Surrealist *menlieu*, to enter into contact with their literature and with their critic sources. Those

---

[‡]Describing Barceló's participation in the occupation of the island, Toni Amengual declared that his interest was more towards "the experience rather than the practice", as stated to Michael Damiano.[3] Originally in the text: "mas por la experiencia que por la causa".

[§]Miró was part of the cultural scene of the island and he sustained economically the new gallery "4 Gats", which had quite a disruptive and anarchic programme (and was managed by Ferran Cano, Àngel Juncosa and Gerard Matas). 4 Gats was the principal point of interest of the avant-garde artists in Barcelona at the opening of the 20th Century, and therefore naming it after that place was already a declaration: they were going back to the avant-garde to promote a new art, mainly exhibiting young painters. Describing the name of the gallery, Ferran declared: "in Catalan, *quatre gats* means 'four drunks', or a place where no one goes: 'only four cats go there'. We liked it, because we thought that no one was going to visit what we were going to exhibit."[4] Originally in the text: "en catalán, *quatre gats* significa 'cuatro borrachos' o un sitio adonde no va nadie: 'allí solamente van cuatro gatos'. Nos gusto este nombre porque pensábamos que, con lo que íbamos a presentar, en realidad no vendría nadie."

were the years in which Barceló started to read Freud, Breton, Debuffet's declarations while he was also working, in order to find his own individual path and research.¶

Trying to create his own style, in fact, the artist went to Barcelona – the closest city to his native island. Barcelona was at that time very much interested in Minimal and Conceptual art and was hosting a series of gatherings, inviting artists of the Italian Movement of the *Arte Povera*. The young artist was there, living in that cultural atmosphere, and yet kept his peculiar interest towards a different aesthetic, towards something much more connected with the natural decomposition. In other terms, the ephemeral was for the artist an aspect of his research, something inevitably connected with transience and decomposition, rather than a simple display. He was not interested in creating something ephemeral in irregular places, but on the contrary wanted to create a sort of permanent reflection upon corruption and decomposition with his art.

The perfect example of his young research is the exhibition *Cadaverina*, held on the island and showing fifteen boxes, each one displaying an organic substance.‖ For the whole duration of the show, which lasted fifteen days, the artist was daily entering the gallery, leaving new boxes with the same matter close to the ones of the day before. The final result was a massive display in which the spectator could re-build the progressive decomposition of the matter, since the box on top was containing an object left on the floor fifteen days before. So the first box was in a state of fifteen days of decomposition, the second one in a state of fourteen days of decomposition and so on and so forth till the very end: the fresh object at the bottom. The boxes of that first installation are still conserved at Barceló's archive today and it is interesting to notice how their process of decomposition is still running, affected also by their storage in a dark place. When discussing the artwork with the artist during one of our meetings, Barceló clearly stated that it is extremely interesting for him to see the final result, because the work of art keeps its own life, it moves and evolves, saving the idea that was at the very base of the intervention. This definition should be seen as a natural process of the work of art, which Barceló defined as "an itinerary of the decomposition".**

¶The deep interest towards the *Informel* movement is testified also by one of his most important trips. When he went to Bologna in 1974, in fact, he assisted with a solo show by Lucio Fontana. Many years later the artist declared: "We attended great contemporary Art exhibitions like Fontana's – I liked so much his work, and the exhibition was placed in a historic building. It was a way to bring modernity to the city."[5]

‖The materials were: one tomato; one banana; one fish; one fried egg; one cake; one flower; one liver; chick peas; one pepper; one heart; an apple; lentils; rice; soup.

**In the original version of the text: "un itinerario de la decomposición".[6]

The interest in decomposition was important to him in his early works,[††] but remained also when he slowly shifted his practice towards painting. After the first experimentation, such as *Cadaverina*, and contrarily to the main trends surrounding him, the artist decided to follow his instinct and started to paint. Pretty soon all the elements that were central in his early years also entered his painting practice and the interest towards the organic corruption of the matter certainly remained with him. But how is it possible to say that the interest towards the corruption, the ephemeral and the waste were still central in the painting practice? How can one associate one of the oldest techniques – the one that gave eternal memory to Giotto, Masaccio and Tiepolo (all of whom Barceló considers as his masters) – with the idea of the corruption? Well to understand this contrast one has to go back to the first paintings and understand how this early interest towards the decomposition was slowly shifted into practice. The first paintings made by the artists were made in conjunction with the last "boxes", which kept the idea of decomposing or rotten materials in the artwork. Once the artist started to paint he kept this practice, and he merged together sand, earth and acrylic. A practice that lasted for several years during the Eighties, as one can see in the series of paintings made in Naples – in 1983 – in which the earth taken from Vesuvius was merged with the painting (something that he started in his series of still lives). Commenting on such paintings, the artist declared:

> I wanted to find an interior and visual correspondence between the materials of painting and the themes painted in them (that mostly referred to foods, to rests after lunch, *etc*) [...] I worked in order to create a coincidence between the solid, the liquid, the white, the hard and the painting matter. What I was interested in, was this metaphoric play.[8,‡‡]

The play between the subject and the subject matter which Barceló is hereby describing started in the Eighties and remained central also in the Nineties, in the series of works he produced in Africa. In the latter works the depiction of the African landscapes was integrated with the insertion of natural materials found on site. In the series of works on paper, the artist mixed the painting with sands and earth, combining the depiction of the landscape with the landscape itself. The process that started with *Cadaverina*, the coexistence of the rotten representation and the rotting in itself, evolved in his

[††]After *Cadaverina*, Miquel Barceló presented a second show at 4 Gats. In this show the artist presented a series of boxes, similar to the ones of *Cadaverina*, that he realized during the winter. Moreover, the show also displayed a chicken in the process of decomposition. Describing the show Ferran said: "As soon as you entered the gallery, the first thing you saw was a rotting chicken. The stink of the decomposition, the corpse of the chicken, all made you want to get out of the gallery instead of getting in!"[7] Originally in the text: "Cuando entrabas en la galería, lo primero que veías era un pollo que estaba putriendo. La peste a descomposición, a cadáver de pollo, ya te daba ganas de... ¡En vez de apetecerte entrar en la galería, te apetecía salir!"

[‡‡]In the Spanish[9] it is possible to read: "Deseaba encontrar a través de los materiales de la pintura una correspondencia a avez interior y visual con los temas que representaban (que se referían, por supuesto, a los alimentos, a los restos de comida, *etc*). [...] Me empeñé en crear una coincidencia entre lo sólido, lo líquido, lo blanco, lo duro y la materia pictórica. Lo que me interesaba era ese juego dc metáforas".

painting giving life to a series of works in which the still life was absolutely matching the materials adopted. Starting from the tradition of the *bodegón* – a Spanish Baroque genre in which the still life was an excuse to represent the transcendence of life – the artist decided to reflect upon the passage of time creating a live painting surface, something in which the natural decomposition of the matters incorporated into the painting was mirroring the subject of the research. The movement of the surface, its progressive changes and decomposition, were therefore already conceived as something that can evolve, transform. There is no decomposition, nor damage, but only the natural evolution of the surface, something that is naturally related to the subject of the painting. The passage of time onto the canvas is the direct consequence of how the act of painting is even imagined by the artist. In his notebook, dated 1992, it is possible to read:

Imagine that they appear with the same naturality as mushrooms in the manure. Some hours after, when waking up on the sofa, he gave a desolate look to the studio: there he would follow the manure, but a track of mushrooms.[§§]

The excerpt is part of his African journals, a series of considerations that the artist was writing while living there, and that were later published in 2003. The same year as the publication, Barceló wrote:

Art is not a metaphor of the world, of the universe. A globular vision of the world, a centrifugal vision of the surface against the eyesight in depth. Art is not a mirror of life – it is a form of life, a good strange way of life at the sea bottom.

Technique advances, with a questionable speed.[¶¶]

In this declaration one can find a close link with the artist's *oevre* and his research. As in the case of the still lifes, in which the daily waste was entering the canvas, in the African production the earth of the river was adopted as a fundamental element, to foster the idea of the subject. The final result takes into account the accident, the decomposition: it has its own life. A life that is different from the one of the real, external world, or, to better say quoting directly the artist, a painting-life that "is a form" of life. For all the abovementioned reasons, the evolution of the painting, differently form the world one, will have its own rhythm. The external world, its science and progress share very little with the one populating the painting: the two might meet, but it is extremely important to understand, before approaching Barceló's art, that the art world keeps its internal laws and time. Even the creation of the work of art is like the natural growth of mushrooms: it has its own spontaneity and it is linked with the natural element, with the cycles of the earth

---

[§§]C.92.11_44 originally in the text: "Imaginas que aparezcan con la misma naturalidad que los champiñones entre el estiércol. Unas horas mas tarde al despertar en el sofá, dejó un vistazo desolado al taller: Ahí seguiría el estiércol, pero un rastro de champiñones."

[¶¶]C.03.02_50 originally in the text: "L'art c'est une métaphore de monde, de l'univers. Une vision global du monde, la vision centrifuge au la superficielle *versus* les vision en profondeur. L'art n'est pas un reflet de la vie – c'est une frome de vie, une bun étrange forme de vie des fond marins. La technique progresse, de forme de marche discutable."

and agriculture. The manure is something linked with that process but is also deeply connected with a practice that would become central few years afterwards in the research of the artist, especially in the African production.

In 1995, a few years after the writings, Barceló approached clay for the very first time and more particularly he decided to work with *dogon* clay, a very ancient material and at the very same time deeply connected with the idea of waste that was always central in Barceló's production, as we saw. The dogon clay is obtained by mixing together clay, waste of precedent pieces and animal faeces. The final result is an impure matter, having a specific texture and usually cooked in wood kiln, in which the temperature and the fire intensity may vary, giving to the final result some discrepancies that were very difficult to predict or to measure. The first experimentations with clay, their link with the waste and the impossible control upon them inspired also the creation of a new series of ceramics, recently presented by the artist. In this latter series, the artist decided to work with a corrupted clay, a matter that had fungus and spores inside it: interestingly enough, when the work of art was situated in a humid environment this spore come out again, transforming the work. The continuous evolution of the work of art, its own life – which cannot be predicted or controlled – is essential to the creation of the pieces and therefore no restoration can be considered possible for such ceramics: their corruption, rather than being a formal problem, is an integral part of the work of art, which is meant to change in time. Due to the natural evolution of the artwork, we have therefore to create a distinctive study for the emergence of fungus in clay and paintings: in the first case, in fact, we are facing something that was already considered by the artist, in the latter – instead – the natural evolution of waste, which was incorporated in the painting matter, was problematic for the overall stability of the painting. Therefore the restorer is called to find the right balance between the correct reading of the artwork and the inevitable corruption of the matter adopted by the artist. The same applies to all the artworks on paper, mainly produced in Mali at the beginning of the Nineties. In this particular production, in fact, the artist decided to use some papers that were attacked by insects in order to foster the overall effect of the composition.

Once again, before approaching the artwork, one has to understand the context that gave life to these creations. Back in the Nineties, Barceló returned to Mali after some months and found his paintings destroyed by some termites. Looking at the painting he noticed that quite often the erosion of the paper gave to the composition new possible readings and decided to somehow "force" the process. He dug a hole and started to make some of the attacks quicker; in other cases he took the "attacked" piece of paper and started to create figures around the insects bites. Quite often, the subject matter of this painting is daily life in Africa, its precarious equilibrium, and therefore, once again, the artist worked upon a distinctive short-circuit between the subject matter and the matter itself. The approach that he started several years before, using earth and food waste and integrating them into the painting, here moved one step forward and arrived at the point at which it is the rotten material that is at the very base of the painting. Obviously that does not mean that such works on paper cannot be restored, but also in that case, the rotten quality of the matter is a distinctive part of the artwork, something that has to remain as an overall effect.

Before entering into the restoration process that will be analysed throughout some case studies, it needs to be stressed that the artist is not against restoration *per se*. On the contrary, even his artistic practice, especially in the latest years, is extremely aware of the restoration process, which sometimes he even integrates into his practice. That is the case of the "black" series of vases in which he creates the ceramic pieces; he cooks them and later positions them on the top of the kiln to smoke them (Figure 12.1). Once the vases are fully black, he then scratches the surface, letting the clay colour come

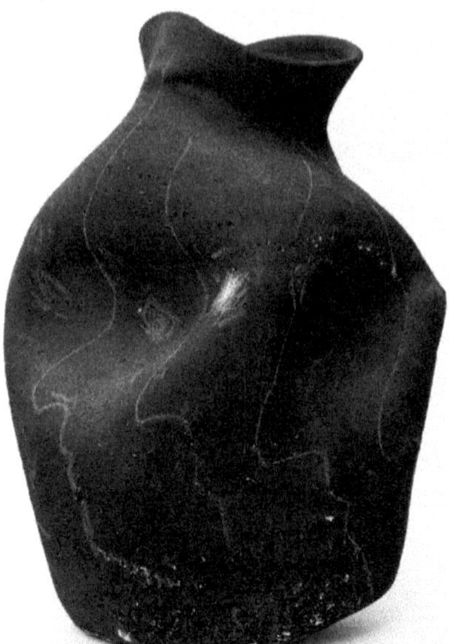

Figure 12.1 M. Barceló, *Les sets edats*, 2014, smoked earthenware. Courtesy of Miquel Barceló Atelier (photo) and Andrea Piffari. Reproduced from Miquel Barceló and Andrea Piffari with permission.

out again and, finally, he fixes the black by working with a team of restorers. This latter process shows how aware the artist is of technology, something that he does not want to fight, but rather is aiming to adopt in order to foster the overall effect of his artworks. The cooperation with the restorers is something that he has developed strongly over the years, taking active part in some restorations, giving indication and then, finally, working with the restorers side by side, even in the creation process.

At this point we can therefore summarise the artist's production with a dichotomy that comprises the two fundamental aspects of his practice: the waste and the rotten as a natural part of the decomposition, which can be limited but does not have to be negated, and the attentive connection with technology in order to "control" the evolution of the work of art and its dynamics. Due to double necessity, therefore, we can summarise Barceló's relation with the restoration as an extremely productive one: he is willing to maintain the possible reading of the artwork but he does not want to impose any idea of overall cleanness, because it never existed. In the following paragraph we will therefore face three different cases that were selected as the most efficient examples of the artist's relationship with restorers. Two of the suggested restorations involve painting: one on canvas and the other one on paper, keeping specific problems mostly connected with the different supports. The third case, finally, was presented in the light of the successful cooperation that the artist established with different craftsmen for the creation of his sculptures and most ambitious projects, in which the adoption of synthetic materials was absolutely important for the final result.

12.2 Case Studies

12.2.1 Painting on Cardboard

The self-portrait was here realised on cardboard, a particularly thick one, and was sent to restoration due to the extremely extended deterioration of the artwork.[‖] More particularly, the artwork is very compromised due to the deterioration of the paper, which is described by the restorer in these terms:

The work shows up as rolled up on itself. The deterioration is really advanced. The support has a vertical break that moves from the top of the artwork to the bottom. It is possible to notice deformations, cracks of considerable dimensions and numerous wrinkles, loss of paintings. There are also several stains of humidity and foxing.[***]

[‖]For all the information in this section my deep gratitude goes to Isabel Adrover Bia and Cristina Planas, painting conservators, who provided me with all the useful information.

[***]In the original report: "Cartulina blanca de grosor medio. La obra se presenta enrollada sobre sí misma. El grado de deterioro es muy alto. El soporte posee una rotura vertical que va desde la zona superior a la zona inferior. Se detectan deformaciones, fisuras de tamaño considerable, y numerosas arrugas y pérdidas de materia que componen la obra. Aparecen manchas de humedades y foxing (oxidación)."

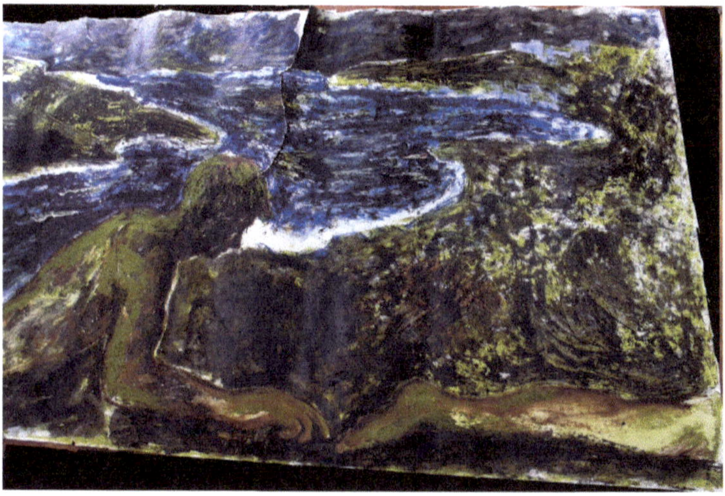

Figure 12.2 M. Barceló, *Autoretrato*, 1986, mixed media on paper, detail. Courtesy of Miquel Barceló, and Adrover Bia and Cristina Planas, Painting Conservators. Reproduced from Miquel Barceló and Adrover Bia and Cristina Planas, Painting Conservators, with permission.

As you can notice from the comments of the restorer, the support shows an advanced state of deterioration that also affected the painting. Taking into consideration this aspect, one has therefore to evaluate the restoration of the piece, stabilizing it and protecting the painting from further damage. In the first section it was underlined, in fact, how the natural evolution of the matter is considered by the artist as an integral part of the artwork, but mostly in those cases where the high *impasto* is going to move and change naturally, keeping the right balance of the composition. In this particular case, there is no natural slow movement of the matter; on the contrary, it is the movement of the cardboard that determines a progressive deterioration of the artwork. For this reason, the intervention in this case is not only necessary, but also highly recommended (Figure 12.2).

Moreover, looking at the causes of such deterioration, one can notice that all the factors determining such changes were mostly external to the artwork, or as the restorers notice: "The deterioration agents are mostly the incorrect storage and bad handling"[†††]. In the balance between the external world and the internal life of the work of art – that we underlined before while examining the artist's practice and his theories – doubtlessly the former had a bigger impact on the painting surface, compromising the stability of the final result. In such cases, therefore, the restoration is called to fix the movements

[†††]In the original report: "Factores de deterioro: Los agentes de deterioro que más le han afectado son un incorrecto almacenamiento, así como mala manipulación. Se observan distensiones en los puntos de rotura, en la zona de cortes y desgarros especialmente. Restos de adhesivo en las zonas de unión."

of the support, whilst bearing in mind the peculiar technique of the artist, *i.e.* the high *impasto*, while intervening.

Looking at the specificity of the object, one can notice that the painting was executed without any preparatory level (Figure 12.3). Miquel Barceló painted directly with the acrylic on the cardboard and the deterioration of the painting is mostly determined by this direct contact between the painting matter and the support: therefore when the cardboard was stored in humid places, it started to move, "rolling up on itself", and during this process even the acrylic painting cracked and started to detach from the surface. Examining the painting surface, the restorer noticed: "Dough acrylic. We find a very deteriorated original polychromy that does not stick to the support, with visible lacunae and loss of colour. It is possible to see also cracks, wrinkles

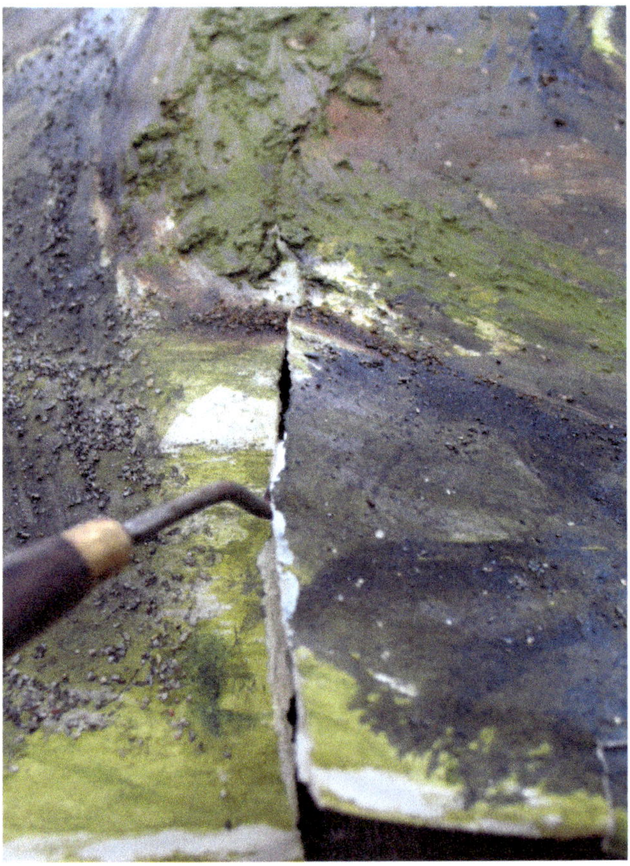

Figure 12.3 M. Barceló, *Autoretrato*, 1986, mixed media on paper, detail of the surface. Courtesy of Miquel Barceló, and Adrover Bia and Cristina Planas, Painting Conservators. Reproduced from Miquel Barceló, and Adrover Bia and Cristina Planas, Painting Conservators, with permission.

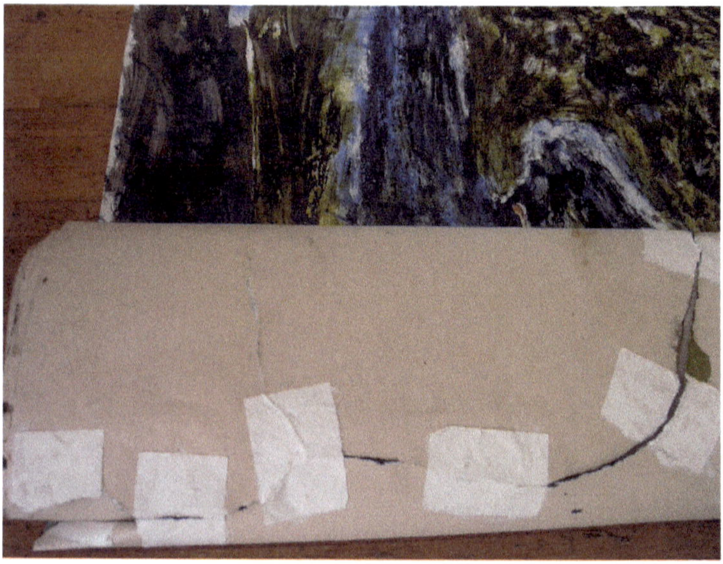

Figure 12.4 M. Barceló, *Autoretrato*, 1986, mixed media on paper, detail of the back. Courtesy of Miquel Barceló, and Adrover Bia and Cristina Planas, Painting Conservators. Reproduced from Miquel Barceló, and Adrover Bia and Cristina Planas, Painting Conservators, with permission.

and loss of pictorial film. The surface shows a thick layer of dirt."[‡‡‡] Due to the extended grade of deterioration, the restoration process had to concentrate mainly on two levels: stabilizing the support and cleaning the painting surface in order to re-equilibrate the final reading of the work.[§§§] The two separate phases of the restoration were conducted following a simple procedure: the surface was cleaned, both in front and on the back (the back of the painting was cleaned using a mechanical approach). In addition, while acting on the back, it was noticed that a previous restoration left some glue stripes that were removed applying some heat and solvents[¶¶¶] (Figure 12.4). The support was then stabilized further by "applying silk gauze and phi-free Japanese paper with methylcellulose in the broken areas, previously lowering

[‡‡‡]Originally in the text: "Acrílico, empastado. Encontramos una policromía original también muy deteriorada, mal adherida al soporte, con importantes lagunas y pérdida de policromía. Se observan cuarteados, cazoletas y levantamientos de la película pictórica. La superficie presenta una capa gruesa de suciedad."

[§§§]In the original report: "La propuesta de intervención va encaminada a la eliminación de deformaciones, intervenciones anteriores (eliminación de colas y papeles) y demás agentes degradantes que inciden en la obra, en la estabilización de la misma y en la intención de recuperación de la imagen cromática que pudiera tener en origen. Así mismo, siempre utilizaremos materiales y técnicas reversibles."

[¶¶¶]Originally in the text: "Limpieza superficial. Limpieza mecánica del reverso. Eliminación de cintas adhesivas: mediante calor, disolventes y mesa de succión. Eliminación de arrugas y deformaciones mediante la aplicación de humedad, calor y presión para conseguir aplanarlas."

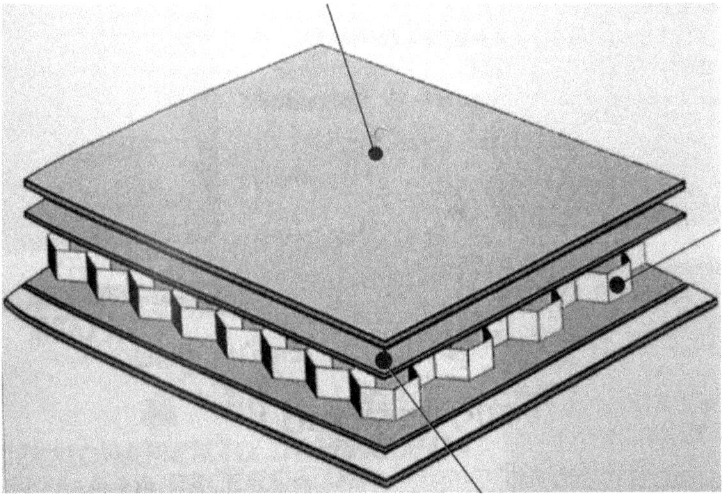

Figure 12.5 Structure of the intervention, as described by Adrover Bia and Cristina Planas, Painting Conservators. Reproduced from Adrover Bia and Cristina Planas, Painting Conservators, with permission.

the seams".[||||] Once it was stabilized, the restorer concentrated on the structure, which was built by laying the work upon an inert support, composed by "two panels of honeycomb of 15 mm of thickness, united by fiber rod glass (of 0.5 diameter) and epoxy resin"**** (Figures 12.5–12.7).

Once the support was fully stabilized, the restorer intervened with a reintegration of the missing colours. The exam of the painting surface revealed that the acrylic was originally mixed with organic elements, a praxis that was typical of the artist, as we saw in the first section. What is relevant to notice is that the integration of other materials into the acrylic did not lead to any specific deterioration and therefore it was decided not to intervene on that surface, keeping also in mind the process that the artist followed while making the painting. In the next case, a painting on canvas, we will notice something different, in the sense that in this second case the numerous materials that were part of the painting layer determined a deterioration that was advancing during the years and that required, instead, a double intervention.

[||||]Originally in the text: "Aplicación de gasas de seda o papel japonés libre de ph con metilcelulosa en zona de rotos y desgarros rebajando previamente las costuras."

****Originally in the text: "Posterioremente, a la colocación de la obra sobre un soporte inerte mediante dos paneles de nido de abeja de 15 mm. de grosor, unidos mediante varillas de fibra de vidrio de 0'5 de diámetro y resina epoxi."

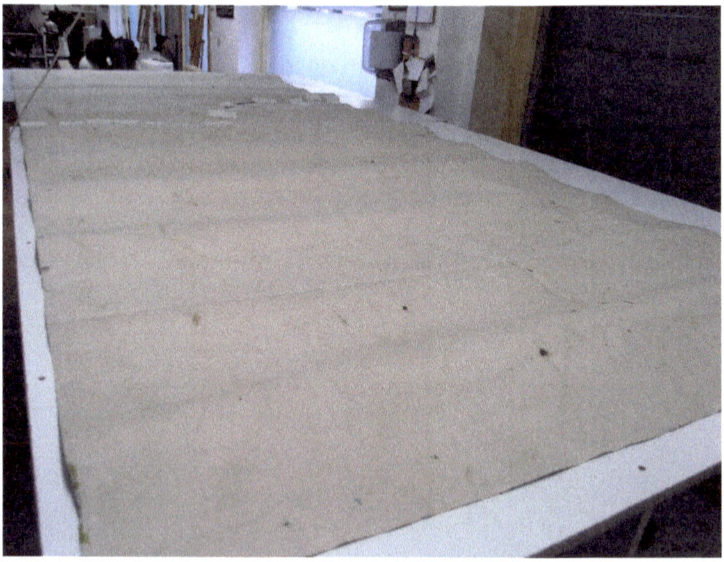

Figure 12.6 Pictures taken during the restoration. Courtesy of Adrover Bia and Cristina Planas, Painting Conservators. Reproduced from Miquel Barceló, and Adrover Bia and Cristina Planas, Painting Conservators, with permission.

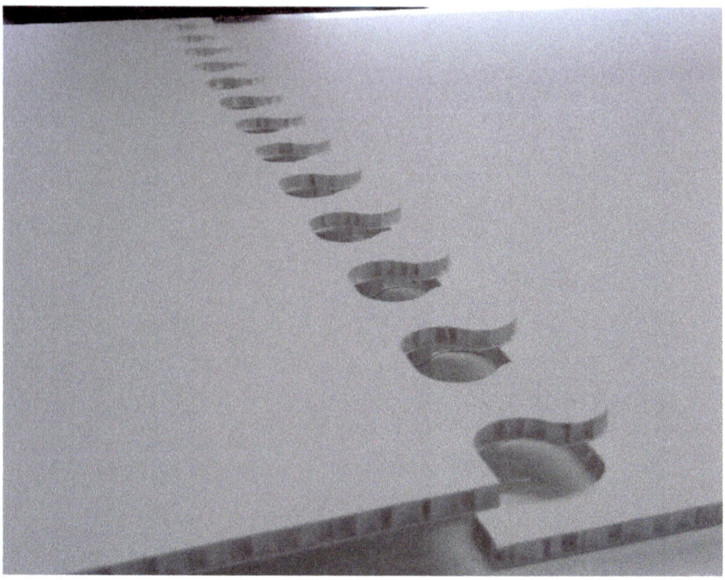

Figure 12.7 Pictures taken during the restoration. Courtesy of Adrover Bia and Cristina Planas, Painting Conservators. Reproduced from Miquel Barceló, and from Adrover Bia and Cristina Planas, Painting Conservators, with permission.

12.2.2 Painting on Canvas

The painting *L'ours blessée* was made in 2000 on canvas and is actually conserved at the Bayerische Staatsgemäldesammlungen in Munich. Contrary to the previous case, the painting did not manifest consistent problems on the support, but the main visible damage was mostly linked with the painting surface, which had visibly deteriorated, showing in 2010 the appearance of spores and mould (Figure 12.8). This case study is particularly close to the question that we examined in the first paragraph, and more particularly to the situations in which spores and mould were an integral part of Miquel

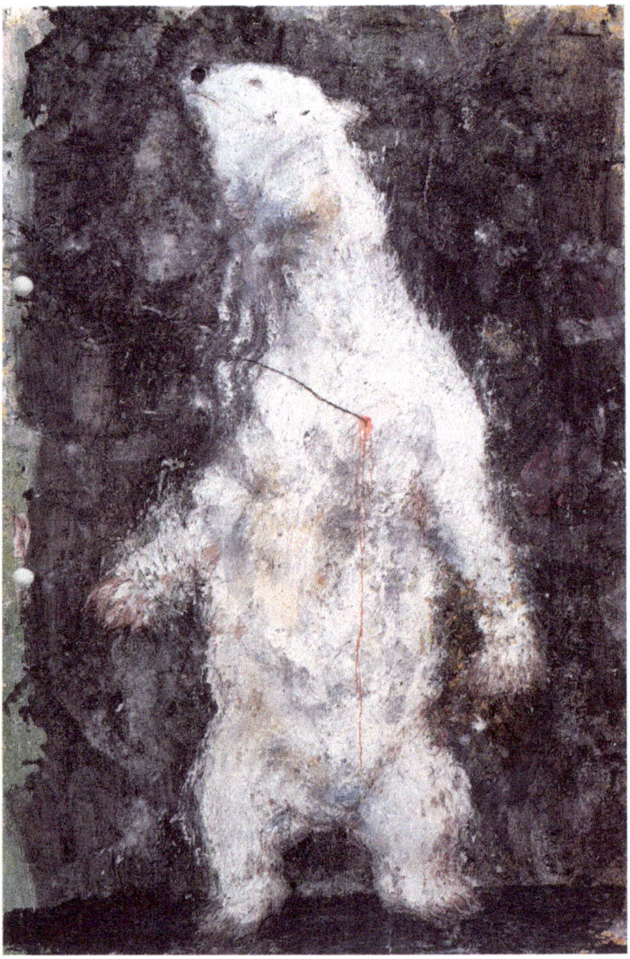

Figure 12.8 M. Barceló, *L'Ours Blessé*, 2000, mixed media on canvas. Courtesy of Udo and Anette Brandhorst Collection, Brandhorst Museum. Reproduced from Miquel Barceló and Udo and Anette Brandhorst Collection, Brandhorst Museum with permission.

Barceló's clays, especially in the series of vases produced in 2014. If in that case the progressive re-birth of such moulds is considered as the natural evolution of the work of art – and therefore no intervention is suggested – in this specific instance the mould was damaging the stability of the artwork and leading to future and consistent problems, therefore the restoration was necessary to block the loss of the painting. The natural life of the different matters present in the painting paste[††††] was therefore here leading to a progressive deterioration, a change in the surface structure that was also interfering with the original balance of colours that the artist tried to keep while painting. Therefore, back in 2010, the restoration of the *L'ours blessée* was made in order to preserve the painting, reaching again the original colour balance. The white spores were therefore treated to limit the danger, but they were not completely killed or eliminated on the whole surface, since in 2019 the painting showed new white spores. Even in this second instance it was decided to clean the artwork, giving priority to the balance of the composition and preventing future damage that might come from the future growth of other moulds and spores.

Therefore, to examine such restoration we have to create a distinction between the two moments of the intervention: we will first see the first cleaning, made in 2010, and then the second one, that took place at the opening of 2019. More particularly, the restoration started with a minimum intervention on the surface: the operation involved a mechanical cleaning (brush) and the adoption of a vacuum cleaner (equipped with HEPA filter). The spores and the mould were treated in order not only to disinfect the affected areas, but also to prevent future problems. The removal of the white spores was made by disinfecting the areas with a mixture of ethanol and water on a cotton swab (70% ethanol and 30% water). In 2010, at the time of this first restoration, the intervention was carried on mainly to eliminate the mould and no other treatments were made on the surface. More particularly, no direct painting reintegration was carried out, mostly because the mould did not affect directly all the layers of the preparation. In 2019, almost nine years after the first cleaning, the painting showed the same problems in other areas: therefore, the first intervention was truly effective, but the painting kept "moving" and "living", to quote Barceló's idea of the work of art (Figure 12.9).

The fact that the areas were different shows that there are no particular problems related to a specific introduction of one organic material, but the opposite – that it is the general mixture and integration of external elements into the painting strata that is determining the "life" of the painting. As was the case in the early works, so in this painting – made in 2000 – the mixed technique is incorporating several matters into the *impasto*, determining a

[††††]Describing the composition of the painting, Elisabeth Bushart (Leitende Restauratorin/Head of Conservation, Museum Brandhorst) wrote: "By bare eye and stereoscope we detected layers of pigmented paper-pulp and charcoal (the black arrow) plus half a cigarette (tobacco and paper). On top of all layers a few areas of glossy paint and possibly as red blood trace a gouache (?)."

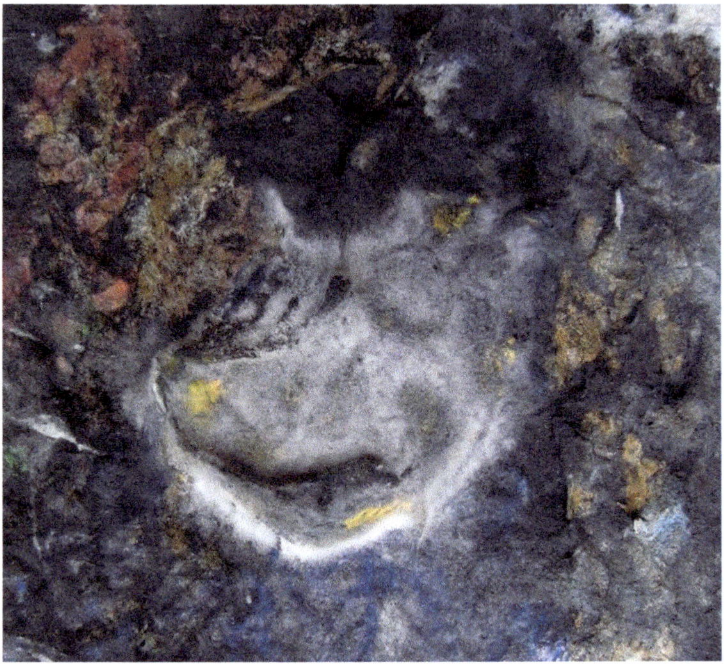

Figure 12.9 M. Barceló, *L'Ours Blessé*, picture of the mould found in 2010. Courtesy of Udo and Anette Brandhorst Collection, Brandhorst Museum. Reproduced from Miquel Barceló and Udo and Anette Brandhorst Collection, Brandhorst Museum with permission.

surface in continuous change and movement, as was shown by the necessity of restoring the painting twice. Nevertheless, the restoration was always of minor entities and was conduced on the surface, without carrying out systematic examinations of the layers.[‡‡‡‡] Pictures of the painting before the restoration show that the spores and mould were at their initial stage and therefore it was decided to promptly intervene, avoiding future growth (Figures 12.10 and 12.11).

The painting was examined in order to show the complexity of Barceló's technique, trying to create a conjuncture between the early work on cardboard and the later one on canvas. As per what was examined in the first paragraph, the technique of mixing and merging several materials in the process is something that is not limited to the painting technique, but – on the contrary – also determined Barceló's ceramic production. Moreover, from the Nineties onwards the artist converted such ceramics into bronze sculptures and then also systematically worked in bronze. The next and final section will examine the process of making his own bronze sculpture in order to understand how the artist works to obtain the *patina* and the most distinctive

[‡‡‡‡]In the declaration of the restorer: "After 8.5 years the mould spores returned in four very small areas and we treated them again with alcohol (70% ethanol, 30% water)."

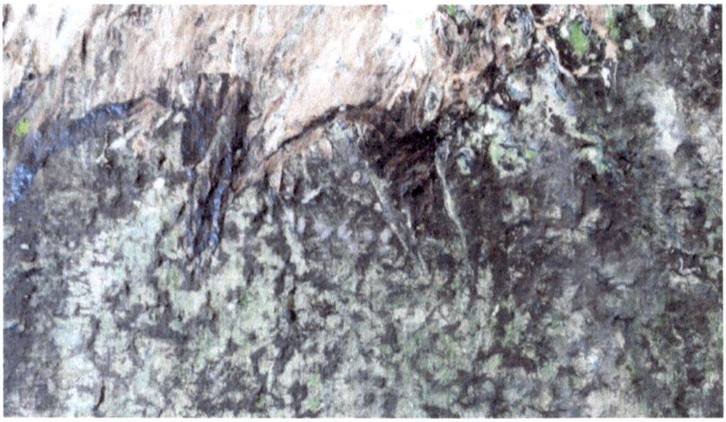

Figure 12.10 M. Barceló, *L´Ours Blessé*, picture of the mould found in 2019. Courtesy of Udo and Anette Brandhorst Collection, Brandhorst Museum. Reproduced from Miquel Barceló and Udo and Anette Brandhorst Collection, Brandhorst Museum with permission.

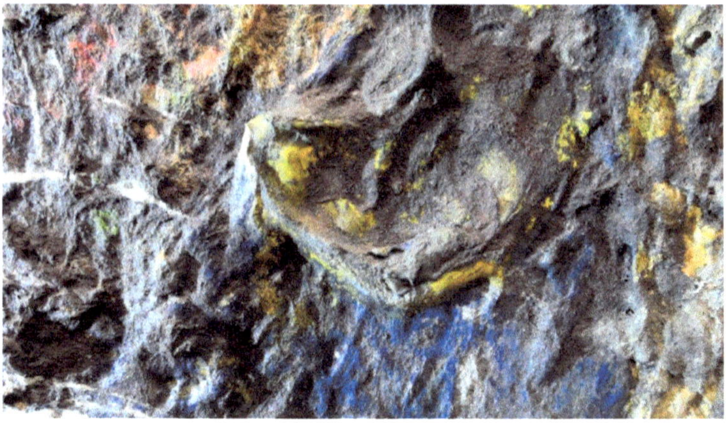

Figure 12.11 M. Barceló, *L´Ours Blessé,* picture after the remotion of the mould found in 2019. Courtesy of Udo and Anette Brandhorst Collection, Brandhorst Museum. Reproduced from Miquel Barceló and Udo and Anette Brandhorst Collection, Brandhorst Museum with permission.

characteristics of his process. Rather than an examination of restoration case studies, therefore, the last and final section of this paper can be interpreted as a guide to his approach, an analysis of how the work is made and how the artist conceives it. The final hope is that by understanding his process one might also trace the basic guidelines towards the heritage and the conservation of such works.

12.2.3 Technique and Construction in Barceló's Largest Projects and Sculpture

For the writing of the present paper, Alfa Arte – the group that worked with Miquel Barceló for the realization of some of his bronze sculptures – was interviewed and asked to share the process of construction. Describing it, the collaborator states:

> Once the bronze is finished, a wire brush is passed over the bronze so that the surface is ready. Then we apply potassium sulfide – very diluted in water, approximately diluted to 5% – and we let it dry, eventually by accelerating the drying with the help of a torch. Then the surface is cleaned with water, the wire brush is passed over the surface again so that in some areas the bronze colour comes out and finally wax is applied.
>
> In some cases – as for example in the case of *Elephant* – the finishing is easier: the surface is sandblasted to equalize it and 10% calcium hydroxide is applied and allowed to dry. Once dry, the residues are removed with a brush and wax is applied on the surface.

The attentive balance between the components was progressively examined by the team and the artist in order to obtain the final *patina*, something that is particularly important in works like *Elephant*, which are usually exhibited for very long periods outside. As per the general principle in Barceló's works, also in the bronze sculpture the life of the artwork is extremely important, and ultimately independent. But the fact that the work might move and change is assumed as a natural element: the only thing that can be done is to present the artwork with the best stability and balance, making it subject to the passage of time.

Also for this reason, the relationship that the artist has with his own collaborators is always based upon a strict connection, mostly due to the importance that the technique and the process have for him. The collaborations were extremely necessary whenever the artist approached materials that required an external expertise: more particularly, ceramics and bronzes were made working in a team structure, discussing all the technical aspects of the work. Despite the centrality of such figures in the technical building of the artwork, it has to be underlined that the collaboration for the artist ends with the brainstorming: once the artistic intervention needs to be done, in fact, he wants to act directly on the matter, working alone. In other words: the teamworking is mostly related to the first tests, to the definition of the correct matters and to the delineation of the working plan. Later the artistic creation always remains for the artist and his solitary work.

That was the case of the huge ceramic mural presented in the Cathedral of Palma, entirely realized in Italy with the collaboration of Vincenzo Santoriello. In that case, Santoriello's laboratory was totally invaded by iron

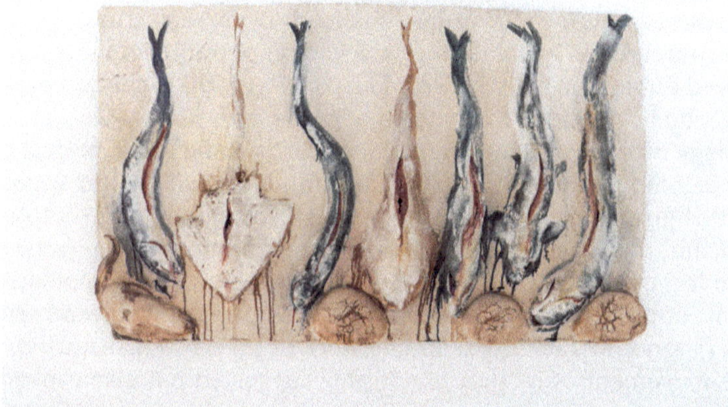

Figure 12.12 M. Barceló, *7 peixos, pans i cap de boc*, 2003, earthenware and glazes, 128 × 212 × 15 cm. Courtesy Atelier Barceló (photo) and Andrea Piffari. Preparatory panel for the work of the Palma de Majorca Cathedral. Reproduced from Miquel Barceló and Andrea Piffari with permission.

structures, built to position the clay almost vertically and helping the artist to work the surface when it was still wet. The balance between glazes was also tested several times and the final result is the production of numerous attempts. In the same way, after some thinking on how to work in the most efficient way, the artist and Santoriello decided to build up the structure, trying to reach also the ideal condition for the clay to dry. Once dry, the work was cooked in parts, more or less the same parts that the artist was working with daily, and only after the pieces were shipped were they finally installed on the wall of the cathedral (Figure 12.12).

The long work made for Palma, which involved a quite extended team for several years, is also the mirror of how Barceló approaches big and challenging projects, such as the huge ceiling in Geneva. For that particular work, in fact, a unit of chemistry experts was involved in order to arrive at the perfect final outcome. In that case, in fact, the artist was aiming to obtain a very natural effect, reproducing with a contemporary technique the consistency of the stalactites. A team of engineers and chemists made several attempts, all having the same problem: the matter was not stable on its own, especially in those cases in which one single stalactite had to be up to one meter tall. For the first two years the team worked upon the idea of shooting concrete upon a metallic structure, covered in a mixture of resins and celluloid. The main problem with such a solution was an aesthetic one: the metallic structure (which had to provide a sort of skeleton to the stalactite) was forcing the matter into a solid scheme and therefore the final effect was not very natural. Then, all of a sudden, the solution came from Ramon López, a biologist, who suggested an epoxy resin, Araldite® Madera, mostly used for the restoration of wood pieces.[§§§§] Once the ceiling was ready from a structural point of view, the artist intervened by "shooting" a selection of pure pigments onto them. He worked directly on the whole surface, only with the exception of the ultramarine, which was applied by a team of restorers.

The huge projects of Geneva and Palma really stand as the perfect example of how technique and chemistry were a fundamental part in Barceló's big and challenging projects, something that is also central in his sculptural production, and more particularly in his bronze sculptures. Finally, the relation between inorganic and organic materials is something that populates all his creations, and once called to intervene the restorer must always remember the co-existence of such matters. The work of art has, for the artist, its own life, the movement of its part is not only suggested but also evoked by the artist. The role of the restorer and all the other parties involved is to preserve its life and uniqueness, defending its life when the movements of its parts might become a threat for the whole composition.

References

1. M. Damiano, *Porque la vida no basta*, Editorial Anagrama, Barcelona, 2012, p. 51.
2. J. M. Mateos, Como Viven los Jovenes Occupantes de la Dragonera, *Diario de Mallorca*, July 1977, p. 18.
3. M. Damiano, *Porque la vida no basta*, Editorial Anagrama, Barcelona, 2012, p. 68.
4. M. Damiano, *Porque la vida no basta*, Editorial Anagrama, Barcelona, 2012, p. 53.
5. *Barceló abans Barceló, 1973–1982*, ed. J. Reus and M. Hevia, Circulos de Lectores, Galaxia Gutenberg, Barcelona, 2009, p. 306.
6. D. Ashton, *Miquel Barceló: A Mitad del Camino de la Vida*, Galaxia Gutemberg, Madrid, 2008, p. 47.
7. M. Damiano, *Porque la vida no basta*, Editorial Anagrama, Barcelona, 2012, p. 65.
8. *Peintures de 1983 à 1985*, ed. J. L. Froment, CAPC Musée d'art Contemporain de Bordeaux, Bordeaux, 1985, p. 14.
9. E. Juncosa, *De Rerum Natura*, Editorial Sintesis, Madrid, 1995, p. 33.
10. R. Rey Rosa, *El mar de Barceló en la Sala de los Derechos Humanos y de Alienza de Civilizaciones de la ONU en Ginebra*, Peninsula, Madrid, 2008.

[§§§§]In the declaration of López: "es el material que usamos para la restauración de piezas de madera y tambien para hacer prototipos de automoviles. [...] se trabaja con las manos, como si fuera una pasta para hacer pizza o goma da mascar."[10]

CHAPTER 13

Explosive Beauty: The Art of Cai Guo-Qiang

RACHEL RIVENC*[a], MICHAEL DOUTRE[b], VINCENT DION[c] AND TOM LEARNER[b]

[a]Getty Research Institute, 1200 Getty Center Drive, Suite 1100, Los Angeles, CA 90049, USA; [b]Getty Conservation Institute, 1200 Getty Center Drive, Suite 700, Los Angeles, CA 90049, USA; [c]Private Practice, Toronto, Canada
*E-mail: rrivenc@getty.edu

13.1 Introduction: From Paint to Gunpowder

Cai Guo-Qiang (born 1957) is one of the most prominent contemporary Chinese artists active today. His polymorph and prolific artistic practice draw on a personal cosmology that freely blends Eastern and Western culture. Having left China in 1986, Cai spent a very formative decade in Japan and in 1995 moved to the US. He has worked and lived in New York City since then. Cai retains very strong ties with China, but while his Asian heritage remains a strong artistic influence, he is first and foremost an international, itinerant artist, who travels the world to realize his many projects. Infused with a strong sense of symbolism, integrating Chinese traditions such as Taoism and Fengshui and reacting to international as well as Chinese history and politics, Cai's work aims at establishing bridges between different cultures and also engages in dialogue with global art histories. His work addresses social, political, and philosophical issues, but also personal and intimate subject matters.

Science and Art: The Contemporary Painted Surface
Edited by Antonio Sgamellotti, Brunetto Giovanni Brunetti and Costanza Miliani

Published by the Royal Society of Chemistry, www.rsc.org

Cai works in many different media, including large-scale installations, daytime fireworks, participative events, and time-based media. He is perhaps best known, however, for his signature use of gunpowder as an artistic medium (see Figure 13.1). Originally a painter, Cai started experimenting with burning his paint layers and combining them with gunpowder in the early 1980s in China. After moving to Japan, he soon dispensed with the paint altogether to explore the sole use of gunpowder as a drawing medium. For years, he experimented with black gunpowder, perfecting his techniques to create drawings that were at once strikingly beautiful and strongly conceptual.

Since 2015, he has been embracing the use of colored pyrotechnic powders to create a series of colored works on canvas, which have become increasingly exuberant and sensual (Figure 13.2 shows one of the first paintings he did in 2015 with colored powders). Although the term is not technically accurate, they are colloquially referred to as "colored gunpowder paintings", and this terminology will be used here. Cai considers this new development in his work to be the rebirth of his painting practice, and through the use of color he has been in the past few years engaged in a fervent dialogue with Western art history. For Cai, the return to painting started with a shift in subject matter which he attributes to the nostalgia and longing he felt during his long absence from home while working on the Beijing Olympics ceremony:

> Before 2008 I was intrigued by the sense of the larger cosmos, the relationship between human and the universe, the earth and land, and the society. During the two years of Olympics preparation time, I started to think a lot about the Chinese culture. For the opening I was seeking the

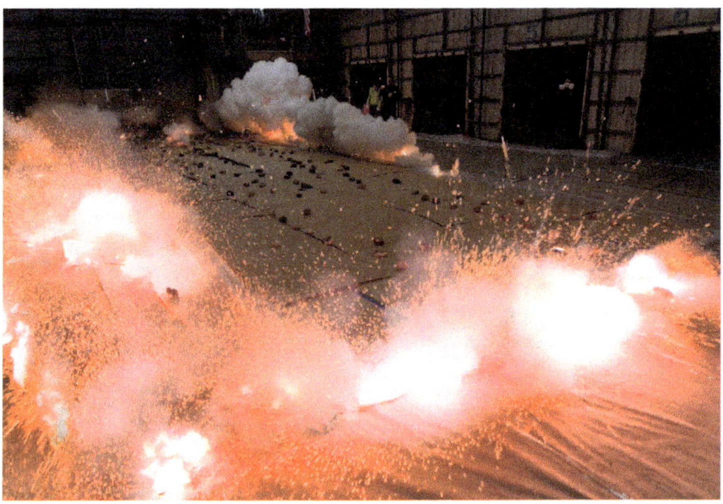

Figure 13.1 Ignition of gunpowder and colored powders to create *White Tone*, Long Island, April 8, 2016. Photo Wen-You Cai, courtesy of Cai Studio. Reproduced with permission.

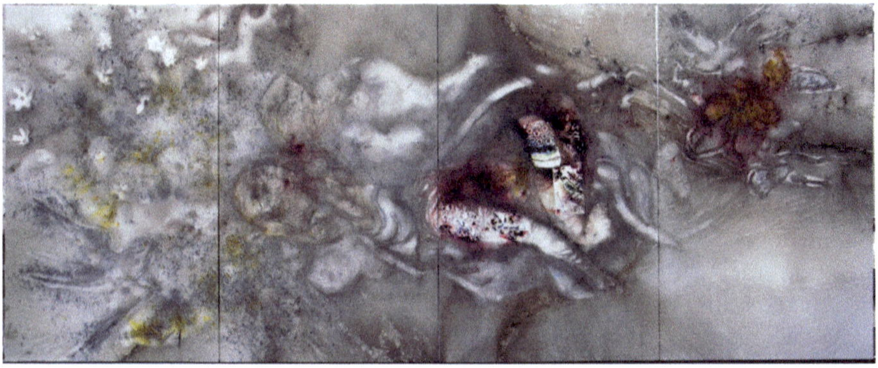

Figure 13.2 *Seasons of Life: Spring*, gunpowder on canvas, 259 × 648 cm, 2015. Photo Yokohama Museum of Art, courtesy of Cai Studio. Reproduced with permission.

> essence of Chinese culture and its significance to the world... The other reality is, during the two years preparing for the Olympics, I was mainly residing in Beijing. On my birthday I received a big poster written by the studio staff and hand carried by Hong Hong [his wife] from New York... This suggests that I have left my family, my studio for a year now. So I had more emotions towards my children; without even realizing, my eldest has graduated from high school and entered college. That's why after the Olympics I had experienced this huge change, which is focusing on florals and nature...this change has continuously evolved to this day, from florals and slowly evolving to colors···Installations and large outdoor explosion events are like social activities and revolutions. But paintings are similar to making love on the bed, it's about intimacy and materiality.[5]

Painting for Cai is not only about the subject, but also about the purpose. For decades, most of his gunpowder drawings were related to ideas that he had for explosion events: they were sketches, or preparatory drawings meant to convey the ideas he had for particular projects and events. Although the materials that he uses are far removed from traditional painting materials, Cai's current artistic practice is very much a quest for the essence or spirit of painting [the title of his solo show at the Prado in 2017], which involves a thorough questioning of its traditional aesthetic tenets and relinquishing of preconceived ideas in favor of material experimentation and complete aesthetic ingenuity – leading to a radically contemporary painting surface. As Cai puts it:

> Right now I'm not rushing into mixing and pairing colors. I'm trying to avoid what I had learned when I studied art, the good color palettes and bad palettes, how colors should be paired together to achieve harmony and beauty. I'm also trying to avoid my impressions of good paintings and

> paintings of the old masters. I'm intentionally trying to put that away. Of course, it's impossible to completely let go, but I don't want to use these as references. I want to use gunpowder colors in ways that are more spontaneous, improvised, discordant, and unrationed. There are several potential objectives for this. The first is the wild and not so rational. The second is to expose the core energy of this material, including its problems. The third is the encounter with the unknown. Since now I'm returning to painting, I'm faced with the history of painting and my own art, so there is this stage where we start from the beginning to historicize. Sometimes it may be difficult to appreciate and it may not feel harmonized, but isn't that what you're looking forward to seeing?[5]

In 2016, Cai's studio sought to understand the durability of the color smokes that Cai had started using and this inquiry brought them in contact with the Getty Conservation Institute (GCI). This was the beginning of a long-term collaboration between the studio and the GCI, the scope of which rapidly broadened to include a large-scale technical study of the materials and processes used by Cai and the aging and longevity of the colored works, and a close examination of Cai's attitude and thoughts towards aging, change, and conservation. The information is being collected through numerous interviews with Cai and some of the professionals assisting him in the creation of his work, as well as extensive and careful examination, sampling, and scientific analyses of a wide range of works. While the entirety of the findings will be published as a book, the present paper focuses on Cai's more recent painting practice using pyrotechnic powders. It describes the materials used, the techniques employed by the artist to create and presenting findings on their aging properties.

13.2 Materials and Process

Cai started using black gunpowder in the early 1980s, first experimenting with mixing the gunpowder into the paint and then by itself, using it to create drawings which he would then ignite to complete the work. He was attracted towards gunpowder first and foremost because of its uncontrollability, because he was seeking to liberate his style and personality from their perceived timidity and cautiousness, and also for its significance in Chinese history and philosophy.[5,11] He now also uses colored daytime fireworks to create ephemeral pyrotechnic displays as well as paintings (see Figure 13.3).

13.2.1 Gunpowder and Fireworks

Gunpowder is mainly composed of black powder, *i.e.* a mixture of an oxidizer, a source of sulfur, and fuels. Saltpeter (potassium nitrate) has historically been the most commonly used oxidizer but various other nitrates, chlorates, and perchlorates can also be found in black powder formulations. Sulfur is

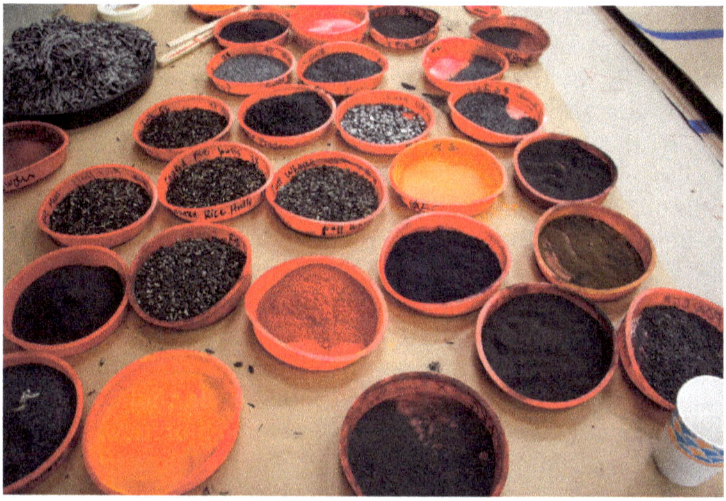

Figure 13.3 Different black and colored powders used to create *White Tone*, Long Island, April 8, 2016. Photo Rachel Rivenc, courtesy of Cai Studio. Reproduced with permission.

found as elemental sulfur or sulfide minerals such as realgar and serves as fuel for the ignition reaction.

Traditional black gunpowder has had a very set composit ion for centuries: 75% potassium nitrate, 15% charcoal, and 10% sulfur.[6,7] To be reactive and properly ignite and detonate ingredients must be brought into intimate contact, which is done by milling them with water then pressing them into a cake that then needs to be broken.[18] Overall, the methods of preparation and conditions of use are crucial to the resulting properties and behavior. Depending on the intended use, fuels can be provided by different forms of carbon in addition to charcoal, such as resins, gums, or carbohydrates. When the powder is used for pyrotechnic displays, additives such as metallic salts and filings, proteinaceous materials, and organic colorants can be employed to control the rate and appearance of the reaction. Main by-products of the combustion reaction include a range of sulfates, carbonates, and carboxylates.[6,14]

It is largely believed that the first fireworks appeared in China sometimes in the second century BC. The first firecrackers were simply bamboo stalks which, when thrown in the fire, exploded with a bang because of overheating of the compressed air. Actual gunpowder was likely discovered fortuitously around the 8th or 9th century, by Chinese alchemists in search of the elixir of life. Having prepared a concoction of honey, sulfur and saltpeter, they saw it abruptly erupt in flames, as the sulfur and honey acted as fuels and were subsequently oxidized by the saltpeter.[6,18]

An early form of rockets propelled by gunpowder was developed in China by the 10th century and precursors to firecrackers containing saltpeter are mentioned in Chinese texts from the 11th century.[6,18] While it is not clear exactly when gunpowder appeared in Europe, it is known that chemists/alchemists by the 13th century were experimenting with different proportions of saltpeter/sulfur/charcoal, and realized that "a loose, open tray of powder was all that was needed to produce a flash, but in order to produce the bang the powder needed to be confined".[18] The use of black powder as a propellant for cannons was widespread in Europe by the 14th century and has been essential to military successes in modern Europe ever since.[6,18]

Actual fireworks were introduced to Europe from China and India *via* the Middle East, reaching Italy first and then, by the 16th century, England.[18] Notable advances in the field of fireworks followed Berthollet's discovery of potassium chlorate in the 1780s, since potassium chlorate was able to produce brilliant flame colors, thereby adding color to the effects of sparks, noise, and motion that were already available with potassium nitrate-based compositions. The latter part of the 19th century marked the beginning of modern pyrotechnic technology, with the manufacture of magnesium and aluminum metals that could produce bright white sparks and white light, and the commercial availability of strontium, barium, and copper compounds which could create vivid red, green, and blue flames.[6]

13.2.2 Daytime Fireworks

The colored powders that Cai Guo-Qiang is using to create his paintings are mixtures designed for daytime fireworks. The compounds used are very different from the compounds used to create color in night-time fireworks. In daytime fireworks, a colored smoke is produced by volatilizing an organic dye which then condenses in air to form small solid particles. The dyes are strong absorbers of visible light. The light reflected off the vaporized particles misses the absorbed wavelengths, so that the complementary hue is perceived by observers. This is the opposite to night-time fireworks, which produce a colored flame. In this case the emitted wavelengths are perceived as color by viewers.

Daytime fireworks have a number of very specific requirements. The mixture needs to produce sufficient heat to vaporize the dye and a sufficient volume and density of gas to disperse the dye into the surrounding space, while at the same time igniting and burning smoothly at a low temperature (well below 1000 °C) in order to avoid decomposition of the dye. The dyes must be volatile enough to convert to the vapor state upon heating without substantial decomposition. For this reason, fairly low molecular weight dyes are generally used. They also need to be chemically stable at high temperature so oxygen-rich functional groups are usually avoided since they easily induce oxidative decomposition of the molecule; instead, amine groups are often used.[6] A crucial requirement is also that the smokes produced be non-toxic, since they are released in the atmosphere in proximity to the audience.

Because of these requirements, potassium chlorate, $KClO_3$, is often used as an oxidizer: it provides a low ignition temperature with reliable burning at low temperatures. Coolants such as sodium bicarbonate or magnesium carbonate are often added to the mixture, in the proportions required to obtain the desired rate of burning and correct reaction temperature. Carbohydrates (*i.e.* sugars such as lactose or sucrose) are often used as a fuel in combination with the potassium chlorate. The amount of heat can also be controlled by adjusting the potassium chlorate to fuel ratio. Lactose is nowadays the most common fuel used for daytime fireworks.[18] Cellulose nitrate is a common component of propellant compositions because the fuel and oxygen are combined within the same molecule.[18]

In terms of color, the range is limited by the number of heat-resistant dyestuffs that are commercially available. Anthraquinones are commonly used and can provide colors from orange to green depending on the level of conjugation – increased conjugation will cause a shift to longer wavelengths.[6] Cai Guo-Qiang works with six colors sourced from Japan and six colors sourced from China. Besides its volatility, the quantity of dye present in the mixture will also affect the burning rate. The dye acts as a diluent, so the greater the quantity of dye used, the slower the burning rate. At the same time, the dye needs to be present in sufficient amount to allow for the partial decomposition that inevitably occurs because of the heat. A typical colored smoke composition often contains 40–60% dye by weight.[6] Cai Guo-Qiang generally adds black gunpowder to the daytime fireworks he uses for his paintings, to offset their slower burning rate and produce a more spectacular deflagration.

13.2.3 Painting Process

The process often starts with lengthy iconographic research, during which Cai and his assistants might gather an abundance of reference images. Cai sketches extensively but the studio also uses computer-assisted drawings.

The whole process of doing the gunpowder drawing or painting is then executed horizontally. Often a stencil is prepared, unless the composition is more abstract, or spontaneous. For the stencil, the drawing is done by Cai himself on brown butcher paper. To enlarge the computer-assisted drawing and to create the stencils a numbered grid system is used, which is transferred to the brown paper, becoming the equivalent of a cartoon in traditional fresco technique. The sheets of brown paper are taped together until reaching the desired dimensions. Cai draws standing up, while the stencil is on the floor. To do this he utilizes a homemade contraption, a marker pen attached to a stick which allows him to draw standing up. His studio assistants, often with additional help from teams of local volunteers for very large drawings, then cut the stencils.

Preparing the actual drawing can take up to several days, depending on the size of the painting. The floor is protected with butcher paper and the substrate laid flat over it. Cai has used a range of substrates. The most common are Japanese paper or stretched canvas, but he has also used porcelain or silk.

For paper, Cai always uses thick handmade Japanese paper from the Kochi prefecture, sometimes marouflaged on thick wooden boards. For canvas Cai now uses commercially prepared canvases, usually locally purchased. The stencil is laid over the substrate, and the process of applying the gunpowder, realgar, and colored powders starts. Black gunpowder is always added to the pyrotechnic powders, otherwise they would produce too weak a deflagration. Cai sprinkles and rubs the different powders slowly, deliberately, almost meditatively. That process looks deceptively simple but relies entirely on Cai's consummate knowledge of the properties, reactivity, and appearance of the different powders. Although the comparison with action painting has limitations, the gunpowder painting process shares with it the deceptive nature of its apparent spontaneity and direct simplicity, in reality supported by hours and hours of practice and reflection. Sometimes, instead of, or in addition to the stencils, Cai uses foliage, branches, flowers, or actual objects or garments. He sprinkles the gunpowder around them, and after ignition most of the material is burnt off while the outline remains.

Once Cai is satisfied, the painting is prepared with the moment of ignition. The method depends on the result that Cai wants to achieve. Sometimes, the whole drawing or painting is covered with glassine; the glassine traps the smoke resulting from the ignition and creates drapery effects. The drawing is then covered with cardboard and weights placed on top of it to contain the deflagration (projected matter could injure Cai and his team, or the audience) and prevent the displacement of matter which would disturb the drawing. If there are a few motifs on vast expanses of white background, only the areas to be ignited get covered. Recently, Cai has experimented with a whole range of additional effects, such as: covering the canvas with a second, blank canvas of equal shape and size, creating a mirror painting; igniting a painting twice, with additional black gunpowder, creating a darker, often more ominous version (see Figure 13.4); using incense sticks or fuse lines dipped into colored powders to create the drawing.

13.3 Findings From Analyses and Interviews

Samples were collected from 20 works by Cai, ranging from his early paintings to his latest paintings using colored pyrotechnic powders, as well as from a range of tests created by the artist himself. Results from the early paintings are not included here. Analyses included FTIR and Raman spectroscopy, ESEM-EDX, and XRF (see Appendix 1 for experimental details).

13.3.1 Black Gunpowder Works

Over the years, Cai has explored the use of many different types of black gunpowder, and expanded his "palette" to include: gunpowder of several different sizes and granulometries; smokeless powders that create a grayer trace; and a few custom-made gunpowders developed specifically for him. In the United

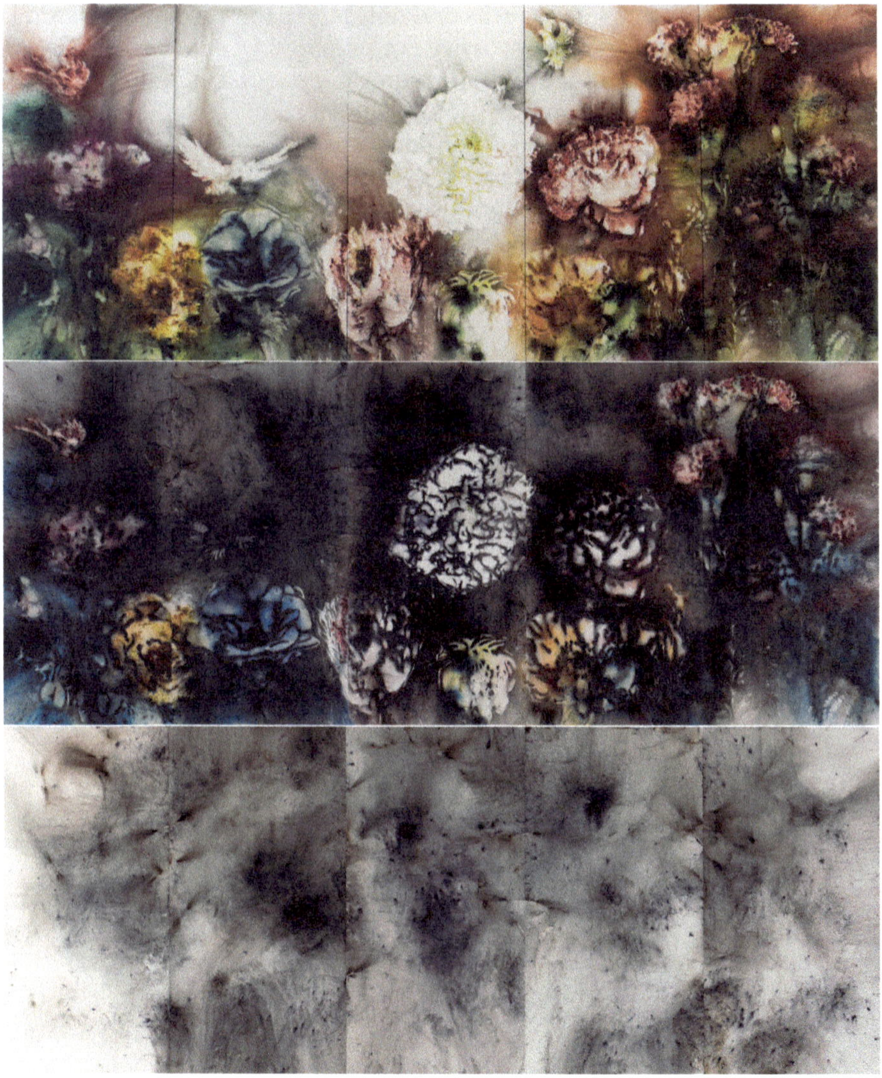

Figure 13.4 *From Top to Bottom*: *Heaven Complex No.1* [Artwork no longer extant]; *Heaven Complex No.1 and No.2* [second ignition]; and *Heaven Complex No.3* [ghost image formed on canvas placed on top of the other]. All: 2017, Gunpowder on canvas, 300 × 750 cm. Photos Wen-You Cai and Yvonne Zhao, courtesy of Cai Studio. Reproduced with permission.

States, Cai has been working for many years with Fireworks by Grucci Inc., whose owner and CEO, Phil Grucci, took a personal interest in the collaboration. Products developed specifically for Cai include gunpowder-coated rice puffs and rice hulls, coated one to three times. These coated materials contribute to producing different effects: the air pockets in the rice puffs make them move very quickly on the substrate creating a very distinctive scattered pattern. The organic material burns and only leaves the gunpowder behind, alongside some carbon residues. They are prepared by spraying on binder to make them tacky and then adding gunpowder and mixing into a tumbler type device – Phil Grucci compares it to a cement mixer[10] (see Table 13.1 for a description of the different blackpowders used by Cai).

When Cai undertakes projects abroad, he also likes to use locally sourced gunpowder, as part of his overall philosophy of creating dialogues between different cultures. He has observed that gunpowders sourced from different

Table 13.1 The main types of black gunpowders used by Cai Guo-Qiang.

English name	Chinese name as used by Cai studio	Description and notes from studio
Quickmatch	导火线	Fast-burning type of gunpowder used for the fuse, comes in two varieties: with paper or waterproof plastic coated casing which burns even faster at 40 m s; often used for igniting multiple points of a large drawing, thus preventing burns.
"Chinese scrap"	中国发射药	Refers to the granulated gunpowder, also known as "lift charge", or propellant black powder, found at the bottom of firework shells and that the Fireworks by Grucci Inc. plant had in surplus (hence "scrap"). Other pyrotechnic vendors would not know it as "Chinese scrap", but would understand "lift charge/propellant black powder from Chinese firework shells". It is used to ignite the less flammable meal powder.
Fuse powder	导火线粉	Black powder of the type used to make quick match, but in batch. When not available needs to be stripped manually from the cotton string inside the fuse.
Meal powder	"臭粉"	Fine black gunpowder. The studio notes that good ones have a pungent sulfurous smell, and generate lots of smoke for charring paper, which Cai uses to produce ink-wash effects.
Meal powder extra fine (XF)	"超细臭粉"	Same as above but with finer grain size. Cai often uses this in large volumes with finely cut cardboard stencils, particularly when making photorealistic works.
Rice hulls		Rice hulls coated with gunpowder, usually three to five times minimum.
Rice puffs		Rice puff cereals coated with gunpowder, specially made for Cai by Fireworks by Grucci Inc. Rice puff cereals must be sugar-free so they wont stick to the paper after deflagration or attract micro-organisms.

locations often behave differently, and in fact always incorporates a period of experimentation in the lead-up to a major exhibition or painting production to get accustomed to the properties of the new products. He has observed in particular that some mixtures burn faster and release more heat than others, leading to increased risks of creating burns and holes in the paper or canvas. This may relate more to differences in preparation (how thoroughly the components are mixed and pressed together, for example) and morphology than differences in chemistry.

Analyses generally show the gunpowder in various degrees of combustion, with organic, carbon-based residues. Elemental mapping with ESEM-EDX helps to visualize the degree of combustion: upon combustion, the potassium nitrate and sulfur recombine as potassium sulfate, as seen in Figure 13.5, while elemental sulfur found in the absence of potassium seems to indicate incomplete combustion. The presence of sulfur can also result from the use of the arsenic sulfide mineral realgar (alpha-As_4S_4), but in that case it is found in combination with arsenic (see also Figure 13.5). Cai has made extensive use of realgar for its orange color, its properties as additional fuel for the deflagration, and its association as a remedy in Chinese medicine (it is used for its antiseptic and antibacterial properties, but always in minute amounts because of its toxicity).[19] Up until 2015, and apart from a brief period of experimentation with colors in the early days, realgar was the only source of color in Cai's gunpowder drawings. ESEM-EDX can also give some indication of the physical form of the gunpowder and its method of preparation. Notably, the amount of organic residue present varies a great deal: for example, rice hulls tend to leave more residues than rice puffs. Tests made by the artist with Chinese gunpowders supplied by Blasting Technology of the Nanjing Institute

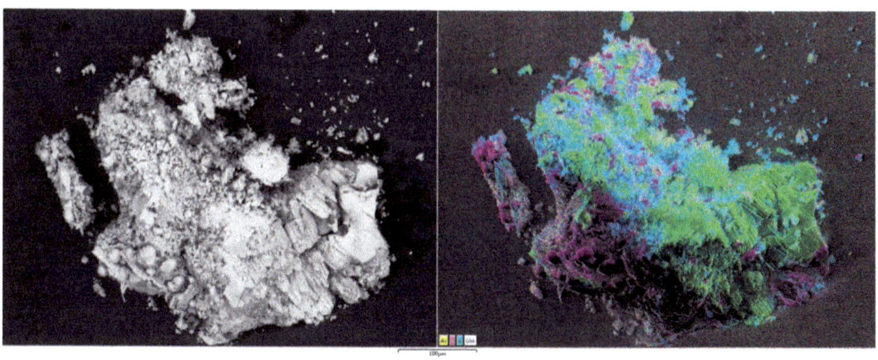

Figure 13.5 Backscatter electron image of a sample of the gunpowder from *Test Panel 1* (2014, left) and the distributions of realgar (green), pure elemental sulfur (purple), and potassium (blue) throughout the material (right). Note that the sulfur and potassium occur almost entirely in separate particles, indicating incomplete combustion. The laminar plates of realgar on the bottom left of the sample indicate that it has converted to pararealgar.

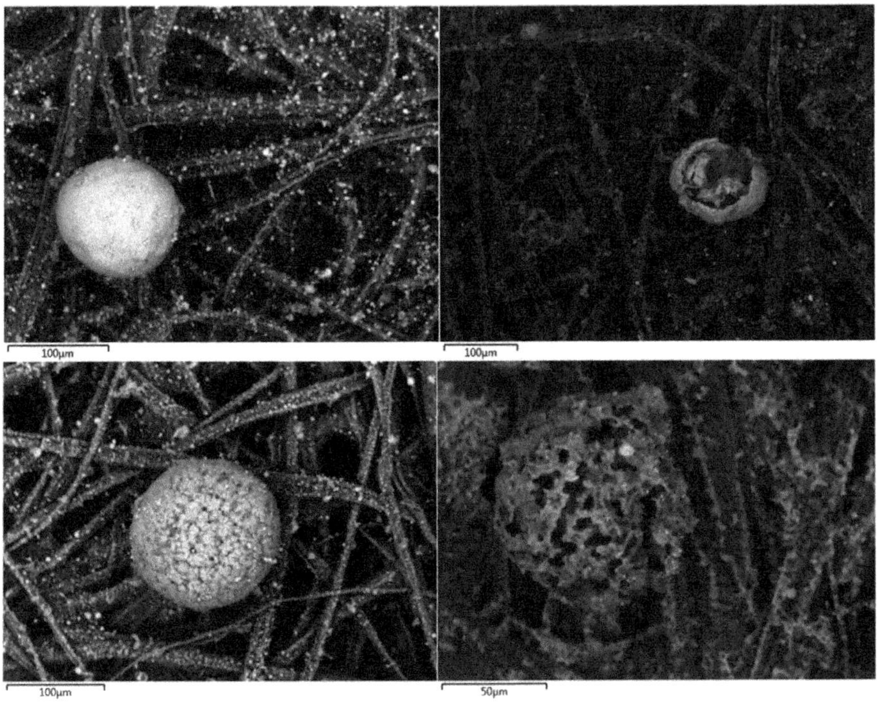

Figure 13.6 *Top*: Smooth pyrotechnic residue spheres formed post deflagration by black powder (potassium nitrate oxidizer) as used in the Japanese yellow on paper (left). Upon artificial aging, the spheres tend to collapse and reveal their hollow interiors (right). *Bottom*: Pebbly pyrotechnic residue spheres (left) formed post deflagration by Japanese green colored fireworks on paper (potassium chlorate oxidized). Upon artificial aging (right), these spheres tend to erode into a porous spherical web. Analyses performed by the Getty Conservation Institute.

of Engineering in preparation for an exhibition in Shanghai in 2014 were analyzed by ESEM-EDX. A coating consisting of a very smooth clay (sodium/aluminum/magnesium silicate) was observed, still intact, on the rice hulls as well as some calcium particles, mixed with black powder. In *Sunshine and Solitude – Pine Tree* (2010), in addition to black gunpowder in varying stages of combustion, siliceous materials were detected (silicon oxide with variable amounts of magnesium, potassium, sodium, calcium, and aluminum) consistent with clays used in granulating black powder. ESEM imaging also allows the observation of the spheres that are typically formed by potassium nitrate and potassium chlorate post deflagration[1,12,14] (see Figure 13.6).

13.3.2 Colored Paintings

As discussed previously, the colorful powders that Cai detonates to create his paintings are mixtures designed for daytime fireworks. A range of analyses were performed on test paintings provided by the artist and including six colored powders from Japan and six from China. The tests were prepared and ignited in the same way that Cai would make an actual painting. The set included:

- Two duplicate sets of five colored powders from Japan on paper (irregularly cut, roughly 14 × 14 cm)
- Two tests with colored powders on primed unstretched canvas (irregularly cut, roughly 23 × 104 cm), referred to as "scroll" canvas test A and B
- 12 samples with colored powders on primed stretched square canvases, six with colors purchased from Japan and six with colors purchased from China (25 × 25 cm)

These represent all the colors that the artist has been using for his colored gunpowder paintings. 20 tests of different types of black gunpowder on square canvases were also prepared by the artist but have not yet been tested.

In the tests made with the Japanese powders on paper, FTIR analyses identified potassium sulfate near the deflagration sites, presumably resulting from the addition of black gunpowder to the colored fireworks to promote a more explosive reaction (this was observed by the author as well as confirmed by the artist in interview[5]). Potassium chlorate was also identified by FTIR, and is the oxidizer for the colored powders which have not been completely consumed by the reaction. Elemental analyses with ESEM-EDX confirmed the presence of potassium, sulfur, and chlorine. FTIR also detected talc and silica, which might be substrates and extenders for the dyes, as well as cellulose ether compounds and proteinaceous material, which could have been used as additional fuels or binders or to alter the burning rate of the powder. The blue colorant was identified as phthalocyanine blue pigment, PB: 15, by both FTIR and Raman spectroscopy (see Figure 13.7). The red and yellow colorants were identified after solvent extraction as an anthraquinone dye, probably Solvent Red 3, and an azo dye, probably Solvent Yellow 56. Both the green and purple powders contain PB: 15 and the color appears to be produced by a mixture of the PB: 15 with another dye. However, analyses of the test marked "yellow" but green in appearance yielded an FTIR spectrum very similar to the one of Solvent Green 3, an anthraquinone dye. This is consistent with Conkling's remark that anthraquinones are commonly used for color smokes and can provide colors from orange to green depending on the level of conjugation in the molecule. The FTIR spectrum of the orange powder showed strong similarities to the spectrum of Solvent Yellow 176, a quinoline dye. A technical date sheet provided by Cai studio technical manager for the orange purchased from 井上玩具煙火 株式会社/Inoue Gangu Enka Co., Ltd mentions the use of CI Solvent Yellow 14 or Solvent Orange R, an azo dye known for being used in orange-colored smoke formulations.

With XRF analyses, sulfur and potassium were found in higher concentration in the fuse and black powder areas. Chlorine is present in higher concentration in the yellow and red areas, suggesting it may be part of the dye

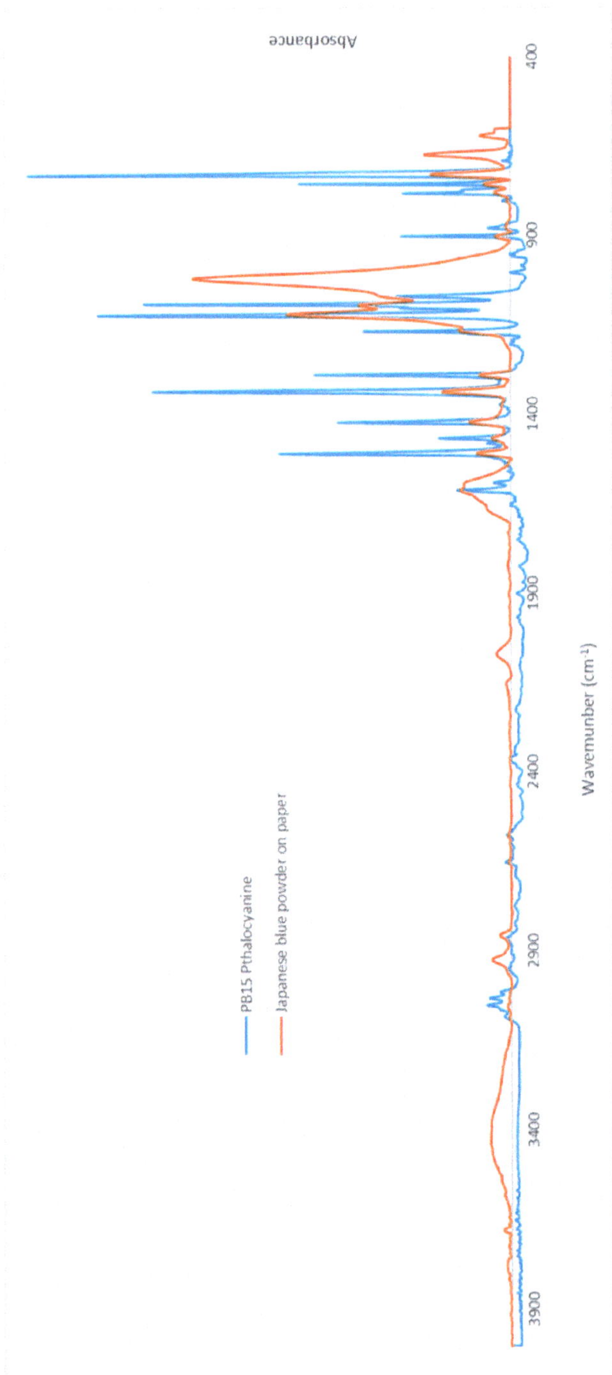

Figure 13.7 FTIR spectra of the Japanese blue gunpowder from the paper samples (orange spectrum) with reference phthalocyanine blue pigment (PB: 15, blue spectrum).

molecules' complex or a compound to which it was precipitated on, or that these colorants used a chlorate oxidizer.

PB15:3 blue is a copper-containing pigment and higher concentrations of this element have been found in blue, green, and purple areas, supporting its speculated use in combination with other colorants to create different colors – this was also confirmed by FTIR and Raman analyses.

In the tests made with the powders of Chinese origin, FTIR analyses also detected the presence of potassium sulfate in the blue, black, and red tests, confirming the use of black gunpowder in addition to the colored powders to magnify the deflagration. FTIR also confirmed the presence of PB15:3 in the blue test, while PV1 (Rhodamine B) was detected in the bright, pinkish-purple test, and an anthraquinone dye in the red test. The white test contained very large amounts of calcium carbonate, which presumably imparts the white color. Observation of the black test under the microscope revealed that a complex mixture of colored particles is present. Finally, efforts to find yellow colorant on the yellow mock-up were unsuccessful. The identification of the Chinese colorants is still ongoing.

13.4 Aging Studies

The aging behavior, in particular the lightfastness, of the colored test paintings prepared by the artist was assessed using both microfadeometry and artificial weathering in a Xenon Arc Weatherometer, in order to be able to compare both and visually assess the results (see Appendix 1 for experimental details).

13.4.1 Microfadeometry

Microfadeometry occurs *in situ* and subjects a microscopic area of a surface to intense illumination (up to five million lux) delivered by fiber optics and the ensuing color change is recorded directly with a spectrophotometer. The spectrophotometer functions according to the three-dimensional CIELAB color space and three distinct coordinates (L^*, a^*, b^*) are recorded as part of the monitored total color change: L^*: the lightness/brightness component (increasing values correspond to increasing brightness); a^*: the red/green component (negative values correspond to a green character and positive values to red); and b^*: the yellow/blue component (negative values correspond to a blue character and positive values to yellow). A reflectance curve of the total color difference $\Delta E94$ observed in time is generated using the changing values of the coordinates during the test. In addition, separate curves for the changing values of each of the L^*, a^*, and b^* coordinates are also generated, allowing one to appreciate their individual contribution to the total color change. The $\Delta E94$ reflectance curves obtained for the tested areas are typically plotted against the $\Delta E94$ reflectance curves of the International Standards Organization (ISO) Blue Wool Standards (BWS) 1 to 3, submitted to the same testing conditions.

The comparison between the fading behavior of the areas tested on the samples and those of Blue Wool Standards is used to give them a Blue Wool Standard Equivalent (BSWE). A preservation target can be defined with the results using the concept of the "just noticeable difference" (JND) benchmark which refers to the minimum color change required for it to be perceptible at least half of the time by a standard observer. Multiple studies have proposed a range of $\Delta E94$ values equivalent to one JND unit; in the field of conservation a $\Delta E94$ value of 1.5 is often used.[3,4,13,15,20] Stefan Michalski suggests that for Blue Wool Standard 1 (or its equivalent), one JND is reached after a light exposure of 300 000 lux-hours, with 900 000 lux-hours and 3 000 000 lux-hours necessary for the same change to happen for Blue Wool Standards 2 and 3 (or their equivalent) respectively.[16]

13.4.2 Artificial Weathering

Using the Arrhenius equation, the total of 576 hours exposure in the Weatherometer to which the tests were subjected can be considered indicative of up to 300 years of exposure in a museum-like, controlled lighting environment with conditions of 50 lux for 8 hours per day, 6 days per week. Of course, outside of museum and galleries, light levels are usually far higher – many typical indoor environments are 500–1500 lux, which would result in this aging protocol simulating something closer to 10–30 years of display in a home or office environment. The 53 °C (±0.5) air temperature in the chamber means that the works are likely to have also experienced a mild thermal aging. This would be most significant for the paper substrate. The colored and black powders, made to withstand high temperatures, are unlikely to have been affected by it. The results were evaluated visually, as color measurements were made impossible by the very friable surface, and by examining changes in the histograms of photographs taken under standardized illumination.

13.4.3 Results

13.4.3.1 Japanese Powders

Overall, both on paper and canvas the blue pigment appeared the most stable in the MFT studies – generally with a BWSE of 3 (see Figure 13.8), followed closely by the orange, green, and red with a BWSE of between 2 and 3, closer to 3 where the powder was thicker. The yellow could only be isolated on the canvas test and appeared more fugitive with a BWSE of 1 and higher in thick areas, but overall showed high variability between the tested areas, making a definitive assessment difficult. The thickness of the application had a significant impact on the lightfastness, with generally a shift towards higher BWSE where the powder is thicker, except for the yellow which is more fugitive when it is applied thicker.

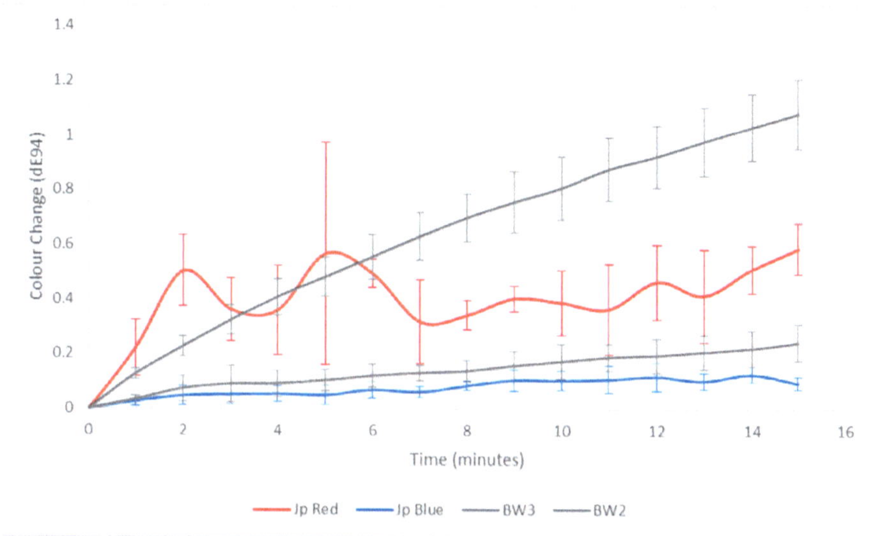

Figure 13.8 Total color change (CIELAB dE94) plotted against time for the pigmented areas on the paper samples of Japanese red colored fireworks (Jp Red) and Japanese blue colored fireworks (Jp Blue). The red colorant's stability rates between BW2 and BW3, whereas the blue colorant has a stability greater than BW2.

The paper substrate was tested in various samples and consistently showed a BWSE of between 2 and 3 while the charred areas showed a BWSE of 3 or above. The behavior of the paper substrate should be considered when looking at the total color change of an area of thin application of colorant or where the colorant has less hiding power. In the scroll canvas tests, the white ground preparation on the canvases displayed a strong fluorescence, possibly due to the presence of optical brighteners, and was given a BWSE of 3.

The results of artificial weathering showed a fairly good correlation with the MFT results. Among the Japanese powders applied on paper, the blue appeared the most stable and was virtually unchanged. In the green and purple, the second (*i.e.* yellow and red) colorants faded, leaving behind only the blue and shifting the overall color balance (see Figure 13.9). In the red test, the red colorant faded to an overall brownish color, with most of the fading occurring within 144 hours and continuing at a slower pace afterwards. The green colorant present on the paper test marked yellow (possibly containing Solvent Green 3) also appeared virtually unchanged. The paper itself appeared to initially darken slightly (after 144 hours) while it got bleached and became whiter upon further aging. The Japanese powders applied on the scroll canvas behaved similarly.

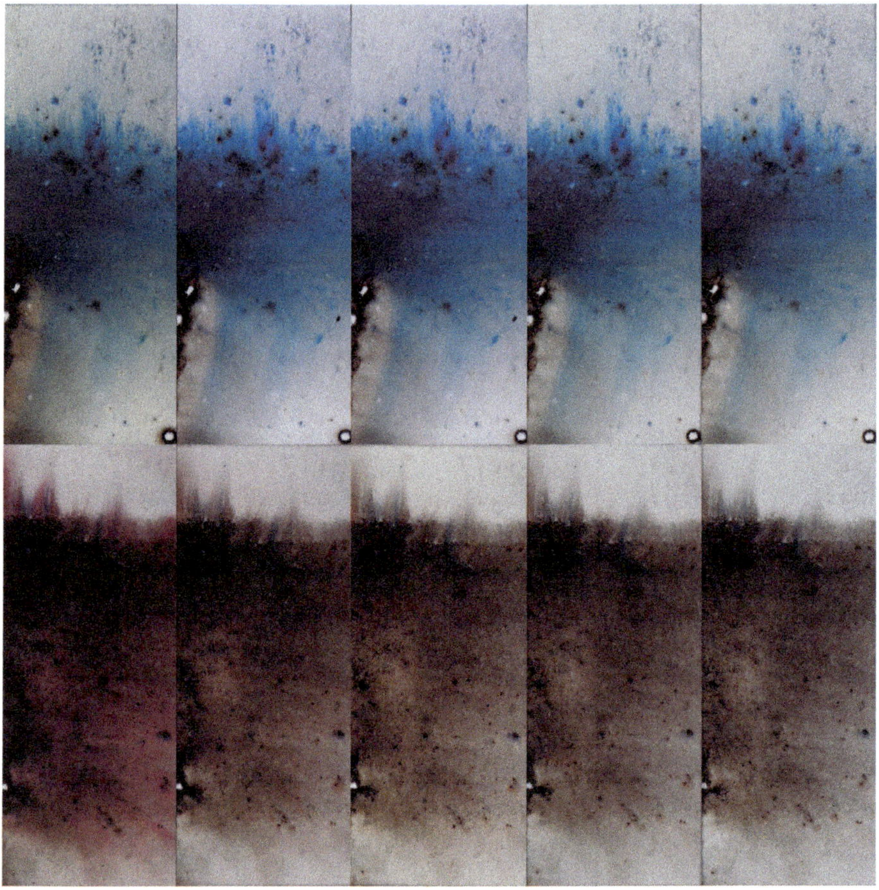

Figure 13.9 Paper samples of the red and blue Japanese colored gunpowders on paper, beginning with the original color and showing the progressive change in color during the artificial aging in 144 h increments (left to right, 0, 144, 288, 432, and 576 hours).

13.4.3.2 Chinese Powders on Square Canvas Tests

The ground preparation on the stretched square canvases ranked a BWSE of 3. The red, blue, and purple colorants ranked a BWSE of between 1 and 3, with the blue closer to 3 and above. The bright pinkish-purple powder showed a light sensitivity much closer to BWSE 1 when thinly applied, and closer to 2 when applied thickly. The white powder also ranked between BWSE 1 and 2. The black powder, on the other hand, was attributed a BWSE of higher than 3. On samples, a metallic sheen can be seen in some areas and the lightfastness in those areas is lower than a BWSE of 3.

Four samples with gunpowder and realgar, square canvases 15, 16, 17, and 18, were also tested. Yellow and orange areas associated with the use of realgar were shown to be very light sensitive, with the majority of the tested areas ranking below BWSE 1. The curves of orange realgar areas suggest a rapid and important change in its appearance during the first few minutes which is indicated by a strong slope in the reflectance curve ending in a peak. Following this abrupt change, the $\Delta E94$ value for the area either progressively slowly increases or decreases through the remaining period of the test. It is known that realgar is altered by exposure to visible light to form pararealgar (beta-As_4S_4).[8] The color change observed in the spectral curves is due to a large increase in the a^* component value (increase in red) and a decrease in the L^* component value (decrease in brightness) which is consistent with the alteration of a compound from an orange color to a red one.

Among the Chinese powders applied on the square canvas, the blue also appeared virtually unchanged, as well as the black and yellow. The red itself also seemed unaffected, except for the areas with metallic sheen exhibiting a dichroic effect (they appear red or green depending on the angle of viewing) which lost the sheen and green color upon aging. The purple, on the other hand, faded spectacularly and rapidly – most of the color was gone after 144 hours and kept fading until only a light salmon pink to white remained (see Figure 13.9).

Some interesting features were observed under optical microscopy, for example the red dye appeared to recrystallize as very thin needles which either sublime or migrate out of the canvas (some were observed on the metal holder which were not present at the beginning of the aging experiment). ESEM-EDX also shows that the spheres typically formed following the deflagration[12] tend to decompose and become hollow upon aging (see Figure 13.6).

13.5 Conclusion

Cai Guo-Qiang uses gunpowder as an artistic medium. This paper reports on the initial findings of a long-term project studying his work and provides chemical information on the compositions of the various types of gunpowders used by Cai, how they alter after deflagration, and how they are likely to change with time. These are all vital pieces of information needed to design appropriate conservation and preservation strategies for his works. This study has shown how the gunpowders have evolved during his career, both in terms of the materials used, and the associated application processes. Cai started using black gunpowder in the early 1980s, first experimenting with mixing the gunpowder into the paint and then by itself, using it to create drawings which he would then ignite to complete the work. After briefly experimenting with color, he limited himself to the use of black gunpowder and realgar for three decades. In 2015, he expanded his painting materials to include colored gunpowders, that are also used to create ephemeral and spectacular pyrotechnic displays. It was particularly interesting to identify the specific pigments and dyes used in the colored gunpowders, including some that are commonly found in modern paints, such as phthalocyanine blue, PB15, as well as realgar, a very traditional pigment that has not been used in paint production for centuries. The fading study of the colored gunpowders showed significant differences in behavior, with the most fugitive pigments being yellow and red, which also impacted on those colors that contained mixtures of pigments, such as green and purple. That said, none of these works showed a light sensitivity that would require any additional precautions to those that are quite typical in a museum environment for works on paper, in terms of light levels and display hours. The ability to work directly with the artist enabled a far deeper level of understanding, providing unprecedented access to his materials, processes, thoughts, and philosophies, including a level of acceptance of likely fading in the colored works. It is rather remarkable how these materials, intended to detonate and light up the sky for just a few minutes, have ended up showing quite reasonable durability when applied to paper or canvas.

Appendix 1. Experimental Details

Optical Microscopy: Samples were examined and imaged in visible light with either a Leica® DMRB analytical microscope coupled to a Canon® E0S 6D digital camera, or a Wild Heerbrugg® M8 stereomicroscope coupled to a Diagnostic Instruments Spot Flex 15.2 64 Mp Shifting Pixel digital camera.

FTIR: For each sample, selected particles were placed on a diamond window and flattened using a metal roller. The samples were analyzed using a 15x Cassegrain objective attached to a Bruker® Optics, Inc. Hyperion 3000 FTIR microscope housing a liquid nitrogen cooled mid-band MCT detector, and purged with dry air, collecting 64 scans at a resolution of 4 cm^{-1}. Reference spectra from the infrared spectral database were utilized in the identification process. Embedded cross-sections were analyzed using a 20x ATR objective, attached to the Bruker® Hyperion 3000 FTIR microscope. The collected spectra are the sum of 64 scans at a resolution of 4 cm^{-1}.

Raman micro-spectroscopy: Raman spectra were collected using Renishaw® inVia™ Raman microscopes (Renishaw, Inc. Hoffman Estates, IL) coupled to a Leica® DMLM microscope with either a 785 or 514 laser excitation. Samples were placed on the instrument platform and all spectra were calibrated using the 520.5 cm^{-1} silicon Raman band. Spectra were collected using an L-50X microscope objective (8 mm working distance, NA = 0.5). Laser power and collection times were varied for each particle examined to optimize the signal while avoiding sample degradation.

In the case of the blue, purple, and green samples, spectra were acquired under the following operational conditions: 785 nm laser excitation using ~0.0087 mW μm^{-2}, 1200 L mm^{-1} grating, Peltier® cooled CCD array detector; integration time 10 s and 1 accumulation over the spectral range 100–2000 cm^{-1}. Additional spectra for red samples were acquired with a 514 nm laser excitation using ~0.054 to 0.0054 mW μm^{-2}, 1200 L mm^{-1} grating, Peltier®

cooled CCD array detector; integration time 10 s and 1 accumulation over the spectral range 100–2000 cm^{-1}.

ESEM-EDX: Embedded cross-sections were imaged and analyzed with an FEI-Philips XL30 ESEM-FEG instrument, coupled to an Oxford Instruments® INCA EDX system, in H2O mode, with a 20 kV accelerating voltage and 0.7 torr water vapor.

XRF analyses: Elemental maps were generated using a Bruker® M6 Jet-stream spectrometer. The instrument consists of a measurement head attached to a motorized *x-y-z* gantry; the maximum scanning area is approximately 800 × 600 mm^2. The M6 uses a Rh-target X-ray tube with a microfocus polycapillary lens and a 30 mm^2 silicon drift detector (SDD) with a maximum energy resolution of <145 eV (Mn Kα) at a throughput of 130 kcps.

Operating parameters are: Voltage (kV) 50, Current (μA) 600, Spot size (μm) 330, Sampling distance (μm) 300, Dwell time (ms/pixel) 20, Matrix size (mm) 584 × 92, Resolution (*x*,*y*) 1947 × 307, Speed (mm s^{-1}) 15, Accel. (mm s^{-2}) 50.

Microfadeometry testing (MFT): The MFT benchtop system used was composed of an Oriel® tower light source with external Newport™ power supply, a Spectra-Physics Digital Exposure intensity controller and a Control Development spectrophotometer (Model PDA-512-USB/400-700/.2/100 um) tethered to a computer for data collection of reflected light from the sample.[9,17] The light source is a 75 W xenon arc lamp with a correlated color temperature between 5500 and 5700 K when UV and IR are filtered. The emitted light reaches the sample at the incident normal angle with a luminous flux of about 850 mlm focused on an estimated area of 0.4 mm diameter through a 450 um diameter silica fiber. The reflected light is collected at a 45° to the incident beam and directed to the spectrophotometer for spectral and colorimetry data acquisition every 60 seconds over a period of 15 minutes. Prior to testing samples, the light source intensity was locked in place using the controller and calibrated using a Labsphere Certified Reflectance Standard. Three reflectance curves were acquired for each of the ISO Blue Wool Standards 1, 2, and 3 to serve as references for the sample. Up to three spots were tested for selected areas of samples and ISO Blue Wool Standards were reevaluated at the beginning of each new testing period. All the colored test paintings were analyzed in areas of both low and high density (thin and thick application) and, based on the results, given a BWSE of between 1 and 3.

Artificial Weathering: Each paper test work was cut into two halves and one half was subjected to accelerated aging while the other was stored in the dark. The tests of Japanese powders were on scroll canvas B and the Chinese powders were on square canvas cut into rectangles of approx. 15 × 6 cm. The aging regime used was based upon test method C of ASTM D4303-10, *Standard Test Methods for Lightfastness of Colorants Used in Artist's Material.*[2] The paper samples were aged for four 144 h segments, for a total of 576 hours in an Atlas® Xenon Arc Weatherometer chamber set to 0.35 Wm^{-2} at 340 nm equivalent to 1793 MJm^{-2} of solar irradiance between 300 and 800 nm. The aging chamber was controlled to 63 °C black panel temperature, 53 °C (±0.5) air temperature, and 30% (±1) relative humidity. The samples were backed with mat board and placed in aluminum holders glazed with 3 mm polycarbonate sheets to protect the friable surface and decrease the UV component experienced by the works to near zero.

References

1. N. Abdul-Karim, C. S. Blackman, P. P. Gill, R. M. Morgan, L. Matjacic, R. Webb and H. N. Wing, Morphological Variations of Explosive Residue Particles and Implications for Understanding Detonation Mechanisms, *Anal. Chem.*, 2016, **88**(7), 3899–3908.
2. American Society for Testing and.Materials (ASTM), *ASTM D4303-10: Standard Test Methods for Lightfastness of Colorants Used in Artists' Materials*, ASTM International, West Conshohocken, PA, 2010.
3. J. Ashley-Smith, A. Derbyshire and B. Pretzel, The continuing development of a practical lighting policy for works of art on paper and other object types at the Victoria and Albert Museum, in *13th Triennial Meeting, Rio de Janiero, 22–27 September 2002: preprints*, ed. R. Vontobel and Icom Committee for Conservation, James & James, London, 2002, vol. 3–8.
4. J. Ashley-Smith, Chapter 12. Light Entertainment, in *Risk Assessment for Object Conservation*, Butterworth-Heinemann, Oxford, Boston, 1999, XIV, 358 p. ill. 24 cm.
5. G.-Q. Cai, *Interview with Rachel Rivenc 17 November 2016*, interview transcript, New York City, 2016.
6. J. A. Conkling, *Chemistry of Pyrotechnics: Basic Principles and Theory*, M. Dekker, New York, Basel, 1985.
7. T. L. Davis, *The Chemistry of Powder and Explosives*, Lindsay Pub, Bradley, IL, Reprint edn, 1990.
8. D. L. Douglass, C. Shing and Ge Wang, The light-induced alteration of realgar to pararealgar, *Am. Mineral.*, 1992, **77**(11–12), 1266–1274.
9. J. Druzik, Oriel Microfading Tester (MFT): a brief description, The Textile Specialty Group postprints of papers delivered at the Textile Subgroup Session: American Institute for Conservation, Annual meeting 2010, 20, pp. 135–143.
10. P. Grucci, Phil,. Interview with Rachel Rivenc, 28 November 2017, interview transcript, Long Island, New York, 2017.
11. P. Kaiser, The Land, the Earth, and the Universe, A Conversation with Cai Guo-Qiang, *Cai Guo-Qiang Ladder to the Sky*, The Museum of Contemporary Art Los Angeles and Prestel Verlag, Munich, London, New York, 2012, pp. 786–882.
12. K. L. Kosanke and B. J. Kosanke, Characterization of Pyrotechnic Reaction Residue Particles by SEM/EDS, *J. Forensic Sci.*, 2003, **48**(3).
13. A. Lerwill, *Microfading: Questioning the Basics,* Image Permanence Institute Bulletin, 2013.

14. C. Martin-Alberca and C. Garcia-Ruiz, Analytical techniques for the analysis of consumer fireworks, *Trends Anal. Chem.*, 2014, **56**, 27.
15. S. Michalski, The Lighting Decision, in *Fabric of an Exhibition: An Interdisciplinary Approach*, Ottawa, 1997, pp. 39–53.
16. S. Michalski, *Agent of Deterioration: Light, Ultraviolet and Infrared*, 2016, http://canada.pch.gc.ca/eng/1444925073140.
17. C. Pesme, A. Lerwill, V. Beltran and J. Druzik, Development of Contact Portable Microfade Tester to Assess Light Sensitivity of Collection Items, *J. Am. Inst. Conserv.*, 2016, **55**(2), 117–137.
18. M. S. Russell, *The Chemistry of Fireworks*, The Royal Society of Chemistry, Cambridge, 2nd edn, 2009.
19. E. H. Schafer, Orpiment and Realgar in Chinese Technology and Tradition, *J. Am. Orient. Soc.*, 1955, **75**(2), 73–89.
20. N. Turner, V. Beltran and C. Patterson, *Cumulative Light Exposure of Illuminated Manuscripts: Steps towards a New Exhibitions Policy*, 2016.

CHAPTER 14

There Are: Fabre's Visual Art

GIACINTO DI PIETRANTONIO*

Accademia di Belle Arti di Brera, Milano, Italy
*E-mail: giacinto.dipietrantonio@gmail.com,
giacintodipietrantonio@ababrera.it

There are alchemists, physicians, philosophers like Paracelsus who pointed out at the height of the Renaissance that: "Each element has its own colour: earth is blue, water is green, air is yellow, fire is red; then there are other colours, chance, motley, barely identifiable. But note the primary colour that dominates, and judge by that."[1]

Basically, everything is colour or has colour, which is why colour plays a key role in our lives. Paracelsus mentions blue and green at the height of the Renaissance, but they had been alternating or absent for some time before that.

But yellow air?

There are, at the start, caves painted in ochres, browns, reds, blacks, and no traces are seen of the blue and green that later become Fabre's colours.

There are mainly propitiatory scenes of the hunting of horned animals, but also sometimes hunter men and free women.

There are hands, many hands, hundreds of hands in Argentina, in the Cueva de las Manos (cave of hands). Hands used as stencils, paint spattered around them, again in the spectrum ranging from ochre to black. A kind of self-portrait dating back 9000 to 13 000 years, sensing the importance of the hand as humanity's natural evolutionary tool. A testimony to the importance of hand and handmade, debated still in art and rendered visible by Fabre in his corpus, especially in the jewel beetle and Bic ballpoint works,

Science and Art: The Contemporary Painted Surface
Edited by Antonio Sgamellotti, Brunetto Giovanni Brunetti and Costanza Miliani

Published by the Royal Society of Chemistry, www.rsc.org

which would be impossible without the skilled hand that assembles them or sketches out the lines.[2–4]

There are – progressing in order and in disorder, rule and unruliness, method and chaos, reason and heart and thus science and art – colours used since the sun first rose on humanity, for knowing and symbolizing the world, with the *What a Pleasant Madness* Bic-Art (1986, Figure 14.1) of being found in art.

There are, however, also the divine Egyptians, Jan Fabre's inspiration, a people who loved both blue and green and used one word to mean colour, being and essence.[2]

There are natural colours and artificial colours; in other words, colours that exist in nature and colours created by culture, especially by art, ranging from the natural hue of the jewel beetle's elytra-shell to artificial Bic-blue.

There are, indeed, people tasked with creating colour, ranging from the alchemists of history to the chemists of today, and artists above all others.

There are artists and poets, like Homer, who told of how the Greeks saw the blue of the Mediterranean as dark red.

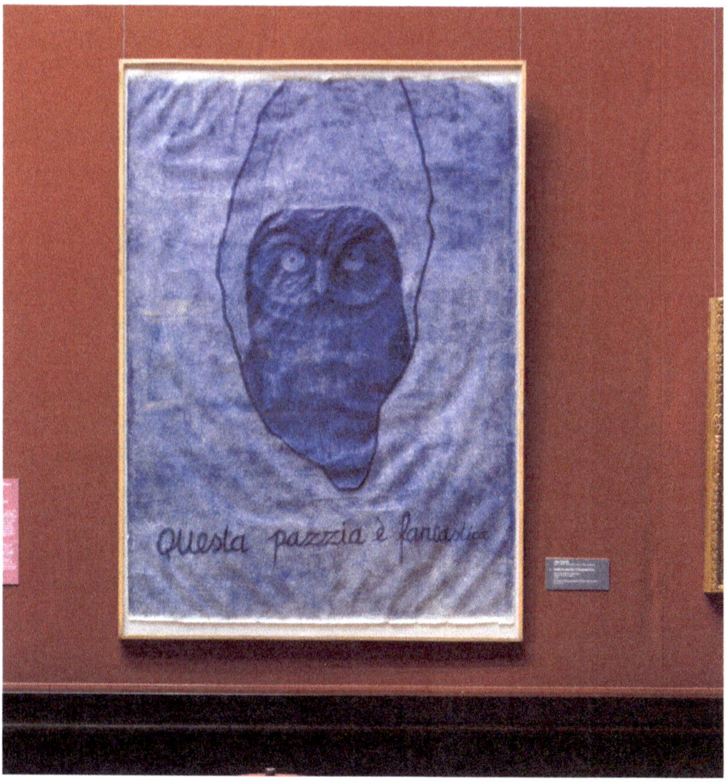

Figure 14.1 Jan Fabre. *Questa pazzia è fantastica/What a Pleasant Madness* (1986). Material: bic balpen op papier/Bic ink on paper. Size: 150 × 200 cm. Photo:/. Copyright: Angelos bvba.

There are, likewise, philosophers, miracle-workers, prophets, physicists and poets like Empedocles. He said that black is the colour of water, white the colour of fire and all other colours are born from their union, anticipating Malevich and his 1915 *Black Square*, the chromatic icon of the twentieth century, obtained by mixing many colours. The idea had already been espoused by many, even Aristotle, who ferried it well past the Renaissance. Because art, philosophy and literature see the world poetically, as Cicero emphasized when he wrote about Empedocles: "(...) those whom the Greeks call physicists are also to be called poets, because Empedocles was a physicist and he wrote a noble poem" (*De Oratore*, I, 217).[5,6]

There are, in literature and in art, colours that are and are not, like the word *bleu*, found nowhere in the tragedies of Corneille or the fables of La Fontaine, but this does not mean the French did not see this colour.

There are Earth, water, finite cosmos and infinite universe, in which everything has a colour that is never its own colour but an attributed colour. Indeed, we call our Earth the *Green Planet* although it is mainly blue, not only and not so much because of the blue-green of the seas and oceans that cover most of the globe's surface, but because the terrestrial globe, seen in its entirety from space appears blue. The blue of the atmosphere that inspired Domenico Modugno's song *Volare* soars into the bluest of skies and might almost have been inspired by Klein, its lyrics painting face and hands blue.

There are planets like Earth, and we call it the *Green Planet*, because as terrestrial animals able to give names to things, and therefore also to colours, we are beings who live on an Earth where green dominates; if we were fish or birds – other beings that Fabre frequents in artistic terms – we might have called it the *Blue Planet*.

There are, therefore, chromatic designations that speak of terrestrial not of celestial atmosphere, which they should, since the Earth is primarily in the atmosphere of the *sky, of the blue sky, the colour of the sea*.

There are, consequently, other questions as to why the two colours that have always been present in the greatest measure on Earth, mentioned and used today in life as in art, do not appear in the chromatic scale of the ancient Greeks. As already mentioned, they saw only red, yellow, white and black, and we shall see that these become the colours of alchemy. Why in the *Iliad* and in the *Odyssey* (especially in the latter, a seafaring tale), we might wonder once more, does Homer state that the sea is red, and not blue?

We end up pondering why the Greek artists like Phidias and Praxiteles, and philosophers like Plato and Aristotle, founders of Western culture – with whom Fabre is conversing through visual artwork like the bronze sculpture *Greek Tragedy & Greek Victory*, 2010 – do not mention blue or green.

Is this why they developed red and black vase painting, while their Egyptian predecessors and neighbours – from whom they inherited so much, including the art embodied by the *kouroi* – loved and used blue and green?

There are answers like this, which make it clear that if a colour is not seen by a society it does not mean it does not exist: it means it plays a secondary role in human activities, in social relations, in religious life, in the world of symbols and imagination.[7–9]

There are the ancient Egyptians, also Fabre's muses, for whom colour was so important that they used the same word to indicate both colour and being.

There are the Egyptians who saw many colours, including blue and green, which they used for painting their divinities.

There are many coloured gods in Egypt, the land where alchemy was born. Here the foremost deity, Amun or Amun-Ra, *unknowable father of one who has no mother*, transitioned from being painted in dark red to lapis lazuli blue, the colour of a mineral more precious than gold. He was also shown wearing the solar disk, sometimes the *ouroburos*, or the circle-scarab-jewel on his head. These are all symbols of the metamorphosis of life, death and rebirth dear to Fabre for shape and colour, meaning and symbology. There are other gods like Osiris, god of death, of fertility and therefore of birth and rebirth, not by chance painted green. The *Great Green of Osiris* and its transient character notwithstanding, it is the quintessential alchemical colour, intermediate fusion of body and soul, nature and spirit.[10]

There are then the ancient Romans also, who did everything they could to be the heirs of the Greeks: "one race, one face". They had little regard for blue and green, colours that from the seventh century BC were already being used by their enigmatic neighbours the Etruscans and, above all, by the barbarians, *e.g.* Celtic druids or the Visigoths commanded by King Alaric – the "peoples of the north" (Fabre's people) who invaded the Empire shortly thereafter.[11–13]

There are these two colours, blue and green, of little interest to the Romans who preferred red, white and yellow for senatorial togas, military uniforms, to the point that blue and green – despite being present aplenty in Pompeii – had no name, because then only red deserved its own name: the *Pompeian Red* of frescoes and therefore of art.

There are symbolic and mythological truths, religious and alchemical truths, and those certified by technology and science, filtered through *the lie of Art that makes us realize truth*.

There are truths and lies that fit into history, application, meaning, destiny, quality and quantity of colour.

There are the four colours of alchemy: *Nigredo*, the black of earth, decay, age and chaos; *Albedo*, the white of water, silver, spring and adolescence; *Citrinas*, the yellow of air, sun, gold, summer, maturity and sublimation; *Rubedo*, the red of fire, sunset, philosophical mercury, androgyne and of the metamorphic phoenix that continues to burn, die and rise again.

But there is also green: only green rises from its ashes, like the phoenix, since the great propitiatory bonfires of spring were lit only from green wood gathered in the early days of the year.

There are also intermediates in the alchemy of green and blue: the Latin *Viriditas* of flowering and budding, alchemically placed between *Nigredo* and *Albedo*, as well as blue as awareness of evil and melancholy.[14,15]

There are alchemies like one found in Isaac Newton's early modern science of colour, initially counting five hues (red, yellow, green, blue, violet). Later, with the prismatic spectrum, he separated light into seven colours, adding orange and indigo, but eventually he reached eleven, considering both alchemy and science. Going back to Aristotle's thinking (accepted without dispute until the Baroque period), there were only three colours in the rainbow, a natural phenomenon and thus observed from the dawn of humanity and of the Blue Hour. The poet Dante, however, in his *Purgatory*, already listed seven, ahead of Newton by more than three centuries, but miscalculating wildly compared to his guide Virgil, who was inclined to count a thousand colours despite urging his reader not to believe in colours. Then came Shelley, who spoke of a million colours.[16–21]

There are sciences, like that of the modern age, in which scientists have quantified 17 000 000 colours; others 2 200 000; and yet others, thriftier still, 346 000.[22] Although, should names be sought: "No language, however, has words for more than about 1000."[22]

There are, therefore, increasing numbers of questions. For instance, why did Titian – undisputed master of colour and of red – consider only six ... like Steve Jobs for his Apple logo?[23]

Apple has only six colours in its apple: indigo (a "shade" of blue) is missing. Perhaps because it saturates the screens of PCs, iPads, iPhones, in short, all the interfaces of electronic devices, namely the virtual face of the new world? Last but not least, television mixes blue, red and green, turning the latter into a primary colour too.

But who can ever say what and how many colours there are?

There are artists from antiquity to modernity in this excursus who create colour, or seize it and possess it with their names.

There are some like Titian's Red, mentioned above, Piero della Francesca's Light-Blue; Veronese's Green; Fabre's Jewel-Beetle Green; Van Gogh's Yellow; Malevich's Black; Ryman's White; "Colour of My Dreams Blue", as Miró called his own; Yves Klein's International Klein Blue; and not least of all, Fabre's Bic-Blue Art.

There are artists like Cézanne and Picasso who created and used some colours for periods of time. The latter is known for his "blue period" and "rose period", two colours, along with green, extensively used by Bosch in the *Garden of Earthly Delights* (*ca.* 1480–90) and another of Fabre's cornerstones, as we will see later.

There are Fabre's two colours, blue and green, less celebrated and used in antiquity but periodically becoming the most loved, used and developed by artists.

Blue was exiled until the early Middle Ages, a time when black and white, the red of martyrdom and the gold of spirituality prevailed. Then, with green, it became the colour of rebirth and began to assert itself in the early Renaissance, so much so that in the fifteenth century Cennino Cennini dedicated entire chapters of his *Libro de l'Arte* to green and blue-green, and then exclusively to blue, including precious lapis lazuli.

There are colours like blue finally considered worthy of appearing in the theory of painting and in the practice of the new art of the celestial Giotto in Assisi (Abbey of Saint Francis) and Padua (Scrovegni Chapel), or in almost all the Virgin Marys by Flemish painter Rogier van der Weyden (another of Fabre's touchstones). Moving on from the sacred to the profane, we see it in the hat worn in van Eyck's *Portrait of a Man* (*ca.* 1429), which paved the way to works with middle-class subjects by converted Catholic Vermeer. From his 1665 *Girl with a Pearl Earring* to the 1688 *Astronomer*, blue is a constant. Impressionists conjure up *plein air* blue. Kandinsky paints his blithe *Blue Rider*: musical and modern, which he likened to the sounds of flute and organ. On the other hand, like Mondrian, he considered green a boring, pointless colour.

There are also the blues woven into the romantic garb of Goethe's Young Werther and the poetry of Rimbaud's vowels with the black: **A**, white, red **I**, green **U**, blue **O**:

U, cycles, divine vibrations of green seas

Peace of pastures scattered with animals, peace of lines

That alchemy stamps on broad scholarly foreheads;

***O**, uplifting Trumpet full of strange sharps*

Silences traversed by Angels and Worlds!

– O the Omega, the violet ray of his eyes[24]

There are new eyes, then, to see new and old colours established in a long, slow process from south to north, from the Mediterranean to Flanders, appearing in much art of the time and of today. Indeed, if it is true that the mystic lamb in van Eyck's Ghent Altarpiece (1426–32) is white because until the Middle Ages white and red were the colours of Christianity and martyrdom, then it is equally true that there came a point when worldly and otherworldly nature was painted in green, like the green lawns composed of thousands of tiny leaves that surround the sacrificial animal.

Then there are saintly celestial women and pious earthly women sanctioning blue and green: the mantle of Piero della Francesca's Virgin Mary and the green gown worn by the bride in van Eyck's *Arnolfini Wedding* (1434).

There are women of bygone and modern times, from the Madonna in Her blue cosmic mantle to the earthly Costanza Trenta Arnolfini in her green wedding dress. The inescapable duality of colours is seen in the unity of birth and therefore of the future. As already mentioned, from the late Roman Empire and Middle Ages, new-borns were wrapped in a green cloth, symbolizing faith, hope and rebirth, qualities still upheld today as we enter the Green Age, after experiencing epochs ranging from stone to reinforced concrete.

There are symbolically new blue and green garments if we observe the Arnolfinis once more: a green wedding dress with blue sleeves, suggesting the modern way of life that was already beginning in the late Middle Ages, when brides chose to wear green as a gesture signifying reforms to repress paganism, ordered and applied by the Church, and replacing the ambiguous red worn both by brides and prostitutes in Roman times.

There are colours like green and blue whose volatility made it difficult for them to become established, and they were not understood until at least 1704, the year of the chance discovery of Prussian blue by Berlin chemists Diescbach and Dippel.

There are blue and green that continue to be ambivalent, accepted then rejected because associated with death, because they are so very difficult to stabilize. Transient, performative colours, just as Jan Fabre's blues and greens are now performative, but not in the chemical sense. Indeed, strange to say and believe, but even in ancient Greek times it was believed that ichor, the pure white blood of the immortal gods, became green and poisonous when a divinity was wounded. Despite the passing of time from ancient to modern, green was again associated with death by poisoning, even in 1814, forsaking mythology for the chemistry of Wilhelm Sattler and Friedrich Ruß, who began to sell *Schweinfurt green*, a pigment made with copper arsenite and therefore toxic. A vivid green that brought death, commonly called *Paris green*, was a favourite with Napoleon, who had the wallpaper of his house at Saint Helena dyed with this copper-arsenic and whose exhalations the exiled emperor absorbed, leading to his death.

There are also the green derivatives of chromium (the oxide and aluminium) created by Antwerp artist–alchemist Jean-Pierre François Lamorinière, "a green never seen before" and thus called *Lamorinière green*.

There are, then, the green and the blue associated with poisonous substances and therefore death, while today we see them as the colours of life. Green and green-blue, blue and blue-green were associated with mortality because the body that leaves this life for the next acquires the green and blue-green hues of death in the process of putrefaction, a process depicted in Andrea Mantegna's archetypal *Dead Christ* (1475–78).

There are, however, yet other reasons, and they can already be seen in Medieval and Renaissance iconography of hell, in which devils, and above all Satan, Lord of Evil, were depicted in green and green-blue, blue and blue-green. For instance, in the paintings of Bosch, already mentioned, or in the *Isenheim Altarpiece* (1512–16) by Matthias Grünewald, whose surname, ironically, means green forest.

There are iconoclastic and chromoclastic reasons, like the battle against blue and green, and blends that distort colours, which Pliny the Elder criticized heatedly in his *Natural History*, because they came from the East and were perceived as part of a more general opposition to what would now be called contemporary art.[25]

There are other reasons that target green alone: "monochromoclasty", revived at the time of the Crusades when it was discovered that green was the colour of Muhammad and therefore of the Muslims. Nor is it a coincidence that Muhammad, in the *Inferno* (1410), frescoed by Giovanni da Modena in Bologna's San Petronio Basilica, can be seen precisely in Hell, above a huge green Satan.[26–44] Then, in the Mannerist period, green began to glow once more thanks to the hues of the emerald stone used to make the table of the alchemist Hermes Trismegistus.

There are changes that are alchemical and others that are not, since the Green Table, already found in pre-Christian times by Sarah, wife of Abraham, becomes influential once more because it is an alchemical colour of faith with the gift of prophecy. This is how green, passing through alchemy, was recovered by the Catholics to thrash the green of the infidels.

There are even more radical chromoclasties, such as that dictated by the Protestant ethic of Luther, Calvin and Zwingli, which allowed more or less only black and white: a chromatic reduction that helped make Rembrandt great.

There are, as already mentioned, colours that change the sense of blue and green, which turns from negative to positive. But there are also things that follow a similar process which we mention because they are to do with Fabre's art: the convex mirror present in van Eyck's *Arnolfini Wedding*.

The Arnolfinis' convex mirror is a symbolic virginal, passionate thing, shining, glimmering like the jewel beetle's elytra-shell and the armour of Fabre's performances.

There are, consequently, colours, things and artists with a special role in the life and art of an artist. For Fabre they are green, blue, armour and Flemish masters.

There are knights of the faith of religious crusades like those of art, or the horse races held in Byzantium, where the Orthodox (blue) Venetian knights represented power and clashed with the (green) monosophist Prasini knights of the plebs. Because throughout history, the poor almost always wear green. The green knight, however, like the people he defended, was always endowed with audacity and insolence, in any case a disturber of established order. For this reason Fabre has long used armoured animals for his art: the elytra-shell of the jewel beetle, "the oldest computer in the world" (Fabre), with which he produces works of great alchemical intensity.

There are armoured scarabs, upside-down animals, because – unlike most of the animal kingdom as a whole – they are hard outside and soft inside. Fabre has addressed the theme of hard and soft, tough and fragile, outside and inside, many times over. A remarkable example was the 2017 *Glass and Bone Sculptures 1977–2017* exhibition in Venice's Abbey of San Gregorio,[3] where the fragility of glass was shown alongside the hardness of bone and the colours jewel-beetle green and Bic-blue.

There are passions, fragilities, emotions, hardships and condemnations like those intrinsic to the 1991 post-colonial work *Canoe* (Figure 14.2), where a typical Congo boat made of hard bones – a metaphor for the victims of Belgian colonialism – has oars of fragile glass with paddle-hands of human

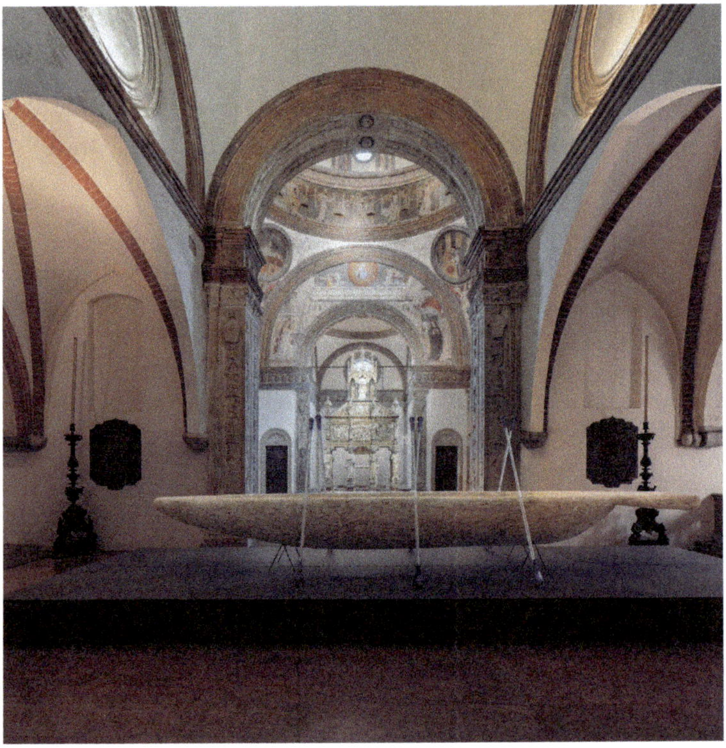

Figure 14.2 Jan Fabre. *Da un' altra Faccia del Tempo/From Another Side of Time* (1988). Material: Glass, Human bones, Bic ink. Size: 587.2 × 66.2 × 4.8 cm. Photo: Pat Verbruggen. Copyright: Angelos bvba.

beings of various non-European races painted Bic-blue in memory of Congolese hands cut off by Belgian colonialists.

There are the Bic hands of the Blue Hour of transition and rebirth, like those of the jewel beetle mosaic series. But the use of bones, especially in this context, also means that while the skin of humans and animals are of different colours, bones are the same egalitarian ivory white for all. And ivory was one of the reasons for the colonialist slaughter of elephants, as we can see in the works *Elephant's Tusks* (Figure 14.3) and *Avant Grade* (Figure 14.4), both 2011, from the *Tribute to Belgian Congo* jewel beetle cycle.

There are condemnations, voiced by Fabre's works, of those beings who have shamefully, racialistically and colonialistically felt the need to define people by colour: Western white, Native American red, Asian yellow, African black. Then there are colours for defining continents: Blue Europe, Yellow Asia, Black Africa. Precisely in the latter, in the Congo, Belgium exercised colonial domination from 1904 to 1960 (Belgian Congo), which Jan Fabre evokes with his cycle of jewel beetle mosaics.

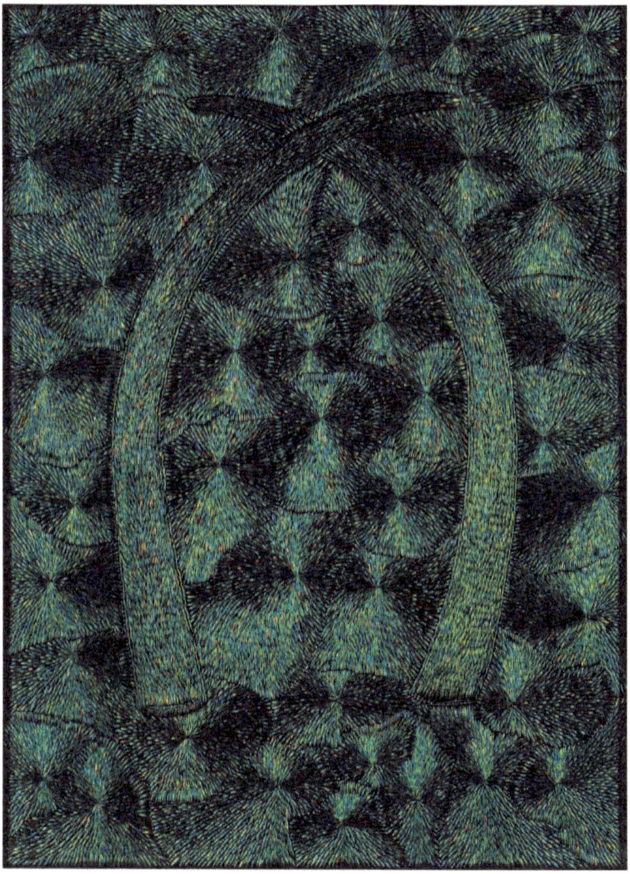

Figure 14.3 Jan Fabre. *Olifantstanden/Elephants Tusks* (2011) (Series: Tribute to Belgian Congo 2010–2013). Material: Schilden van de juweelkever op hout/jewel beetle wing-cases on wood. Size: 227.5 × 173 × 8.1 cm. Photo: Lieven Herreman. Copyright: Angelos bvba.

There are many questions about colours: why does ethnic colouring not include blue or green? Has some sort of cosmic colonialism shifted them to the other world of science-fiction: green Martians from the red planet of Mars, blue people from the green planet of Avatar, for a chromatic pastiche of Indian gods?

There are planetary chromographies, from finite earth to infinite universe, as we can read in many books, including one already mentioned by Cennino Cennini, or in Mario Equicola's 1525 *De Natura de Amore*,[45] which indicates seven hierarchical colours:

White, Moon and Monday

Red, Mars and Tuesday

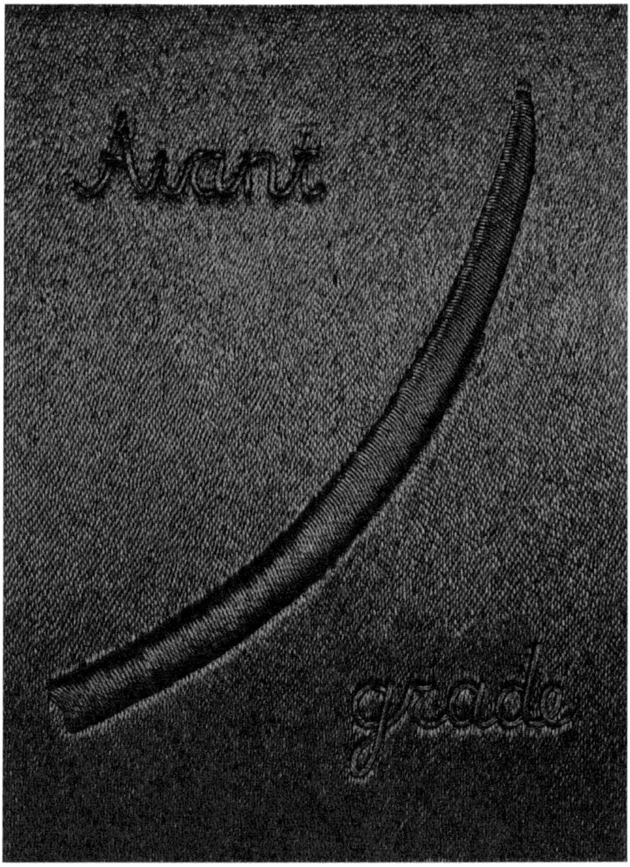

Figure 14.4 Jan Fabre. *Avant Grade* (2011) (Series: Tribute to Belgian Congo 2010–2013). Material: Schilden van de juweelkever op hout/jewel beetle wing-cases on wood. Size: 227.5 × 173 × 8.1 cm. Photo: Pat Verbruggen. Copyright: Angelos bvba.

Violet, Mercury and Wednesday

Blue, Jupiter and Thursday

Green, Venus and Friday

Brown, Saturn and Saturday

Yellow, Sun and Sunday

There are, returning to Earth and down to earth, Overseas Territories, *Territoire d'outre-mer* (TOM), which means those lands beyond the waters of Europe's sea, now no longer dark, but the ultramarine blue of colonial conquests, the location of real and imaginary lands of conquest, of *hic sunt leones*, and from which thousands of migrants now arrive.

There are three green animals in traditional symbology: the parrot, the lizard and the jewel beetle – symbols of beauty, joy and love.

There are animals that help us to live and to relive, like the beautiful bird in Fabre's 2011 *Parrot* (Figure 14.5) from his *Tribute to Belgian Congo* in the 2010–13 Congolese cycle. The artwork is made with the elytra of jewel beetles, beings of love, the love for life of the metamorphic scarab that is born, dies and is reborn, so much so that even early Christians used it as the quintessential symbol of resurrection, rather than the cross.

There are jewel beetles chosen by Fabre because they are metamorphic, alchemic beings, as alchemical is the iridescent colour of their elytra that shines, glitters, reflects and chromatically performs, shifting from green to blue to the gold of perfection and immortality.

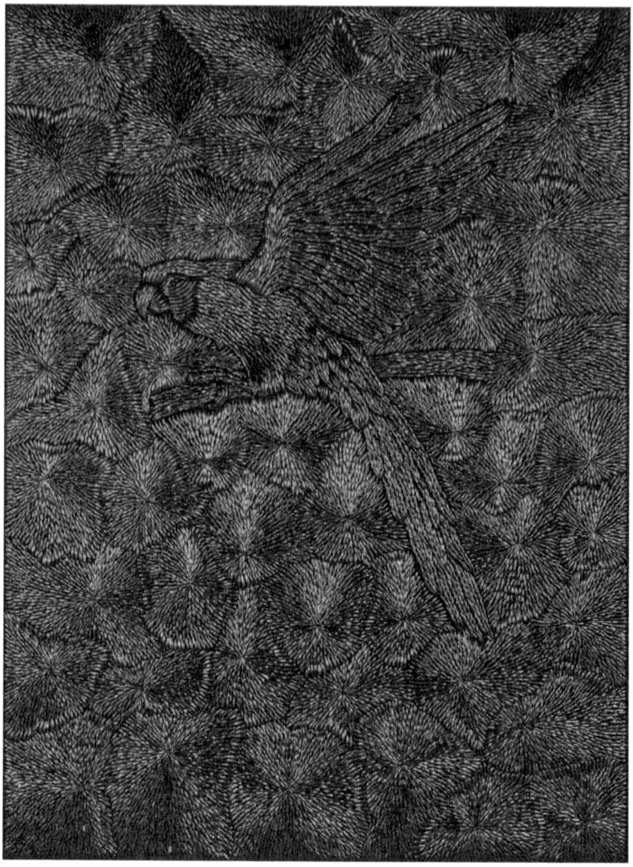

Figure 14.5 Jan Fabre. *Papegaai/Parrot* (2011) (Series: Tribute to Belgian Congo 2010–2013). Material: Schilden van de juweelkever op hout/jewel beetle wing-cases on wood. Size: 227.5 × 173 × 8.1 cm. Photo: Attilio Maranzano. Copyright: Angelos bvba.

There are works that Fabre creates with jewel beetle elytra that have a metallic tinge, a sort of natural animal armour that mimics the artificiality of metal, the alchemical–visual quality of the transformation of body elements, because – as we continue to repeat – green and blue are unstable colours that contrast one another to turn one into the other because they are alchemically active.

There are colours that Fabre has created, choosing them from what already exists: the jewel-green of the beetle and the Bic-blue of the ballpoint pen. We can then consider them ready-made colours, just as Duchamp – another of Fabre's icons – considered objects to be ready-made.

There are these colours, that Fabre found, and which bring together two traditions: one ancient in the action of Flemish painters; one in the quest of modern conceptualists.

There are two artists, Duchamp and Fabre, also inspired by alchemy and astrology, to the point that we cannot fail to notice that astrologically Sagittarius (Jan Fabre) is of a shade of blue that lies between the green-blue of Scorpio and the blue-violet of Capricorn and Leo (Duchamp) is tinged with yellow-gold.

There are conceptualisms and dexterity, with which Fabre seems to hold together thought and hand, thinking and manual making, because he never simply leaves these colours on show: he uses them to create shapes and images made by his hands "in action that takes, creates and sometimes seems to think";[46] the mental hand, the philosopher and spirit hand that begins and ends (...) on fingertips, (...) the capital organ of the possible.[47]

There are, therefore, actions with which Fabre creates bas-relief paintings as he gradually assembles the jewel beetle elytra-shells, inspired on one hand by his grandfather entomologist Jan Henry Fabre and, on the other, by the *Garden of Earthly Delights* by his artistic forebear Hieronymus Bosch.

There are thence unique works such as that of the stunning ceiling in Brussels' Palais Royal. A 2002 opus in the Hall of Mirrors, even its title – *Heaven of Delight* (Figure 14.6) – was inspired by Bosch's *Garden of Earthly Delights* and rightly defined a membrane of light by Laurente Busine and Jan Hoet. It is a monumental work that bears comparison with Michelangelo's "expressionist" Sistine, but – in my opinion – is also tinged with Raphael, the beautiful artist of beauty and alliance. Indeed, rapport, resilience and conciliation are approaches and questions that Fabre poses in his art, as he calls upon many co-workers, as Raphael did for his frescoes in the Vatican Rooms.

There are – returning to jewel beetle elytra material – transformations that go from the ideology of material to the matter of art, because Fabre does not exhibit for himself, but for what he can give figuratively.[48]

There are more than 1 400 000 elytra in place to create the Hall of Mirrors ceiling–sky, which echoes the thousands of tiny painted leaves creating the lawns for van Eyck's *Ghent Altarpiece*.

There are also other traditions to attract our attention, like that of the gleaming mosaic, because Byzantine mosaic vaults are composed of gold tesserae that create, reflect and diffuse light as does the Hall of Mirrors.

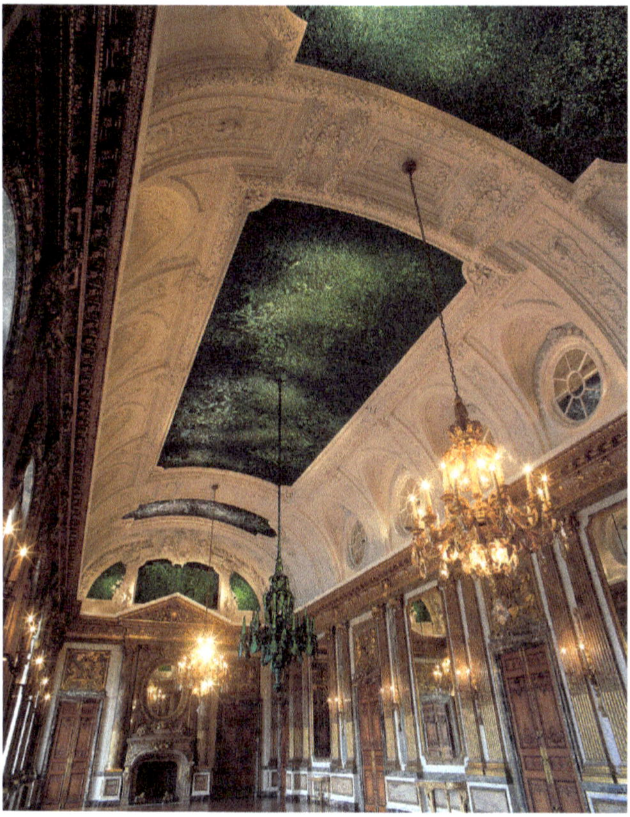

Figure 14.6 Jan Fabre. *Heaven of Delight* (2002). Material: Schildjes van de prachtkever op plafond/wing-case of buprestids on the ceiling. Hall of Mirrors, Royal Palace, Brussels. Photo: Dirk Braeckman.

A tradition Fabre measured up to last summer in Monreale Cathedral, home to one of the largest mosaics in the world.

There are actions and works whose intention is to show that Fabre never backs away from a strong challenge. In addition to those already mentioned, he also took up the gauntlet cast by Rubens, van Dyck and Jordaens. So, his works now fill the gaps left in Antwerp's church of Saint Augustine when paintings by the Flemish Masters were moved to the Royal Museum of Fine Arts.

There are techniques to choose from and Fabre relied again on that of the jewel beetle with which he produced works on equal footing with Peter Paul Rubens' *Mystic Marriage of Saint Catherine* that once soared on the high altar; Jacob Jordaens' *Martyrdom of Saint Apollonia* from the right aisle; van Dyck's *The Ecstasy of St Augustine* in the left aisle.

There are sacred and profane; Christian and alchemical symbolism; past, present and future deities; ranging from the reinterpretation of van Eyck's *Mystic Lamb*, in this case with Antwerp's philosopher's stone and economic cornerstone diamond placed on its back, upon which Fabre grafts the primitive sanctity of the tree of life encircled by the angel wing ring of faith.

There are changes in colours, figuration and spatiality that Fabre uses to channel the polychromatic, spiralling Baroque of the three aforesaid artists into van Eyck's vertical Flemish Renaissance. Similarly, in *The Ecstatic Recording* (2018) (Figure 14.7) verticality and musicality are the spatial and poetic cores. In *The Monastic Performance* (2018) (Figure 14.8) music once again rules, evoked by the image of the saint, also vertical but no longer gazing skyward (as van Dyck has him posed in *The Ecstasy of St Augustine* (1628)),

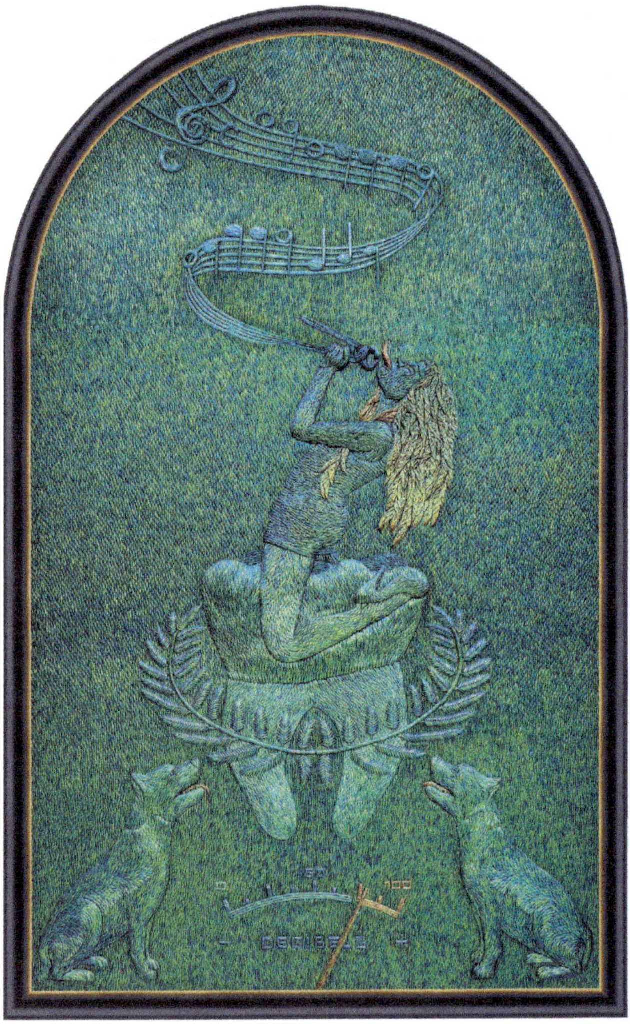

Figure 14.7 Jan Fabre. *De extatische opname/The Ecstatic Recording* (2018). Material: Schilden van de Juweelkever op dibond/jewel beetle wing-cases on dibond. Size: Approx 383 × 244 cm without frame. Photo: Attilio Maranzano. Copyright: Angelos bvba.

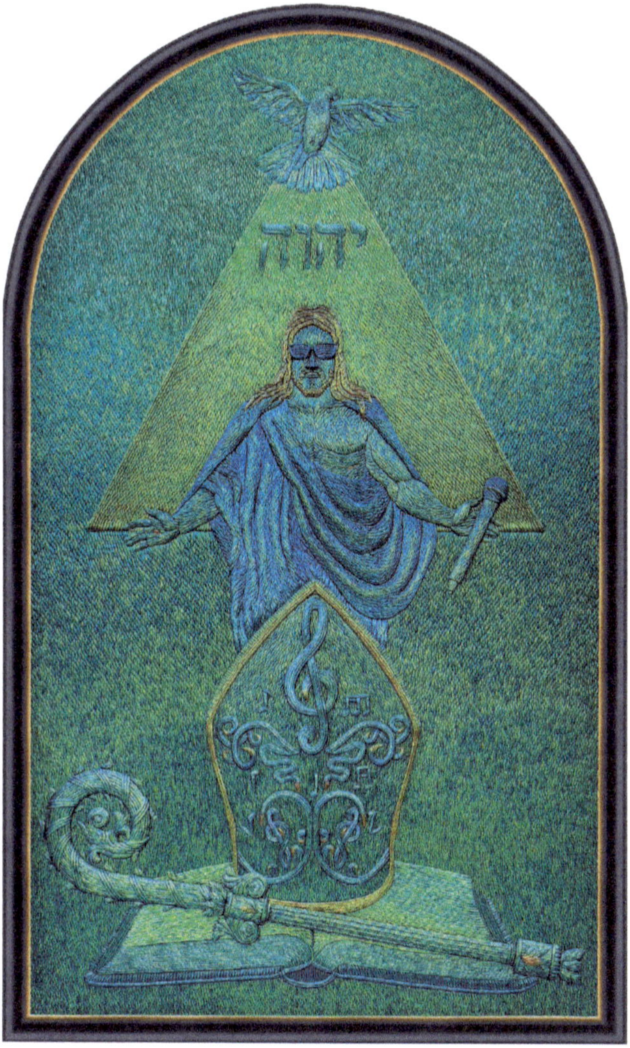

Figure 14.8 Jan Fabre. *Het mystieke contract/The Mystic Contract* (2018). Material: Schilden van de Juweelkever op dibond/jewel beetle wing-cases on dibond. Size: Approx 566 × 398 cm without frame. Photo: Attilio Maranzano. Copyright: Angelos bvba.

instead long-haired and looking at us from behind sunglasses with a microphone in his hand, suggesting not so much ecclesiastical sermons as a pop star during a concert or a rock opera like *Jesus Christ Superstar*.

There are iconographies brought into the modern age, given that the monastery of Saint Augustine is now an international music centre where the swirling, all-enveloping colour of Baroque artwork is replaced by Fabre's monochromes, where everything is green and where we see everything as evergreen, its iridescence notwithstanding, to the point that we think the transparent green diamond is an emerald.

There are gems – diamonds, emeralds, rubies, the wealth of Antwerp, the city of diamonds, which drove the colonial enterprise embodied so precisely by Fabre in the two cycles *Tribute to Belgian Congo* (2010–13) and *Tribute to Hieronymus Bosch in Congo* (2011–13), works inspired by the *Garden of Earthly Delights*, where everything is monstrous and for this reason even Paradise seems hellish.

There are, indeed, works that speak of this, like the 2011 *Slave Shitting Pearls* (Figure 14.9), or the 2012 *The Innocent Hand* (Figure 14.10), to mention just two of a corpus for working through the post-colonial guilt of Belgians – and not only Belgians.

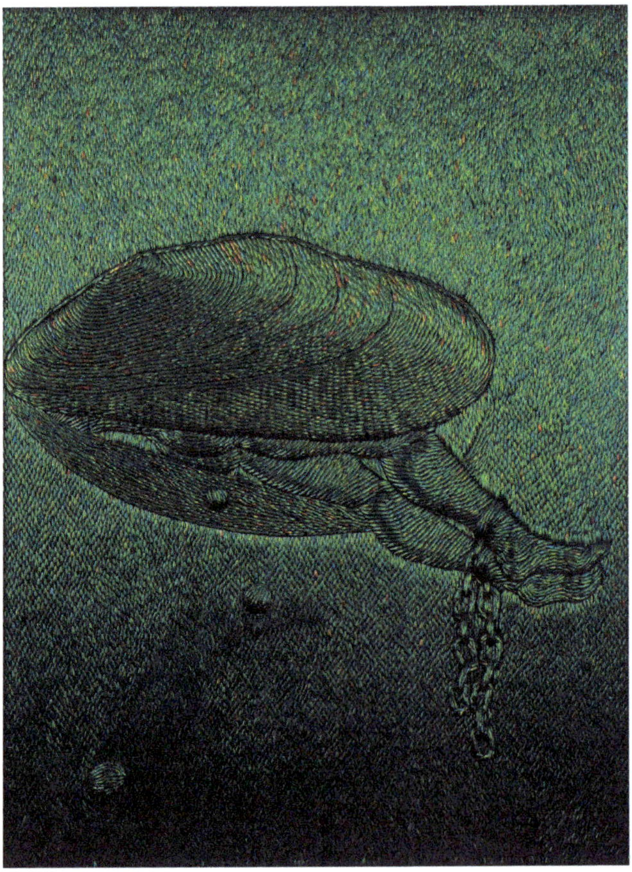

Figure 14.9 Jan Fabre. *Slaaf schijt parels/Slave Shitting Pearls* (2011) (Series: Tribute to Hieronymus Bosch in Congo 2010–2013). Material: Schilden van de juweelkever op hout/jewel beetle wing-cases on wood. Size: 227.5 × 173 × 8.1 cm. Photo: Pat Verbruggen. Copyright: Angelos bvba.

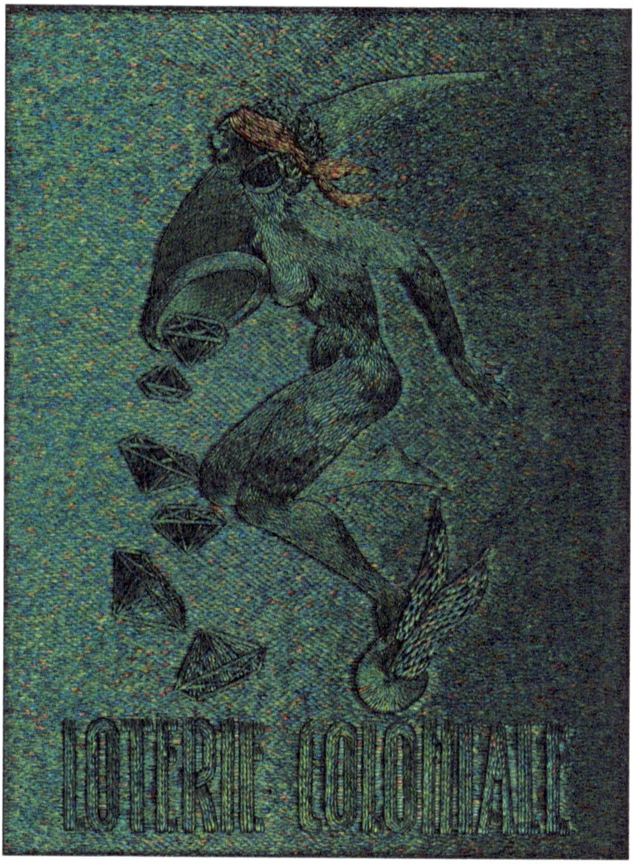

Figure 14.10 Jan Fabre. *De onschuldige hand/The Innocent Hand* (Series: Tribute to Hieronymus Bosch in Congo 2010–2013). Material: Schilden van de juweelkever op hout/jewel beetle wing-cases on wood. Size: 227.5 × 173 × 8.1 cm. Photo: Lieven Herreman. Copyright: Angelos bvba.

There are other coloured thoughts that appear and lead us to blue, because blue is the colour of the soul and therefore of the blues, the blues–soul that came from Africa too, the blue that "transcends the solemn geography of human limits".

There are more reflections that lead us to say that Fabre does not use brushes, acrylics, tempera, whatever: we know he uses pencils, ink, blood, semen, saliva, sweat and the blue of a Bic ballpoint, an ink born as a colour for writing and only later becoming a colour for painting. Thus, its destiny is similar to that of the ancient blue used by miniaturists in medieval books, after the scriptor had produced the text.

There are first writers then the miniaturists who illuminate the book with images, because in the ancient debate between the Liberal and Mechanical Arts, writing always came before painting in the hierarchy.

There are, therefore, reflections that have to do with Fabre, because he is also a scriptor, a man of letters who writes everything by hand, never using machines or computers of any kind. He used blue for the first time with a Bic ballpoint, with which he gave life to a series still in existence, which he entitled Bic-Art. The first opus was the 1979 *Bic-dweil* (Figure 14.11), a white cloth on which he drew, but above all wrote texts like

ONLY

ACTS

OF POETICAL

TERRORISME

or *Iliad of the Bic-Art*, (1980 self-portrait), a half-length nude black-and-white photo with the chest inscribed with the title, also in Bic-blue. The same

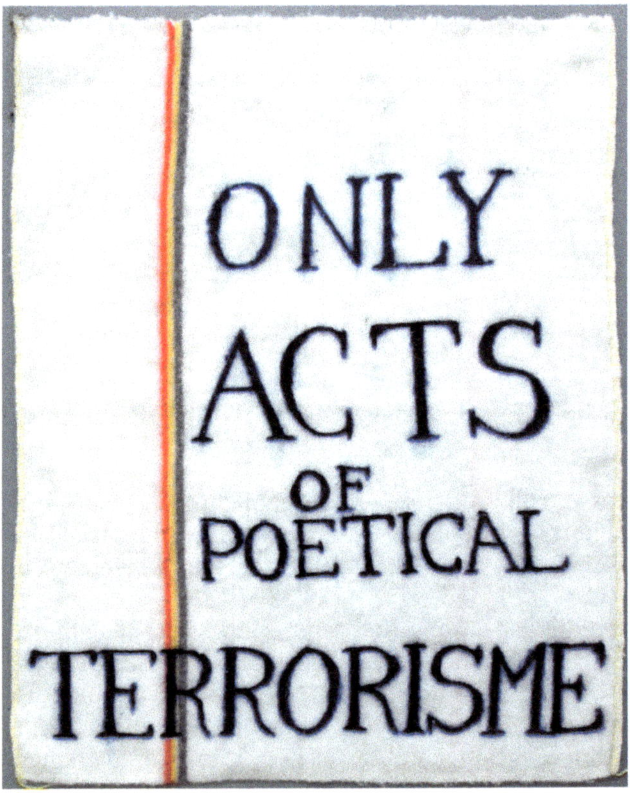

Figure 14.11 Jan Fabre. *Biccloth/Only Acts of Poetical Terrorisme* (1979). Material: Bic-ink on textile. Size: 80 × 64.5 cm. Photo: Christine Clinckx. Copyright: Angelos bvba.

occurred the following year, 1981, with *the Iliad of the Bic-Art, The Bic-Art Room* performance, lasting seventy-two hours, during which Fabre shut himself in a room painted white, with white furniture, writing and drawing only with his Bic ballpoint.

There are memories that make us think – at least as a temporal reference – of Joseph Beuys, of his 1974 performance *I like America, America likes Me*, in which the German shaman artist locked himself in a room in the René Block Gallery in New York, living and fraternizing with a coyote, the symbol of true America, that of the Native Indians.

There are interpretative necessities taking us so far back to explain Fabre's Bic-Art corpus which, from writing and drawing gradually becomes painting through a performative process, just as Bic-blue is performative, because like jewel beetle elytra it is a shimmering material, shining and able to mimic oils, a gift from Flemish Art to humanity. So, Bic-blue is a colour for writing that becomes a sign, then drawing, then painting. It is an A–A = ancient and alchemic process, so we say that painting is born from drawing, actually a challenge defiant to the last, like that between Apelles and Protogenes, considered the greatest painters of their era. Pliny the Elder wrote of Apelles' visit to the studio of Protogenes in Rhodes. In the latter's absence he drew a very fine sign on a blank panel. When he returned, Protogenes drew a line finer than that left by Apelles, cutting it in two. Later Apelles traced another, dividing it into three and forcing Protogenes to admit defeat in a battle to the last sign, because "the human is a sign".[49]

There are writing, signs, drawings that evolve into paintings, a Bic-blue painting in which the challenge seems to target the IBK of Klein, who had said that in the future people would start to paint images with just one colour and nothing but colour.

There are colours, sky blue, sea blue, ultramarine blue, Prussian blue, Bic-blue, Picasso blue, IBK blue and – not least of all – Fabre's Bic-blue, a colour obtained not by spreading or dipping, but by drawing, tracing increasingly intense Bic-blue signs to obtain a coloured surface that can be perceived and glimpsed, made of signs.

There are signs of signs, signs of drawings, signs of painting with which Fabre wants to cover the world, like his 1990 *Tivoli* (Figure 14.12), where he coloured with Bic-blue painting strokes the entire surface of Mechelen's Tivoli Castle (on an even larger scale than the ceiling of the Hall of Mirrors). He filmed the 24-h transition from day to night until the sun rose on the new day, in the metamorphic passing of the Blue Hour, which separates dark from light and of which Fabre emblematically takes possession, becoming the Blue Hour artist. The Blue Hour that he carries above everything and everywhere, like in *The Gaze Within (The Hour Blue)* (2011–13, Figure 14.13), a permanent work commissioned in 2009 from Fabre for a staircase in Belgium's Royal Museums of Fine Arts, originally used only by the king and queen to enter and leave the building unobtrusively.

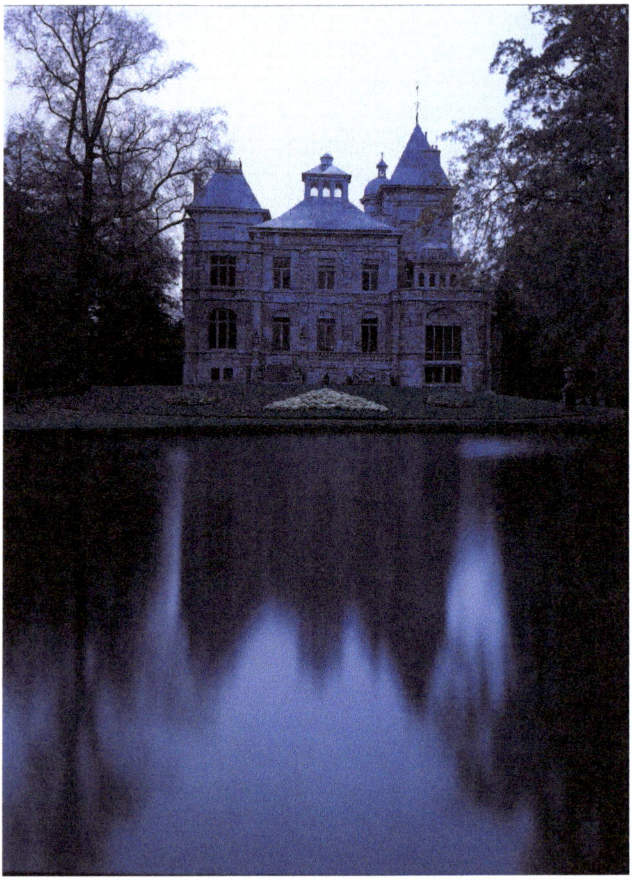

Figure 14.12 Jan Fabre. *Tivoli* (1990). Material: Cibachrome. Size: 122 × 170 cm. Photo:/. Copyright: Angelos bvba.

There are, of course, stairs and staircases, and this points to a unique collaboration, one of the rare occasions in which a museum offers a contemporary artist the chance to install a work permanently, aware that in choosing Jan Fabre there is a guarantee of an artist who has been capable of conversing with classical tradition for over forty years, as can be seen in the work on the Hall of Mirrors ceiling of the Palais Royal in Brussels, which we have already discussed.

There are seven large panels at the top of the staircase, suspended just beneath the ceiling coloured in intense Bic-blue. A saturated dark blue surface, apparently uniform but which glimmer red or purplish-green as spotlights shift.

There are four pairs of eyes drawn, painted, represented: a woman, an owl, a butterfly and a jewel beetle. Eyes that look at each other and at us. Between semblance and reality, thcy show the metamorphosis of nature born from

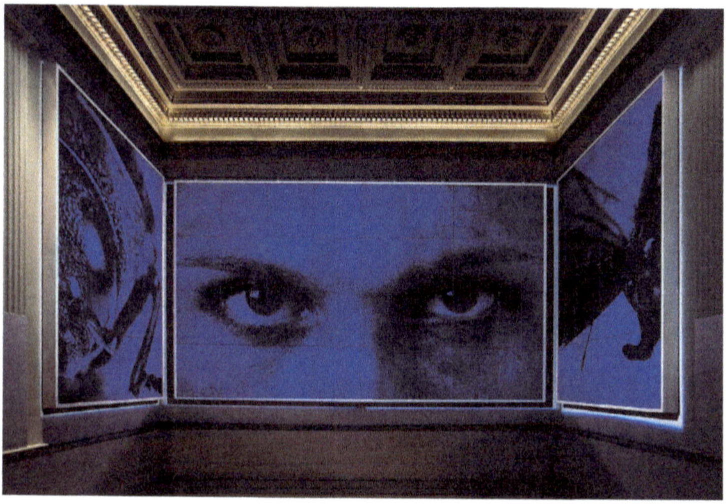

Figure 14.13 Jan Fabre. De blik binnenin (Het Uur Blauw) (2011–2013)/The Gaze Within (The Hour Blue). Materiaal: Bic balpen op cibachrome foto/ Bic ballpoint pen on cibachrome. Afmeting: 7 panels: 2×(4 × 1.5 m); 4×(4 × 3 m); 1×(4 × 6.85 m). Photo: Attilio Maranzano. Copyright: Angelos bvba.

alchemy of the Bic-blue colour, the Blue Hour, being and non-being, because "colour must be seen".[50]

There are, finally, conclusions that are never conclusive, and therefore posed only as questions like: after all this, have we really understood the essence, the being of Jan Fabre's art? If the mind is unknowable, is that of the artist yet more unknowable?

There are, moreover, my words. And my words say that Fabre is an alchemist–artist because with only two colours – jewel-beetle green and Bic-blue – he casts a multicoloured spell.

Acknowledgements

The author thanks Angela Maria Arnone for the translation from Italian to English.

References

1. Paracelsus, *7 libri dei supremi insegnamenti magici*, ed. D. Meldi, Giunti-Demetra, Firenze, 2007.
2. J. Fabre, *Les annés de l'Heure Bleue*, Musée d'Art Moderne, Metropole Sain Etienne, curated by Lorand Hégy, Silvana Editoriale, Milano, 2012.
3. J. Fabre, *Glass and Bone Sculptures, 1977–2017*, ed. Forma, Exhibition catalogue, Abbazia di San Gregorio, Firenze, 2017.

4. J. Fabre and E. Schneider, *Jan Fabre, Tribute to Hieronymus Bosch in Congo/ Tribute to Belgian Congo*, Skira editore, Milano, 2015.
5. E. di Agrigento, *Frammenti e testimonianze. Origini, Purificazioni, con i frammenti del Papiro di Strasburgo*, curated by Angelo Tonelli, Bompiani, Milano, 2002.
6. F. Montevecchi, *Empedocle d'Agrigento*. Liguori, Napoli, 2010.
7. M. Pastoreau, *Blu. Storia di un colore*, Ponte alle Grazie, Firenze, 2010.
8. M. Pastoreau, *Rosso. Storia di un colore*, Ponte alle Grazie, Firenze, 2010.
9. M. Pastoreau, *Verde. Storia di un colore*, Ponte alle Grazie, Firenze, 2013.
10. J. Hillman, *Psicologia alchemica*, Adelphi, Milano, 2013.
11. J. Fabre, *Progetto Roma*, Catalogue of the exhibition at Accademia Belgica di Roma, 2001.
12. L. Luzzato and R. Pompas, *Il significato dei colori nelle civiltà antiche*, Bompiani, Milano, 2001.
13. A. Kondratiev, *Il tempo dei celti. Riti e miti: una guida alla spiritualità celtica*, Apogeo, Milano, 2005.
14. C. G. Jung, *Psicologia ed Alchimia*, Bollati Boringhieri, Torino, 2006.
15. E. J. Holmyard, *Storia dell'Alchimia*, Odoya, Bologna, 2009.
16. R. Iliffe, *Isaac Newton: Il sacerdote della natura*, curated by Franco Giudice. Hoepli, Milano, 2019.
17. M. White, *Isaac Newton: L'ultimo mago*, Rizzoli, Milano, 2001.
18. I. Newton, *Scritti sulla luce e sui colori*, curated by Franco Giudice, BUR: Biblioteca Universale Rizzoli, Milano, 2006.
19. Aristotle, *I colori e i Suoni*, curated by Maria Fernanda Ferrini, Bompiani, Milano, 2008.
20. D. Alighieri, *Divina Commedia*, Hoepli, Milano, 1988.
21. M. Shelley, *Frankenstein*, Feltrinelli, Milano, 2013.
22. D. S. Kastan and S. Farthing, *Sul Colore*, Einaudi, Torino, 2018.
23. J. Elliot and W. L. Simon, *Steve Jobs. L'uomo che ha inventato il futuro*, Hoepli, Milano, 2012.
24. A. Rimbaud, *Opera completa*, curated by Diana Grange Fiori, Arnoldo Mondadori Editore, Milano, 1984.
25. Pliny the Elder, *Storia Naturale*, Einaudi, Torino, 1997.
26. C. Iggulden, *Il sangue degli dei*, Piemme, Segrate, 2018.
27. P. Ball, *Colore. Una biografia*, Rizzoli, Milano, 2002.
28. M. Pedrini, *Verde di Schwheinfurt, sul pigmento, materia prima di ogni arte pittorica*, Sebino Colori, 2007.
29. V. Villavecchia and G. Eigenmann, Rame acetoarsenito, in *Nuovo dizionario di merceologia e chimica applicata*, Hoepli, 1976, vol. 6.
30. N. Bevilacqua, L. Borgioli and I. Adrover Gracia, *I pigmenti nell'arte dalla preistoria alla rivoluzione industriale*, The Prato editions, Villatorta, 2010.
31. M. Clemenza, E. Fiorini, E. Previtali and E. Sala, *Misure con attivazione neutronica sulla presenza dell'arsenico nei capelli di Napoleone Bonaparte e dei suoi famigliari*, Bulletin of the Italian Physical Society, il Nuovo Saggiatore, Bologna, 2008.

32. A. Benhamou, *L'autre Saint-Hélène. La captivité, la maladie, la mort et les médicins autour de Napoléon*, Broché, Paris, 2010.
33. A. Goldcher, *Napoléon Ier. L'ultime autopsie*, Broché, Parigi, 2012.
34. J. Pierre and F. Lamorinière, Ressources relatives aux Beaux Arts, *Dictionanire des peintre belges*, Musée d'Orsay, Paris, 1994.
35. N. Eastaugh, *Pigment Compendium: A Dictionary of Historical Pigments*, Routledge, London, 2004, vol. 1.
36. G. Agosti, *Su Mantegna*, Feltrinelli, Milano, 2005.
37. Mantegna, *Catalogo della mostra (Perigi, Musée du Louvre, 2008–2009)*, curated by Giovanni Agosti and Dominique Thiébaut, Officina Libraria, Milano, 2008.
38. R. Fry, *Mantegna*, Abscondita, Milano, 2006.
39. S. Fischer and H. Bosch, *Opera Completa*, Taschen, Colonia-Milano, 2014.
40. E. Villata, *Grünevald, Pittore e mistico tra Lutero e Hindemith*, Hapax, 2018.
41. E. Villata, *I Chiodi di Grünevald*, EDUCatt: Università Cattolica, Milano, 2010.
42. F. R. Martin, M. Menu and S. Ramond, *Grünewald*. Jaka Book, Milano, 2012.
43. D. Benati and M. Medica, *Giovanni da Modena, Un pittore all'ombra di San Petronio*, Silvana, Cinisello Balsamo, 2014.
44. A. B. Venier and C. Fabre, *Il Cristo del Mantegna e oltre*, Morgana Edizioni, Firenze, 2008.
45. M. Equicola, *De Natura de Amore*, Aracne, 2018.
46. H. Focillon, *La vita delle forme – Elogio della mano*, Einaudi, Torino, 2002.
47. P. Valéry, *Eupalinos o l'architetto*, curated by Barbara Scapola, Mimesis, Sesto San Giovanni, 2011.
48. *Jan Fabre – Homo Faber*, ed. G. di Pietrantonio, Mercatorfonds, Brussells, 2006.
49. J. Derrida, *La mano di Heidegger,* Laterza, Roma-Bari, 1991.
50. W. Benjamin, *L'opera d'arte nell'epoca della sua riproducibilità tecnica*, Einaudi, Torino, 2000.

CHAPTER 15

Hockney, Hume and Chandra: Surface, Change and Conservation

BRONWYN ORMSBY*[a], RACHEL SCOTT[a], HELEN BRETT[a], JUDITH LEE[a] AND REBECCA HELLEN[a]

[a]Tate Britain, Conservation Department, Millbank, Pimlico, SW1P 4RG, London, UK
*E-mail: bronwyn.ormsby@tate.org.uk

15.1 Introduction

Contemporary artists embrace paint types for a wide variety of reasons; some prefer precise technical approaches, some desire specific colours and optical or textural effects, some follow an intuitive response, while others choose specific paints that relate to memories, inspiration or other practices that hold meaning. On a pragmatic level, artists' decisions as to which paint(s) are used can also be influenced by availability and numerous technical factors, such as purchase volumes, drying time, application methods, supports and paint diluent type.

Waterborne synthetic polymer paints, such as acrylic emulsions, poly-vinyl acetates (PVAc) and others which have been popular since the early to mid 1960s, were widely adopted, sourced as either house, student grade or artist quality paints. Many contemporary painted works are now showing signs of

Science and Art: The Contemporary Painted Surface
Edited by Antonio Sgamellotti, Brunetto Giovanni Brunetti and Costanza Miliani

Published by the Royal Society of Chemistry, www.rsc.org

change, which naturally prompt discussions on conservation and a range of questions including how the works were made, how the materials used evolve with time and how artists' views may change. In addition, many of the changes noted are proving to be chemically complex. Acrylic emulsion paints have now been the subject of considerable research exploring their use and manufacture, paint properties, ageing behaviour and approaches to conservation treatment. Oil-containing paints such as drying oils and oil-modified alkyds have also remained popular due to their inherent saturation and wide range of textural and drying effects. Oil paints used since the mid to late 20th century (often known as modern oils) have recently been the subject of research aimed at understanding the effects of both paint formulation and the environment.

This material complexity also needs to be considered within the context in which a given artist creates and then considers the effects of ageing on their works, in addition to the environment in which works are displayed. Where possible, the artist is included in an ongoing dialogue which informs how we understand works of art and their surfaces, directly influencing how these works are interpreted and conserved. This chapter explores these complexities *via* investigations into two key paint classes – synthetic polymer and oil-containing paints – through works drawn from the Tate collection by British artists David Hockney and Gary Hume and the Indian artist Avinash Chandra.

15.2 Acrylic Paints and Varnishes: David Hockney

David Hockney's (b. 1937) consistent desire to embrace innovative technologies has always been evident and was no less keen in the 1960s than it is now. On moving to California from the United Kingdom in 1964, he switched from oil to acrylic emulsion paints which, like many artists, led to a period of experimentation. "I'd tried acrylic paint in England before and I hadn't liked it at all; [the] texture [and] colours weren't so very good. But the American ones I liked. It was a superior paint."[1] Key differences in the properties of oils and acrylics, such as handling, drying times and finished effect also informed his subsequent changing between these two paint types, in addition to conceptual needs, people and places.[2,3] One of the key advantages in his use of acrylic paints was that, unlike oil paints, the new acrylic media dried quickly, allowing Hockney to layer paint – wet over dry – without picking up any under-layers. This allowed him to focus on one painting at a time, and to produce more works in any given period. The following paintings form part of his output from this transformative period, including an early example – *Man in Shower in Beverley Hills*, 1964 – and a later, arguably more technically adept larger work, *Mr and Mrs Clark and Percy*, 1970–71.

15.2.1 Man in Shower in Beverley Hills, 1964 (Tate, T03074)

During 1964, Hockney is known to have used New Masters® acrylic-vinyl copolymer emulsion paints, produced in California.[4] His first work using these paints was aptly entitled *Plastic Tree Plus City Hall*,[1] and *Man in Shower in Beverley Hills* (Figure 15.1) was painted in the summer of that year. Analysis of paint samples (see Table 15.1) confirmed that all paints tested are an acrylic-vinyl blend, a poly ethyl acrylate/methyl methacrylate/vinyl acetate [pEA/MMA/VAc] terpolymer with a phthalate plasticiser,[4] apart from the ground, which is a pEA/MMA copolymer with some of the hallmarks of Liquitex® formulations.

The New Masters® paint line had a higher gloss than existing acrylic emulsions, and a consistency more viscous than the new Liquitex® bodied paints which came onto the market in 1963.[5] Leonard Bocour, producer of Magna, and later Aqua-tec™[6] paints, took over the company and recalled of New Masters® that: "the reason they added vinyl is that it's slow-drying and very cheap...[but] it also gets very gummy...".[7]

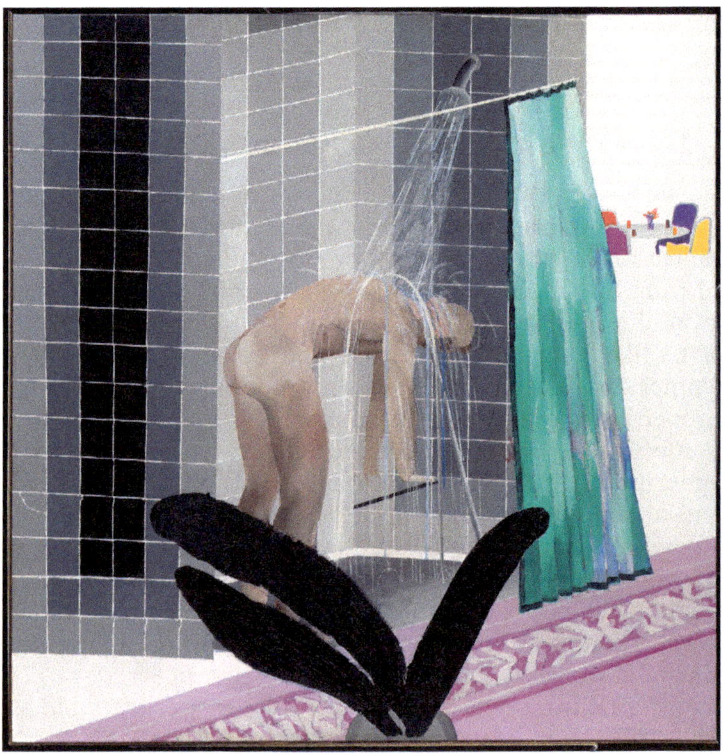

Figure 15.1 David Hockney, *Man in Shower in Beverley Hills*, 1964, T03074, Tate, © David Hockney, photo © Tate.

Table 15.1 Combined analysis results from *Man in Shower in Beverley Hills.*

Sample	Binding medium (FTIR, PyGCMS)	Pigments/solids (FTIR, EDX)	Notes
Grey	Vinyl acrylic, [EA/MMA/VAc] phthalate plasticiser	Titanium white, probable carbon black	n/a
Green	Vinyl acrylic, [EA/MMA/VAc] phthalate plasticiser	None	n/a
Black	Vinyl acrylic, [EA/MMA/VAc] phthalate plasticiser	Probable carbon black	n/a
Priming (gesso)	pEA/MMA acrylic copolymer	Chalk, kaolin, titanium white	Possible Liquitex priming formulation
Coating	Acrylic resin	n/a	Water-soluble acrylic resin

As described in Table 15.1, Hockney first applied the thick acrylic gesso priming layer consisting of a pEA/MMA copolymer, with chalk, titanium white and kaolin, forming the white background and grouting between the tiles. The tiles were then painted with the aid of masking tape and extended under the figure, curtain and leaves. The acrylic gesso priming has a slightly dry, porous appearance which is full of small air bubbles, in contrast with the glossy, lightly impasted texture of the New Masters® acrylic-vinyl paint. Hockney then enhanced the sheen of the "wet" bathroom tiles further, by glazing each one individually with a clear acrylic varnish coating, as is clearly shown in Figure 15.2.

New Masters® sold their paints in distinctive transparent tubes, with the promise of a "viscosity" akin to oil paint,[8] which may have attracted Hockney, as – despite his interest in colour at the time – he was clearly still interested in textural effects. For example, the glossy black leaves at the front centre have a distinctive texture, achieved by pulling the bodied paint up into peaks and ridges with a palette knife or the flat surface of a brush. Even thick applications of these paints dried within a few hours and couldn't be re-dissolved once dry. Unlike oil, they didn't allow for easy wet-in-wet blending of colours, and the painting of the shower curtain shows that Hockney was perhaps still mastering this aspect. Choppy gradations of adjacent light and dark greens are also visible, as well as the deep pink of a second figure, begun under the curtain, which was then painted out. In raking light, the scattered lumps of dried paint embedded in the surface suggest Hockney attempted to scrape off applications of dried paint.

Hockney later mentioned the difficulty in making changes when using early acrylics, which challenged spontaneity, making him more cautious: "once acrylic is down, you can't get it off. It's that permanent",[9] whereas slow-drying oil paint was easy to scrape down and remove. With *Man in Shower*, Hockney admitted difficulty with modelling the feet, which led him to cover them with one of the leaves.[10] Like many artists, Hockney preferred

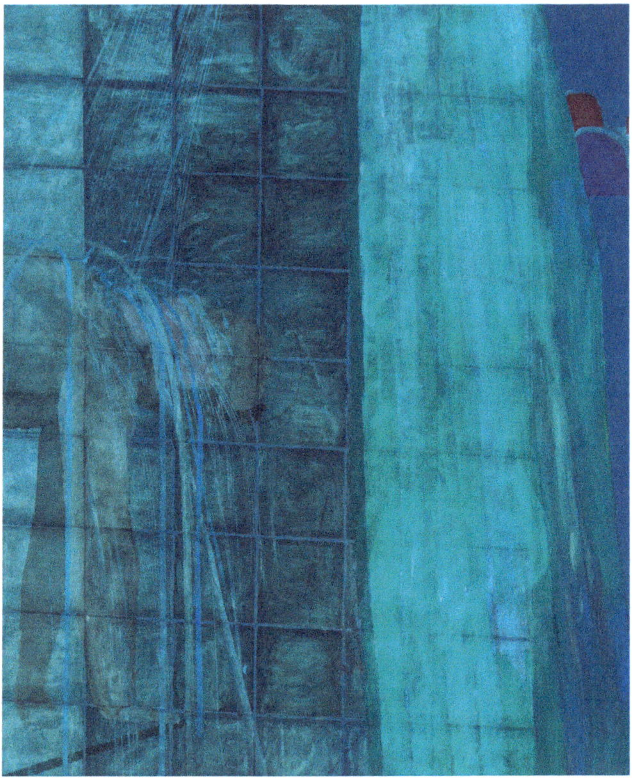

Figure 15.2 David Hockney, *Man in Shower in Beverley Hills*, 1964, T03074, Tate; ultraviolet light (UV) detail of applied coatings on shower tiles, © David Hockney, photo © Tate.

the relative stability that acrylic paints offered, stating: "the colours are very intense, and they stay intense, they don't alter much. Also, people said they were very safe to use – permanent."[9]

In addition to varnishing each tile, Hockney also applied a thin, overall unpigmented, presumably acrylic emulsion varnish, and both coatings have now become slightly yellowed, as shown in the detail image (Figure 15.3). Where varnish has accumulated in paint crevices and ridges, the yellowing is more obvious, however this is also combined with imbibed sand – presumably originating from early in the painting's life – and migrated surfactant from the combined emulsion materials,[11] in addition to accumulated dust and grime.

How this knowledge informs the conservation of this work is an ongoing conversation. Interestingly, recent discrete cleaning tests quickly revealed that the layers of acrylic coatings are adversely affected by water. This degree of water sensitivity for unpigmented acrylic resin layers is unusual, hence it is possible that these coatings may have been designed as water-soluble

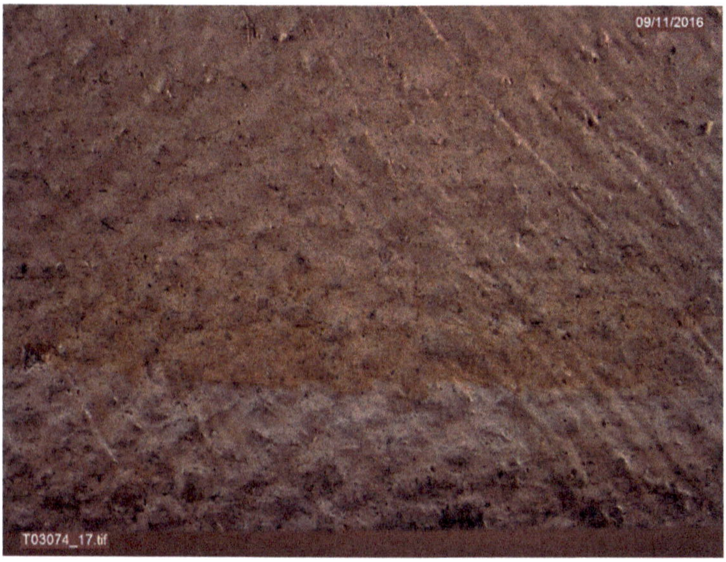

Figure 15.3 David Hockney, *Man in Shower in Beverley Hills*, 1964, T03074, Tate; detail of yellowed varnish from bottom edge, © David Hockney, photo © Tate.

materials, and/or perhaps the water solubility may arise from excess surfactant used in relatively early acrylic dispersion formulations. Despite recent advances in cleaning systems for acrylic painted surfaces, removing deposited soiling from water-sensitive layers remains a difficult challenge, hence it is advantageous at this point that the yellowed coatings have a relatively minor impact on this work. Interviews with Hockney where he explores his use of materials are relatively rare. On viewing the painting in 2016, during recent preparations for a retrospective exhibition at Tate, he commented that he was both surprised and pleased that this work, among his other earlier works, was still looking good.[12]

Leaving this work in its current condition is a relatively easy decision, given both the appearance of the work and Hockney's thoughts. However, the coatings will continue to become yellowed, further distorting tonal relationships within the composition and possibly further retaining deposited soil. In any case, the removal of the two varnishes would not be a straightforward decision, given that Hockney individually varnished each tile before applying his overall varnish layer, and that both layers are similarly sensitive to water. If removed – as is widespread practice for varnished historical works when the coating has discoloured – these layers would also need to be replaced to regain the desired aesthetic effect. This would involve very careful consideration about application, gloss, thickness and reversibility. However, it seems more appropriate at this stage to argue that these original varnish layers should remain part of this work, and that effort would perhaps be best diverted to reflecting on our expectations for the appearance and ageing of contemporary painted art; particularly with so many inherent limitations to the materials used.

15.2.2 *Mr and Mrs Clark and Percy*, 1970–71 (Tate, T01269)

The iconic acrylic painting by Hockney, *Mr and Mrs Clark and Percy*, (Figure 15.4) was bought directly from the artist and finished by him in the galleries just prior to its display at the Tate Gallery, now Tate Britain. Hockney began the process in 1969 by taking photographs of the sitters, Celia Birtwell and Ossie Clark, and making drawings. In the spring of 1970 he started painting, and completed the work in 1971.[13]

For this work, a pencil-drawn grid was used to assist in transferring the design based on drawings to the cotton canvas. The acrylic paints were diluted significantly to produce thin washes, with a build-up of several layers in some areas. Demonstrating greater success with his manipulation of acrylic paints, Hockney made several attempts to paint features such as Clark's head, while removing the paint between attempts. He kept the surface of the painting wet by spraying it with water, aiding the blending of paints. As shown in Table 15.2, samples of the white ground, grey paint and yellowed varnish were taken for analysis in addition to paints from Hockney's palette used *in situ* at the (then) Tate Gallery. The binding medium in all colours was identified as an acrylic emulsion of the poly ethyl acrylate methyl methacrylate (p(EA/MMA)) type, typical of early acrylic emulsion formulations.

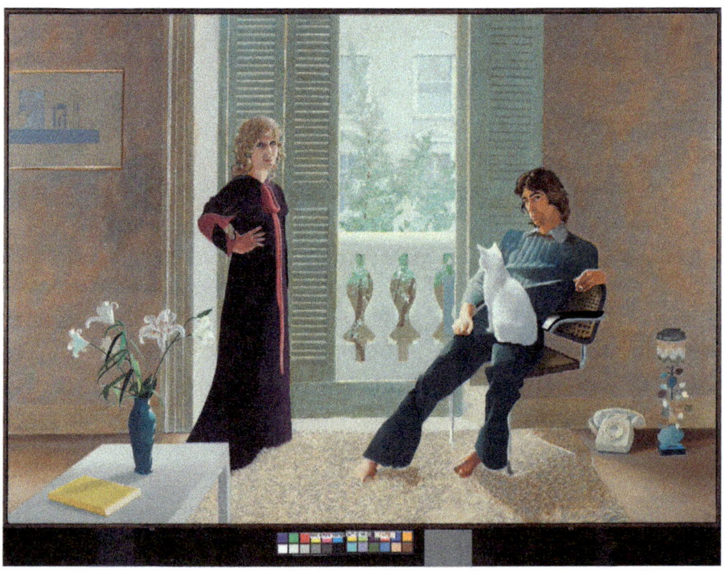

Figure 15.4 David Hockney, *Mr and Mrs Clark and Percy*, 1970–71, T01269, Tate, © David Hockney, photo © Tate.

Table 15.2 Combined analysis results from *Mr and Mrs Clark and Percy*.

Sample	Binding medium (FTIR, PyGCMS)	Pigments/solids (FTIR, PyGCMS, EDX)
White ground	pEA/MMA acrylic copolymer	Titanium white
Grey	pEA/MMA	Titanium white, carbon black
Yellowed coating	Styrene-acrylic (Sty-BA-MMA)	None
Palette red (light)	pEA/MMA	Naphthol red, azo PY73/PY74 and chalk
Palette yellow	pEA/MMA	Cadmium yellow, barium sulphate, chalk
Palette blue	pEA/MMA	Cobalt blue, zinc white, chalk, kaolin
Palette green	pEA/MMA	Phthalocyanine green (PG7), titanium white
Palette white	pEA/MMA	Titanium white (rutile), chalk, silica extenders
Palette earths	pEA/MMA	Natural ochres, umbers, chalk, silica

As stated above, Hockney completed his finishing touches *in situ*, where photographic documentation of the process shows Hockney with either a paint brush and water, or an acrylic varnish containing spray bottle in hand.[14] Recent technical examination using ultraviolet (UV) light confirms this, as the dripping and uneven application is clearly visible, as shown in Figure 15.5. Analysis of the yellowed varnish confirmed a styrenated acrylic emulsion (see Table 15.2). Styrenated acrylic polymers tend to be cheaper than pure acrylics and are thought to be more susceptible to yellowing than pure acrylate containing equivalents.[15] It is noted that the non-styrenated acrylic varnishes applied to *Man in Shower* are older and remain less yellowed. How much this varnish has changed colour since the execution of the painting is not known, but the styrene component suggests that the yellowing may continue to develop slowly.

There are two key issues for this painting's care regarding Hockney's choice of materials. The acrylic emulsion paint medium and coatings attract and imbibe dust and the artist-applied varnish and drips have continued to darken and change colour over time. Indeed, the heavily discoloured drips often caused alarm in the galleries, as worried invigilators mistook them for splashed fizzy drinks or coffee and reported them as potential incidents.[16] Additionally, the complex layer structure of the paint surface and its sequencing of "paint, varnish, paint" mean that varnish removal may not be ethically desirable, as Hockney clearly added paint on top of his varnish applications. In tandem with concerns around the preservation of the uppermost layers of Hockney's paint is the technical challenge of designing a system which facilitates the safe removal of a varnish layer comprised of the same polymer as the original paint; for which there is currently no successful method.

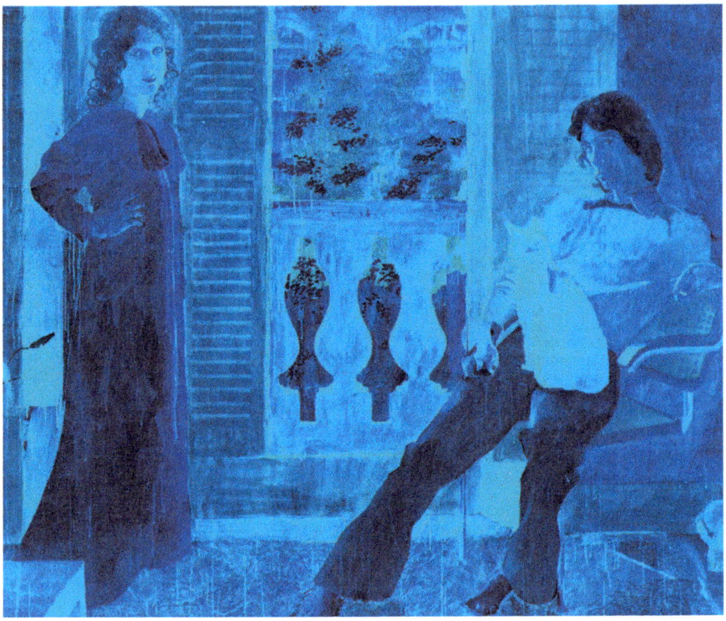

Figure 15.5 David Hockney, *Mr and Mrs Clark and Percy*, 1970–71, T01269, Tate; UV detail of fluorescent varnish drips and absorbing final paint additions on top of varnish, © David Hockney, photo © Tate.

Interestingly, unlike in *Man in Shower*, the varnish used on *Mr and Mrs Clark* is not water-sensitive; this may be afforded by the inclusion of hydrophobic styrene monomer, using a different product, and/or by improvements in coating formulations. Similarly to *Man in Shower*, recent discussions with the artist highlighted his consistently relaxed and pragmatic approach to his older works, stating: "Look how well they are doing...not bad for a 50 year-old painting."[17]

15.3 Oil-modified Alkyd Paints: Gary Hume

Gary Hume (b. 1962) came to prominence as one of a group of artists known as the YBAs – or Young British Artists – graduating from Goldsmiths Art College in the late 1980s and exhibiting at Damien Hirst's celebrated *Freeze* exhibition of 1988. Hume rejected traditional artists' materials when he began making his seminal *doors* series in what was to become his trademark medium, household gloss paint. In these literal representations of doors, this choice of paint fitted with themes of appropriation and the abstract invention prevalent in art, and particularly sculpture.

Hume's paintings are characterized by flat planes of paint on aluminium sheets; intense colour fields which are joined by low, linear ridges creating boundaries between colours and rendering subtle embossed relief effects which draw attention to their physicality. Hume's choices of materials and techniques have a subtle but important impact on the surfaces of his paintings and close examination and technical analysis helps to reveal the significance of his choices. He remains strongly committed to his chosen medium, oil-modified alkyd-based house paint, and is fully engaged in exploring its technical possibilities.

15.3.1 *Water Painting*, 1999 (Tate, T07618)

Water Painting (Figure 15.6) is one of a series of paintings Hume worked on for the 48th Venice Biennale in 1999. The outlining of the figures marked a new stylistic departure for Hume and contrasts with his more typical hard-edge juxtaposition of colour shapes used for the door series. The image is derived from a collection of drawings on acetate, where he traced the drawing from the projection onto a prepared aluminium support using pencil.

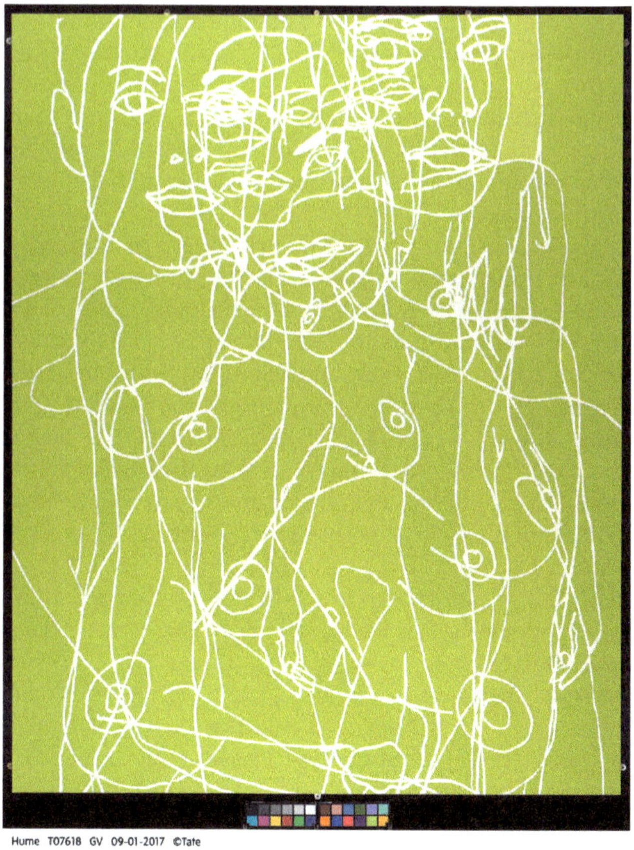

Figure 15.6 Gary Hume, *Water Painting*, 1999, T07618, Tate, © Gary Hume, photo © Tate.

The aluminium panel is made up of two sheets joined with epoxy glue and supported by an aluminium box framework which is glued with the same adhesive, to the reverse. Aluminium supports had superseded the Formica and MDF boards he used in the early 1990s. Aluminium was proposed in response to Hume's need for a support that would mimic a traditional canvas and stretcher, but be smooth, inert and light, without risks of scuffing corners on the studio floor. His panels are still made in this format, although now with the improvement of a double-sided pressure tape rather than glue, used to attach the box frame to the reverse. The bars on the reverse are positioned at the correct height to allow him to lift the works easily.

After peeling off the protective film, the surface is degreased with naphtha petroleum spirit and roughened with wire wool or SCOTCH-BRITE® scourer. For *Water Painting*, Hume drew the outlines and then painted directly onto the degreased panel, although subsequent panels were first primed with a coloured two-part acid etch primer. Hume generally uses either a white or brown coloured primer: "The white primer was used when I felt that I was "projecting" the image rather than the brown for "growing" the image."[18] Speaking of his chosen medium, Hume has said: "Only when I started using it did I realise how beautiful gloss paint is: the nature of it is so perfect...gloss paint has its own unique qualities; it's very different to oil paint, which can mimic different surfaces. For example, with an oil painting of a vase you're aware the image is constructed out of paint but feel you know the material of a vase. Whereas with gloss paint you are always aware that the picture is just a thin film of solidified liquid."[18]

Oil-modified alkyd resins are the main binder used in fast drying gloss "oil" household paints. In order to be used as a paint binder for household paints, alkyd resins are modified by adding a large proportion of oil at 60–80% by weight – known as long oil alkyds – which imparts a painterly quality and makes the brittle alkyd resin retain sufficient flexibility once the paint has dried.[19] Alkyd paints dry by the same oxidative polymerization process as pure oil paints, although their higher molecular weight means that they require fewer cross-linking reactions to harden, thereby speeding up drying time. Designed as exterior coatings, these paints also include additives which affect properties, such as gloss, self-levelling, resistance to ultraviolet light and weathering.

Working with house paints requires different application methods to artists' oils, acrylic or even oil-modified alkyd artists' paints. One of Hume's chief pleasures and motives in using gloss house paint is that it "carries light" without creating the illusion of light. "There's no illusion of three-dimensional space in the works, but there is plenty of time: light time, because they change throughout the day. I never paint light, but I wanted to be able to have light in the pictures. Gloss reflects the light that we have – the sun and artificial light – it can hold the light by reflecting it."[20]

For *Water Painting*, the panel was laid on the floor and Hume lay across the surface to paint sections in stages, starting with the greens first, with a reserve left for the white lines. The paint was applied free hand. The self-levelling property of alkyd house paint is designed not to hold brushstrokes, and thus in the greens it is only possible to see faint, rippling echoes of brush marks forming subtle mounds which describe how the loaded brush followed the pencil lines, paint pooling within the confines of each green island. Hume recalls the difficulty in painting the fluid white lines starting from one corner, painting one and then catching another, before it started to get "claggy".

Hume paints directly from the tin but also adjusts the tones by mixing paints. In *Water Painting* Hume painted using subtly different shades of green. As described in Table 15.3, analysis of a series of paint samples characterized this work as containing Mars yellow (synthetic iron oxide), a synthetic organic azo yellow pigment PY74, ultramarine blue and titanium white, although there appears to be a greater proportion of blue in the darker green.

Table 15.3 Combined analysis results from *Water Painting*. The organic pigments PY74 (Hansa yellow) and PB15 (phthalocyanine blue) are classed as grade I, 'excellent lightfastness' by the ASTM (American Society for Testing and Materials), though stability may vary in different paint media.

Sample	Binder (FTIR, GCMS)	Solids (FTIR, EDX)	Notes on relative proportions
White drip paint	Oil-modified alkyd – long oil	Titanium white, some chalk/gypsum	Well-bound paint
Unpigmented drip	Poly-vinyl (acetate/butyral) polymer	Talc	Dominated by talc
Lighter green paint	Oil-modified alkyd – long oil	Titanium white, synthetic organic pigment PY74 (azo), Mars yellow? (Fe), ultramarine blue	High PW6, high PY74, moderate medium
Darker green paint	Oil-modified alkyd – long oil	Titanium white, synthetic organic pigment PY74, Mars yellow? (Fe), ultramarine blue, chalk, possible bone black (P)	Moderate PW6, high PY74, moderate medium
Surface efflorescence	Good match for palmitic acid	n/a	Dominated by fatty acids
Darker green, drip, top right corner	Oil-modified alkyd – long oil	Minor titanium white, synthetic organic pigment PY74. Possible chalk, ultramarine blue, kaolin? (Mg), phthalocyanine blue? (Cu), bone black? (P), Mars yellow? (Fe)	Moderate PW6, moderate PY74, high medium
Darker 'all over' green drip, top edge	Oil-modified alkyd – long oil	Minor titanium white, synthetic organic pigment PY74. Titanium white, chalk, ultramarine blue, kaolin? (Mg), bone black (P), Mars yellow? (Fe)	Moderate PW6, moderate PY74, high medium

Hume recalls that he may have chosen two different types of alkyd house paint for this work: an eggshell or satin paint for the lighter, yellower tone, and a gloss for the darker green. In some areas he has applied more than one layer to change the colour.[20] At the top right corner, a lighter green is overpainted with a darker shade and the adjacent passage of hair to the left was changed from dark to light green, as the overlapping drips at the side of the panel make clear.

Water Painting is unusual in that it had developed a significant amount of fatty acid efflorescence within a short timeframe, as shown in Figure 15.7. While this can be common to oil-containing paints, two of the factors known to promote efflorescence were not relevant in this case – the painting has not been framed and glazed or somehow enclosed since painting,[21] nor has it been exposed to high humidity.[22] As Hume used household gloss paints directly from the can with no alteration apart from mixing with each other, the source of the fatty acids cannot be due to the artist incorporating additional materials to modify paint properties. However, additives used in paint manufacturing can include the addition of fatty acids as stabilisers and pigment dispersants.[23]

There is a noted occurrence of fatty acid efflorescence in oil paints dominated by organic pigments, primarily due to the lack of inorganic metal (pigment) ions that aid in the oil drying process through cross-linking and the formation of stabilising divalent metal fatty acid soaps.[24,25] It is likely, therefore, that the relative lack of inorganic pigments in *Water Painting* has resulted in higher relative proportions of unreacted fatty acids, which have

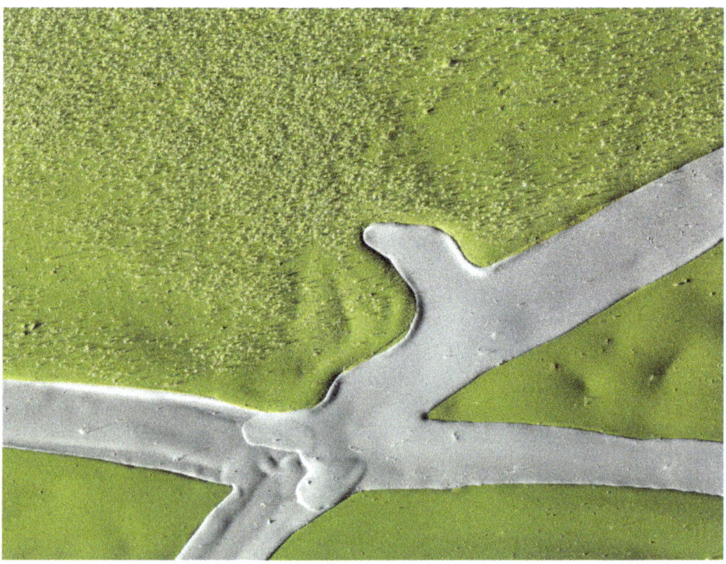

Figure 15.7 Gary Hume, *Water Painting*, 1999, T07618, Tate; detail of fine fatty acid efflorescence crystals on a green area © Gary Hume, photo © Tate.

migrated to the paint surface.[26] Interestingly, the efflorescence was also noted as heavier in areas where Hume used the eggshell finish lighter green paint and where he had overlaid different layers, giving an increased thickness of paint. Although leaner paints such as eggshell formulations have lower relative medium content (and hence lower potential fatty acid content) the higher solids content can impart a more porous structure, where volatile materials such as fatty acids may move more freely to the surface.

Discussion with the artist[27] established that Hume removes similar efflorescence layers himself, and, upon inspecting *Water Painting*, he expressed that he would like the surface gloss reinstated through the removal of the accumulated efflorescence. Conservation treatment was carried out in 2016, where options for removing the efflorescence crystals were explored using gloss measurement and microscopy. A range of aliphatic hydrocarbon solvents and dry-cleaning systems were explored. The method eventually chosen involved the application of petroleum spirit (boiling point range 100–120 °C) with a cotton wool swab, which would dissolve and pick up as much of the efflorescence as possible. This was followed by gently wiping and buffing the area with dry cotton wool to remove uneven areas before the solvent had completely evaporated. After treatment, the greatest improvement was seen in the top right passage which had the greatest efflorescence and equally, the least difference was seen in the white paint, which had minimal efflorescence.

It is clear from discussions that Hume wishes to retain the surface gloss of his paintings, however he appears possibly more ambivalent about other potential changes to his works due to ageing. In the same discussion he stated: "Age comes to us, things aren't stuck in aspic and there ought to be change. How attractive that change is, as an autonomous being...that's up for grabs...I can live with it as long as it looks good." When thinking of paintings by Mondrian which have "aged in such a beautiful way" Hume states, "they've got to have a bit of ageing but like damage – it's just such a "suck it and see", because we're at the beginning of them hitting a certain age, when they start to show themselves. It's an ongoing, interesting question...but I have chosen to paint in these paints that haven't got a history of their own ageing process yet, I can't be sure about it." Hume is also aware of other chemical changes that can occur to paint films, such as the yellowing of the oil component or the slow fading of pigments, and that the pigments in household paints are potentially more likely to fade than those in artists' paints. He states: "Every colour will change at a different rate to the other colour. Most of my paintings have between two and four colours, so it might be interesting. I know that in the universe nothing is fixed. I paint things in the universe, so I accept it and I accept it positively, but I'm intrigued by it." He also recalls looking at a painting by Sidney Nolan who worked in oil-modified alkyd house paints from the 1940s onwards: "That gave me encouragement, because it looked really great. But then I thought, I've no idea what that is. What he really used and what I'm really using because it's all chemistry."

As with Hockney, our course of action when caring for with Hume's paintings is clear for the foreseeable future. We have Hume's sanction and approval of treatments aiming to maintain his high-gloss surfaces and we will continue to explore and contextualise changes to the paintings over time and with conservation treatment using imaging and surface measurement tools in parallel with studies into the properties and ageing of these types of non-conventional paints. How close we can get to the original appearance is a somewhat redundant challenge, as paints can change markedly from their original appearance and the chemistry of oil-containing paints is, on a molecular level, impressively dynamic. Hume's own pragmatic view is that future decisions will be made through instinctive, albeit subjective judgements of the work's appearance; as he says: "I can live with it as long as it looks good."

15.4 Modern Oil Paints: Avinash Chandra

Avinash Chandra's *Hills of Gold* is a member of a group of modern oil paintings in the Tate collection that are beginning to show signs of being sensitive towards water as they age. This is a growing phenomenon that has received some attention in recent years.[28] Since many modern oil paintings are left intentionally unvarnished for aesthetic reasons, and glazing is not necessarily practical nor appropriate, surface dirt (or soiling) is naturally deposited onto painting surfaces over time. This can have a significant impact on the colours and saturation of the surface, altering tonal relationships and in some cases becoming increasingly difficult to remove. The removal of surface dirt traditionally involves the skilled application of aqueous cleaning agents, however the presence of mildly to severely water-sensitive paint passages can render water unusable. Recent research has aimed to investigate both the causes and characteristics of water sensitivity in modern oil paintings, as well as to develop low risk, effective methods for surface cleaning.[29] Thus far, the causes of water sensitivity have been shown to relate to paint formulation, including pigment type, type(s) of oil, additives such as driers, a lack of formation of metal soaps and the inclusion of specific extender pigments. In addition, ambient environmental conditions, *i.e.* temperature and relative humidity, artists' modifications to paints and application techniques are all influencing factors.

15.4.1 Avinash Chandra, *Hills of Gold*, 1965 (Tate, T00724)

Avinash Chandra (1931–99) was an Indian artist whose significant creative output received international recognition, whose works form part of public collections in Europe, America and India. *Hills of Gold*, 1965 (Figure 15.8) was exhibited in the *Indian Painting Now* exhibition at the Commonwealth Institute, London in 1965 and was acquired by Tate in the same year.

It is said that Chandra's work was influenced heavily by that of the Impressionists and post-Impressionists,[30] however during his first years as a painter, Chandra was also consciously trying to extricate himself from past traditions

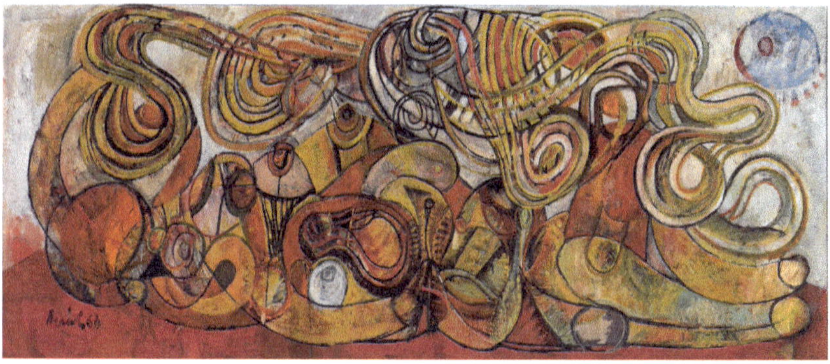

Figure 15.8 Avinash Chandra, *Hills of Gold*, 1964, T00724, Tate, © The Estate of Avinash Chandra, photo © Tate.

and find his own means of expression, working hard against the influences of Soutine and van Gogh. Similarly, while Chandra stated that he was familiar with the work of contemporary British and American artists working in the 1950s and 1960s, citing friendships with both Alan Davie and Willem de Kooning, he claims that they did not influence his artistic development. In 1956, Chandra moved to England with a series of unfinished landscape paintings that incorporated imagery inspired by the mountainous landscape of his home town. Chandra remarked that during this period he painted "memories of landscapes with trees which had warmth"[31] to contrast the frosty reception he received from 1950s London. During his early days in London, Chandra could not afford to buy canvas and used hardboard as a support, however he was careful to use "good quality paint" and commented that this is why "the colours are still bright" when reflecting on his works from the 1960s.[32]

Hills of Gold exemplifies Chandra's working technique of the 1950s and 1960s, where the composition was reworked over a prolonged period. Chandra did not try to disguise this process, and the surface of this work has deliberately retained ridges and textures that are not reflected in the final composition. In addition, Chandra would apply thin washes of diluted oil paint over dried passages of more thickly applied paint, which created translucent veils of colour through which the complexities of the underlying composition can be glimpsed. In other areas, pure blocks of colour contrast with paint that is mixed directly on the surface, creating dynamism.

Hills of Gold is painted in oil, on a medium-weight plain-weave canvas with a commercially applied priming layer that contains lead and zinc white. Like many modern and contemporary oil paintings, the surface was left unvarnished by the artist, who was interested in achieving a varied surface rather than uniformity, where subtle modulations in gloss associated with different paint passages are retained. Consistent with Chandra's painting technique, *Hills of Gold* was accomplished in more than one session, with evidence of reworking and repeated painting over earlier compositions. A tiny cross-section of paint taken from a bright red passage (not shown) indicates the presence of a second priming layer between two paint layers, the uppermost red and an earlier pale yellow, suggesting the canvas might have been repurposed. There are also many areas of wet-in-wet painting, where paint is blended directly on the canvas. Forms are delineated using a glossy, diluted black paint in continuous fluid lines that have dripped in some areas, albeit obscured by the growing complexity of the surface. Orange and yellow paints were applied within and blended with the black lines, as both thick buttery paint and in a diluted form, contributing to the growing dynamism of the surface.

As listed in Table 15.4, Chandra's palette included alizarin crimson, cadmium red and orange, chrome yellow, an organic green pigment, bone black and Prussian blue pigments. Extenders including chalk, barium sulphate, alumina hydrate and magnesium carbonate were also identified. The presence of magnesium carbonate in the red and yellow paints suggests the use of British paint manufacturer Winsor & Newton® oil paints for those colours. Fatty acid efflorescence was detected in areas of medium-rich paint and passages containing organic pigments such as the alizarin crimson organic red pigment (Figure 15.9). As noted for Hume's *Water Painting*, this phenomenon is frequently observed in modern and contemporary oil-containing paintings, originating at least in part from additives such as metal soaps and free fatty acids included as part of the paint formulation.

In *Hills of Gold*, rod- and hexagonal-shaped crystals were also identified at the surface of paint samples taken from cadmium orange, cadmium red and chrome yellow paints, as shown in Figure 15.10. Elemental analysis of the paint surface identified that these crystals (Table 15.4) were rich in magnesium and sulphur, suggesting the formation of magnesium sulphate heptahydrate (epsomite), which is a water-soluble salt known to form on paint surfaces as a degradation product of magnesium carbonate.[33] While not visible to the naked eye, the formation of epsomite can affect paint gloss levels and complicates surface-cleaning treatments due to the extreme water solubility of this salt and the fact that these crystals can be partially embedded in paint surfaces. As a result (at least in part) of the formation of epsomite, *Hills of Gold* is now sensitive to water, and particularly so in the chrome yellow and cadmium red passages most affected. The surface is also sensitive to mechanical action such as swabbing with cotton wool, where the crystals can be disturbed, risking abrasion and pigment pickup during cleaning.

There are several other degradation phenomena that contribute to the fragility and conservation treatment challenges associated with water- and solvent-sensitive oil paint surfaces, such as medium skin formation, powdery paints and even liquifying paints. When combined with the formation of water-soluble epsomite, all these possible occurrences render the safe removal of accumulated soiling from water-sensitive oil paint surfaces one of the most difficult challenges for the preservation of unvarnished painted modern and contemporary works of art.

Table 15.4 Combined analysis results from *Hills of Gold*.

Ample	Water sensitive (Y/N)	Epsomite detected (Y/N)	Solids EDX: Elemental composition [* = major element(s)]	Binder and solids (FTIR)	Inferred sample composition
Orange impasto	Y	Y	C, O, Mg, Al, S, Cr*, Pb*, Ba	Oil, magnesium carbonate, barium sulphate, possible alumina hydrate	Lead chromate, magnesium carbonate, barium sulphate, aluminium stearate and/or alumina hydrate
Red	Y	Y	C, O, Mg, Al, Si, S, Cl, Fe,	Oil, organic red pigment (possibly PR2/PR10), magnesium carbonate	Organic red pigment, magnesium carbonate, silica, possible alumina hydrate and/or aluminium stearate, silica, iron oxide
Black	Y	N	C, O, Pb*, Zn	Oil, magnesium carbonate, organic red (possibly PR83, alizarin crimson), possibly bone black	Organic red pigment, magnesium carbonate, carbon black, lead and zinc driers
Yellow	Y	Y	C, O, Na, Mg, Al, S*, Cr*, Zn, Pb*, Ba*	Oil, magnesium carbonate, barium sulphate, possible silica, possible epsomite	Lead chromate, magnesium carbonate, barium sulphate, zinc stearate, aluminium stearate, magnesium stearate/carbonate
Yellow	N	Y	C, O, Al, S, Cr, Zn, Pb*, Ba*	Oil, magnesium carbonate, barium sulphate, crystalline zinc soaps, amorphous metal soaps	Lead chromate, zinc chromate, magnesium carbonate, barium sulphate, zinc stearate, aluminium stearate
Dark green + white	Y	N	Mg, Pb, Cr, Fe, Mg	Oil, magnesium carbonate, Prussian blue, titanium white	Lead chromate, Prussian blue, magnesium carbonate, titanium white

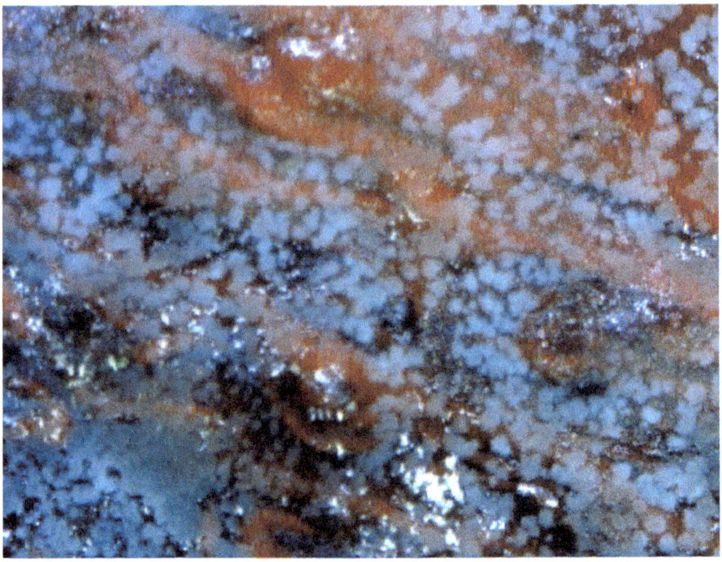

Figure 15.9 Avinash Chandra, *Hills of Gold*, 1964, T00724, Tate; photomicrograph (215x) of fatty acid efflorescence over cadmium orange and black paint, © The Estate of Avinash Chandra, photo courtesy of Lucia Bay, photo © Tate.

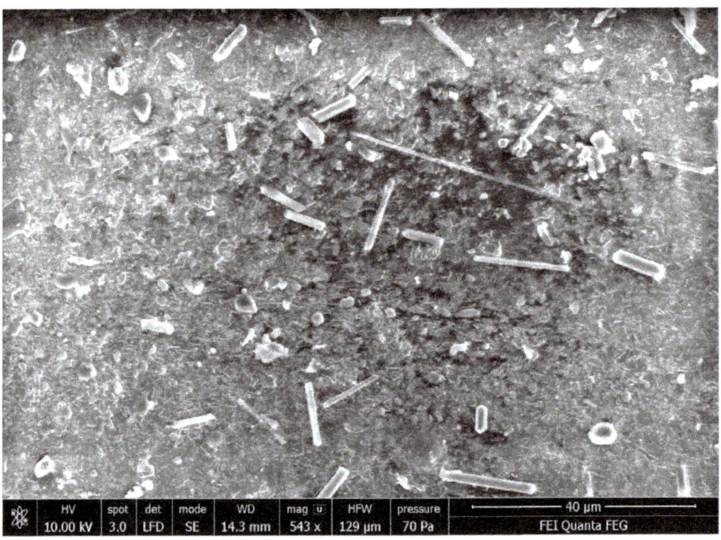

Figure 15.10 Avinash Chandra, *Hills of Gold*, 1964, T00724, Tate; scanning electron microscope (SEM) image of epsomite crystals on chrome yellow paint. © The Estate of Avinash Chandra, photo courtesy of Lucia Bay, photo © Tate.

15.5 Conclusions

Holistic studies into modern and contemporary works of art involving scientific, technical and art historical information, combined with ongoing discussions with artists, produce valuable information that can directly enhance our understanding of the complexity and nuance of the materials and processes used to create a work of art. Furthermore, they provide insight into motivations, collaborations and meaning that may have never been voiced, inform new ways of looking and contribute to much-needed research on change and the degradation of modern and contemporary art materials. Although the paints used by Hockney, Hume and Chandra are primarily stable and relatively young, these and other contemporary art materials can change in ways which can significantly alter appearance and potentially compromise the integrity and longevity of a given work. Technical investigations into the detail of changes noted on works of art form a vital part of this ongoing, necessarily diverse dialogue through enriching our understanding of the material character of painting surfaces, providing insight into the degradation of materials through contextualising visual changes and minimising risk associated with conservation treatment. This is particularly valuable when conserving a work for the first time, and when the artists' intention is difficult to discern or unknown. Particularly relevant for modern and contemporary art is the artist's "right of integrity", which must be placed at the forefront when reflecting on change and conservation, and for this, discussion with artists, preferably in front of their works, proves invaluable in balancing the need for preservation with the desire to view a work in its most "authentic" state.

Acknowledgements

The authors would like to thank the following people and departments for their contributions to this work: Tate staff including Joyce Townsend, Andrew Wilson and Bernard Horrocks; Tate Photography and Tate Conservation Departments; former Tate staff and collaborators Lora Angelova, Anna Cooper, Lucia Bay, Aviva Burnstock and Tom Learner.

References

1. D. Hockney, in *David Hockney by David Hockney: My Early Years*, ed. N. Stangos and H. Geldzahler, Thames and Hudson, London, 1988, p. 98.
2. D. Hockney and N. Stangos, *Hockney by Hockney*, Thames and Hudson, London, 1976, pp. 125–128.
3. J. Crook and T. Learner, *The Impact of Modern Paints*, Tate, London, 2000, pp. 12–31 and 84–97.
4. J. Crook and T. Learner, *The Impact of Modern Paints*, Tate, London, 2000, p. 89.

5. J. Crook and T. Learner, *The Impact of Modern Paints,* Tate, London, 2000, p. 29.
6. C. E. Rogge and B. A. Epley, Behind the Bocour Label: Identification of Pigments and Binders in Historic Bocour Oil and Acrylic Paints, *J. Am. Inst. Conserv.*, 2017, **56**(1), 1–28.
7. P. Cummings, *Interview with Leonard Bocour, June 8 1978*, Archives of American Art, www.aaa.si.edu/collections/interviews/oral-history-interview-leonard-bocour-12884#transcript, Accessed 17 September 2018.
8. J. Crook and T. Learner, *The Impact of Modern Paints,* Tate, London, 2000, p. 30.
9. D. Hockney and N. Stangos, *Hockney by Hockney*, Thames and Hudson, London, 1976, p. 125.
10. J. Crook and T. Learner, *The Impact of Modern Paints*, Tate, London, 2000, p. 90.
11. B. Ormsby and T. Learner, The Effects of Wet Surface Cleaning Treatments on Acrylic Emulsion Artists' Paints – A Review of Recent Scientific Research, *Rev. Conserv.*, 2009, **10**, 29–41.
12. R. Hellen, *Personal Communication with David Hockney*, Tate Britain, July 2016.
13. D. Pearce, *Portrait of David Hockney*, 1972, www.bfi.org.uk/films-tv-people/4ce2b6bf0f7c1, Accessed 15 October 18.
14. Anon, CN 9.1 David Hockney, T01269 Tate Conservation file, unpublished report.
15. E. Jablonski, T. Learner, J. Hayes and M. Golden, Conservation Concerns for Acrylic Emulsion Paints: A Literature Review, in *Tate Papers*, no.2, Autumn, 2004, https://www.tate.org.uk/research/publications/tate-papers/02/conservation-concerns-for-acrylic-emulsion-paints-literature-review, accessed 15 November 2018.
16. R. Hellen, Tate Britain, November 2016, personal communication.
17. Anon, 1972, CN 9.1 David Hockney, T01269 Tate Conservation file, unpublished report..
18. D. Barret and L. Head, *Gary Hume, New Art Up-Close 1*, London, 2004, pp. 4–5.
19. T. Learner, *Analysis of Modern Paints*, Getty Publications, 2005, p. 19.
20. R. Scott and H. Brett, Interview with Gary Hume: TATE Stores, 20 April 2015.
21. R. Mulholland, PhD thesis, Royal College of Art, London, 2010.
22. L. Akerlund, *Efflorescence: An Investigation of Selected Paintings from the 19th to the 21st Century with a Preliminary Experimental Study of the Role of Moisture in the Development of Efflorescence*, 2013, http://collectiewijzer.nl/2013/01/18/efflorescence-in-paintings-and-the-role-of-moisture-by-luciana-akerlund/#_ftn1, Accessed 15 November 2018.
23. E. Ordonez and J. Twilley, Clarifying the Haze: Efflorescence on Works of Art, *WAAC Newsletter*, 1998, **20**(1).

24. B. Singer, J. Devenport and D. Wise, Examination of a Blooming Problem in a Collection of Unvarnished Oil Paintings, *The Conservator*, 1995, (19), 3–9.
25. M. Mecklenburg, C. Tumosa and E. Vicenzi, The Influence of Pigments and Ion Migration on the Durability of Drying Oil and Alkyd Paints, *Smithsonian Contrib. Mus. Conserv.*, 2013, (3), 59–67.
26. A. Sawicka, A. Burnstock, F. Izzo, K. Keune, J. J. Boon, K. Kirsch and K. J. van den Berg, An Investigation into the Viability of Removal of Lead Soap Efflorescence from Contemporary Oil Paintings, *Issues in Contemporary Oil Paint*, 2013, pp. 312–313.
27. R. Scott and H. Brett, *Interview with Gary Hume*, TATE stores, London, 20 April 2015.
28. Cleaning Modern Oil Paints project, www.tate.org.uk/about-us/projects/cleaning-modern-oil-paints-0, Accessed 5 November 2018.
29. K. J. van den Berg, *et al.*, *Issues in Contemporary Oil Paint,* Springer Publications, 2014.
30. R. Garfield, Avinash Chandra: Towards a Reappraisal, *Avinash Chandra A Retrospective*, Osborne Samuel LLP Berkeley Square Gallery, London, 2006, pp. 7–13.
31. R. Araeen, *Conversation with Avinash Chandra*, Third Text, vol. 2, 1998, pp. 69–95.
32. R. Alley, *Catalogue of the Tate Gallery's Collection of Modern Art Other than Works by British Artists*, Tate Gallery and Sotheby Parke-Bernet, London, 1981, p. 117.
33. G. Silvester, *et al.*, A cause of water-sensitivity in modern oil paint films: The formation of magnesium sulphate, *Stud. Conserv.*, 2014, **59**, 38–51.

CHAPTER 16

The Deep Meaning of Poetry: Eduardo Kac's Art of the Fundamental Processes

PIER LUIGI CAPUCCI*

Accademia di Belle Arti Reggio Calabria, Italy
*E-mail: plc@noemalab.org

16.1 Networks, Interactive Arts, Telepresence, Robotics

In his wide activity, Eduardo Kac has been working with digital technologies, telepresence and networks. In these realms he has been especially interested in the possibilities of a "dialogue based on telecommunications, to the extent that it overcomes local boundaries and makes intersubjective experiences possible across the network on a global scale".[1] It is the art of networking, opposed to the "monological ideologies"[1] embodied by the traditional one-way television transmission. This "dialogism", this reciprocal information exchange that is a critical topic in networks and telecommunications, pervades all Kac's artistic research, including outside the digital and networks realms. Along this path he identifies the potentialities of participation, of the interactive artistic forms (that he calls "dialogic electronic art"[2]) which tend to limit "the emphasis of visuality to give precedence to interrelation and connectivity".[2] A path that goes beyond the traditional idea of an art to contemplate, that introduces the use of unusual tools and techniques for art and

Science and Art: The Contemporary Painted Surface
Edited by Antonio Sgamellotti, Brunetto Giovanni Brunetti and Costanza Miliani

Published by the Royal Society of Chemistry, www.rsc.org

goes beyond their rules and constitutive technical aims, in order to develop a critique of the increasingly saturated and oppressive daily infosphere.

Telepresence and robotics can reveal an important artistic potential. Kac extends the meaning of the art of telepresence by defining it as "the integration of telecommunications, robotics, human–machine interfaces and computers",[3] enclosing it in the wider framework of interactive electronic art.[3] In recent decades, robotics has experienced overwhelming growth, and in its many applications and forms – industrial robotics, social robotics, biorobotics, drones, autonomous vehicles, *etc.* – it is going to become a pervasive and essential field in everyday life. The tendency to delegate repetitive or unattainable activities and operations to external devices created by humanity, and to which humanity has transferred its own knowledge, has existed for a long time. Automata have been historically present in many cultures, from Europe to Islam,[†] to China, to Japan, as well as the attempts to construct lifelike entities, mythological simulacra of human and animal life. These ideas pervade the whole history of humanity: the Greek Talos, the Jewish Golem, the Medieval,[5] Renaissance and eighteenth-century automata, the nineteenth-century Mary Shelley's *Frankenstein*, the twentieth-century robots, the androids, cyborgs and replicants in literature and cinema, the sophisticated contemporary social and industrial robots, the explorer robots sent to Mars...

Robotics is a transdisciplinary territory which ranges from the mythological and historical traditions of many cultures to the literary and theatre narratives (in fact the word "robot" was born in theatre[‡]), to cinema and television, to industrial applications. But it is also a transdisciplinary research field, where both humanistic and scientific disciplines collaborate: biology, physiology, psychology, philosophy, automation, electronics, artificial life, artificial intelligence, physics, computer science, mathematics, mechanics... Since the end of the 1950s and the experiences of kinetic art, robotic art (or "cybernetic art", as it was initially called) has often hybridized with traditional art forms like performance, installations, dance, theatre. And also telepresence and telecommunication, like in Eduardo Kac's artworks *Ornitorrinco* (1989) and *Rara Avis* (1996).[6,7]

The art of telepresence highlights the aspects of relationship, the dialogical opportunities. It can have a critical and social value, being

> [...] a means for questioning the unidirectional communication structures that mark both traditional fine arts (painting, sculpture) and mass media (television, radio). I see telepresence art as a way to express on an aesthetic level the cultural changes brought about by remote control, remote vision, telekinesis, and real-time exchange of audiovisual information. I see telepresence art as challenging the teleological nature of technology. To me, telepresence art creates a unique context in which participants are invited

---

[†]From 31 October 2015 to 28 February 2016 at the ZKM in Karlsruhe an exhibition on Medieval Arab automata was performed.[4]

[‡]"Robot" is from the Czech word *robota*, which means "heavy labor" or "forced labor", introduced in 1920 by the writer Karel Čapek in the three-act drama *R.U.R. (Rossumovi univerzální roboti)*.

to experience invented remote worlds from perspectives and scales other than the human.[8]

According to Kac, this "space of reciprocity", which due to digital technologies and telepresence undergoes a profound transformation, constitutes a key element. The "real space" and the very notion of distance become progressively irrelevant,[9] while the image, and more generally the narration, replaces the real by implementing the predominance "of the effect of reality on the principle of reality".[10]

Recently, many digital-based technologies have emerged. Among them are: Augmented Reality, which interactively and in real time adds symbolic information layers to the environment; Big Data, which are technologies to analyze or extract information from data sets that are too large to manage using common data-processing applications; the Internet of Things (IoT, and its evolution, the Internet of Everything, IoE), which interconnects objects, tools and devices, allowing them to exchange data and take an active role; the development of industrial, social and personal robotics, with increasingly autonomous and economic devices and vehicles. Moreover, a number of applications can combine these technologies with Artificial Intelligence, Artificial Life, Deep Learning and Machine Learning, with increasingly sophisticated algorithms, multiplying the possibilities and at the same time raising troubles of a social, cultural and political nature: let's think, for example, about the security- and privacy-related issues.

16.2 The Art of the Living

I was directly involved for the first time in Eduardo Kac's activity in 2000. I had known about his long and multifaceted activity before that, but at that time I got directly in touch with it. In fact, in 2000 I was involved in Cafe9.net, a pilot project supported by the European Community and the Youth for Europe Project that interconnected, in a creative physical and Internet-based network, the European Cities of Culture for the year 2000: Avignon, Bergen, Bologna, Brussels, Helsinki, Krakow, Prague and Reykjavik. I was the director and content manager for the activities of the city of Bologna, and we organized and shared many cultural events, concerts, discussions, exhibitions – many of them about art. In September 2000 we organized an event dedicated to bioart, in particular to transgenic art, in teleconference with Avignon. Participants included Louis Bec – the great *zoosystematicien*, scientist and artist engaged in the realm of Artificial Life, who died in 2018. He was my French colleague in Cafe9.net as well as at that time the director of the partner cultural venue of Avignon – and one of the two scientists who technically created *GFP Bunny*, named *Alba* Figures 16.1 and 16.2, the green fluorescent rabbit-*chimera* that Kac publicly revealed. Louis Bec was deeply involved in this project, and we had a wide discussion about bioart, and in particular on transgenic art. We also debated on the fact that, although Kac's bio-media performance had become very famous, it was hardly criticized, and *Alba* was never publicly shown in any exhibition or event because of the protests of

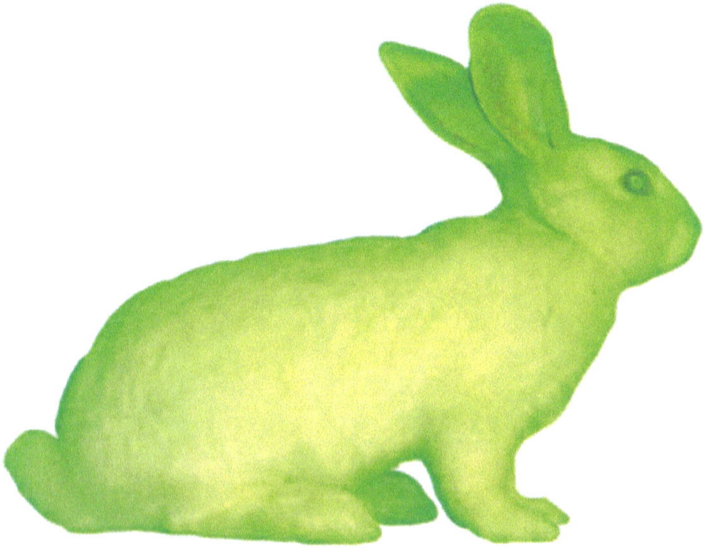

Figure 16.1 Eduardo Kac, *GFP Bunny*, 2000, transgenic artwork. Alba, the fluorescent bunny. Reproduced with permission from Eduardo Kac.

many organizations. We also discussed why art is generally considered as inappropriate and inadequate to discuss science in the biotech realm, and to creatively act inside this field.

Indeed, by the end of June 2000, just two months before the beginning of Cafe9.net, I received an e-mail from Louis Bec reporting that *Alba* had been censored, and was forbidden to participate in the public exhibition "Artransgénique" in Avignon, the French partner city in Cafe9.net. The e-mail also contained an invitation to collaborate and share this information and the topics that bioart raised. So some days after, along with a text by Eduardo Kac about *Alba* and transgenic art[11] and with the call by Louis Bec,[12,§] I published an article in *Noema*.[13] Those were probably the first articles about bioart published in Italy.

> [...] *Alba* (or *GFP Bunny*) opens up new art horizons but also raises new questions. Given that its life is respected, as it seems evident from Eduardo Kac's text, we could ask: do we have the right to do so? And, more specifically: art, which by definition is mainly considered as having no scientific, practical or utilitarian purposes that can justify this kind of experimentation, but cultural scopes, has the right to participate in this discussion?
>
> You know what we think: art has always used technologies, even against the culture that opposed them. Why there should there be a censure for some technologies? Where is the contradiction?
>
> *Alba*, like all interesting art, actually leads away from art, with its provocative existence it contributes to undermining some fundamental concepts and highlights contradictions, hypocrisies and questions on which our

§The texts by Eduardo Kac and Louis Bec published in *Noema* can be reached online: https://noemalab.eu/org/sections/ideas/ideas_3.html.

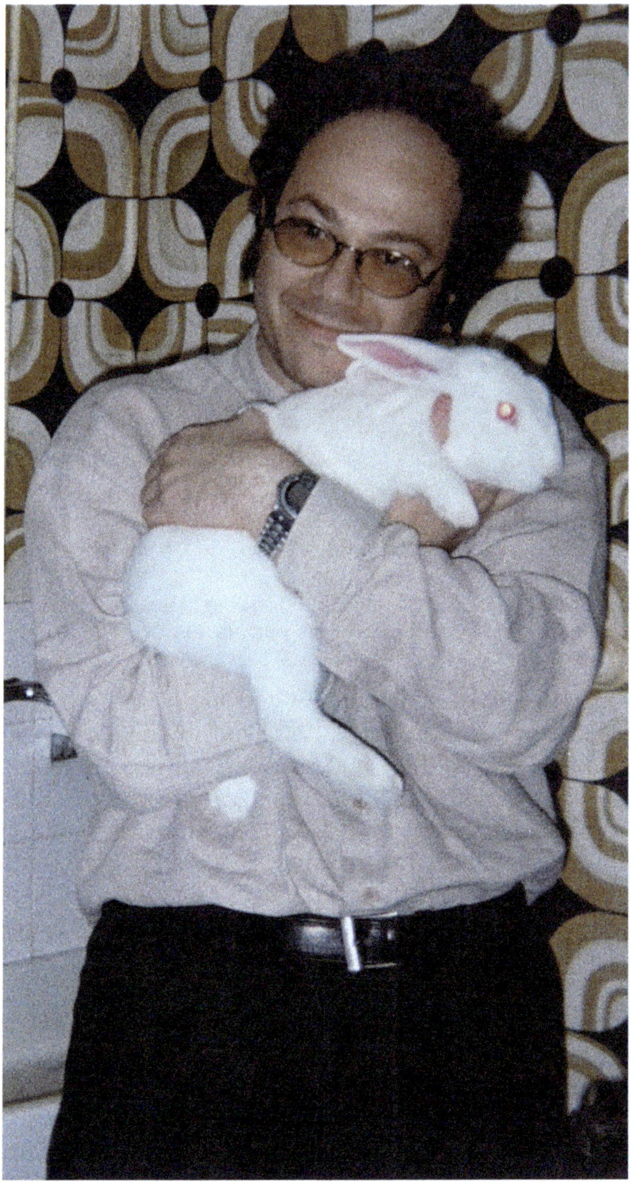

Figure 16.2 Eduardo Kac and Alba, 2000. Reproduced with permission from Eduardo Kac.

culture is based on (including the idea of "artificial" as opposed to "natural", the centrality of human culture, ethics, the idea of "life", the meaning of "nature" and "art"...). *Alba* points out to an uncomfortable reality, destined to grow in the future, at the end of which for now there are only nuanced distinctions that sound more as prejudice, sophistry, ideology or religion rather than reason or science. So let's open this discussion, that in the past has engaged us and will engage our culture in the years

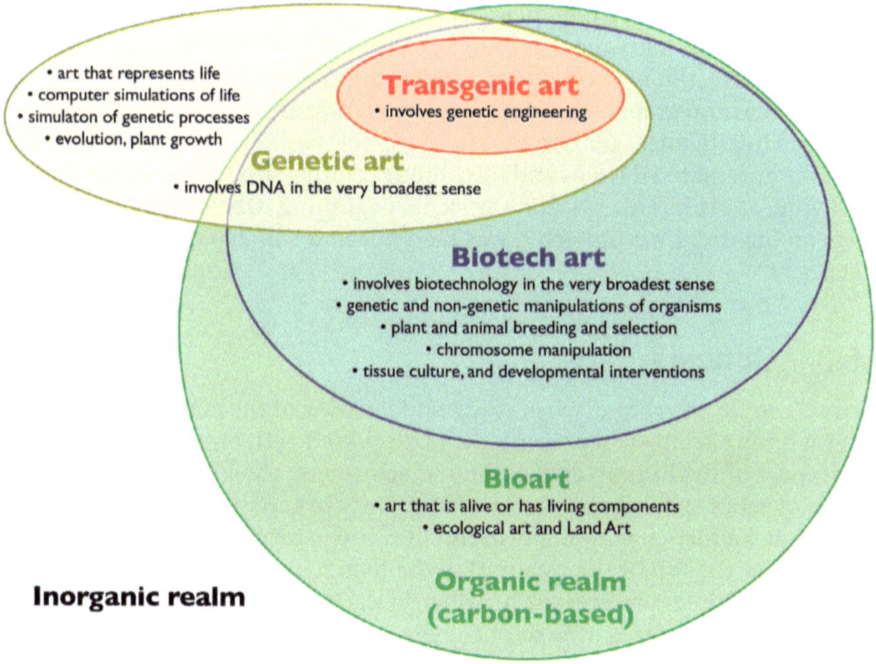

Figure 16.3 Pier Luigi Capucci, *The Bioarts Realm*, 2007, based on a text posted by George Gessert to Yasmin, mailing list for art–science–technology interactions around the Mediterranean Rim, on 25 March 2006 [Image created by the author of the article, originally published in Ivana Mulatero (ed.), *Dalla Land Art alla Bioarte/From Land Art to Bioart*, Turin, Hopefulmonster, 2007, bilingual (Italian/English)]. Reproduced from Mulatero, 2007 with permission from P. L. Capucci.

> to come. We believe it would be very interesting if *Noema* readers could express their thoughts.¶ [...]

Eduardo Kac's activity on bioart, and especially the *Alba* affair, promoted a wide discussion worldwide about biotechnologies and genetic engineering. *Alba* even appeared on the first page of *Le Monde*[14] and it was cited in many international magazines. In 2000, biotechnologies and genetics were under suspicion and their use in art was considered as bizarre and even nasty, an experiment to reject and lock up in the room of horrors. Biotechnologies and genetics evoked awful ghosts such as eugenics and bacteriological war. As reported in many texts of that period, the artists working in this field by modifying and even creating living organisms made of the very matter and dynamics of the living organisms, (humans included), were "playing God", were playing with life.

These art forms also us forced to reconsider art history and the art forms who used living beings and organic matter as tools and substance to create, like performance, land art, ecological art for example Figure 16.3. Bioarts were apart from the classic tools and matter of art, like brushes, canvases,

¶From ref. 13. Translation into English by the author of this chapter.

colors, stones, metals, wood, plastics. They were also apart from the digital-based art forms, like the computer-generated arts, the algorithmic and interactive arts, the network art processes, although sometimes they can be involved and integrated in the bio-creations. And they were apart from the Artificial Intelligence art constructs, from the Artificial Life applications, from the automated devices and the robotic entities simulating or emulating the living, as art has always done since its beginning. One of the first exhibitions on bioart, "L'art biotech'", was held in Nantes in 2003, curated by Jens Hauser.[15,16]

16.3 Genesis

Bioarts can be a tool to discuss many nodal issues of the contemporary that cannot be as effectively represented and discussed with other art forms. Like the impact of the human culture on the ecosphere; the killing and extermination of other species; the ability of humanity to genetically modify the living, as has happened with many tools since prehistory; the historical idea of humanity at the top of the pyramid of the living; a critique of anthropocentrism. These topics are present in Eduardo Kac's artworks. In *Genesis* Figure 16.4, 1999, he criticizes a verse of the *Book of Genesis* which puts humanity at the top of the pyramid of the living, giving it a supremacy over all living beings:

> Let us make man in our image, like us: and let him have rule over the fish of the sea and over the birds of the air and over the cattle and over all the earth and over every living thing which goes flat on the earth.[17]

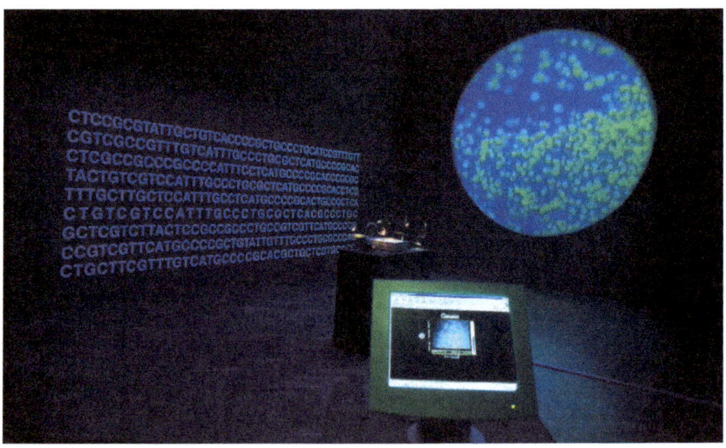

Figure 16.4 Eduardo Kac, *Genesis*, transgenic telematic installation, 1998–1999. Reproduced with permission from Eduardo Kac.

Kac creates a synthetic artist gene, non-existent in nature, transducing that verse into the Morse code – which in Kac's idea represents the dawn of the information age – and then in the DNA code, which is inserted into a population of bacteria through the four nitrogenous bases (Cytosine, Thymine, Adenine and Guanine). In the same Petri dish coexist cyan (which also contains the synthetic gene) and yellow fluorescent mutant bacteria, which may or may not generate a mixture of colors (green). The synthetic gene – and therefore the verse from the *Book of Genesis* – can mutate both as a result of the normal multiplication process, both because of the interaction between bacteria and for a human intervention. The latter can be done by activating *via* a Web interface an ultraviolet light source targeted to the bacteria. The installation allows participants to monitor the evolution of bacteria locally, in the gallery and remotely through a video camera Figure 16.5.

The set – consisting of a Petri dish with the bacteria, a video camera and a source of ultraviolet rays – is connected to a video projector and two networked computers. One of them acts as a Web server streaming live audio and video and manages the remote requests for activating the ultraviolet source. The other computer coordinates the DNA musical synthesis (the music is generated live in the gallery and sent online through multiplication

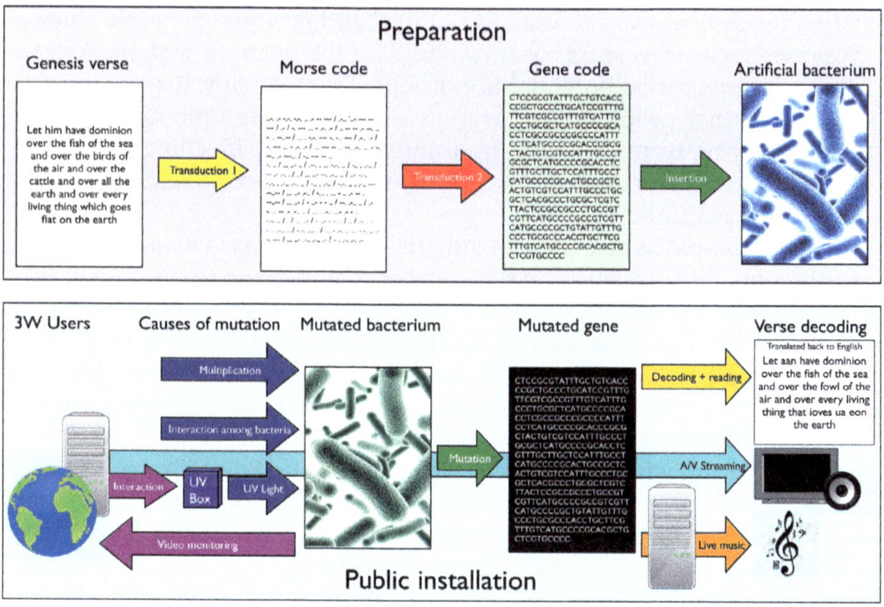

Figure 16.5 Pier Luigi Capucci, the scheme of Eduardo Kac's *Genesis* installation, 2017 [Image created by the author of the article, originally published in *The New and History. art*science 2017. Conference Proceedings*, Pier Luigi Capucci, Giorgio Cipolletta (eds), Ravenna, Noema Media & Publishing, 2018]. Reproduced with permission from P. L. Capucci.

and mutation algorithms). Thanks to the video camera the video projection in the gallery shows the interactions among the bacteria and their multiplication. The ultraviolet light, remotely activated by the users, causes the bacteria to emit visible cyan and yellow light, and disrupts the DNA sequence by accelerating the mutation rate. According to Kac, the triple system of *Genesis* codes – natural language, DNA code and binary logic – explores the fact that today biological processes have become writable and programmable, able to store and process data similarly to digital computers.

At the end of the exhibition the verse of the *Book of Genesis* is decoded, transduced and read in English, and the mutations that have deconstructed the initial text are highlighted.[18,19]

16.4 Looking into the Future

In the last ten years many important acquisitions have emerged in the field of life sciences. A growing importance has been given to the role of bacteria, in the environment as well as inside the organisms, to an extent that they can be considered as peculiar markers of the individuals and able to influence their behavior. Synthetic Biology has become a widespread practice to modify and create living organisms for many purposes, for example in medicine.[20,21] Genetic engineering has considerably accelerated thanks to CRISPR-Cas9,‖ an easy-to-use, economical and powerful genomic editing technique that in a few years has revolutionized life sciences, with the related ethical concerns for its potential use on humans**. Finally, for a few years, De-Extinction has promised to revive extinct animals and plants with technologies ranging from the more traditional ones, like breeding and selection, to genetic engineering.[24,25] A sort of Jurassic Park-related idea but with more recent creatures.

Today the bio-based disciplines – and their applications in medicine, in the environment, in life sciences, in the study of evolution, up to the merge with robotics and digital devices – are among the most active fields of research, and biology has emerged from being an exclusive scientific dimension to a cultural paradigm. Although sometimes they are still considered with suspicion, genetics and biotechnology give birth to many useful and often vital practical applications.

In the art realm what two decades ago was considered as bizarre or monstrous is now a wide and complex field through which to look into the future. There are many artists, events and exhibitions, many books and essays have been published, there is an important international discussion. The bioarts have emerged from the fence of curiosities to reach museums and galleries and are also in dialogue with other art forms, with philosophy, with scientific, cultural and social issues. In the last years many important exhibitions,

‖On CRISPR-Cas9 see the dedicated section of *Le Scienze*.[22]

**In 2015 in China for the first time this technique was used to intervene on a "non-vital" human embryo.[23]

events, conferences and publications worldwide have raised new perspectives in the fields of philosophy, art and culture. Like "NeoLife" at the University of Western Australia in Perth[26] organized by Oron Catts and SymbioticA, on the emergence of new life forms, which included a conference curated by the Society for Literature, Science and the Arts and some workshops and exhibitions on the sciences and the arts of life. Like the events "Limits of the Human: Human, Inhuman, Overhuman, Antihuman", at the National Centre for Contemporary Arts in Moscow, curated by Dmitry Bulatov and Madina Tlostanova,[27,28] a series of conferences and exhibitions on human nature and the meaning of the human beyond the idea of being a measure of all things, and on recognizing within the human things that are normally considered outside it. Like the special issue of the French magazine *mcd – magazine des cultures digitales*[29] dedicated to life-based art forms, curated by Annick Bureaud, or the exhibition "La Fabrique du vivant",[30] in Paris at the Centre Pompidou, on the relationships between the living and the artificial, on the processes of artificial recreation of life, on the manipulation of the living matter and on the hybridization of organic and industrial matter, of animal and plant cells.

The topics and fields of investigation have expanded, multiplied and deepened, connecting apparently distant areas: for example, by scrolling through the "NeoLife" lectures there are topics such as anthropology, design, nutrition, ethology, ethics, biopolitics, patentability, copyright, law, literature, theatre and De-Extinction.

The bioarts have also entered the world of traditional art, which has always used the organic matter, the body and the environment to create. And since biotechnologies are becoming increasingly common, maybe in a near future the word "biotech" might disappear from the name of these art forms. Moreover, the biotech art forms can hybridize with the Net, the storage, archive, management and visualization systems, with Artificial Intelligence, Big Data, the Internet of Things, Artificial Life, Robotics, Synthetic Biology. This evolution is going to increase the complexity of the field, and contribute to the rise of new life forms created or managed by human culture: a sort of "Third Life",[31–33] with the "First Life" being the organic life and the "Second Life" being the life in the symbolic realm.

16.5 Interview with Eduardo Kac

Pier Luigi Capucci (PLC): I think it is very interesting that in your work you have used many different techniques, technologies and tools: poetry, photography, performance, holography, computer, networks, telepresence, robotics, biotechnologies, creating inorganic and organic-based art forms. Is it possible to draw some kind of a poetic and artistic line across all your ways of intending and making art?

Eduardo Kac (EK): Yes, because the question I am interested in is not technology itself. In fact, everything is technology: pencil is technology, oil painting is technology, every tool that we employ is technology. So, as a poet, as an artist, I create form, I create form that gives itself to the senses, which is to say form that provides an aesthetic experience. That is what I do and that is what art is supposed to do. What is particular in what I do, that is unique and underpins all of my interests, is that early on in my life I understood that life could not fold into haze: either we would find ourselves in a condition that we would be free or we would find ourselves in a condition that we would not be free. We could rebel against that lack of freedom and we could seek to change reality as we found it. What struck me was the analogy that I perceived among artworks that come to you completely close and leave you the possibility of accepting or refusing to engage, but the form itself came completely pre-packaged, so it is basically a question of accepting or rejecting.

That reminded me of modes in which you, on the receiving end of the experience, lacked greater freedom. That led me to the idea of developing forms of art that were dialogical: that acknowledge that you are there, engaging in the work right now and that mattered that it is *you*, not a human, not a person, but *you*, because only *you* will make the decisions that *you* will make right here, right now, decisions that might be different tomorrow. This mode of working in which you do not have a subject to object relationship but a subject to subject relationship, which I called dialogical, became a lifelong pursuit.

PLC: Communication is at the heart of your artistic research and poetics, at many levels, from the inner existence of living matter to the reason for making art, to the way the artwork can dialogue with its environmental and human contexts.

EK: From a larger philosophical, materialist perspective, what that means to me ultimately is that when we talk about communication most people do not realize that what they mean is a subset of the world of communication. When they say "communication" they usually mean "persuasive communication". This is why you have the expression "Do you know what I mean?". So, basically, everything that is involved in the communication process – the sender, the channel, the noise... – becomes irrelevant because what is important is that you get what I am saying to you, and as a result that you behave in the manner that I expect. Well, in reality that is a minute aspect of the communication processes. Instead, what I am really interested in is creating communication processes that remain open ended and producing works that create conditions for subject to subject to engage, and, ultimately, that reveal my understanding of communication as, quite possibly, the most fundamental of the living processes. Because a rock does not communicate anything to a piece of plastic, that is my perspective, and any attempt to read any communication in a rock or a piece of plastic is a human projection, in my view, over inanimate objects. There are certainly bacteria that live inside the rock and on the surface of the plastic, but I am talking about a hypothetical communication by the plastic and by the rock itself.

Communication in my view is a fundamental material, physical, underlying process of reality. Let us imagine the moment when the first single cell emerges: if that single cell could not communicate to another single cell there would not be the formation of the first complex creature. And not all communication is about coming together, some communication is unfortunately lethal: when a virus communicates its contents to a bacterium, the bacterium dies. But the possibility of commingling, and coming together, and exchanging information, in my view, is the fundamental substrate of reality, so that is what I seek to explore and capture.

PLC: In 2007 you published the Space Poetry *manifesto,*[34] *and in 2017 you created* Inner Telescope[35] Figure 16.6, *an artwork aboard the International Space Station specifically conceived for zero gravity, which was made in space following your instructions.* Inner Telescope *is an instrument of observation and poetic reflection, which leads us to rethink our relationship with the world and our position in the Universe. This is a dream that you long pursued, why is "space art" interesting?*

EK: Again, the question is not technology, the question is more profound. I made my first digital work, which was a digital poem, in 1982. I started very early, at the age of 17, I formed a performance group in 1980 and I worked in mixed media, multimedia, with the performance from 1980 to 1982, but when I finished that project I realized that everything that had been done up until then – everything from the first cave paintings to Pollock, to Warhol, including my old work up until that point – was analog, early computer art was on paper, everything was analog. And I understood that this emerging form which we call "digital" was more than a technique. I understood that it was the beginning of a new culture. I lived in an environment in which I was surrounded mostly by painters and poets that published books. Although critics were interested in what I did, almost nobody else was. I explained that in the 1960s and 1970s we had gone through something that Oscar Masotta

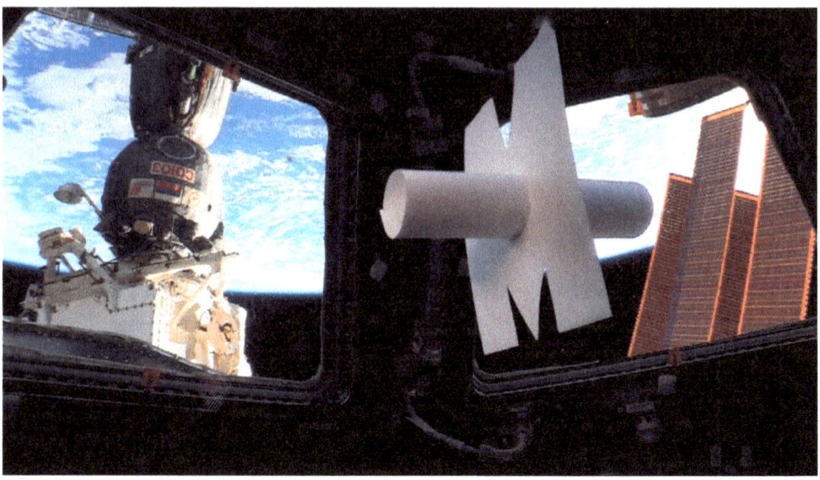

Figure 16.6 Eduardo Kac, *Inner Telescope*, artwork floating in the International Space Station (ISS), in orbit around Earth, 2017. Reproduced with permission from Eduardo Kac.

from Buenos Aires had described as dematerialisation, which he actually got from Lissitzky. He puts an epigraph in his essay on dematerialisation in which he quotes Lissitzky, who said: "We live in an era of telegraphy, radio, *et cetera*, these are dematerialized forms". And then he wrote that dematerialisation was the order of the day in the late 1960s with newspapers, magazines, television and all that. So, what I am saying is that we – for me starting from 1982, with my first work – are no longer experiencing dematerialisation, we can now create works that are themselves directly immaterial, because they are digital. It is no longer a question of diminishing the importance of matter, it is a question of creating directly work that is information. That signalled, in my view, a new culture. As a young poet and artist, I wanted to participate in this new culture, contributing to the development of the art and poetry of this new culture, which I did for many years.

With space there is a similar situation: we are in the beginning of a new era in which space will progressively become accessible to the individual, to the civilian. Space tourism will become ordinary in the next decades. It is quite conceivable that there will be a space hotel eventually as well. We will build a base on the Moon which will serve as a preliminary area for study about living outside of the Earth because, as we know, our body did not evolve for that purpose, and eventually we will land on Mars. The landing on Mars will be significant for us as a culture, but we are not going to start colonising Mars anytime soon. We will, however, have a space station around the Moon in about fifteen to twenty years, and we will have a Moon base where people would be going to on a regular basis, like they go to the International Space Station right now. But, most importantly, we will have suborbital flights as an ordinary event, we will cross the hundreds of kilometres from the surface line that separate us from space, and we are going to do that on an ordinary basis. At that moment you have lots of people having similar experiences and sharing them, describing them, and creating works for those experiences: a new culture emerges. As that new culture emerges, more people will want to participate with new forms of cuisine, of theatre, of performance, of poetry...

My *Inner Telescope* is a work created for zero gravity. Creating work that is specific for zero gravity is not a trivial thing, because we have no tradition, we have no landmarks. You know what Beethoven did, what Bach did, what Chopin did, so you have landmarks that stay as a reference: you can go in or against that direction, but we have no *Mona Lisa* of the space age. One thing that is significant about *Inner Telescope* is that the work was actually made in outer space. It is the first poem, the first artwork that was crafted by hand outside of our home planet, and that is the landmark, like the flag on the Moon: we have arrived, here we are, we can only develop from here. That is why I think that *Inner Telescope* is both a visual work of art and a poem, because for me poetry is an art form, and I pursue poetry in a way that is visual, sensorial, multimodal, I have always done that since I started with digital poetry. I have always pursued poetry in this malleable, transformable way that involves visuality and other senses, because I see a *continuum* between word and image, between visual art and poetry. I know that for many people poetry is contiguous to music, the beginning of poetry was like that and many people explore this way, which is wonderful. For me personally, as a practician, I explore this *continuum* between visual art and poetry, and I have always done that. Conceiving something for zero gravity, where the rules of Earth do not apply, is a particular challenge but many people will tackle this challenge in the future, and the new space culture will emerge.

PLC: The relationship among art, science, technology and society is at the basis of many of your artworks and your making art. How do you see the relationship between art and science?

EK: I think the question is: why haven't we recognized that art has influenced science, when everybody thinks that science has influenced art? I think that is the real question. Science has always been influenced by art, it is just through our prejudice that we have not acknowledged that. I think there is a book that has to be written about how science borrowed from art. It has always been reciprocal but we have not recognized science borrowing from art. We will give some examples. The metaphor is a tool of the poet and science has always borrowed this instrument but pretended that it did not, and by pretending that it is not borrowing anything from poetry, science produces facts. Let us give an example: the concept of the gene. The concept of the gene was speculated upon already in the 1930s, and before the discovery of the double helix structure there was speculation on what it might look like, but also what it might be, how it might function, and the result of that is that this metaphor was coined: that the gene works like Morse code. It was an analogy with something that you do not know because you have not seen it. You did not know that this thing really existed, you speculated on its existence, you had not seen its structure, you had not really proved its existence, but it was speculated on the notion of a unit.

It was Schrödinger that coined this analogy when, as a physicist, he looked at biology from the perspective of a physicist, and in his book *What is Life?*,[36] published in the 1940s, he coined this metaphor. So, it is by collapsing the metaphor and pretending that they did not employ that tool. So, you have two elements of an analogy and then you have the link through the metaphor, you collapse the metaphor and then you link the elements of the analogy into a fact. This is the basic mechanism of the rhetoric of science, and many people have written long essays in exposing how this logic works. But beyond the instrument of the poet, perspective was invented by artists, photography was invented by artists, and can you imagine science without photography? The debt that science has towards artists is fundamental. You look at an X-ray picture of the cosmos, those pictures are raw data, there is no color and then scientists, in order to communicate an idea, employ false colors. We invented that: painters invented false colors to communicate an idea or an emotion. The list goes on, it is a very long list that we would have to track down in history and different cultures to demonstrate that it has always been a two-way road, it is just that out of prejudice the society has only recognized one side.

PLC: Some people today consider that the recent acquisitions of Artificial Intelligence, Big Data, Deep Learning and Machine Learning, Robotics, are leading to some kind of dishumanization, are a threat for humanity. This is a long-standing idea; what do you think about it, and how do you see the future?

EK: This feeling of being inadequate because traditional modes of existence have been disrupted by technology is quite old in itself. Most people, when they complain about technology in one way or another, do not realize that they stand at the end of a line that goes back in time, since technology was not controlled by a small group but started to be part of society. Of course this phenomenon became more accentuated during the Industrial Revolution, when the ordinary folk started to be touched in a more direct way by technology. The earliest example that I casually have found – I have not really stopped to research this and do an extensive examination of people responding negatively to technology – is in Edgar Allan Poe, who lived in the first half of the nineteenth century. In fact, I am currently reading an anthology of Heinrich Hein, the German poet that operated this transition between Romanticism and Modernity (he precedes Baudelaire, he is a contemporary of Poe), and he too has a passage in which he complains about the "humanizing technology", which for us today would be the object of laughter.

It is a feeling that has been shared socially since technology started to touch society, and it is often a feeling that the young people do not have. It is the older folks, who knew the world before these technologies and who feel nostalgic about the way the world was, that express that feeling. The young folks, that do not know the older world, are perfectly happy in this new one. That is one thing. The other thing is that, as we know, the elephant in the room is power. Everybody dances around all kinds of issues, but the elephant in the room is always power. The question is not abstract, it is not technology itself, the question is whether those who are in power will use these technologies to perpetuate themselves in power. Those who consolidate wealth will use these technologies to increase their wealth. This may be less apparent in Europe, but it is certainly the case in the rest of the world. The ultimate challenge is not to stop these technologies, but to find ways in which they are used in a manner that assures a democratic distribution of power and a fair distribution of wealth. That is the real question, it is not the technology itself.

PLC: Through sciences, technologies and the resulting tools, our species is deeply pervading and modifying the world it lives in, it is influencing the environment and threatening the other species, transforming the meaning of "Nature". According to some thinkers and philosophers it is fundamental to go back to being human, the solution would be "more humanity". Is it really so? Or, since just "being humans" has led to this situation, is it necessary to go beyond humanity?

EK: I do not think that we can "go back to being human", because there is nothing specific in my view about being human. The human is an entity that continuously transforms itself, evolution has not stopped, we have not arrived at anything, we are just here as a part of a *continuum* that will develop and transform indefinitely, and there is nothing to go back to because there was nothing there to begin with. For example, we were talking about bio-art, and it is really interesting that the Human Genome Project revealed that we all have bacterial DNA in our genome. What do you call a creature that has a DNA of another creature? Transgenic. We have always been transgenic, we just did not know it until the Human Genome Project. So, going back to being transgenic again? Well, we have always been, we just did not know. So, what does that mean, going back to what? I do not think there is anything to go back to, because we are malleable and transformable.

The problem is the same: the problem is power. It is not that humans have created global warming, it is power, it is the wealthy, it is countries that put their own interests first and not the interests of humanity as a whole, that pollute immensely. We know who are the top polluters in the world and that there is no remaining body of water that is not contaminated by plastic. Any form of water that you drink will have some plastic. Not to mention the islands of plastic in the ocean. It is power, it is greed, it is the accumulation of wealth that has led to this, so those are the real issues. If those issues could be addressed, everything else would follow.

PLC: What is your position on contemporary issues that are transforming the natural, geopolitical and economic balance of our planet, like Climate Change, migrations, economic crises? May an art exist that can deal with these topics?

EK: In the 1960s an art form emerged that was called "ecological art". Artists were already dealing with ecological problems: they were, for example, looking for plants that could absorb pollutants in an area, *etc*. It is wonderful, it is great, but art can talk about any topic: you could make a painting about global warming, but the planet itself, reality, life, those are too large for a single artist to resolve. That has to be resolved at the level of United Nations, by elected politicians that represent the ultimate aspirations of the world's planet, which are survival, happiness, fair wealth distribution, access to education, access to hygiene, access to hospitals... The solution for the planet, the real solution for the planet, is not to be looked for in art, it is to be looked for at a political level. But artists can raise issues, they can make us aware of these issues, they can produce insight that motivates us to action.

So, art can play its part. But we should not look for solutions directly in art, because these problems are too large and they have to be resolved politically. For that we need democracy, and for that we need ways in which we can be vigilant. So that we protect ourselves from the most dramatic effect of our era, which is beginning to be known as the "information world war". We have started a new phase of history that is likely to be remembered as the "information world war". In the wars of the past you knew exactly your enemy and where they were, even though you had coded messages that had to be cracked, but the war of today is one in which people hide behind IP addresses. Shell after shell after shell, and they exert influence in other societies at a distance.

So, democracy alone, without protection against the effects of the information world war, will suffer, as we have seen. I think it is a time in which a lot has to be rethought, reconfigured, re-examined, and then we need to start anew. We need to understand what is coming to us from all these fronts, and we need to ensure that the possibility of a genuine free exchange and communication takes place, so that we can elect democratically politicians that will in fact make changes to protect the planet and to ensure fair access to all the human rights. I think that this is the real issue.

(Interview conducted on 26 December 2018)

References

1. E. Kac, *Telepresence & Bio Art. Networking Humans, Rabbits and Robots*, The University of Michigan Press, Ann Arbor, 2005, p. 104.
2. E. Kac, *Telepresence & Bio Art. Networking Humans, Rabbits and Robots*, The University of Michigan Press, Ann Arbor, 2005, p. 103.
3. E. Kac, *Telepresence & Bio Art. Networking Humans, Rabbits and Robots*, The University of Michigan Press, Ann Arbor, 2005, p. 138.
4. *Allah's Automata. Artifacts of the Arab-Islamic Renaissance (800–1200)*, ed. S. Zielinski and P. Weibel, Hatje Cantz, Ostfildern, 2015.
5. E. R. Truitt, *Medieval Robots: Mechanism, Magic, Nature, and Art*, University of Pennsylvania Press, Philadelphia, 2015.
6. E. Kac, *Ornitorrinco* and *Rara Avis*: Telepresence Art on the Internet, *Leonardo*, 1996, **29**(5), 389–400.
7. E. Kac, *Telepresence & Bio Art. Networking Humans, Rabbits and Robots*, The University of Michigan Press, Ann Arbor, 2005, ch. 2, pp. 125–214.
8. E. Kac, *Telepresence & Bio Art. Networking Humans, Rabbits and Robots*, The University of Michigan Press, Ann Arbor, 2005, p. 139.
9. E. Kac, *Telepresence & Bio Art. Networking Humans, Rabbits and Robots*, The University of Michigan Press, Ann Arbor, 2005, p. 142.
10. E. Kac, *Telepresence & Bio Art. Networking Humans, Rabbits and Robots*, The University of Michigan Press, Ann Arbor, 2005, p. 144.
11. *Eduardo Kac: Telepresence, Biotelematics, and Transgenic Art*, ed. P. T. Dobrila, Kibla, Maribor, 2000.
12. Louis Bec, Le lapin fluorescent, *Noema*, June 2000, https://noemalab.eu/org/sections/ideas/ideas_articles/bec.html, (accessed December 2018).
13. P. L. Capucci, *Alba*, arte transgenica, arte del vivente, *Noema*, June 2000, https://noemalab.eu/org/sections/ideas/ideas_articles/alba.html (accessed February 2019).
14. Lapin Fluo, *Le Monde*, 5 October 2000, p. 1.
15. *L'art Biotech'*, ed. J. Hauser, Filigranes Editions, Nantes, 2003.
16. An expanded Italian edition of the catalog, in *Art Biotech*, ed. J. Hauser, CLUEB-mediaversi, Bologna, 2007.
17. Genesis, 1, 26.
18. E. Kac, *Genesis*, pp. 249–263.

19. P. L. Capucci, Art as a philosophy of contemporaneity. Poetics of complexity, Third Life, locality and universality, in *The New and History. Art*-science 2017 Conference Proceedings*, ed. P. L. Capucci and G. Cipolletta, Noema Media & Publishing, Ravenna, 2018, pp. 49–62.
20. W. W. Gibbs, Synthetic Life, *Sci. Am.*, 2004, **290**, 75–81.
21. D. Baker, *et al.*, Engineering life: Building a fab for biology, *Sci. Am.*, 2006, **294**(6), 44–51.
22. *Le Scienze*, April 2016, n. 572, pp. 28–47.
23. P. Liang, Y. Xu and X. Zhang, *et al.*, CRISPR/Cas9-mediated gene editing in human tripronuclear zygotes, *Protein Cell*, 2015, **6**, 363–372.
24. B. Shapiro, *How to Clone a Mammoth: The Science of De-extinction*, Princeton University Press, Princeton, 2015.
25. http://reviverestore.org (last accessed March 2019).
26. http://www.symbiotica.uwa.edu.au/activities/symposiums/neolife-slsa-2015 (last accessed February 2019).
27. http://www.ncca.ru/en/events.text?filial=2&id=1500 (last access January 2019).
28. http://thehumancondition.ru/page?en&mid=209&id=78 (last access January 2019).
29. mcd - magazine des cultures digitales, *Arts&Sciences*, March–May 2016, n. 81. Available online: http://www.digitalmcd.com/mcd81-arts-sciences/ (last access August 2016).
30. https://www.centrepompidou.fr/cpv/agenda/event.action?param.id=FR_R-0e1ca28a3e1128d624b2d8375beead0&param.idSource=FR_E-0e1ca28a3e1128d624b2d8375beead0.
31. P. L. Capucci, From life to life. The multiplicity of the living, in *New Realities: Being Syncretic*, ed. R. Ascott, G. Bast, W. Fiel, M. Jahrmann and R. Schnell, Springer-Verlag, Vienna, 2009, pp. 56–59.
32. P. L. Capucci, Declinations of the living: Toward the Third Life, in *Evolution Haute Couture. Art and Science in the Post-Biological Age*, ed. D. Bulatov, BB NCCA, Kaliningrad, 2013, pp. 50–63.
33. P. L. Capucci, Art as a philosophy of contemporaneity. Poetics of complexity, Third Life, locality and universality, in *The New and History. Art*-science 2017 Conference Proceedings*, ed. P. L. Capucci and G. Cipolletta, Noema Media & Publishing, Ravenna, 2018, pp. 49–62.
34. E. Kac, *Space Poetry*, Available online http://www.ekac.org/spacepoetry.html (accessed January 2019).
35. E. Kac, *Inner Telescope*, Available online http://www.ekac.org/inner_telescope.html (accessed January 2019).
36. E. Schrödinger, *What Is Life? the Physical Aspect of the Living Cell*, Cambridge University Press, Cambridge, 1944.

CHAPTER 17

When Surgery Meets Conservation: The Treatment of the Multi-material Sculpture Love Me *by Sarah Lucas*

FABIOLA ROCCO*[a], TOMMASO POLI[b], OSCAR CHIANTORE[b] AND ANTONIO MIRABILE[c]

[a]SUSCOR – University Structure in Science for Conservation, Restoration, Exploitation of Cultural Heritage, University of Turin in Cooperation with the Conservation – Restoration Center *La Venaria Reale*, Via XX Settembre 18, Venaria Reale, 10078, Turin, Italy; [b]Department of Chemistry, University of Turin, via Giuria 7, 10125, Turin, Italy; [c]Paper Conservator, 11, rue de Bellefond, 75009, Paris, France
*E-mail: fabiola.rocco1991@gmail.com

17.1 Introduction

Since the mid-nineteenth century artists have had access to a wide range of innovative materials, which has facilitated their ability to renovate traditional aesthetic values.[1,2] Among them, sculptors have progressively abandoned bronze and marble for modern materials, embodying a leading role in the definition of contemporary art.[1] The industrial development of plastics has prompted artists to use second-hand objects and industrial materials as part of their sculptural language, instead of more traditional methods such as casting and modelling. Moreover, artists started to investigate the

Science and Art: The Contemporary Painted Surface
Edited by Antonio Sgamellotti, Brunetto Giovanni Brunetti and Costanza Miliani

Published by the Royal Society of Chemistry, www.rsc.org

expressive potential of acrylic, polyester and epoxy resins for the creation of artistic, abstract or fanciful shapes.[2] More recently, the constant research of an inedited way of expression has pushed contemporary artists to the limits of the concept of sculpture itself, including transient and edible materials as sculptural items.[1] The work of the English artist Sarah Lucas (born in 1962) is no exception. Along with many of her contemporaries and compatriots (*e.g.* Damien Hirst, Angus Fairhurst, Tracy Emin), Lucas became synonymous with the phenomenon of the Young British Artists (YBA) who dominated the Brit Art scene of the nineties.[3] Besides the shared themes and vocabulary in common with the other YBAs, Lucas developed her own individual and unconventional artistic language from the beginning of her artistic enquiry. Her work focuses on the relationship between sexes, social identity, death and destruction.[3–5] Lucas makes sculptures with ordinary "things"[6] – items, found objects, vegetables, cigarettes – all quotidian materials which bear the traces of use and become "epitome of commonness".[7] The employment of simple materials underlies the provocative and gruesome qualities of her work where she plays with social cliché, verbal and visual representations of sexuality and masculine stereotypes.[5]

One of the peculiar expressions of Lucas' artistic research is the *Love Me* sculpture representing the lower part of a woman's body sitting on a chair (Figure 17.1a). The sculpture surface is completely covered with images cut from magazines, showing repeated female mouths (on the right half) and eyes (on the left half). Due to its exquisite artistic value, *Love Me* has been widely exhibited in several temporary shows in Italy and abroad.[†] The frequent transportations combined with handling necessities during installation/de-installation, together with the intrinsic fragility of the constituent materials, has produced localized damage in a few places, the most severe of which was a deformation on the right leg at the level of the artwork's "ankle" (Figure 17.1b). Therefore, in 2016 *Love Me* was transferred to the Conservation and Restoration Center "La Venaria Reale" for restoration. The main focus of the work was securing the damage and re-establishing the correct leg inclination in order to achieve structural stability, thereby limiting the risks of further damage and allowing the artwork to be showed again.

Analytical investigation was combined with the historical-artistic research and information provided by the artist herself *via* an interview, which allowed the characterization of the constituent materials and the discovery of the uncommon techniques employed by the artist. The information obtained proved to be fundamental to the development of an experimental program and the conservation treatment. Due to the nature of the damage, a new

[†]"Lei. Donne nelle collezioni italiane"/"She: Women in the Italian Collections", Fondazione Sandretto Re Rebaudengo/Sandretto Re Rebaudengo Foundation (Turin), 2003; "Glowbowl. Opere dalla collezione Sandretto Re Rebaudengo"/"Glowbowl: Works from the Collection of Sandretto Re Rebaudengo", Salle Quai Antoine/Quai Antoine Room (Principality of Monaco), 2007; "Espíritu y Espacio Colección Sandretto Re Rebaudengo"/"Spirit and Space: Sandretto Re Rebaudengo Collection", Ciudad Grupo Santander/Santander Group City (Madrid), 2011 and "Tuttovero"/"All True", Galleria Arte Moderna/Modern Art Gallery (Turin), 2015.

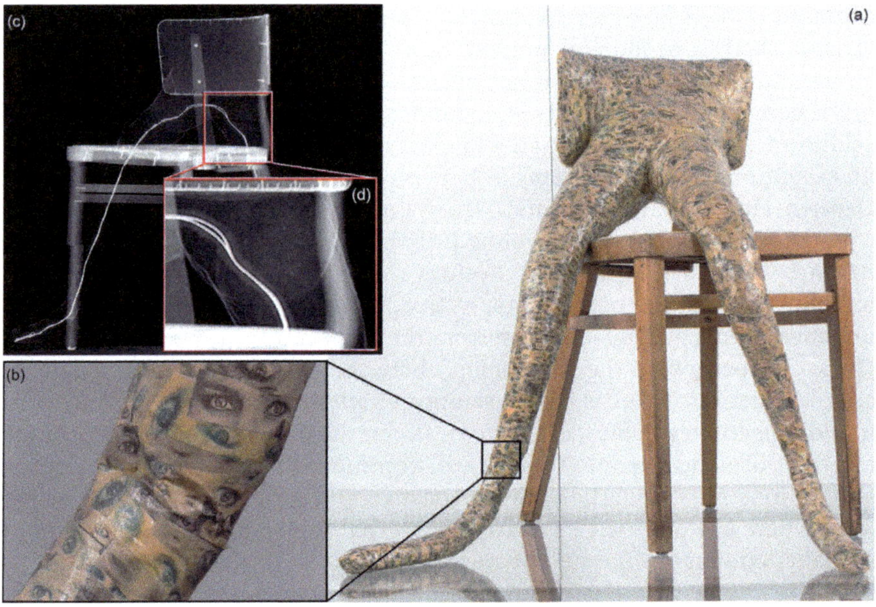

Figure 17.1 *Love Me*. (a) Front side view; (b) Detail of the dent located on the sculpture's right leg; (c) X-ray radiography of the sculpture which highlights the iron wire used by the artist as internal armature; (d) Detail of the newspaper used to cover the iron wire.

restoration product based on nanocellulose particles was tested. Moreover, an innovative conservation strategy, more respectful to the original material than traditional methods, was successfully developed.

This contribution highlights the complex problems faced by conservators of twentieth-century artworks made of non-traditional materials.

17.2 Love Me

17.2.1 Description of the Artwork and Artistic Meaning

Love Me represents a headless female figure lying on a small secondhand chair in an unseemly manner, with wide-open legs. The sculpture's torso, devoid of arms, is fused with the chair-back, composing an inseparable shape.

The appearance of *Love Me* is reminiscent of the iconic Sarah Lucas' *Bunny* series of 1997. The *Bunnies* are life-sized, decapitated sculptures realized by stuffing nude-colored ladies' tights with cushion padding and wire that were then gracelessly clamped on office chairs, provocatively suggesting sexual fatigue and female submissiveness.[4–6] Even if correlations between the artworks seem obvious, the images used in *Love Me's* collage express something more deep and concealed. In order to completely understand the artwork's meaning, *Love Me* has to be investigated considering also two later works, *Hysterical Attack (Mouths)* and *Hysterical Attack (Eye)*. These artworks, realized in 1999 for the solo show *Beyond the Pleasure Principle* (Freud Museum, London), present the same form as *Love Me* and, like their relatives, Lucas' *Bunny* sculptures, represent a feminine figure in a languishing way.[4] The *Hysterical Attack* sculptures are evenly covered with a mosaic-like collage reproducing only female mouths (*Hysterical Attack (Mouth)*) or only female eyes (*Hysterical Attack (Eye)*). The Freudian meaning of each *Hysterical Attack* seems to be 'scopophilia',[‡] regarding the eyes-covered sculpture, and oral fixation in the case of *Hysterical Attack (Mouth)*.[4] With this in mind, *Love Me*, where the two Freudian meanings are represented together in an allegorical artwork, could be read as a progenitor of the later works. In *Love Me*, Sarah Lucas focuses on the interchangeability between "mouths and sexual orifices, the gaze and the sexual act"[§] in a celebration of erotic symbolism. In this way, the three works express a deeper psychoanalytical conception of the female body compared to the *Bunny* sculptures, because "they are caught between physicality and symbolism".

17.2.2 Executive Technique: the Artist's Interview

The characterization of *Love Me's* constituent materials and the identification of the technique employed by Lucas proved to be highly difficult. With the exception of the visible surface, all the elements used to create the sculpture's shape were hidden and impossible to characterize at first glance. Therefore, in order to provide insight into *Love Me's* technique, it was decided to directly contact the artist.

Lucas emphasized the closeness of *Love Me* to the *Bunnies* "[*Love Me*] may be regarded as a discrete subgroup or variant of the *Bunny* series", suggesting the use of a woman's nylon tights filled with cushion padding as stuffing material. Moreover, the artist explained in general her sculpting technique: "an internal armature was created using chicken wire covered with newspaper" (see Figure 17.2c); the wire was then placed inside the tights, which were filled and firmly secured to the back of the chair. Subsequently, several layers of images, cut out from "colour glossy magazine and black-and-white images from newspaper newsprint", were glued directly to the "surface of the work, using wallpaper paste". The artist has a long history of using newspaper images as an artistic device.[3,4] At the beginning of her career, Lucas cut out pornographic photos from tabloids like *The Sunday Sport* and *The Daily Mirror* and exposed them on the gallery wall, to the embarrassment and annoyance of the public.[4] Tabloids were used for two reasons: firstly, they are mass-produced materials available to a wide range of consumers

[‡]Sexual pleasure derived chiefly from watching others when they are naked or engaged in sexual activity.

[§]Information about the artistic meaning, the manufacturing process and the restrictions about the conservation treatment were given directly by the artist *via* interview.

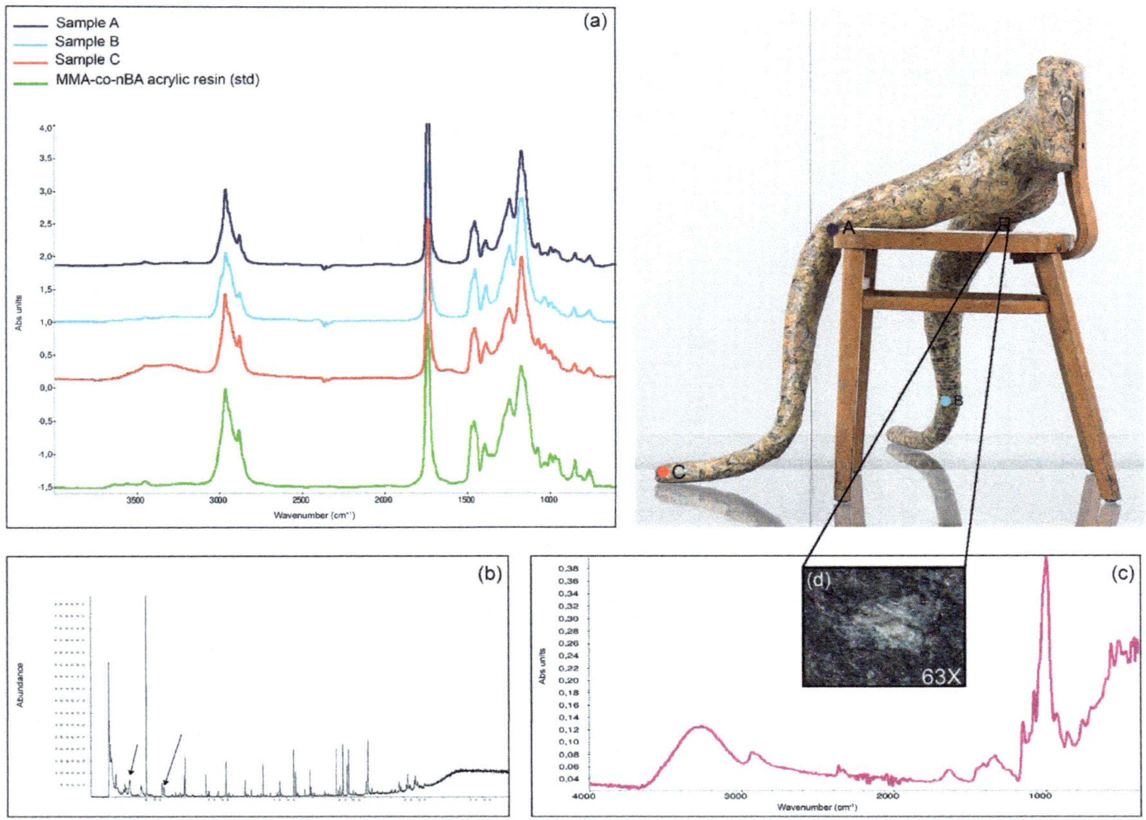

Figure 17.2 (a) Analyses performed on the sculpture: Infrared spectra of varnish from three samples A, B, and C (blue, light blue and red lines) taken from different areas of the sculpture compared with the reference spectrum of a MMA-co-nBA acrylic resin (green line); (b) Pyrolysis spectrum of a sample coming from the varnish. The arrows show the signal referable to the presence of the two co-monomers: MMA and nBA; (c) Infrared spectrum of the cellulose-based adhesive sampled on the back of a superficial paper piece (d).

and secondly they are tremendously cheap.[3–7] Lucas' characteristic technique of featuring the re-utilization of existing and discarded things, all coming from her near surroundings, is visible also in *Love Me's* chair, which, as the artist pointed out, is "readymade and well-used".

Finally, to level out the artwork surface and fix the collage, *Love Me* was covered with a clear industrial acrylic varnish which is almost undetectable on the back and more consistent on the front because "this is the side most often viewed".

17.3 Materials and Characterization

In order to confirm the information about the artwork's materials provided by the artist, selected samples collected from the sculpture were analyzed by spectroscopic and chromatographic techniques, namely infrared spectroscopy[¶] (FTIR) and pyrolysis-gas chromatography/mass spectrometry[‖] (py-GC/MS). The industrial varnish employed by the artist was investigated by means of FTIR and py-GC/MS analyses. Micro-samples of coating were removed from the artwork using a scalpel and immediately evaluated to avoid contamination. Multiple samples, collected from different areas across the artwork, were analyzed in order to obtain a representative investigation. In all of the infrared spectra acquired from different areas of the sculpture, see Figure 17.2a, it is possible to see the exact correspondence between the reference spectrum (green line) of a methyl methacrylate-co-*n*-butyl acrylate polymer and three spectra of the varnish coming from the sculpture (blue, light blue and red lines). Typical absorption bands of an acrylic backbone are clearly present. In the region between 3100 and 2800 cm^{-1}, it is possible to observe both symmetric and asymmetric stretching of the methylenic groups (CH_2 and CH_3) at 2954 and 2874 cm^{-1}. The intense peak at 1734 cm^{-1} is due to the carbonyl C=O stretching which is a characteristic peak for an acrylic compound. Two sharp peaks at 1240 and 1169 cm^{-1} were recorded. Those bands were attributed to the stretching of the C–O group, the former, and the

---

[¶]FTIR transmission spectra (64 scans) recorded using a diamond anvil cell (High Pressure Diamond Optics, Inc.) were obtained on a Bruker® Vertex 70 spectrophotometer coupled with a Bruker® Hyperion 3000 IR microscope equipped with an MCT detector (InfraRed Associates, Inc.), working in the spectral range from 4000 to 600 cm^{-1} with an average spectral resolution of 4 cm^{-1}. The silicon wafer paints were analyzed in the optical bench in transmission modality in the range 4000–400 cm^{-1}.

[‖]Py-GC/MS analyses were carried out with a Pyroprobe 1500 (CDS Analytical Inc., USA) filament pyrolyzer directly connected to a GC/MS system. The GC was a 6890N Network GC System (Agilent Technologies, USA) gas chromatograph with a (5%-Phenyl)-methylpolysiloxane (30 m, 0.25 mm i.d., 0.25 μm) capillary column. The temperature of pyrolysis was 600 °C, the temperature of pyrolysis interface was 300 °C and the temperature of the injector was kept at 280 °C. The carrier gas was helium (1.0 mL min^{-1}) and split ratio was 1/20 of the total flow. The mass spectrometer coupled to the GC apparatus was a 5973 Network Mass Selective Detector (Agilent Technologies, USA). Mass spectra were recorded under electron impact at 70 eV, scan range 40–600 *m*/*z*. The interface was kept at 280 °C, ion source at 230 °C and quadrupole mass analyzer at 150 °C. All instruments were controlled by Enhanced ChemStation® (ver. 9.00.00.38) software.

stretching of the C–C group, the latter. The nature of the copolymer has also been confirmed by py-GC/MS analyses, as shown in Figure 17.2b, where the characteristic ions of MMA and nBA monomers were detected.

In order to investigate the nature of the adhesive employed by the artist for adhering the cut-out images, the decision was made to detach a newsprint fragment from a non-visible area of the sculpture. A spectrum was acquired from a micro-sample collected from a tiny dense accumulation of adhesive located on the back of the fragment (Figure 17.2b). Figure 17.2c shows the infrared spectrum of the adhesive. The signals are compatible with a product based on polysaccharides (starch, wheat paste) or a modified cellulose-based one. This identification is critical due to the difficulty of the sampling of the adhesive. Moreover, the possible presence of paper as a contaminant in the sampling can alter the results. Paper, which also consists of cellulose, presents a spectrum characterized by the same typical absorption bands as other saccharides, namely a strong C–O stretching in the region ranging from 900 to 1100 cm^{-1}. As reported during the interview (see Section 17.2.2), Lucas glued the images with an ordinary wallpaper paste. Wallpaper pastes are normally commercialized in two formulations: as a dry mixture of cellulose ethers or as an aqueous dispersion of vinylic/acrylic compounds and cellulose ethers. Since none of the bands identified can be related to a synthetic material, it was concluded that the artist had brushed the cut-out newsprint and tabloid images with a pure cellulose-based paste.

The position and form of the internal armature was detected by examining the sculpture by means of X-ray radiography.** As can be seen from Figure 17.1c, the radiography allowed the visualization of the internal armature. The artist employed a single, 1.5 mm thick wire which was twisted several times in order to provide more support to the sculpture. The armature runs across the two legs and forms a semicircle in the lower part of the sculpture's torso. Moreover, contrasting the image it was possible to identify the presence of a second material, which, as the artist explained, was connected to the newsprint used to cover the armature (Figure 17.1d).

17.4 Conservation Concept

Establishing a conservation/preservation plan for contemporary artworks is always a compromise between many different considerations, which can sometimes conflict with one another.[8,9] Discussions concerning ethics in contemporary art restoration need to take into account the artist's idea and the owner's will.[8] Regarding the case study presented in this paper, direct consultation with the artist highlighted her preference for a treatment that is the "[least] interventionist as possible", focusing on the "respect of the

**X-ray investigation was performed with the General Electric® Eresco 42 MF4 X-ray generator with a tube voltage of 90 kV, anodic current set at 10 mA and 3 mm focal spot. Five digital images were recorded as 12-bit TIPP and transformed using an image software program.

original materials" using conservation products which "are similar to those employed in the creation of the original sculpture".

Moreover, Lucas emphasized that the treatment "should be invisible – not affecting the stability or outward appearance of the work". Our primary concern was to a large extent the stabilization and protection of the piece and, only secondly, re-establishing its aesthetic coherence. Based on the restrictions posed by the artist, paper conservation literature and relevant case studies were extensively reviewed.[10–21] The survey of globe and papier-mâché literature provided treatment options for the restoration of dents. In his review of globe restoration, Bayes-Cope[15] proposes to split the cover paper and fill the dent with a plaster compound of papier-mâché, chalk, zinc and parchment. Alternatively, another approach is cutting out neighboring areas and pushing out the dent using scalpels and blades.[16,17] Both methods were considered to be much too invasive for the restoration of *Love Me*. After a careful weighting of the possible options, it was decided to develop an innovative external traction system to ease out the dent without operating on the original material.

Moreover, it was realized that for restoring the dent it would also be necessary to reinforce the area to prevent the reformation of the damage. For the strengthening of aged paper, different materials are available. Natural polymers, such as wheat starch and cellulose ethers or synthetic polymers based on polyvinyl acetate (PVAc), are especially useful.[10,11,18] Alternatively, a widely used conservation method consists of laminating paper with a stronger material to strengthen the object.[13,14] The material most used is Japanese paper, due to its low basal weight and long fibers.[10,11,13,14,19] Despite the extended range of possible materials and methods, reinforcement of deteriorated papers on a large scale has proved to be problematic.[16] Recently, the range of possible products has been increased with the introduction of nanomaterials based on micro- and nanocellulose particles[22–25] and bacterial cellulose.[26] Cellulose nanocrystals and microcrystals have been added as reinforcing filler to the water-soluble thermoplastic adhesive Aquazol® 500 with the aim of improving its properties as lining adhesive for oil painting restoration.[22,23] Nanocellulose, in the form of aqueous and water–ethanol suspension, and bacterial cellulose have recently been employed for consolidation of paper[24,26] as well as commercially prepared cotton canvas.[25] Cellulose nanomaterials show specific properties that made them particularly suitable for the restoration of cellulose-based artworks: they have high strength and degree of crystallinity, as well as good optical and thermal properties.[22–26] They are non-toxic and biocompatible with cellulose, hence they should not alter the chemical composition of paper,[22–26] which is in line with the minimally interventionist principles of restoration[8,9] and the restrictions stated by the artist. Therefore, for all the above-mentioned reasons, it was decided to test novel materials based on cellulose nanocrystals and nanofibrils as newsprint reinforcing agents.

17.5 Experimentation

Considering what was discussed in the conservation concept, a research experiment was carried out. The aim of this study was to evaluate the best, least interventionist method for easing out the dent and to test the behavior of nanocomposite materials as strengthening agents for papier-mâché. The three-dimensionality of the work, the position of the damage and the soft internal core of the sculpture that could be easily deformed even by light pressure complicated its treatment. Taking these issues into consideration, a fundamental requirement was that the fixator would be self-supporting and not rest on the sculpture. Regarding the use of nanocellulose, the central issue was discovering if these new materials could offer an alternative to traditional paper adhesive and consolidant.

17.5.1 Reinforcing Agents

Two different types of nanocellulose were tested: cellulose nanocrystals (CNC) and mechanically isolated cellulose nanofibrils (CNF). CNC in powder form was purchased from CelluForce® (Canada). It was produced from bleached craft pulp by sulfuric acid hydrolysis. CNF in the form of aqueous suspension was provided by the Finnish company UPM, in collaboration with the Finnish Centre for Nanocellulosic Technologies. The CNF was produced from birch pulp containing *ca.* 20% of hemicellulose and traces of lignin. CNF had a thickness of 4–10 nm and a length of several micrometers, whereas CNC had a similar diameter, 5–10 nm, but was smaller in length, *ca.* 100 nm. The reinforcing agents were prepared by dilution with a 1 : 1 w/w hydroalcoholic solution of water : ethanol. In total, five formulations were tested: (a) cellulose nanocrystal 0.5–1–3% (CNC0.5, CNC1 and CNC3); and (b) cellulose nanofibrils 0.2–0.4% (CNF0.2 and CNF0.4).

17.5.2 Mock-up Preparation

Reinforcing treatment tests were conducted on newspaper samples cut into strips, 10 × 80 mm, using common newsprint paper (45 g m^{-2}). A set of specimens was cut in a cross direction (CD) fiber direction perpendicular to the direction of the web, whereas the second set was cut in a machine direction (MD) fiber direction longitudinal to the direction of the web. Three-dimensional samples were employed to evaluate the potential of the external traction system.

Based on the results obtained from scientific analyses, three-dimensional mock-ups were prepared by stuffing nylon tights with kapok (natural cellulose fiber) and iron wire as internal armature. Subsequently, the surface was covered with black-and-white newsprint pieces and colorful tabloid images glued with the Metylan®, a methylcellulose adhesive supplied by Henkel®.

The specimens were varnished with Plextol® D498 (nBA/MMA copolymer), acrylic water based resin provided by Kremer Pigmente. Dents were created by bending the samples manually, simulating the sculpture's damage.

17.5.3 Application of Cellulose

Nanocellulose solutions were applied on newspaper samples with a 10 mm wide brush. The brush was dipped in the reinforcing agents and applied onto pre-specified surface areas using slight and controlled pressure. The solutions were brushed on the specimens' surface two times. Every brushing is defined as a forward–back movement. Successive nanocellulose applications were done six hours after the previous.

17.5.4 Testing Method

Tensile tests were performed using a Discovery Hybrid Rheometer HR-3, equipped with two grips for performing linear Dynamic Mechanical Analysis (DMA). The measurements were carried out at a constant extension rate of 94 $\mu m\ s^{-1}$ with the initial distance between the grips of 40 mm. The ultimate tensile strength (UTS) before a break was measured by dividing the average breaking force of the specimen width and then expressed in kilonewtons per meter ($kN\ m^{-1}$).

The specimens were conditioned at 21 ± 1° and 50% relative humidity in a closed environment, with a solution of glycerol and demineralized water for 24 hours. Six measurements were performed for each series and the average values were calculated.

17.6 Results and Discussion

17.6.1 Realization and Tuning of the Traction System

As mentioned above in Section 17.4, it was decided to not follow the common restoration procedure for re-shaping dents but to construct an innovative external traction system inspired by surgical operations. The objective of the external traction system was to provide a homogeneous lengthening of the area in order to mechanically tackle the damage without cutting the original material.

After a review of medical limb-lengthening surgery literature, an external circular fixator similar in shape to the surgical instruments used in leg-lengthening surgery was constructed. The circular fixator was composed of four iron rings, two movable and two stable, four 8 mm threaded connecting rods and bolts (Figure 17.3). By rotating the bolts, the upper and lower rings can be pulled apart very gradually. Moreover, an external arm was constructed. The arm allows the cylindrical fixator to be oriented in the x, y and z dimensions, permits the structure to be self-supporting and improves the fixator stability.

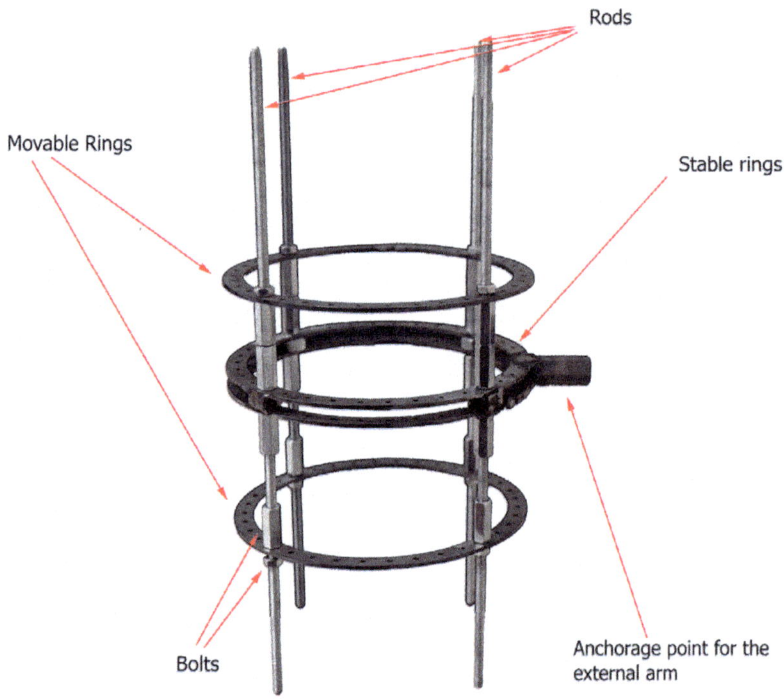

Figure 17.3 Schematic representation of the external traction system. The arrows show the different elements of the system.

Different methods for connecting the artwork leg to the circular fixator rings were studied. These included: (a) fish hooks and nylon thread, (b) safety pins and nylon thread, (c) 10 mm × 100 mm *Kozo* Japanese paper tie rods (machine-made Japanese mulberry paper, 10 g m^{-2}) and bulldog clips, (d) 10 mm × 130 mm 70% polypropylene–30% cellulose pulp (PLP5) tie rods and bulldog clips. Additionally, two methods for securing the tie rods to the artwork surface were evaluated: protecting the surface with a cotton medical dressing (a, b) or adhering a 30 mm × 180 mm Japanese paper sheet with wheat starch directly onto the area (c, d) (Figure 17.4). The areas above and below the sample dents were protected by these two methods and the tie rods were firmly secured to the movable rings. Subsequently, the two rings were pulled apart very gradually until the damage was recovered.

17.6.2 Evaluation of the Traction System

By each of the lengthening test methods it was possible to ease out the dent and re-establish the correct leg inclination. However, systems (a) and (b) were not wholly satisfactory due to the dissimilar forces applied on each tie rod during the lengthening of the movable rings. Moreover, since the medical dressings were not firmly secured onto the specimen surface they slightly moved during the tests, jeopardizing the operation. Systems (c) and

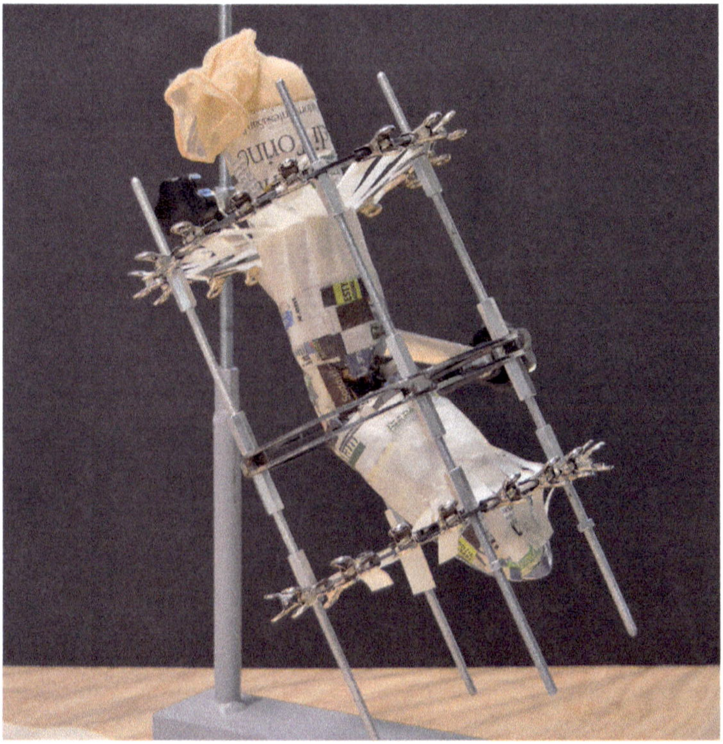

Figure 17.4 System (d) before lengthening. System (d) is characterized by the use of two Japanese paper sheets as a protecting layer and 10 mm wide by 130 mm long, 70% polypropylene–30% cellulose pulp (PLP5) tie rods.

(d) provided the best results. The increased number of points of fixation between the artwork leg and the movable rings allowed a homogeneous distribution of the traction force and assured a safer operation. Furthermore, the bulldog clips considerably improved the anchorage to the circular rings and allowed all the tie rods to be put under the same initial tension. Nevertheless, some of the Japanese paper tie rods broke during the lengthening, while the PLP5 tie rods perfectly supported the stress, permitting a complete re-shaping of the damaged area.

17.6.3 Mechanical Properties of the Reinforced Newspaper Samples

Figure 17.5 shows tensile strength for newsprint specimens reinforced with different nanocellulose-based formulations measured in cross direction (CD).

Mechanical properties of pristine samples, as well as those of samples wetted with demineralised water, are also given as reference.

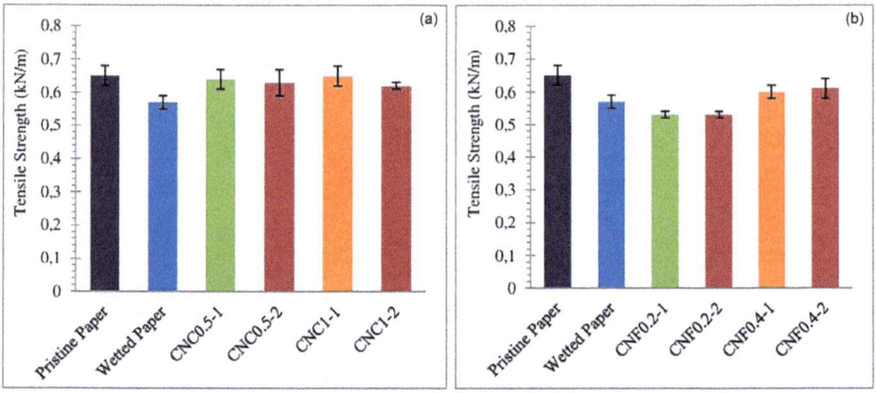

Figure 17.5 Tensile strength results from CNC (a) and CNF (b) tests measured in cross direction.

It can be seen from the figure that none of the reinforcing agents were able to significantly improve newspaper strength. Moreover, it was noticed that multiple nanocellulose applications decreased the overall newsprint strength as compared to untreated samples. These observations indicate that the treatment or the high moisture content of the nanocellulose suspensions have an overall effect on the tensile strength of the samples. It was supposed that the poor reinforcement provided by the nanoparticles was due to excessive water uptake by the cellulose fibers, leading to paper swelling and the loss of mechanical strength.[26] Cellulose-based materials are especially sensitive to water absorption which not only causes a reduction in size, but additionally can influence the formation of intra- or intermolecular bonds by moistening and drying.[10] The measurements also reported a higher decrease in tensile strength for CNF rather than CNC. As mentioned above (see Section 17.5.1), cellulose nanofibrils consist of longer particles than CNC. The dimensional differences among the nanocellulose particles may influence the film-forming properties on paper substrates, hence it was assumed that CNF may not be able to penetrate much into the material. Similar observations have been reported when coating cotton canvases with aqueous dispersion of CNF and CNC.[25]

In order to confirm this assumption it was decided to repeat the test again with newspaper samples cut in machine direction (MD), since longitudinally oriented fibers are less sensitive to water plasticizing action. Moreover, it was decided to treat the specimens with only a single application of reinforcing agents and to also test CNC at 3% (CNC3).

Comparative evaluation of the values obtained from the reference samples, pristine and wetted, and from the reinforced newsprint specimens, showed that samples treated with CNC3 slightly increased their tensile strength compared to reference specimens (data not shown). The mechanical behavior of CNC3 samples seemed to assess the capability of cellulose nanoparticles to act as effective reinforcing agents in newspaper specimens. Even if the measurements were not conclusive, the results show that the tensile strength is directly affected by the amount of nanocellulose material applied on the surface and that CNC, which possesses a lower aspect ratio and can penetrate more in depth into the porosity of paper, can slightly improve the final mechanical properties. Therefore, paper consolidation treatments with nanocellulose particles will be a compromise between the highest concentration of nanoparticles and lowest water content of the suspension. Furthermore, it is worth noting that the substrate did not distinctly change in the visual appearance, which is imperative in art conservation. However, since the mechanical behavior of CNC and CNF were not completely known, it was not considered safe to use the nanocellulose particles for this restoration treatment. The research was consequently focused on traditional paper lamination methods.[10,11,13,14] Three different *Kozo* Japanese papers, namely, RK 00 (3.6 g m^{-2}), RK 0 (5 g m^{-2}) and RK 2 (8 g m^{-2}), were evaluated. The Japanese papers were cut into small strips, glued onto the specimen surface with 5% w/w Culminal® MC200 in water and let dry at ambient temperature (21°, 50% RH). Under direct visual observation, the Japanese paper RK 00 appeared completely invisible to the naked eye, while both RK 0 and RK 2 were slightly visible. Tests on a three-dimensional mock-up confirmed the structural consolidation provided by the Japanese paper RK 00, which was able to give enough support to contrast the reformation of the damage and, at the same time, be invisible after drying. The results obtained from these tests supported the use of *Kozo* Japanese paper RK 00 for the restoration of *Love Me*.

17.7 Treatment

As the mechanical results obtained from the nanocellulose experimentation were not so promising as to lead to the use of those products in the artwork conservation treatment, the restoration was done following a more traditional technique.

Undamaged areas located above and below the dent were protected with a thin sheet of Japanese paper (RK 15, 10 g m^{-2}). The Japanese papers were cut into two strips 30 mm wide by 180 mm long and adhered onto the surface with wheat starch paste. The adhesive was prepared by diluting wheat starch powder (Zin Shofu®) in a proportion of 1:4 w/w (powder to water), stirred and cooked for approximately 1.5 hours, stirring until the proper consistency was achieved. After cooling for 24 hours under cold water, the paste was passed three times through a sieve of horsehair and then diluted in water. The Japanese paper protecting strips were pasted with the adhesive with a horsehair brush and then placed on the artwork (Figure 17.6a).

A *Noribake* brush was employed for gently hitting the strips to remove bubbles and possible wrinkles. Each PLP5 tie rod (10 mm wide by 130 mm long) was pasted for 30 mm with wheat starch and applied on the protecting papers for a total amount of 40 tie rods, half on the upper sheet and likewise

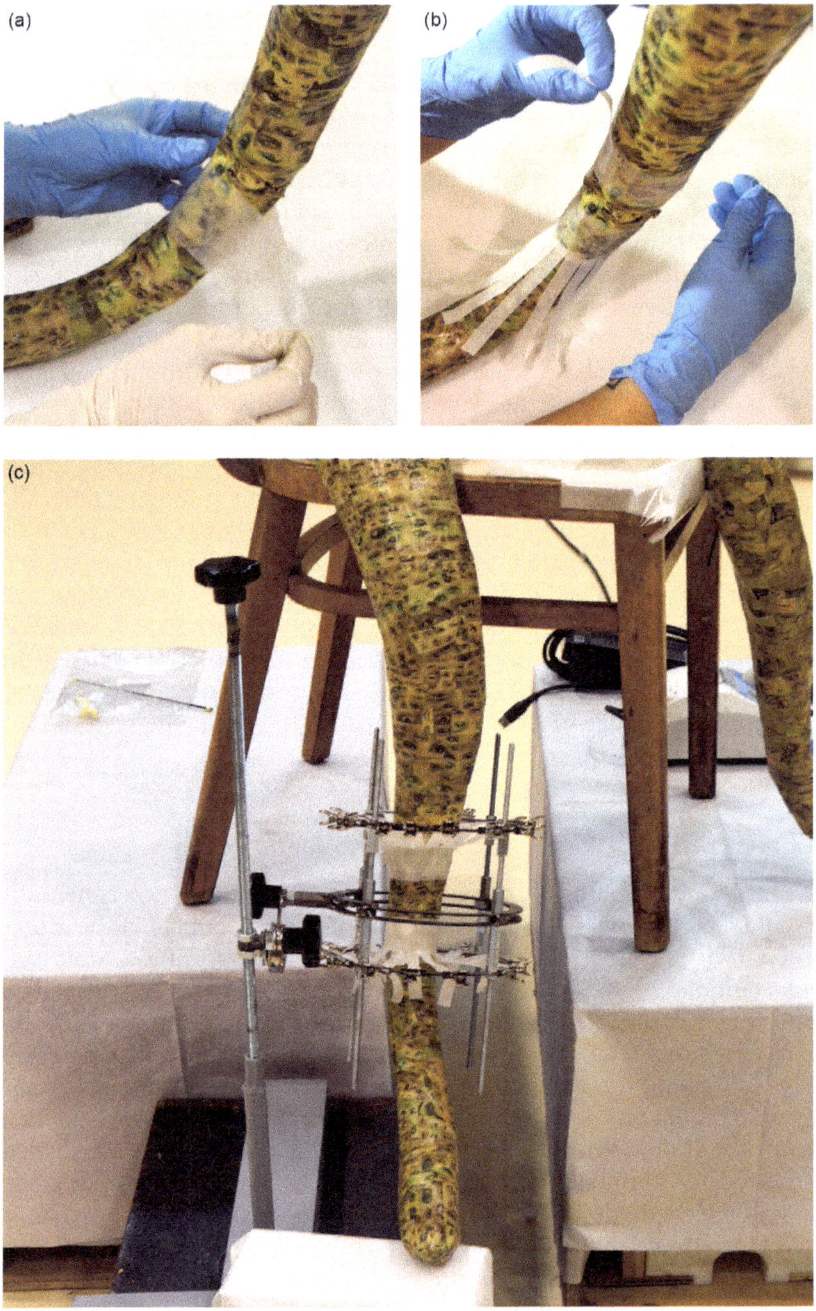

Figure 17.6 Set-up of the traction system: (a) Placing of the two Japanese paper protecting strips on the artwork; (b) Positioning of the forty PLP5 tie rods, half on the upper strips and likewise on the lower; (c) Positioning of the artwork inside the external traction system and anchoring of the tie rods to the movable rings with bulldog clips.

on the lower (Figure 17.6b). Afterwards, the tie rod rings were further reinforced with a second layer of Japanese paper to secure the adhesion.

The artwork's leg was positioned inside the external traction system and the tie rods were anchored to the two movable rings with the bulldog clips (Figure 17.6c). The rings were slowly pulled apart until the dent was eased out. Localized humidity, often suggested to soften the papier-mâché layers,[16,19,20] was not adopted and the treatment was carried out in ambient conditions (21 °C, 50% RH). Finally, a thin sheet of Japanese paper (RK 00, 3.6 g m^{-2}) laminated with 5% w/w Culminal® MC200 in water was positioned where the dent was located in order to prevent the reformation of the damage. The sculpture remained in traction for a week. Subsequently, the force was diminished daily until it was possible to unlock the bulldog clips and remove the traction system. After treatment the artwork's leg regained its characteristic curvy shape and the dent was almost completely recovered and firmly secured.

17.8 Conclusion

Working on a multi-material sculpture was an interdisciplinary experience that required the expertise of contemporary art and paper conservators, of conservation scientists and art historians. Scientific analysis combined with the direct information provided by the artist allowed the precise documentation of the materials used for the creation of the artwork and led to an overall understanding of the artist's technique. Moreover, the dialogue with the artist turned out to be a crucial occasion for revealing the connection between *Love Me* and the two later works, *Hysterical Attack (Mouths)* and *Hysterical Attack (Eye)*, as well as its relationship with the *Bunnies* series, and proved to be fundamental for the comprehension of Lucas' ideas about conservation. The restriction posed by the artist stimulated vibrant discussions which led to the experimentation of novel reinforcing agents based on nanocellulose particles and the development of an innovative system to ease out dents in sculpture. The experimental mock-ups proved that it would not be safe to use nanocellulose for the restoration of *Love Me*. Mechanical analyses of the increase in strength of the newspaper were difficult to assess, due to the incomparable results obtained in machine direction (MD) and cross direction (CD); this was likely caused by the solvent swelling effect. However, the results of the preliminary trials laid the foundation for the use of these products as reinforcing agents for aged paper. Treating paper with cellulose nanomaterials would be in line with the principle of minimal intervention, since nanocellulose is similar in nature to cellulosic paper. Finally, through this research it was possible to demonstrate that one of the crucial issues in three-dimensional artwork conservation, such as dent restoration, can be tackled without adopting invasive and dangerous methods like cutting the neighboring areas or splitting out the paper. Thanks to the construction of an external fixator, the dent located on the artwork's right leg was completely eased out, re-establishing the correct leg inclination. Furthermore, the strengthening provided by the Japanese paper will protect and stabilize the area from future damage.

This contribution highlights the complex problems that contemporary artwork restorers face when working on objects made with unusual materials and techniques, which can, however, lead to innovative ideas and solutions.

Acknowledgements

This contribution is part of the Master's thesis of the first author, developed at the Conservation and Restoration Center "La Venaria Reale" in fulfillment of the Degree in Conservation–Restoration of Cultural Heritage of the University of Turin (Turin, Italy). Fabiola Rocco wishes to thank the people and institutions that helped her during her Master's degree research: Sandra Vazquez Perez, head of the contemporary art lab at "La Venaria Reale" for the quality of her supervision; and Francesca Comisso and Sara Abram, contemporary art historians, for the support during the artist's interview; the private collector who entrusted her with this fascinating and amazing artwork; Professor Piero Baglioni, head of the Center for Colloid and Surface Science (CSGI, Florence) for having hosted her in the CSGI labs during the nanocellulose experimentation; and Federico Demeo for having prepared the nanocellulose solutions and for having helped her during the mechanical tests.

References

1. J. Heuman, *From Marble to Chocolate: The Conservation of Modern Sculpture*, Archetype Publications, London, 1995, p. 172.
2. O. Chiantore and A. Rava, *Conserving Contemporary Art: Issues, Methods, Materials, and Research*, Getty Pubblications, Los Angeles, 2013, p. 332.
3. J. Stallabrass, *High Art Lite: The Rise and Fall of Young British Art*, Verso, London, 2006, p. 356.
4. M. Collings, *SL: Sarah Lucas*, Tate Publishing, London, 2002, p. 125.
5. Y. Dziewior, C. Grunenberg and B. Ruf, *Sarah Lucas : Exhibitions and Catalogue Raisonné: 1989–2005*: Kunstahalle Zürich, Kunstverein in Hamburg, Tate Liverpool, Tate Publishing, London, 2005, p. 195.
6. L. Le Feuvre, Ordinary Things, in *Sarah Lucas: Ordinary Things*, ed. L. Le Feuvre, Henry Moore Institute, Leeds, 2012, pp. 9–15.
7. G. Muir, It Must Be A Camel (For Now), in *In-A-Gadda-Da-Vida*, ed. G. Muir, Tate Publishing, London, 2004, pp. 78–94.
8. G. Ryan, Conservation of Outdoor Painted Sculpture at the Hirshhorn: Taking Lessons from the Preservation of Variable Media Art, in *Conserving Outdoor Painted Sculpture*, ed. L. Beerkens and T. Learner, The Getty Conservation Institute, Los Angeles, 2014, pp. 5–12.
9. I. Hummelen, D. Sille and M. Zijlmans, The decision-making model for the conservation and restoration of modern and contemporary art, in

Modern Art: Who Cares? An Interdisciplinary Research Project and an International Symposium on the Conservation of Modern and Contemporary Art, ed. I. Hummelen, D. Sille and M. Zijlmans, The Foundation for Conservation of Modern Art, Amsterdam, 1999, pp. 164–177.
10. S. Zervos and I. Alexopoulou, *Cellulose*, 2015, **22**(5), 2859.
11. I. Alexopoulou and S. Zervos, *J. Cult. Herit.*, 2016, **21**, 922.
12. R. Teygeler, Preserving Paper: Recent Advances, in *Managing Preservation for Libraries and Archives. Current Practice and Future Developments*, ed. J. Feather, Ashgate, Aldershot, 2004, p. 194.
13. D. G. Suryawanshi and O. P. Agrawal, *Restaurator*, 1996, **17**, 229.
14. H. Bansa and R. Ishii, *Restaurator*, 1997, **18**, 51.
15. A. D. Bayes-Cope, *The Study and Conservation of Globes*, Internationale Coronelli-Gesellschaft, Wien, 1985, p. 80.
16. D. van der Reyden and D. Williams, *J. Pap. Conserv.*, 1988, **2**, 21.
17. K. E. Leyshon, *Pap. Conserv.*, 1988, **12**, 12.
18. C. Marchal, *Support/Tracé*, 2013, **13**, 72.
19. G. Fairclough and R. Harrison, *Icon News*, 2007, **10**, 22.
20. L. Dessennes, *J. Pap. Conserv.*, 2011, **12**(3), 30.
21. E. Nijhoff Asser, B. Reissland, B. J. W. Grob and E. Goetz, Lost fingers, scurfy skin and corroding veins – conservation of anatomical papier-mâché models, in *ICOM Committee for Conservation, 15th Triennal Conference, New Dehli, 22–26 September 2008*, ed. J. Bridgland, Allied Publisher, New Dehli, 2008, vol. I, pp. 285–292.
22. A. Cataldi, A. Dorigato, F. Deflorian, L. Berglund, A. Pegoretti, *Polymer Composite with Micro- and Nanocellulose for Artwork Protection and Restoration*, 2014, http://www.escm.eu.org/eccm16/assets/0550.pdf. Accessed 21 August 2018.
23. A. Cataldi, L. Berglund, F. Deflorian and A. Pegoretti, *Nanocomposite*, 2015, **1**(4), 195.
24. S. M. Santos, J. M. Carbajo, N. Gómez, E. Quintana, M. Ladero, A. Sánchez, G. Chinga-Carrasco and J. C. Villar, *J. Mater. Sci.*, 2016, **51**, 1541.
25. O. Nechyporchuk, K. Kolman, A. Bridarolli, M. Odlyha, L. Bozec, M. Oriola, G. Campo-Francés, M. Persson, K. Holmberg and R. Bordes, *Carbohydr. Polym.*, 2017, **194**, 161.
26. L. Völkel, K. Ahn, U. Hähner, W. Gindl-Altmutter and A. Potthast, *Herit. Sci.*, 2017, **5**, 23.

CHAPTER 18

Diagrams, Technology and New Materials: The Scientific Nature of Loris Cecchini's Artworks

MATTIA PATTI*[a], LUCA BINDI[b], LORIS CECCHINI[c] AND BIANCALUCIA MAGLIONE[a]

[a]Università degli Studi di Pisa, Italy; [b]Università degli Studi di Firenze, Italy; [c]Loris Cecchini Studio, Milan, Italy
*E-mail: mattia.patti@unipi.it

18.1 Introduction

Born in Milan in 1969, Loris Cecchini was aware right from the outset that the only path that would bring him lasting satisfaction lay in the world of art. After obtaining his diploma from the *Istituto d'Arte di Siena* in 1989, he frequented the academies in Florence and then Milan, and these years left an indelible mark on the young Cecchini. His studies in the areas of draughtsmanship and painting, while removed from the areas that he would subsequently excel in, are only superficially so. "When I think of a work, I begin with some sketches in my sketchpad or on sheets of paper",[†] Cecchini has stated, as if to point to a process of mental elaboration as the generative spur behind each of his works of art, extemporaneously accompanying the act of drawing as a synthesis and materialization of this. Cecchini's first artistic projects, exhibited at the earliest shows the artist participated in, are made

[†]Statement taken from <https://www.youtube.com/watch?v=bqsfreIYDeQ> (last accessed March 2019).

Science and Art: The Contemporary Painted Surface
Edited by Antonio Sgamellotti, Brunetto Giovanni Brunetti and Costanza Miliani

Published by the Royal Society of Chemistry, www.rsc.org

of plastic and synthetic resin or other materials, "Anxiety-inducing objects",[1] as Cecchini himself has defined them, containing probable echoes of Rosenberg,[‡] which were soon cast aside in favour of different expressive forms. Subsequently, indeed, and contemporaneous to his fascination with the world of cinema and photography (the latter reinforced by a stint working in a studio in Monza), the artist began to make use of digital modelling as a means of creating photographic works of art, such as the series *Pause in Background* and *No Casting*, digital collages where, as the titles show, Cecchini's interest in the theatre is a catalyst.[§] The title of a renowned subsequent series, *Stage Evidences*, also alludes to the preparation of a play. In this case, rubber casts on a scale 1:1 of objects relating to the 'day-to-day affections', remodelled[¶] in such a way as to 'collapse' on themselves, evoke the performance of a play that was never actually staged, but for which a stage was carefully set up and the various elements put 'in the spotlight'. As far as the influences underlying these works are concerned, the artist himself has often made reference to some philosophical elements from the beginning of the Nineties, which may be traced back in particular to Baudrillard and Virilio, dealing with reflections on loss and the consequent redefinition of the concept of the 'real'.

The artist's talent for investigating space in an almost architectural sense, already noticeable in some of his rubber sculptures, came to the fore at the beginning of the new millennium, when Cecchini dedicated himself to the creation of what may be termed instances of 'micro-architecture', which can be accessed yet also enjoyed individually.[‖] As well as representing a space themselves, many of these works of art establish a definite dialogue with their locations, in accordance with site-specific dynamics that will constantly underpin the artist's output from this point on. It is precisely this mode of reflection, in addition to scientific and natural studies, that all come together to act on the artist in finalising and defining his early works as modular in character. This indeed is one of the categories that, despite variations in form and material, Cecchini has continued to adopt with systematic frequency right up to the present time. In these works all Cecchini's talent for construction, along with the dialectic between the natural and the artificial (which are pillars of Cecchini's vision**) can clearly be seen: the artist uses nature and its mechanisms as a useful starting point for an individual module, which will then take shape through complex technical and industrial procedures, while each new work also takes into account the changing nature of the

---

‡The reference is to *L'oggetto ansioso*, in the Italian translation of the book by H. Rosenberg.[2]

§This interest has been pointed out several times by the artist himself, who traces it back to biographical experiences.[3]

¶Over the course of his career, what has been crucial for the artist has been precisely this investigation into the concept of the model and then, as shall be seen, the module.[4]

‖As may be clearly understood from the title of the series that includes these works: *Monologue Patterns*.

**In an interview from 2004 the artist stated that if he had to summarize the pillars of his poetic vision in only three points, they would undoubtedly be: his desire for transfiguration, his inclination for construction, which for him carries a symbolic tension, and finally the diverse approaches he takes to observing nature.[5]

construction materials in question. In the final phase of the work the modules are assembled by the artist almost along the lines of an installation, with an approach that here again focuses more on what is natural.

Loris Cecchini has had numerous exhibitions in prestigious venues, among which are: the MoMA, New York; the Palais de Tokyo, Paris; and the Shanghai Duolun MoMA. He has also participated in important shows such as the Venice Biennale, the Rome Quadriennale and the Shanghai Biennale. His permanent works are kept in important public and private collections: Collezione Gori – Fattoria di Celle (Santomato, PT), MAXXI – Museo nazionale delle arti del XXI secolo (Rome), Arts and Medicine Institute, Cleveland Clinic Foundation, and many others.

18.2 Dialogue

Mattia Patti (MP): *In considering the different phases of your work and the different areas where you have carried out your research, the scientific dimension seems to play an important role on several levels.*

On the one hand, scientific and technological research acts at the design and production level of the work, in the choice of materials to use and the technology to adopt in order to assemble the various parts of the body of the sculpture; on the other hand, science seems in some way to inform the very foundation of your work: the relationship with naturalness that characterizes many of your sculptures and the way they interact with the space containing them seems to evoke systems and structures that belong to the world of science – diagrams and frameworks with their origins in chemistry or biology.

Do you acknowledge a scientific character to your work?

Loris Cecchini (LC): In recent years scientific imagery has come to represent a sort of conceptual material that I find useful for my speculations on concepts, providing a wider cultural area than that linked purely to the language of art, and more in general that of aesthetics. I do not believe that my work has in and of itself a scientific character; rather, it seems to me that in a decidedly empirical way science enables me to explore a wider territory, which draws on analogy and similes, natural complexity and artificial systems, subject and language. What I definitely seek is a poetic dimension that continues to permeate the artwork, and in this sense (in some cases) the scientific character of phenomenological reality is very interesting if it is shifted onto the level of the imagination and paradox in the construction of a simultaneous subject; that is, a work wherein an implicit order may be perceived that opens up to multiple interpretations. I have always seen technology as an integral part of the cultural landscape: having the opportunity to utilise it in various forms is a further opportunity for knowledge.

MP: *Have you ever had the opportunity to look into scientific theories that you feel are close to your work or that you have used, if you will, to elaborate your work? Do you ever happen to read material of a scientific nature?*

LC: As far as the scientific material is concerned, I have several interests there, which range from looking at books on botany from different eras, to

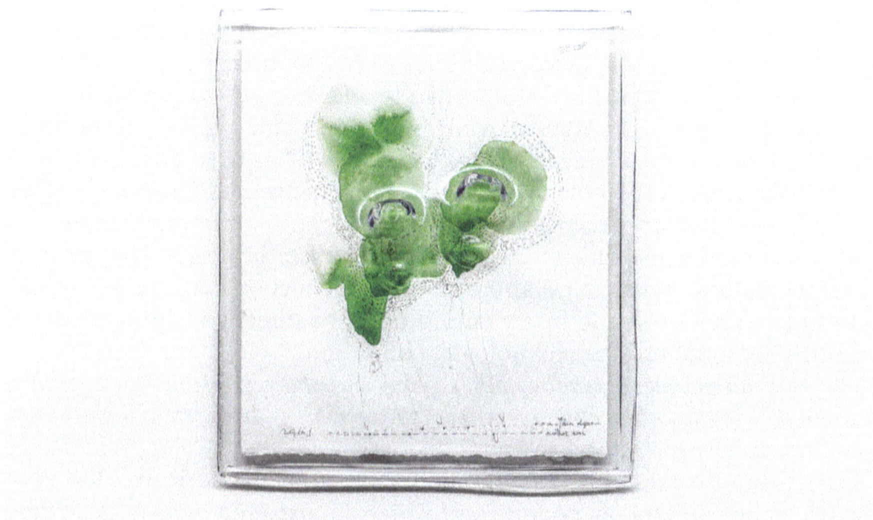

Figure 18.1 Loris Cecchini, *I Must Have Seen Things nr.7*, 2018. Watercolour on Canson paper 300 gr in thermoformed PETG, 38.5 × 34 cm. Photo Ela Bialkowska. Courtesy of Galleria Continua, San Gimignano – Beijing – Les Moulins – Habana - Roma.

herbals – especially at the moment, since I'm working with watercolour (Figure 18.1) – to material on physics or the chemistry of the elements. Digitalized online libraries are important sources of information and all this material, whether of a scientific bent or the humanities, is all over the Internet, affecting knowledge *tout court*. I'm intrigued by abstracts dealing with chemistry, biology or neuroscience on account of their specificity, but also for a sort of linguistic abstraction characteristic of autonomous branches of learning. I have a passion for scientific notebooks, what were once – during the pre-printing press days – extremely important work instruments in the dissemination and circulation [of knowledge] carried out through manual copies.

Alongside these types of publications, I read other material regarding phenomenological research interpreted and translated by means of specific software: for instance, my wave-based works in relief[††] applied to the 'skin' of the

[††]LC is referring to a series of artworks entitled *Wallwave Vibration*, created from about 2009 onwards. Starting from the observation of vibration movement and the scientific diagrams describing it, the artist makes some polyester resin discs in his studio that reproduce the crests of waves. Subsequently, counter-panelling is made, giving continuous coverage of the wall which will house the artwork, or, less frequently, an independent wall is specially made. At this point the artworks are hung on the walls: these elements are fixed by means of tassels placed along the edge, so that the extraversion of the whole surface of the wall may be perceived, which has been turned into a sort of visual continuum. After this, all the elements and the wall are painted the same colour white, thus achieving an even more homogeneous surface.

architecture originate in the study and use of software that enables an analysis of the phenomenon of vibration, whether it be liquid, sound or something else, and an evaluation of its sources, frequencies and intensity (Figure 18.2). These observations may represent an approach that spills over into other disciplines such as geology, or engineering or botany...all these disciplines lead to drawings! I would also like to add, in what I still hold to be a scientific dimension, the study of software dealing with translations of parametric material: for example, for many years now it has been possible to use a plugin from the well-known digital modelling software, *Rhinoceros*, called *Grasshopper*. This tool enables you to write – organize – various parametric functions, in the sense that within the software itself the design is organized by mathematical, behavioural and constructive functions. All these aspects have links to my work and in my opinion may be directly connected to the scientific area, not just the technological one.

MP: *Has this manner of reading things, along with this type of approach – which is also linked to the design and execution of the artworks – been strengthened over time? Can they be traced back to your formative experience?*

LC: I think that this interest simply derives from the curiosity that permeates us as subjects, as people, and which in artists manifests itself frequently yet to a varying extent in a continual overlapping of nature, poetry and language. In my own case this curiosity is nurtured by different kinds of things I observe and experience, in a continual process of regeneration and discovery. Observation is part of an analytical process, an investigation. This is what drives me to examine certain factors, which I then try to make use of

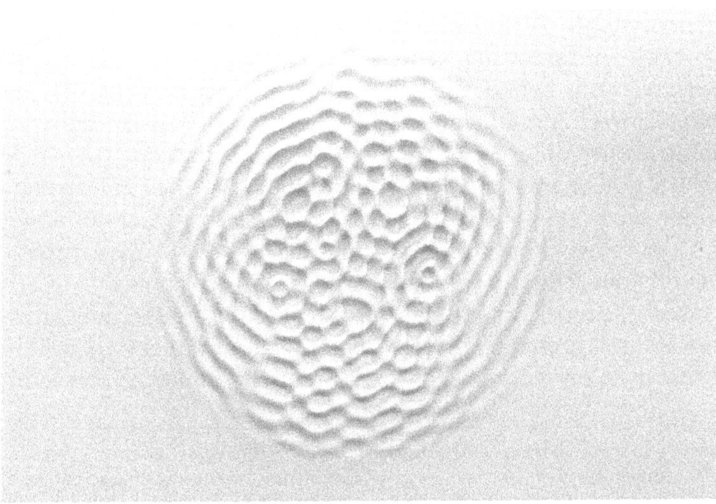

Figure 18.2 Loris Cecchini, *Wallwave Vibration (ear particle tingles)*, 2012. Polyester resin, wall paint, diameter 100 cm. Photo Oak Taylor-Smith. Courtesy of Galleria Continua, San Gimignano – Beijing – Les Moulins – Habana - Roma.

in my work. The final outcome of this process may of course be very different from traditional scientific observation, but only in the outcome, though. Indeed, in some processes, study, observation, stratification and the identification of key information is very important. At the moment I am developing sculpture and environmental installations by following an idea of the parcelling up of matter, which is almost a form of the molecular explosion of sculpture, whereby scientific phenomenology becomes an intimate structure and go-between for one's vision.

Biancalucia Maglione (BM): *Apart from your reading activities, during your formative period were there encounters – with artists from the previous generation, or with people from outside the world of art – or exhibitions you went to, or works of art and images that reinforced your interest in the scientific?*

LC: Definitely. I visited places and exhibitions that were very important for me, which struck a deep chord with me, influencing my work. I remember, for example, an amazing exhibition in the museum La Specola in Florence about ten years ago, which focused on a huge collection of crystals – it was an exceptional exhibition, highly evocative, and conveying an incredible sense of the wonder of nature.[‡‡] Being able to see these elements concretely, as opposed to photographic reproductions, is very important: the physicality of the objects overwhelms you, as you're not dealing with an image or a reproduction, so that the physical dimension of the objects, their weight, colour and transparency, all transmit an important memory, the type of memory that makes a deep impression.

I would often visit the Museum of Natural History in Berlin when I lived there, as well as the beautiful Botanic Gardens, and I still have about a dozen large seeds from some of the plants in the greenhouses there (Figure 18.3). I also remember the Museum of Anthropology in Rotterdam,[§§] where the then director showed me a vast array of objects, some of which were not on display but kept in storage. Here, too, the material presence of the objects gathered from the various cultures of the Dutch colonial empire – a semantic stratification between artisan manufacture and the intrinsic strength of the natural element in which the internal forces continuously overlap – introduced me to a type of scientific observation produced in the past and thus, in a certain sense, brought me back to the level of history: the powerful physicality of these objects had to be perceived in a very different manner compared to the more abstract, contemporary dimension, which is characterized more by digital technologies, virtualization and the superficiality of the Internet.

For that again, I also am greatly intrigued by what has been going on in the world of architectural research over the past twenty years concerning the idea of techno-organic culture. Speaking of science in the widest possible sense, I feel I have to mention architecture as well because when you come

[‡‡]The exhibition was *Cristalli – La più bella mostra di minerali al mondo, (Crystals: the most famous mineral show in the world),* Museo di Storia Naturale La Specola, Firenze, 1st April–1st October 2009 (extended until 30th June 2012). The exhibition showcased material from Adalberto Giazotto's collection.

[§§]The Wereldmuseum in Rotterdam.

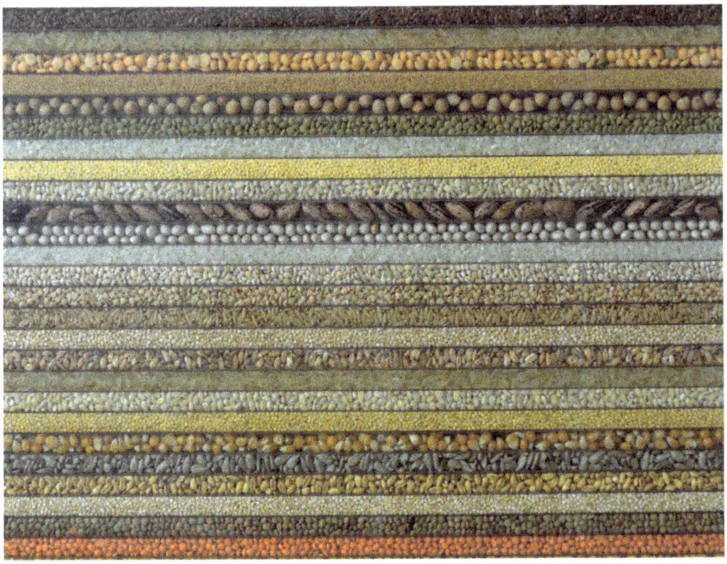

Figure 18.3 Loris Cecchini, *The Polychromesandsessions*, 2015. Natural seeds encapsulated in UV-protected polycarbonate, wood and metal frame, 201 × 151 × 5 cm. Photo Loris Cecchini Studio. Courtesy of Galleria Continua, San Gimignano – Beijing – Les Moulins – Habana - Roma.

into contact with spaces and projects that are strongly organic, such as those by Greg Lynn, Zaha Hadid, Santiago Calatrava or Frank Gehry, you realize just how pervasive science and nature are also in this particular context, transforming cultural tokens into functional aesthetic codes that can be appreciated also by the public.

Luca Bindi (LB): *Have you ever seen a conference of scientists who talk about new materials, their features and applications and gained some inspiration for your work?*

LC: On the web there are many material presentations which also, thanks to the suggestions of friends, artists, engineers or others, we often find ourselves commenting on... Specifically, in a recent work made of ceramic I had the opportunity to collaborate with a great Tuscan company experimenting with the use of glass ceramic slabs and a nanotechnological film applied to traditional baked clay. Such a thin film, through an ionization process, is able to filter pollutants into the air: an active process, but not visible... I liked this idea very much in relation to the fact that it was a public sculpture put permanently in a square. I know that in recent years there have been various attempts at using this material in architecture.

BM: *Right from the outset of the modern era, the relationship between art and science has experienced some particularly lively and intense periods. In your opinion, which aspects of scientific research have had the greatest impact on contemporary artistic production, and to what extent?*

LC: Nowadays I believe that a certain type of phenomenological idea of science has permeated the work of most artists. The best known in this sense are Anish Kapoor and Olafur Eliasson, but I'm also thinking of James Turrell and Georges Seurat, all of whom are artists mediating a conciliation between the aesthetic environment and phenomenological observation, whose thought- and method-based approach, specifically philosophical but clearly connoted by a scientific basis, are evident. The same is true for Italian artists: Davide Boriani, for example, as well as all of the group linked to the *Cinetica e Programmata* art movement: in the Sixties they used to make series of artworks linked to light and movement phenomena, in eccentric compositions that had a lot in common with scientific practice and the observation of matter in its purest state. Using magnets and iron dust and making them act on a surface, thus giving rise to an aesthetic experience with a great deal in common with scientific investigation. In a similar way now, in the area of cymatics (the study of the morphology of sounds and their energy), some behavioural patterns concerning the structure of matter are being brought into focus. Cymatics produces visible energy, which is normally not so, and this leads to areas such as quantum physics or naturally occurring fractals. I recall an intense piece of work by Carsten Nicolai at the Biennale in Venice, which used two powerful subwoofers and volumes of water to produce beautiful crests of water in a rhythmic movement.[¶¶] As far as the pervasiveness of science in the field of art is concerned, I also recall the last Triennale in Milan: this current exhibition is based essentially on our relationship with nature, and showcases approaches and stances that are very diverse aesthetically, yet possess a shared conceptual responsibility. Having said that, it is almost inevitable that nowadays artists, designers or more in general any agent of creativity, should find themselves interacting with the scientific field.

BM: *In the interest you have towards the world of science in its broadest sense, do you think that, at least partially, a 'Tuscan matrix' is recognisable? I'm thinking of the great characters that, since the fifteenth century, have bridged the confines between the two areas – the artistic and the scientific.*

LC: I don't think so. It is true, however, that when I was a child my house was full of reproductions of Leonardo's drawings: the swirling water one in particular, and I also remember etchings and watercolours with sections and cross-sections of enormous hothouses made of glass and iron from the end of the nineteenth century. I suppose I believe that what I'm interacting with now is the result of the memory of certain images seen in the past, which sediment in us and leave a mark. But I don't think that from this point of view it is important to talk of a Tuscan matrix. I'm thinking, for example, of other artist friends, whose biographies I know better, such as Michele Guido, a native of Lecce, who often works by using a sort of gold-standard botany as a basis for research, recognizing it and stratifying it in the culture of the architectural designs of the past and re-elaborating this whole idea of botany,

[¶¶]The 49th edition of the Venice Biennale in 2001, during which C. Nicolai presented the artwork *Frozen Water*.

architecture, sense of order and wonder in a different key – which is new and nearer to the current context in terms of perception. He too used to tell me that he had been struck by certain aesthetic elements as a child, and this definitely affects the dynamics of one's subsequent formation. However, I don't think that belonging to a particular region is decisive in this; it is not a necessary factor in developing one's cultural sensibilities in a wider temporal framework.

BM: *From the middle of the Seventies on, the debate regarding the artistic use of technology took off internationally. This involved names such as Susan Sontag, Lucy Lippard and Jack Burnham. Against those who were critical of the aesthetic validity of the most recent technological means and rejected them, other approaches attempted instead to establish a new train of thought around the dialectic between art and technology,*[6] *giving rise to what can only be called 'technophilia'.*[|||] *What is your own relationship with new technology, and how can it be reconciled to the manual aspect of your creative process, which in some phases is still present? How has it changed over time?*

LC: I have a pretty empirical way of relating to the fascination I have for nature as a subject, and I circle round it continually. This comes from the observation of the morphology of certain elements such as leaves, seeds or growth and expansion structures. After an initial approach made through drawing, watercolour or other manual compositional techniques, I then proceed almost more like a designer, elaborating through 3D software a base element, a module that I generally use in my sculptures in order to make them into aggregates of different shapes and sizes. The 3D design enables me to rapidly visualize if there are any changing morphologies, thus helping me to register any opportunities for links or connections among the shapes themselves: thus, from a singularity it is possible to progress to the plurality that makes up the sculpture. In any case, the technological aspect is always present in my work process, since a phase from drawing to prototype is necessary. I could print the prototype in 3D by going to an outside server, but this would not simplify the subsequent phase, which is the actual production where the pieces are created using metal, and in quite a large quantity: when an element is designed to be reproduced in thousands of copies, it's clear that you need to avail of different techniques and technologies. In my work this occurs when I use metals such as steel and aluminium, which are not easy to work with in the studio and so necessitate a very demanding production system. Naturally, after the prototypes and several trials, the moment of assembling the first sculpture is always a surprise, and so I am always amazed to see the shape that an object may take on; at the same time I also reflect on the structural character and the 'durability' of the pieces, as well as on the forces that are created in the composition, such as any problems intrinsic to the physical 'performance' which, in the empiricism of the

[|||]Pamela Lee, mentioned in ref. 6.

designed project, I have not managed to resolve. I'm very aware that all these aspects could be dealt with and partially resolved with the help of an engineer.

Last week an engineer came to visit me in my studio who has a specialization in parametric calculation in three-dimensional modelling, as I briefly mentioned above. He brought a script based on modules he had seen installed as exhibits and which he had reproduced in 3D, even if they were not quite exact, and he showed me some procedural solutions where the sculpture is autogenerated in relation to various constraints, or else it can be induced to act upon a volume or a surface such as, for instance, the façade of a building. As many architects and engineers tell me, in a highly empirical yet substantial way I have touched on what is nowadays often studied by means of computers, and when they see the installations on a large scale and the reiterated module, they think that my starting point is the parametric calculation; instead, this is not so, as everything is assembled by hand – in a certain sense, I'm 'winging it'. It is my personal attempt to set up the sculpture as if it were a virus, an expanding organism, a connecting network capable of dancing gracefully in the space housing it, but which at the same time manages to maintain its independence from the context.

Engaging with the technological landscape, succeeding in impacting it through design, is for me an important way of acquiring knowledge of the world. It overlaps with a certain poetic narrative wherein technique *per se* is secondary to the non-informative image; this, nevertheless, strikes a chord with our sensibilities, which are formed instead through our direct experience of the work of art.

LB: *I noticed that you make a lot of use of technology for choosing your starting materials. How important is it for you to stay up to date on new materials and their characteristics?*

LC: Over the years the curiosity towards materials has developed a lot: I have to say that it also takes a long time to acquire a certain mastery in the design and use that can be done; every time the technical interlocutors that I meet for a project become a real wealth. Working with petroleum derivatives, as with various types of rubbers and resins, or marble, as well as glass or leather, sintered prototyping or certain types of steel, always represents special journeys to different worlds. The technical operators, the artisans, teach a lot with their experience; even if often the will of an artist is also to test the limit of craftsmanship, literally forcing the hand, and looking for the limit of the material itself. The updates become almost consequent when you enter into the specific problems and meet the so-called 'state of the art' of certain technical researches. I add that the material itself is almost never the end of the work.

LB: *The use of modules that present a sort of periodic repetition ('translational') is very evident in your latest works. One expects an artist like you to give free rein to his fantasies by not following canonical rules dictated by the laws of mathematics (such as those of translational symmetry), but instead it seems to happen. Can you explain it to us?*

LC: I really like this precise definition of the periodic behaviour of the sculptures. As I mentioned in another answer, I found great freedom in this kind of work: in the use of a basic form and its subsequent repetition/multiplication, I find on the one hand a sort of relationship of *filiation* with the historical minimalism of the Sixties – I am thinking of Sol LeWitt, Donald Judd, Robert Morris – on the other hand, the implicitly organic meaning of the sculptures, the compositional variable and the dynamic relationship in the space in which they are placed, allow me a great freedom of improvisation in the spatial distribution and morphology that I intend to give to the sculpture. The sculpture in its manifestation must have, beyond the mechanical function that determines it, a musical character, the elegance of a plant, the lightness of an idea.

BM: *Apart from functioning as an initial spur, in some specific cases the connection with the world of science becomes a constituent part of the work itself. I am thinking, in this sense, of a work which, in its form as a complex system, has so far represented something unique in your career: Waterbones (Green Sponge + L System)* (Figure 18.4).*** *In this case, the installation dialogues with a sound created by Alessio de Girolamo, starting from the reduction of the modular structure to a mathematical function. Can you explain the genesis of this project?*

LC: The idea for the work came about a couple of years ago. Alessio De Girolamo is a visual artist, but also a musician, and it was he who suggested we collaborate in this direction: observing the tripolar connection of the module in *Waterbones*, he noted an analogy with the so-called *Lindenmayer System*, known as 'The L-system', a theoretical model that arose in the Sixties in the USA in order to resolve the graphic function and representation of plant growth diagrammatically, along with the branching out of fractal structure. Alessio associated it with the features of my module, creating 'The L-system' as a composition of generative music brought about by reducing this particular modular installation to a mathematical function. Alessio's research into sound, 'Nn, NomenNescio' bears similarities with Bohr's atomic model and the pianoforte invented by Busoni (Bosendorfer Imperial 290), and it is the compositional structure that introduces the theme of the pianoforte. The chords garnered from chemical formulae, as well as the melody, are the result of an exploration of the frequency of notes compared to the elements on the periodic table.

We asked an IT engineer to formulate an algorithm that could register the mathematical indications produced by the size of the module, the tripolar connection and the fact that consequential assembly can develop in infinite vector variables. Basically, the software algorithm works on whatever samples the musician decides to make. Once all these elements have been

***This work was created for a personal exhibition.[7]

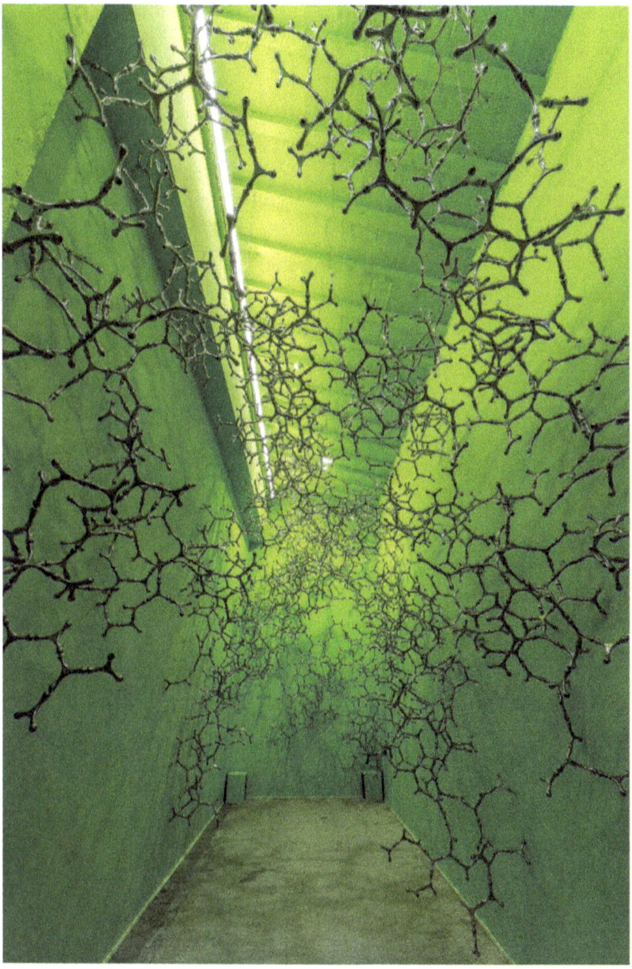

Figure 18.4 Loris Cecchini, *Waterbones (Green Sponge + L System)*, 2018.2000 stainless steel modules + sound produced by Alessio de Girolamo. Photo Ela Bialkowska, OKNO studio. Courtesy of Galleria Continua, San Gimignano – Beijing – Les Moulins – Habana - Roma.

inserted, the algorithm continuously reproduces itself, or rather autogenerates itself, parallel to what is formally and visually taking place in the sculptural aggregate.

Clearly, the purpose of the project was to bring together this twofold aspect in tandem, visual and auditory, in order to produce an immersive situation that could act on a double composition.

LB: *In your last works I noticed the presence of a repetition of metallic modules with ternary symmetry with spherical globule termination (remembering an atom). These modules are then more or less linked to form a three-dimensional framework (Figure 18.5). This type of work is very reminiscent*

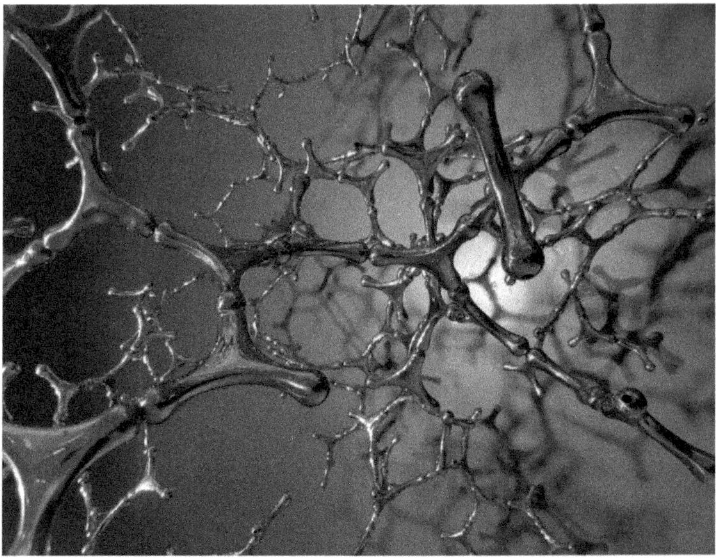

Figure 18.5 Loris Cecchini, *Waterbones* (detail), 2012. Welded stainless steel modules, variable dimensions. Installation view Gallery Diana Lowenstein, Miami, USA. Photo Loris Cecchini Studio. Courtesy of Galleria Continua, San Gimignano – Beijing – Les Moulins – Habana - Roma.

of the atomic structure of certain materials. In the specific case, the planar triangular grouping is typical of the carbonate ion $(CO3)^{2-}$. *I therefore wonder if even the chemical sciences could have influenced your evolution as an artist, as well as the almost maniacal engineering attention in the selection of building materials.*

LC: I started looking at chemical composition diagrams – even knowing very little about chemistry – as projects for three-dimensional structures, somehow attributing to them a scale that can be practiced on a personal scale, a bit in the utopian dimension of certain architecture by Buckminster Fuller...and in the meantime I was drawing shapes similar to seeds, but with a strong geometric component, with the idea of constructing colonies of aggregates independent of any objective reference morphology, looking for an abstract formulation completely independent of recognizable contexts. Going back to your question, I think that more than the chemical sciences, now considered fundamental through the centuries in art and with which language technologies have evolved, a certain idea of alchemical science is part of my profound education, in the sense of a parallel form of transformation that from material becomes philosophical and spiritual. Art has the great power to make one read beyond visible matter.

LB: *In addition to chemical-engineering contaminations, I believe you have had, perhaps unconsciously, also strong physical contaminations, as is clearly evident from your magnificent 'Wall reliefs'. In fact, a detailed study of how the waves propagate is evident, which, added to your desire to connect translational modules in three dimensions, makes you call yourself a real 'artist-scientist', very attentive to what is around you.*

LC: In the works you mention, I use a simple software that allows me to evaluate the undulatory phenomenon – which acts on sounds, liquids, light – working on various sources, intensity and consequent interference. The idea was to go beyond experimental verification and, in a somewhat surreal and paradoxical way, to directly relate this phenomenon of vibration to the 'skin' of architecture, again on a different scale, displaying something on the walls that generally is not visible to the naked eye and as a scientific phenomenon it pervades all matter and all bodies. In this sense the sculpture welded to the wall loses the contours of the object to become a sort of 'emotional point' in the architectural box, a mystery that becomes visible.

BM: *In a very recent dialogue*[†††] *you stated that even works you created at the beginning of your career based especially on philosophical premises (the rubber sculptures collapsed down on themselves) have been reinterpreted by you now partially in a scientific key, in the sense of an explosion of matter into particles. Over the course of your career, do you feel that there may be other cases where a scientific interpretation could be put forward for the ex post works, so to speak, depending on the avenue of research being pursued?*

LC: This statement emerged from the observation of matter as it was going into the creation of the sculptures (Figure 18.6);[‡‡‡] very simply, utilising polyurethane, silicone and aggregates of various kinds, and working with bi- and tri-component materials, has enabled me to witness transitional states of matter. When I mentioned the particles, it was because having observed these transitional modalities of matter for a long time, I was giving myself a sort of interpretative key for my subsequent career, in the sense that today the material in the sculptures is particle-based and in a certain sense explodes in space... Initially I used to observe how liquid transforms into a solid, as with the rubber, which takes on many forms: it is solid yet flexible and elastic, it can go from a shore durometer of 15 to 90...with a paradoxical sense of the solid compared to common opinion. Against that, however, in recent times the use of more evident macroscopic structures such as spheres and lozenge shapes, cones and prisms and other geometrical mixed shapes means that the works of art contain a great deal of 'empty' space within themselves. Obviously, the emptiness is an important, constituent part of the sculpture: it testifies to how the two states of matter are correlated in the space continuum.

LB: *I would very much like you to try combining planar pentagonal metallic modules, classic geometric figures that are not able to cover the two-dimensional space according to the classical mathematical rules, so as to form, thanks to your immense creative imagination, new two- and three-dimensional patterns. Do you think it could be an inspiration?*

LC: And you ask me? I'm going to look at Penrose's tessellations...

[†††] *Conversazioni d'autore – Loris Cecchini*, conversation between Loris Cecchini and Bartolomeo Pietromarchi held at MAXXI in Rome, 25th January 2019.[8]

[‡‡‡] LC is referring in this case to the artworks in the series *Stage Evidence*.

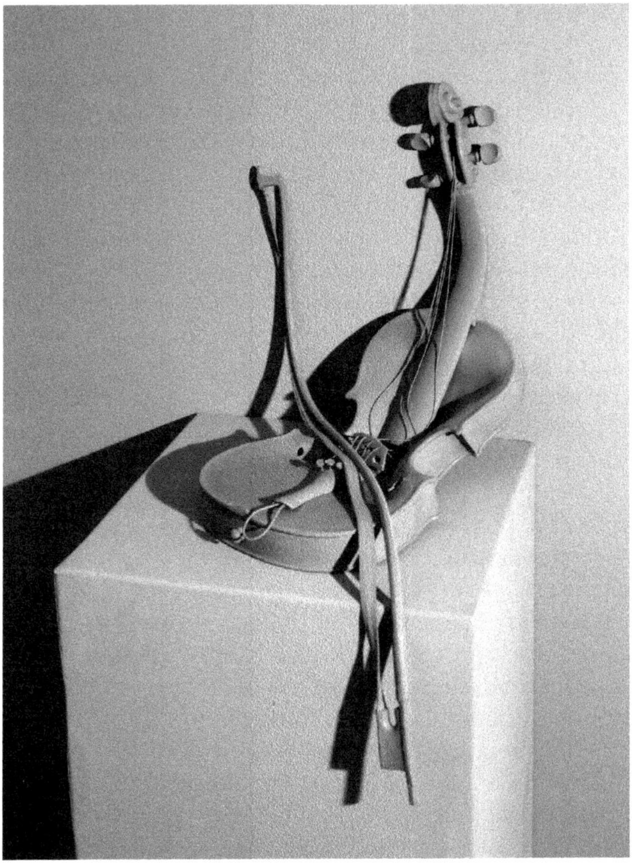

Figure 18.6 Loris Cecchini, *Stage Evidence (violin)*, 2005. Urethane rubber. Courtesy of Galleria Continua, San Gimignano – Beijing – Les Moulins – Habana - Roma.

References

1. http://www.artext.it/Cecchini_Loris.html (accessed March 2019).
2. H. Rosenberg, *The Anxious Object. Art Today and its Audience*, Horizon Press, New York, 1964.
3. L. Cecchini, *Twister Magazine*, 2009–2010, 1.
4. L. Cecchini, *Modulo e modello*, exhibition catalogue, ed. M. Meneguzzo, Fondazione Arnaldo Pomodoro, Milan, 18 March–27 June, 2014.
5. M. Galbiati, *Espoarte*, 2014, 41–45.

6. R. Lami, *Ricerche di storia dell'arte*, 2009, **98**, 11–22.
7. L. Cecchini, *The Ineffable Gardner*, Galleria Continua, San Gimignano, 20 January–01 May, 2018.
8. https://www.youtube.com/watch?v=sAIT8SUwztI (accessed March 2019).

CHAPTER 19

Spectral Revisions: A Colourful Expert Conversation

TIM OTTO ROTH*[a] AND KONRAD SCHEURMANN*[b]

[a]Imachination Projects, Germany; [b]Historische Farbstoffsammlung, TU Dresden, Dresden, Germany
*E-mail: info@imachination.net, konrad.scheurmann@tu-dresden.de

19.1 Introduction

The intense examination of colour brought together Dr Konrad Scheurmann and Dr Tim Otto Roth more than ten years ago. The research interest of curator and art historian Scheurmann oscillates between the fields of arts and sciences. The conceptual artist Roth, who holds a doctorate in art and science history (Academy of Media Arts Cologne), collaborates regularly with scientific institutions and in his research tackles in a novel way intrinsically artistic questions, such as those about the phenomenon of colour, which obviously involve also physical aspects. So, it's no wonder that Scheurmann commissioned Roth to create several works for the colour-related exhibitions he curated. For "Colour continuo – System und Kunst der Farbe", which was presented in 2009/2010 at the Altana Gallery in Dresden, Roth created *Chromology* (Figure 19.1), an installation consisting of a matrix of 25 rotating colour discs. The work links questions on the dynamics of colour relations, as formulated, for instance, by Paul Klee, with recent mathematical models of self-organisation.

Science and Art: The Contemporary Painted Surface
Edited by Antonio Sgamellotti, Brunetto Giovanni Brunetti and Costanza Miliani

Published by the Royal Society of Chemistry, www.rsc.org

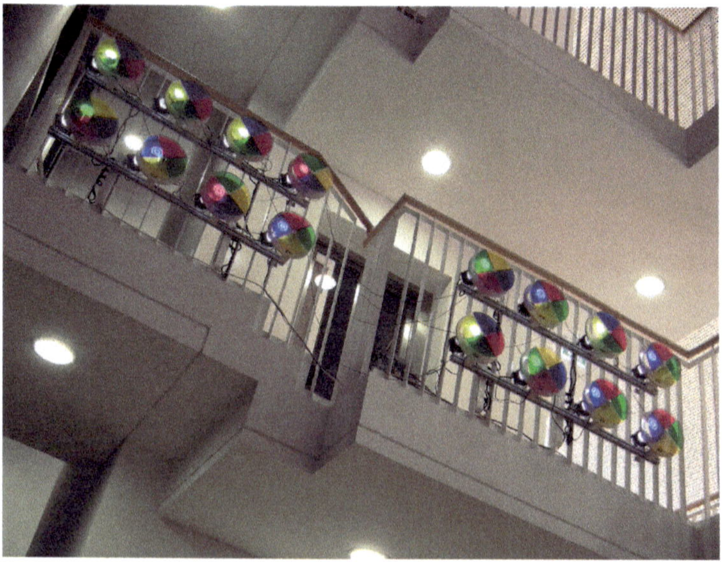

Figure 19.1 Tim Otto Roth, *Chromology* (2009) is a kind of cellular automaton calculating with four colour states: the colour disco wheels function like interconnected colour clocks, changing their "colour time" by adding clockwise the values of the neighbouring wheels connected *via* wires, Städtische Galerie Wetzlar 2010. Image courtesy of imachination projects. Reproduced with permission.

In 2008 he created the light sculpture *Colour of the Universe* (Figure 19.2) for the exhibition "Red.Green.Blue. – Experiments in Light and Colour" using the most abundant element in the Universe, hydrogen, as a luminous colour source. Thanks to a glass specialist in Thuringia, who is experienced in the reconstruction of historic Geissler tubes, a special glass capillary was twisted to form a spherical glass sculpture reflecting the sphere of the Universe. Filled with hydrogen and excited at extremely high voltages, the sculpture emits a luminous colour characterized by very specific lines in the spectrum, related to the intimate nature of the atom. In astrophysics, the light emitted by objects in the Universe (*i.e.* the spectrum) is used as a means of identifying chemical species in interstellar gases. This possibility of colour analysis makes astrophysics a modern colour discipline where the investigated matter, the colour (*i.e.* the spectrum), is simultaneously phenomenon and instrument.

Roth's profound knowledge was one of the reasons to take him on board as an associated member of a joint research project funded by the German Federal Ministry of Education and Research (BMBF), which was initiated and directed by Scheurmann: the joint project "FARBAKS – Farbe als Akteur und Speicher" (Colour as Agent and Memory) brings together different

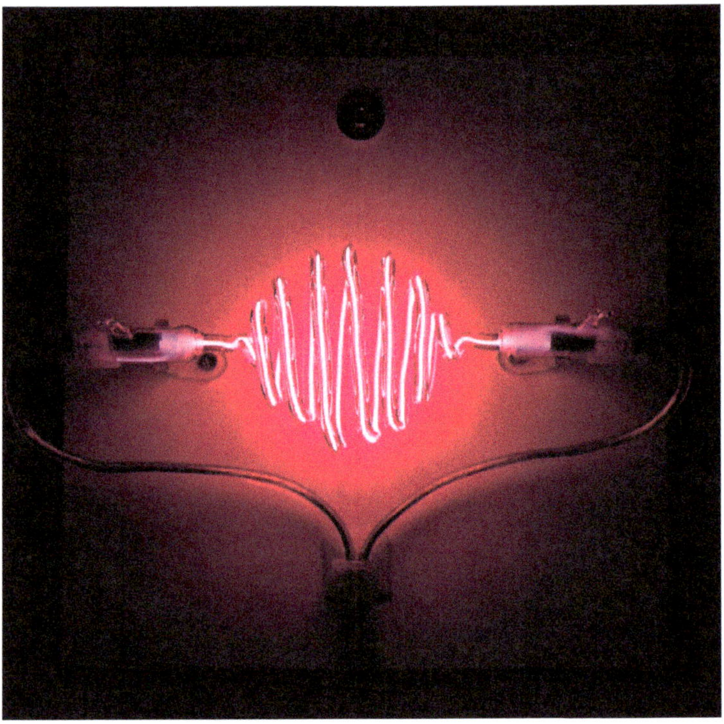

Figure 19.2 Tim Otto Roth, *Colours of the Universe* (2008), light sculpture, *ca.* 9 × 30 cm, soda-lime glass, elementary hydrogen, metal, cable, high-voltage electronics, Fischerhütte Ilmenau 2008. Image courtesy of imachination projects. Reproduced with permission.

approaches of the humanities and the sciences, examining the materiality and the cultural encoding of colour. Here the members of the eight project branches rely, among other things, on colour as a material and immaterial expression of objects selected among German university collections to reconstruct the historical development of colour research. At the end of the joint project, Konrad Scheurmann had an extended conversation with Tim Otto Roth covering the spectrum of their collaboration. The conceptual artist's various approaches to colour as phenomenon and epistemic object, based on his sound historical knowledge, turned the interview into a survey on the contemporary *status* of colour research.

19.2 The Interview

Scheurmann. Mr Roth, you are known as an artist creating big light installations which are made, for example, of coloured light projections in combination with sound as their essential elements. Many of these huge projects take place in collaboration with research institutions, using urban spaces as their stage: sometimes large *façades*, sometimes even whole buildings; and you, by means of such *mise en scéne*, are time and again addressing questions about the universe and cosmology. Why are you interested in such an examination of light, colours and cosmos, which appear quite scientific? What fascinates you about a universe whose man-made unit of measurement is the light-year, a measure that for laymen is a mere metaphor of infinity?

Roth. You refer to my projects which deal mainly with astronomy or, better, with astrophysics. The essence of astrophysics is its involvement with a peculiar and simultaneously ordinary compound: light – not only as a mere shading of black and white, but also as coloured light. Indeed, certain physical and chemical properties of observed objects can be inferred by their colour. For a long time, astronomy, whose procedures consist mainly in measuring and mapping the universe, was the prime discipline for the observation of the sky. The construction of larger and better telescopes marks a crucial *caesura*: the objects not only moved closer, but also much more light was collected. As a consequence, the astronomers not only saw points with different degrees of brightness in the sky but, for the first time, they appeared coloured to them, because the increased amount of light stimulates the less sensitive colour cells on the eye's retina. This discovery was taken into account in developing diverse colour reference systems and colorimeters. In particular, the colorimeter by Friedrich Zöllner allowed a very reliable colour determination in the second half of the 19th century by the comparison of the observed starlight with monochromatic light, so that people – having different colour perceptions – could realize a reliable colour assignment. The real breakthrough occurred indeed with spectroscopy. The splitting of light into its coloured spectral components made it possible to analyze the different spectra of cosmic objects and phenomena and to ultimately individuate them. In this case, colour/light is not a measuring instrument but a physical self-expression. The optician Joseph von Fraunhofer was the extraordinary forerunner who in 1814 discovered with his spectroscope the Fraunhofer lines that were named after him (Figure 19.3). He compared in sequence light from different sources and established that all spectra were not continuous and moreover interrupted by very different types of lines.

40 years later the chemists Gustav Robert Kirchhoff and Robert Wilhelm Bunsen gave an explanation thereof. They demonstrated that the occurrence of these lines is due to chemical elements absorbing light at well-defined wavelengths, when subjected to continuous radiation, or emitting light – when excited – just in correspondence with these lines. Spectroscopy revolutionized our gaze at the sky, since it is now possible to precisely make physical and chemical assertions about objects which are very far away. The new possibilities of colour analysis founded a new discipline – astrophysics

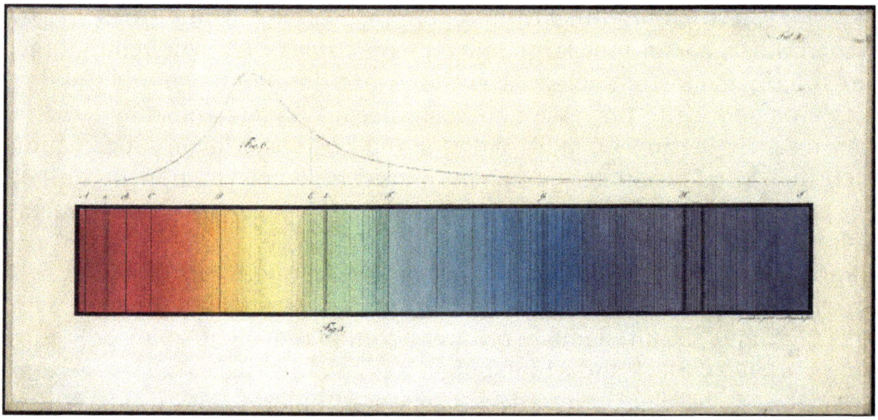

Figure 19.3 Joseph von Fraunhofer, absorption lines within the solar spectrum, 1814. Image courtesy of Deutsches Museum, Munich, Archives, BN43952b.

– which is the ultimate modern colour discipline,† and the peculiarity of its research technique – spectroscopy – consists in the fact that the investigated matter, the colour, *i.e.* the spectrum, is twofold: it is simultaneously phenomenon and instrument.

Scheurmann. Remarkably, in astronomy, colour photography as an analysis and interpretation tool was introduced quite recently, in the 1960s. Is this due to the fledgling stages of colour film at that time? Or does it trace back to the conviction that black and white as pictorial document is more precise and more adequate for reproducing contrasts than colour photographs?

Roth. In a certain way, colour information in astronomic images was already processed before, but in a virtual way: two different black-and-white images recorded with different coloured filters were simply compared by putting them side by side or by superimposing them by a stereoscopic method. There are also copying techniques copying two photographic plates subsequently – consequently the distribution of definite elements could be made visible according to the applied filter.

But astronomers only reluctantly relied on three colour image techniques. I showed in a composite of three photographic plates of the *Orion Nebula* recorded by different observatories between 1896 and 1914, that in principle it was already possible at the beginning of the 20th century[2] (Figure 19.4). The composite appears like a modern astronomical colour picture and it could have been produced, for example, with a carbon print, an established colour printing process at that time. Indeed, some attempts to produce spectroscopic recordings were made using an early type of slide film, the so-called autochrome technique.

†A summary of this "colour revolution" and its consequences is given in ref. 1.

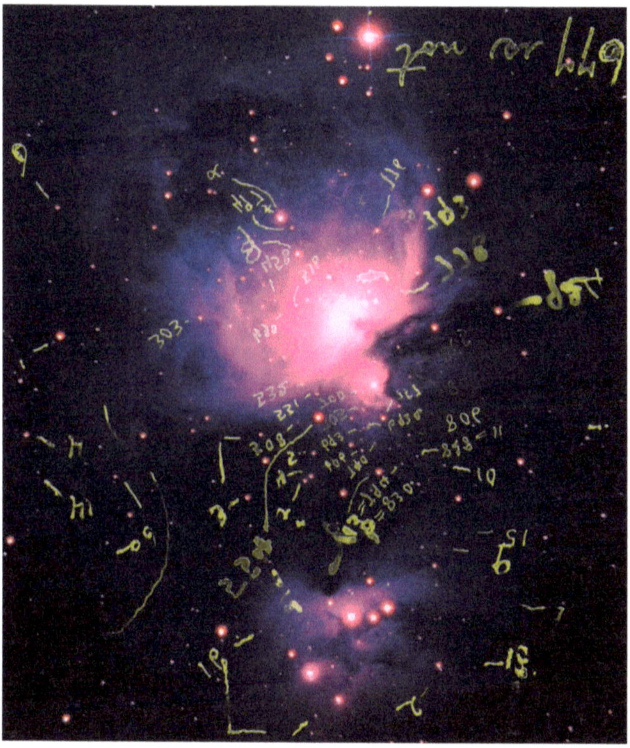

Figure 19.4 Tim Otto Roth, *Orion Nebula* (from the series "Triple Sensation"). The colour composite of both gas nebulae was elaborated from three glass-plate photographs recorded in Arequipa (Peru) 1896, at Landessternwarte Heidelberg-Königstuhl 1906 and in Cambridge, MA (USA) in 1914. The exposure time of those relatively faint celestial objects ranged between 30 and 256 minutes. The last plate, which was exposed in combination with a yellow filter during Halloween night, shows the subsequent inscriptions of the "computers", namely the women, processing the image analysis. The original plates of the composite are from the Heidelberg Digitized Astronomical Plates (HDAP) and from the Harvard College Observatory's DASCH Project. Image courtesy of imachination projects. Reproduced with permission.

Unfortunately, the three colours red, green and blue in the emulsion did not reproduce the spectrum to the same extent. It is the same issue you have been confronted with in black-and-white photography until the end of the 19th century. The light-sensitive emulsion was not uniformly sensitive throughout the whole spectrum and so it was only partially suitable for astronomical colour determination. Until a few decades ago, the astronomers were really forced to manually draw the spectral lines into the wavelength diagram just as they saw them in the spectroscope. And this has very much to do with colour, even though we see only lines there. This is an extremely conceptual and "abstract" way to deal with colours that until now had not found a wide resonance apart from in astrophysics.

Scheurmann. Astrophysical images record phenomena not only within the range of visible light. They also visualize phenomena recorded at wavelengths invisible for humans translating them into so-called "false colour" images. What limits do our visual faculties impose on astro-research?

Roth. Astrophysics reveals something fundamental concerning colours: the limitedness of human colour perception. The semiologist Charles Sanders Peirce highlighted the primitiveness of colour vision in his early photometric publication comprising his own observations of the colour brightness of stars. In the introduction he states: "Light is a triple sensation"[3] – each colour impression thins out in the retina to only three values. This restrictedness becomes particularly evident by comparing colour vision with audition. With our ears, we process an acoustic spectral analysis and we can distinguish individual frequencies within a sound. But in vision most of the rich spectral information about colour gets lost. As a consequence, two colours which appear as identical to us may instead have a totally different spectrum. Only when the illumination changes is it possible to note a difference.

Scheurmann. The so-called colour metamerism.

Roth. Exactly. Metamerism implies a gnosiological problem, because it reveals that we are not able to see the "right" colour but, going back to your previous question, we basically always see false colours. Our eyes reduce not only the real colours to false ones, but this fundamentally occurs also with every image. The art historian Ernst Gombrich was already aware of this: the reduction of the image brightness involves the reduction of the colour palette.[4] From a spectral point of view, the pigments on a canvas or the colour layers of a colour photograph have nothing to do with the objects they represent – consequently, every picture must be considered as a false-coloured one. Astrophysicists belong to those few scientists who are aware of this inadequacy. Thus, they can deal quite self-confidently with coloured depictions, which are intrinsically false.

Scheurmann. But this does not correspond to general knowledge. In their everyday dealing with colours people are normally guided by the impression produced by their physiological vision and this is supported by different mathematical methods and systems for colour classification, as those provided by producers, which are themselves built on the same physiological colour perception. Colour coordinates such as the CIELab, RGB, CMYK or colour codices such as RAL or NCS are exemplary and they were developed for this purpose both for private and professional use of colours.

Roth. For a designer or a graphic designer, colour classifications such as RAL® or Pantone® could provide a sufficient orientation for the daily creative routine. However, if I want to understand why these systems function reliably only according to the standard illumination, or why the green of a tree in CMYK appears totally different from how it does in RGB, then I must have a look at their spectra. Just to underline once more the gnosiological issue: since Kant's critical philosophy, we know that the world as we see it cannot be reduced to a mere sensation.‡ What we perceive is always an interpretation of our sensation in accordance with well-defined representations [*Vorstellungen*]. This enables us to see things totally anew – as, for example, did the young Kant with the Milky Way. As a mere phenomenon, we see a wide bright band in the nocturnal sky; the Königsberg philosopher instead, with a little kind of geometrical imagination inspired by Thomas Wright, catapults us into an expanding space, interpreting the band as a side view of a disc-like stellar system. This notorious turn revolutionized our concept of the universe. Revolutionary to the same extent – concerning the colours – could be considered Joseph von Fraunhofer's discovery of the spectral lines which indeed opened to us a totally new access to the phenomenon of colour. This was the final cut for Goethe's *Farbenlehre* [Theory of Colours], published four years before.

Scheurmann. You must explain this once again in detail, please. The observations made by Goethe, concerning above all the psychological influence of colour, were indeed not entirely wrong. But even the optical-physical phenomena are today rediscussed in relation to names like Olaf Müller, Ingo Nussbaumer, Johannes Grebe-Ellis and Matthias Rang.[6–9]

Roth. Goethe's experiments excel through their peculiar view on colour phenomena, which on closer inspection everybody can discover in daily life. I don't question his power of observation at all. The problem relies not on the observed phenomena but on their interpretation: the privy councillor (Goethe's appointment at the court of Weimar during 1779–1786, Ed.) wanted to deduce from his observations a general theory of light and colours – thus he takes over the interpretative control. But his theory runs out of gas with regard to Fraunhofer's spectrum, simply because it can't explain such a phenomenon. Yet the problem is even more fundamental: Goethe's claim to universality must necessarily fail, since he considered the human colour perception – we already discussed its inadequacy – as the measure of all things. What can be well retraced about Goethe is a kind of rearguard action into inwardness that is not open to a counterintuitive natural science, but even heavily attacks it: so, he polemized not only against Isaac Newton, but also against Joseph von Fraunhofer.[10] In a way he opposed his theory to the natural science, as, to say it in current jargon, an "alternative fact". This counterfactual position is a throwback to Kant and thus to modern and enlightened thinking.

Scheurmann. But isn't this judgement, at least partially, expressed *post festum* from our contemporary point of view, which is based on up-to-date knowledge for instance about the physiology of perceptual apparatus or about the distinction between light and reflected colours?

Roth. My argument refers to the science-historical context of that time, and it is not based solely on Fraunhofer's discovery. We must think that some years before Goethe's *Farbenlehre*, some researchers had already observed that the spectrum transcends its visible part: for example, William Herschel, who

‡The impact of Kant's *Theorie des Himmels* [Theory of Heavens] is examined in depth in ref. 5.

measured with a thermometer the spectrum in the area where the red gets lost in the dark and ascertained an increase in temperature – an indication of the existence of infrared radiation. The existence of an invisible radiation at the opposite part of the spectrum was realized by Johann Wilhelm Ritter, as he observed that silver nitrate, beyond the visible blue-violet area, blackened.

Scheurmann. Let's return to your work: what motivates the artist-scientist [*Künstler-Wissenschaftler*] Tim Otto Roth to translate the above-discussed themes, scientific concepts and procedures into art?

Roth. I am always astonished by expressions like artist-scientist or "scientartist". Within the artistic field, it appears that the idea of two supposed cultures still dominates, as C. P. Snow already problematized in 1959 within his very much cited, yet scarcely read, essay.[11] You would never bring to the foreground in such a way the nexus to science in the case of a composer doing research also as musicologist or concentrating on mathematics. Art has always been connected with its contemporary technologies and its consequent intellectual developments. However, today, within defined art trends it does not seem to be like that, for reasons that cannot be discussed here.

Scheurmann. Please let me formulate it otherwise: How does the art correlate with high-tech science and research? Nowadays, do artists like you hark back to Renaissance art? Peter Weibel, head of the ZKM Center for Art and Media Karlsruhe, talked about a 2.0 Renaissance on the occasion of the "Globale" opening (an art event that took place in Karlsruhe from June 2015 to April 2016), alluding to our digital-based changes in the present time and their future potentialities.

Roth. I rather see a technological continuity therein. Let's return once again to early modern times: if we consider Baroque painting through the eyes of those days, then it was the high-tech of its time. The artists were acting at the "pictorial forefront": what can I figuratively represent as a painter and how should I handle the colours or how should I model light and shadow, people, space and atmosphere? The contribution of Rembrandt & Co. in their times could be considered today as the high school of rendering. I simply follow that tradition. This is why I am interested in the leading image generating and processing disciplines. I do locate such an imaging avant-garde more in the realm of natural sciences than in classic artistic disciplines. I am convinced that today both Rembrandt and Rubens would no longer paint.

Just to connect this concept to a concrete project: I was stimulated, for example, when developing *From the Distant Past*,[12] by a peculiar instrument of the Hubble Space Telescope that is able to deduce a spectrum from minimal light information. For technical reasons many photons are lost by performing classical spectroscopy, so it is very difficult to obtain reasonable spectra from very faint objects. By the combination of a prism and a diffraction grating to form a so-called "grism", a blurry image of overlaying spectra of faint objects can be obtained. But thanks to subsequent calculations a meaningful spectrum can be reconstructed. Consequently, such a technology makes it possible to extract colour information from objects which are faint and thus implicitly very distant. The fascinating thing about spectra is that depicted as a colour intensity curve they give a very precise colour description. In the case of *From the Distant Past*, we are not simply dealing with colours, but we can also truly say we are dealing with the oldest colours currently recordable in the universe. Here I see a chance to give a fresh impetus to the reflection of colour as one of the big themes in the tradition of visual art besides pictorial representativeness.

Scheurmann. So you have dealt with the most distant stars, from which we are still getting light: a sort of light-greeting from the "edge" of the Universe?

Roth. The light sources are stellar systems, like galaxies or the so-called quasars. The distances are too big to resolve pictures of individual stars using contemporary telescopes.

Scheurmann. And thus, you finally caught their fingerprints and translated them into this artwork of laser light drawing?

Roth. "Translated" is the right expression – whereas the intervention was relatively minimal: I more or less adopted the plots, which were then animated. Thus, there was no static image, but a wave-diagram in motion that I projected onto façades in Venice, Baltimore or New York (Figure 19.5).

There is also a little symbolic game associated therewith. Depending on the context the observer came from, and where the diagram was projected, various associations are possible: be it a heartbeat, a brainwave or the last

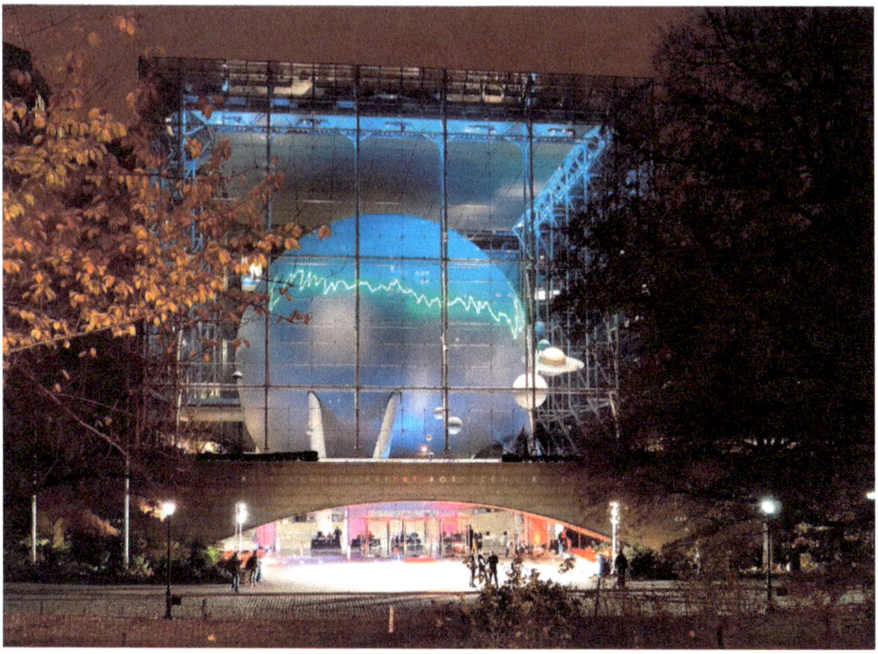

Figure 19.5 Tim Otto Roth, *From the Distant Past*, laser projection on the Hayden Planetarium Sphere, American Museum of Natural History, New York, NY, 2011. Image courtesy of imachination projects and AMNH. Reproduced with permission.

values of the stock market. In this regard, my cooperation partners of the Space Telescope Science Institute, Baltimore[13] have yet showed themselves to be very prone to letting this indefiniteness stretch out for as long as possible without imposing immediately an explanation. Of course, also the open space and the architecture also come into play here. In New York we projected with two lasers, at almost 180°, onto the 30 m measuring sphere of the Hayden Planetarium. Depending on the observer's position, you could barely recognize if there was something moving on the surface or if the whole sphere was rotating.

Scheurmann. What makes the difference of such a project comparing it with the normal public relations activities of these big international astronomic institutions? Almost everyone today knows the colourful images of the Hubble Space Telescope.

Roth. We must be grateful for the enduring campaign by the Space Telescope Science Institute – by the way, joined also by the European Space Agency. They made a wide public aware that processes occurring in the heavens are always accompanied by the emanation of colour. That heaven is colourful is the widespread message and merit of the circulating Hubble images.

My project goes beyond the scope of pictorial representations of astronomical results. It picks up on an essential issue of astrophysics that was never addressed up to now by classic public outreach: the separation of light into its coloured components, as a spectrum. Today each of us can easily produce a spectrum: everyone knows the iridescent colours on the surface of a DVD or a CD. This can be refined a little by bringing a thin slit before the self-diffractive or refractive medium to obtain the above-described lines. You have just referred to spectra as fingerprints, which appears a little bit misleading. A fingerprint marks a certain variability along with the genetic code. And it's exactly this codex metaphor which does not fit to a spectrum in terms of physical effect. A code operates with symbols, which can be interpreted in different modes due to convention. Within a spectrum such arbitrariness is lacking: if I excite a hydrogen atom to a defined energy level in defined conditions of temperature and pressure, I will invariantly obtain, at the same wavelength, one or more lines. This is a completely different approach towards colour. Normally colour is related to a bunch of associations. Such a symbolism resonating various levels of meaning is absolutely irrelevant when dealing with a spectrum, having the pure effect as an outcome. This is a complementary access to colour which due to its purism is for me, as an artist, very stimulating (Figure 19.7).

Scheurmann. Is this fascination for such a "purism" based solely on the physical phenomenon or is it also triggered by the experimental precision?

Roth. Mere precision is not in the forefront, instead you encounter the physical phenomenon everywhere. Just an example: although a CD is a simple "optical instrument", it has a rudimentary precision. If you are lucky, you can see within the red, at sunset, a couple of expanded spectral bands if you obliquely look at the CD against the light. Alternatively, it can be observed with an energy-saving lamp that the colour bands are clearly distinguished and do not form any continuous spectrum. This is the maximum that can be observed with such a simple instrument, but it is an incontrovertible physical effect. I deliberately work with such effects, because I consider my art as a plea for a physics of art. A physics of art not in the sense of physical formulas, but like that which resonates in the German expression "Physik", namely the physique and physis, thus a certain corporeity. This physicalness is more than a phenomenon, an effect, but a kind of incontestable disclosure that forms a counterpoint against the dominance of the little symbolic games in 20th century art. Hence, I mean the gesture in the lineage of a tradition triggered by Marcel Duchamp, which by now has run on for 100 years, with the shift of meanings boosting objects from popular culture to art objects simply by presenting them within the altered context of an exhibition. Since then, all sorts of things have been carried into the white cube. This is basically a pure gesture politics.

I conceive my works, as mentioned before with respect to Rembrandt, as a physical art. I consider my work as a plea to return again to such physical roots. This is reflected also by the feedback of visitors who experience my artworks. They are exactly searching for such a material reconnection.

Scheurmann. The 20th century brought forth also other trends in art addressing colour very materially, for example colour field painting.

Roth. More or less parallel to colour field painting, another art movement emerged: conceptual art. Taking language as a basis, this form of art also dealt with colour, but in the end the two movements had little in common. In the cycle "Spectral Revisions"[14] I shorten to some extent colour field painting and conceptual art, comparing colour as a physical phenomenon with its exact description – not as a verbal expression, but as a physical description in the form of a spectral curve. This shape is cut into the coloured image support. Consequently, colour appears in a twofold manner. I have worked so far with different materials, with industrial powder-coated composite panels of aluminium, with cobalt-based pigments (Figure 19.6) or recently with standard RAL-colours from spray cans.

The spectral revision performed in this way gets a peculiar dynamics by a formal arrangement: originally I presented these spectra separately or well staggered, like a mountainous landscape. But then I started to display always two colours as a complementary couple. Combined in this way, you can watch immediately the spectral colour relationships. Above all the colour duo with its necessarily unequal curves generates its own form-dynamics which imparts in the colours a new kind of liveliness. In such a way, physics and emotion find common ground.

Scheurmann. Mr Roth, I am grateful for this conversation.

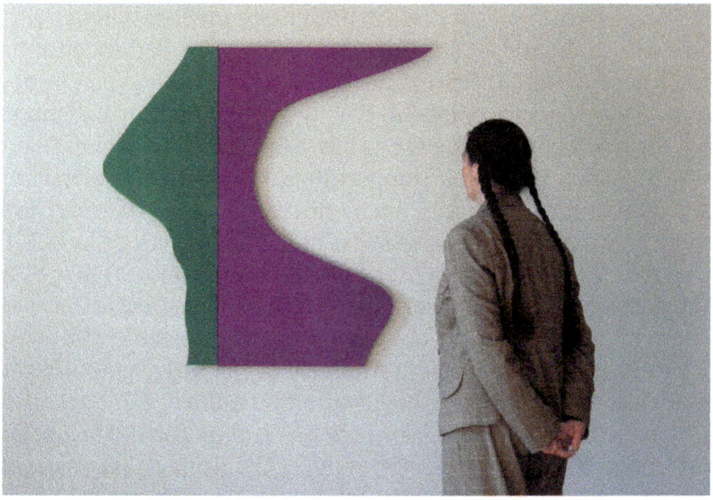

Figure 19.6 Tim Otto Roth, *Spectral Revision* (Cobalt Series), cobalt-based pigments and binder on aluminium-dibond, *ca.* 100 × 120 cm, 2016. Image courtesy of imachination projects. Reproduced with permission.

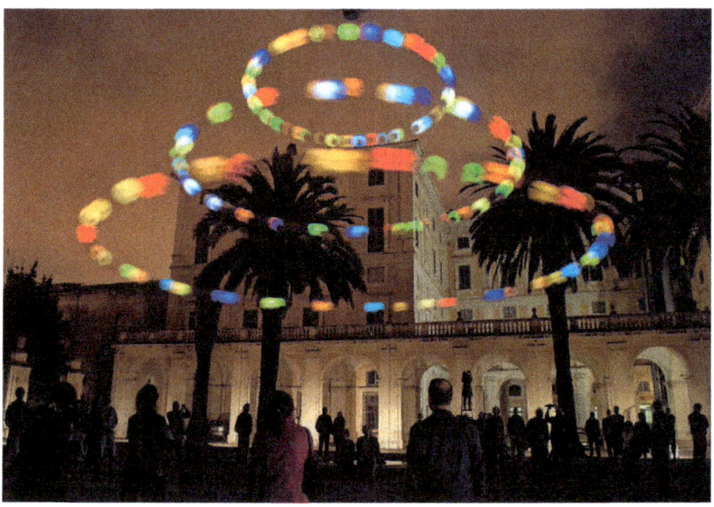

Figure 19.7 Premiere of the *Heaven's Carousel* at the Accademia Nazionale dei Lincei, Rome in 2014. The kinetic light and sound sculpture with 36 rotating illuminated loudspeakers translates, among other things, spectral emissions into a new kind of harmony of the spheres. Image courtesy of Pam Jeffries/STScI. Reproduced with permission.

References

1. T. O. Roth and R. F. Fosbury, Colour beyond the sky. The chromatic revolution in astronomy, in *Light, Image, Imagination – the Spectrum Beyond Reality and Illusion*, Martha Blassnigg, Amsterdam, 2013, pp. 242–268.
2. T. O. Roth, R. Hook and A. Doane, Das Mißtrauen gegenüber der Farbe im Bild – Zur späten Ankunft der Farbenfotografie in der Astronomie [The distrust towards colour in images – On the late arrival of colour photography in astronomy], *Rundbrief Fotografie*, 2013, **20**(4), 3–4.
3. C. S. Peirce, *Photometric Researches*, Leipzig, 1878, p. 1.
4. E. H. Gombrich, *Art and Illusion: A Study in the Psychology of Pictorial Representation*, Phaidon, Oxford, 1977.
5. H. Blumenberg, *The Genesis of the Copernican World*, MIT Press, Cambridge, MA, 1997.
6. O. L. Müller, *Mehr Licht: Goethe mit Newton im Streit um die Farben* [More light: Goethe and Newton fighting for colours], S. Fischer, Frankfurt, 2015.
7. I. Nussbaumer, *Zur Farbenlehre: Entdeckung der unordentlichen Spektren* [Theory of Colours: Discovery of disordered spectra], Edition Splitter, Wien, 2008.
8. J. Kühl, N. Löbe and M. Rang, *Experiment Farbe: 200 Jahre Goethes Farbenlehre* [Experiment Colour: 200 Years of Goethe's Theory of Colours], Verlag am Goetheanum, Dornach, 2010.
9. J. Grebe-Ellis and M. Rang, *Experimentum lucis: Ausstellungsinstallation im Lichthof der Humboldt-Universität zu Berlin* [*Experimentum lucis*: Installation exhibition in the atrium of the Humboldt-Universität of Berlin], October 2010.
10. H. Zehe, "Gott hat die Natur einfältig gemacht, sie aber suchen viel Künste." Goethes Reaktion auf die Fraunhoferschen Entdeckungen ["God has made nature very simple, but they are looking for lots of arts." Goethe's reaction to Fraunhofer's discoveries], in *Sitzungsberichte der Heidelberger Akademie der Wissenschaften*, Mathematisch-Naturwissenschaftliche Klasse, Berlin, 1990, pp. 359–392.
11. C. P. Snow, *The Two Cultures and the Scientific Revolution*, Cambridge University Press, New York, 1959.
12. T. O. Roth, K. Sembach, A. Nota and B. Staude, From the Distant Past, *Leonardo*, 2013, **46**(5), 492–493.
13. http://www.stsci.edu/institute/ (accessed April 2017).
14. T. O. Roth, *XX oder der "Mummelsee in der Pfanne"* [XX or the "Mummelsee in the pan"], Städtische Galerie Offenburg, 2016, pp. 32–36.

CHAPTER 20

Painting with Acrylics: José Gutiérrez, Gunther Gerzso and the Material Innovation in Mexican Contemporary Painting

SANDRA ZETINA*[a], JOSÉ LUIS RUVALCABA-SIL[b], REBECA BARQUERA[a], NITZINÉ OCAMPO-ÁVILA[b], ALEJANDRO MITRANI[b], MIGUEL MAYNEZ[b], EUMELIA HERNÁNDEZ[a], EDGAR CASANOVA[c], ADRIÁN MEJÍA-GONZÁLEZ[d] AND NURIA ESTURAU[d]

[a]Laboratorio Nacional de Ciencias para la Investigación y Conservación del Patrimonio Cultural (LANCIC) Instituto de Investigaciones Estéticas, Universidad Nacional Autónoma de México, Circuito Mario de la Cueva s/n, Coyoacán, 04510, Mexico City, Mexico; [b]LANCIC Instituto de Física, Universidad Nacional Autónoma de México, Circuito de la Investigación Científica s/n, Coyoacán, 04510, Mexico City, Mexico; [c]CONACYT-LANCIC Instituto de Física, Universidad Nacional Autónoma de México, Circuito de la Investigación Científica s/n, Coyoacán, 04510, Mexico City, Mexico; [d]LANCIC Instituto de Química, Universidad Nacional Autónoma de México, Circuito de la Investigación Científica s/n, Coyoacán, 04510, Mexico City, Mexico
*E-mail: sandra.zetina@gmail.com

Science and Art: The Contemporary Painted Surface
Edited by Antonio Sgamellotti, Brunetto Giovanni Brunetti and Costanza Miliani

Published by the Royal Society of Chemistry, www.rsc.org

20.1 Early Use of Industrial Paints in Mexico and Development of a Mexican Brand of Artists' Acrylic Paint

Mexico stands out for its early artistic experimentation with industrial paints created for other purposes (*e.g.* automobile and furniture paints), and with modern mechanical tools (such as airbrushes, spray guns, projectors and cameras) since the 1930s, years before synthetic paints were produced specifically for artistic use. The testing of modern painting media and techniques intensified in the Mexican artistic circles during the 1940s, and by the 1950s Politec®, a Mexican artists' acrylic paint brand, came into the market. Two artists were involved in the early interest in modern materials and tools in Mexico: David Alfaro Siqueiros (1896–1974) and José López Gutiérrez (1900–1968). Siqueiros' intense activity throughout the world (he travelled to the United States and South America) and his theories about modern art expressed as manifestos and conferences were the point of departure for the creation of an artist culture of modern techniques in Mexico and the United States. He was a pioneer in the use of synthetic media with a strong theoretical position on the agency of materials in the production of art. On the other hand, José López Gutiérrez, or José Gutiérrez, was a less visible figure, but his role as teacher, technical advisor, promotor and manufacturer of synthetic paintings was crucial for the application of innovative industrial materials and techniques in Mexico, Canada and the United States, uniting artists of all tendencies and developing new procedures customized for each artist. The first part of this research explains the background of the use of industrial materials in Mexican modern art, and the particular agency that modern techniques acquired during the first half of the twentieth century, where the artists had the intention to introduce significant innovations in art history. The second part explores the use of acrylic paintings in Gunther Gerzso's *oeuvre*; he was a painter that contaminated the long-lasting figurative stance of the muralist movement with a challenging abstraction, rooted in geometry, surrealism, and the uncanny experience of the Mexican pre-Hispanic ruins.

20.1.1 Siqueiros and the Agency of Industrial Technology and Collectivism in Modern Art

In 1932 Siqueiros for the first time articulated his ideas about a technical revolution in painting at a conference given in the John Reed Club of Los Angeles: "Los Vehículos de la Pintura Diálectico Subversiva [The Vehicles of Subversive Dialectical Painting]".[1] The conference was presented in the form of a manifesto where he claimed that artists were using outdated materials and techniques that were not suitable with the recent discoveries of science and technology: they were anachronical and did not respond to social conditions.

Siqueiros departed from Marxist and communist notions to advocate for an update in technology and the collectivization of artistic work. The conference summarized his experiences during the creation of two murals (*Workers' Meeting* and *Tropical America*, 1932) produced by The Block of Mural Painters, a Los Angeles artist collective. By suggestion of the modernist architect Richard Neutra, Siqueiros started to use airbrushes and developed a mural fresco technique over white Portland cement. The subversive proposal of the Mexican painter was to produce painting that could be multi reproductible, an outdoor mural painting as a tool for revolutionary agitation and propaganda for the masses. He pronounced himself against the traditional paintbrush, fresco, oil, watercolour and pastel techniques which he defined as archaic modes of production and he proposed the use of drill guns, cement guns, spray guns, blow torches, white water-proof cement, electrical projectors and photographic and cinematographic cameras for sketching. The importance of this early manifesto is that Siqueiros first stated that only modern techniques could produce revolutionary art and he pushed the agency of the industrial materials to create modern and revolutionary aesthetic values.

By the end of 1932 in Buenos Aires along with the Equipo Poligráfico (another painters' collective) Siqueiros painted *Plastic Exercise*, a "plastic cube", a room fully covered with images painted with ethyl silicate.[2] But it was 1935 when Siqueiros found cellulose nitrate automotive lacquers and, a year later, created the Siqueiros New York Experimental Workshop where his ideas about modern art were fully developed.

He again formed a collective of artists, this time based in New York, that experimented in the production of political art and propaganda for the Communist Party during the US presidential elections of 1936. Among the participants were José Gutiérrez and Jackson Pollock, who would later use some of the ideas developed at the workshop for creating his abstract expressionist technique. Philip Guston also attended some meetings.[3–5]

The Siqueiros New York Experimental Workshop established new interesting ways of using fast-drying cellulose nitrate lacquers that were dripped, splattered and poured over supports placed horizontally. Siqueiros developed the idea of the controlled accident, an automatist technique, near to surrealist modes of painting. He also insisted on the importance of collective creation: he used stencils and friskets that helped to unify the shapes and to avoid the personal touch. New modern panels such as plywood or metal sheets were also tested, as well as many mechanical tools: the electric saw, airbrushes, spray guns and photographic cameras to reproduce paintings. Easel painting was presented as a photogenic matrix for the creation of posters by photographic enlargement that were colored with lacquers and spray guns.[5,6]

It is possible that José Gutiérrez during the New York Siqueiros Experimental Workshop introduced Siqueiros to cellulose nitrate lacquers as a medium, but certainly he would later follow the experimentation with other modern materials.[5]

20.1.2 José López Gutiérrez: the Artist's Technical Advisor

Biographical information about José López Gutiérrez (1900–1968) is scarce and often contradictory. Possibly because of the Anglo-Saxon habit of using just one last name, he is usually mentioned as José Gutiérrez. He came from humble origins in Miacatlán, Morelos, a Zapatist region during the Mexican Revolution. Escaping the violence of the Revolution, he emigrated alone to California when he was very young and made his living as a farm labourer. Later, he travelled with a journalist across the United States, and around 1918 he established himself in New York, although apparently he made a few short trips back to Mexico City around 1920. In New York he worked at a restaurant and at a paint manufacturing company in New Jersey, a chemical laboratory for painting and a Westinghouse plant. There are no direct sources about his early life, but his daughters recalled that he worked at aeronautic companies, possibly Kaiser Aeronautics where guncotton explosives were manufactured. This is the raw material for producing nitro-cellulose lacquers, also called pyroxylin. He also worked as a tool designer during the war for Brewster Aeronautical Corporation in Long Island.[8] When Gutiérrez started his involvement in art as a part-time activity, he had not received any formal training, only briefly attending the Escuela al Aire Libre de Santa Anita with Alfredo Ramos Martínez in Mexico, before leaving for the United States. In New York he took classes at the Pratt Institute, the Art Students League, the Greenwich Settlement House and possibly at an art institute in Harlem.[7]

During the 1930s he was regarded as an advisor on synthetic painting materials, probably after his contact with industrial manufacturers and laboratories. At a conference on the work of José Clemente Orozco in 1932, Gutiérrez met the American journalist Alma Reed, with whom he established a lifelong friendship. He claimed that he used industrial paintings before meeting Siqueiros and attending the New York Experimental Workshop. Gutiérrez also participated as a mural painter for the Works Progress Administration and he stated that his formulations were used in some of the program's art projects.[7]

In 1944 he travelled to Havana where he painted a mural for a military hospital, possibly recommended by Siqueiros who had stayed in the city a year before. Gutiérrez returned to Mexico in 1945 and immediately opened the Taller de Ensaye de Materiales Plásticos [Workshop for the Research of Plastic Materials] at the Instituto Politécnico Nacional (IPN). The workshop was created by mandate of the Minister for Education Jaime Torres Bodet. José Clemente Orozco, Diego Rivera and especially David Alfaro Siqueiros supported Gutiérrez in the creation of the workshop and were part of the advisory committee; by then, they were already known as the three great Mexican muralists. The creation of the workshop was highly publicized at press conferences. Gutiérrez was the director and researcher on artistic techniques and Manuel Jiménez Rueda was the technical chemist. Both taught the courses of the IPN's Escuela Superior de Ingeniería y Arquitectura [Higher Studies School in Engineering and Architecture] located at the Pre-vocational School in Tacubaya, Mexico City.[9]

The IPN workshop was designed with two main objectives: firstly, to research the properties and capacities of new industrial materials for their application in modern painting; secondly, to provide young students with a technical education, offering them the possibility to acquire a certificate as experts in painting materials or as industrial painters, after two years of education. Gutiérrez was aware that this kind of teaching on synthetic materials was new; he affirmed that there was no other place in the world that offered such a program.[10] Also, based on research, the workshop advised engineers, architects, archaeologists and artists on how to solve specific technical problems and disseminated information about the use of synthetic materials in modern art, as finishing materials for architecture or for conservation purposes in archaeological objects.[10]

A bulletin from the IPN workshop, printed around 1946, indicated the novel painting techniques that were part of the curricula: "silicon mural painting technique" (ethyl silicate applied over Portland cement); "vinilita" (Vinylite®, the trademark name for polyvinyl acetate PVAc and polyvinyl chloride, PVC); "pyroxylin lacquers" (nitro-cellulose); "catalinas" (Catalin®, a trademark for cast phenolic resins in semi-transparent colours).[10] The brochure also suggested that the permanence of these resins was "certified by research", and highlighted their luminosity and transparency, "that made colours vibrate". Along with the industrial techniques, the workshop offered technical knowledge on traditional techniques, such as fresco, oil, watercolor, pastel, *etc.*[10]

Although some hardships came soon after the founding of the workshop, it survived, serving mainly as an advisory center. Gutiérrez advised many important official mural commissions, such as José Clemente Orozco's ethyl silicate outdoor mural, painted for the open auditorium at the Escuela Normal Superior in 1947 and Rufino Tamayo's use of Vinylite® for the mural *Mexico de hoy*, painted at the Palacio de Bellas Artes in 1952. Gutiérrez was very close to Siqueiros and designed for him all his painting formulations. For instance, he created a mixture of pyroxylin especially for him, mixing pigments, lacquer and asbestos to obtain thicker impastos with matte finishing.[11]

Gutiérrez became well known as a technical advisor not only in Mexico, but also in Canada, in the United States and, to a lesser extent, in Europe and Latin America. He was invited by the National Gallery of Canada to teach about painting materials and he travelled the country teaching at six universities. Based on these courses, he wrote his book *From Fresco to Plastics*, published in Canada in 1959.[12] In the US, he went to Columbia University in New York, Southern California University in LA and San Francisco Art Center in California. He also sent written professional recommendations to the Museum of Portugal, the Art School at Carrara, the UNESCO in France,

Maidstone College in Kent and the Art Schools in Lima, San Salvador, Havana and Guatemala City.[13,14] He entered into contact with North American experts in artists' materials, such as Rutherford Gettens and Ralph Mayer, and his formulations were quoted and used in other handbooks from the 1960s.[7]

Gutiérrez established a large network of collaboration with institutions, artists, paint companies and laboratories. Although he never had a professional education in chemistry, he corresponded with chemical manufacturers close to New York, where he had previously worked. During the 1940s he used products from Union Carbide, and as soon as the Rhom and Haas laboratories offered the acrylic emulsion medium to the market Gutiérrez sent them letters and decided to create his own artists' paint brand. Although Gutiérrez always considered himself an artist, and exhibited his artworks in important galleries, such as for example the Ines Amor Galería de Arte Mexicano, the main purpose of those events was to present the properties and techniques of modern media. He was never regarded as an important artist; the critique emphasised the material composition of his paintings, but never his aesthetic values.[7]

20.1.3 Painting With Acrylics and Politec®, the First Mexican Acrylic Artists' Paint

In 1956 Gutiérrez designed his own acrylic paint formulation "by an artist for artists" using Rhom and Haas acrylic emulsion and resins. The brand was named Politec® after the IPN, the institution where it was developed.[15] Gutiérrez marketed several products: Politec® 100 was an emulsion acrylic paint for professional use; Politec® 300 was also an emulsion acrylic paint, a tempera formulated for commercial artists, schools and hobbyists.[16] He also offered an acrylic emulsion, named varnish-sealer, designed to be used for glazing, as final matte varnish or to be mixed with pigments. Luzitron, marketed as varnish-medium, was a glossy medium that was not water-based: it was composed of acrylic resin dissolved in xylol and toluene. The "Plastilita" extender was a kind of gesso composed of acrylic emulsion, marble dust and silica.[16] The brand is still on the market, with several of the described products still being produced. A recent study of yellow and red Politec® acrylic paints showed that the organic paint constituents were the acrylic base, as the copolymer of EA–MMA–EMA, the surfactant, as OPE compound 4-(1,1,3,3-tetramethylbutyl) phenyl-polyethylene glycol, and the pigments Hansa Yellow and Toluidine Red, respectively. The inorganic aggregates identified were diatoms, talc, Mg and Ca carbonates.[17]

In 1965, after ten years of experimentation with acrylic-based paints, José Gutiérrez, along with Nikolas Roukes, wrote *Painting with Acrylics*, a manual for artists who wanted to explore the "unique character" of the new industrial products of the modern age.[15] The first part of the book summarized the history and the main acrylic products in the world market (Acqua-tec,

Hyplar™, Liquitex™, Magna, Reeves®, Shiva, and many others) but also detailed formulas for customizing self-prepared paints with different media: pyroxylin, ethyl silicate, polyvinyl acetate and acrylics.[15] The second half of *Painting with Acrylics* presents studio tips accompanied by photographs of actual modern artworks that Gutiérrez advised on: the caption of the images describes the precise technique or medium used by the artists. The list of artists is very heterogeneous and expresses quite well his relationships and tastes; it does not only comprise artists living in Mexico, Canada or the United States, but also creators of all tendencies, many of them contradictory. Some, as José Clemente Orozco and David Alfaro Siqueiros, were the leading figures of muralism, that pertained to a kind of social realism related to the official culture that sprang in the 1920s and became the mainstream in the Mexican art field during the 1940s and 1950s. Other artists, such as Rufino Tamayo and Carlos Mérida, were interested in formal experimentation with a less figurative approach, closer to cosmopolitan tendencies. Gutiérrez also included younger and lesser-known artists, such as Gunther Gerzso or Arnold Belkin.

The preface of *Painting with Acrylics* was written by David Alfaro Siqueiros from the Lecumberri prison in Mexico City, where the artist was incarcerated for political reasons.[15] Confined in a small space, he started experimenting with acrylics to avoid the fumes of pyroxylin; he had previously disliked acrylics, because they lacked the impasto, and also the fast-drying properties of pyroxylin. In the preface, the renowned muralist recalled his 1932 manifesto about the revolution in painting, and presented the New York Experimental Workshop from 1936 as the precedent for the use of synthetic paintings in art. For Siqueiros, the research of modern synthetic paintings was the result of concerns the Mexican artists had about early murals, resolved when they entered into contact with the American industry. In *Painting with Acrylics*, synthetic paintings were regarded as a sign of modernization, of industrial development, although it must be said that Gutiérrez' contribution was that he was up to date in the developments of the chemical industry and in contact with manufacturers, so he introduced the products to the art field, found new applications and imported raw materials to Mexico.

20.1.4 The Mexican Art Scene During the 1950s and 1960s: Gunther Gerzso and the Rupture

The Mexican state that emerged from the Revolution in the 1920s endorsed a cultural project that favored mural paintings and the development of a local, socially committed realism close to communist views and expressed in official commissions of murals. After World War II, the state became more powerful and centralized and during the 1950s and 1960s the Mexican economy grew duc to the creation of a local industry and a close relationship with the United States.[18] Before, the Mexican art scene was configured around state commissions and was devoted to the preservation of a national identity. In 1945 David Alfaro Siqueiros published "No hay mas ruta que la nuestra [There is No Other Route than Ours]", a manifesto where he compelled artists to follow socially committed realism and questioned North American and Parisian abstract tendencies, which he claimed were dominant cultural strategies of capitalism and imperialism.[19]

Cold War tensions manifested in ideologically charged debates around art that polarized abstraction and realism, tensions that were resolved around 1956 in the *Ruptura* [Rupture] movement, signaled by the emergence of new artists and the openness to other tendencies such as surrealism or abstract painting. This change in the art field was also related to the creation of an art market and changes in the media: television was launched in Mexico, new artistic interdisciplinary groups and cultural magazines proposed a more experimental and cosmopolitan approach.[18]

Gunther Gerzso (1915–2001), through his painting practices related to cubism and surrealism, in the isolation of his studio a few years before the Rupture, started to quietly undermine the solidity of muralism. He started out as a scenographer, and in 1950 had his first solo exhibition at the Galería de Arte Mexicano. Gerzso was born in Mexico and from Hungarian and German origins, he lived in Europe during his infancy and returned to his homeland when he was still young. In the early 1930s he studied scenography at the Cleveland Play House and developed a successful career as a set designer for Mexican cinema. It was during this time that Gerzso started to paint.[20]

Gerzso developed a very personal artistic vision in which he explored the otherness of pre-Hispanic ruins, his emphasis on polished surfaces and decorative rhythms. After departing from surrealism, he attempted to highlight the tensions between the emotional otherness felt in the Mexican pre-Hispanic ruins and the space and color expressed his uncanny experiences in the tropical landscape through a quite sophisticated technique. His painting is concerned with producing exquisite surfaces, and although he explored the values of space and color, he never considered himself an abstract painter but a surrealist. He also proclaimed that his work was a modality of cubism manifested in geometric planes, inspired by the aesthetics of the golden section.

20.2 Material Study of Gunther Gerzso Paintings

Gunther Gerzso was interested in the materials and techniques of the Old Masters; in his studio he kept modern treatises such as *The Materials of the Artist and Their Use in Painting with Notes on the Techniques of the Old Masters* (1934) by Max Doerner, *The Secret Formulas and Techniques of the Masters* (1948) by Jacques Maroger, *The Mastery of Oil Painting* (1953) by Frederic Taubes and *The Materials and Techniques of Medieval Painting* (1956) by Daniel Thompson, along with monographs of Bosch, Brueghel, Canaletto, Piero della Francesca, Tintoretto and Tiziano, as well as the Flemish Primitives.

The painter obsessively controlled all the details of his painting production and he annotated them in the sketches. There, it is possible to analyze the careful geometry of the compositions; he added notes on the materials and even the painting brands he used (see Figure 20.1). Gerzso, as well as the surrealist circle in which he participated (Remedios Varo and Leonora Carrington, as well as Julio Castellanos, Antonio Rodríguez El Corcito and Juan O'Gorman), was interested in the revival of traditional painting techniques (such as tempera) to produce clean and translucent finishing to his paintings in the manner of the Flemish painters, but he also recurred to the modern materials. Once he said he liked to read catalogues and painting technique magazines in the same way that novels are read.

It is possible that José Gutiérrez advised Gerzso for *Paisaje de Papantla*, a painting from 1955 that appears as an example of mixed media in *Painting with Acrylics*, and could be one of the first Mexican paintings created with Politec® paints. Gerzso experimented with acrylic paints from the 1950s and he continued to use them as one of his favorite media, testing and describing the brands he used in his sketches (some from the 1970s describe the use of Liquitex® gesso and acrylic). Alongside *Paisaje de Papantla*, whose technique was described by Gutiérrez, *Muro azul* was also chosen to be analyzed through a non-destructive approach comparing results obtained with UV and IR imaging, surface microscopy and spectroscopic techniques (FORS, FTIR, XRF and Raman). While they were produced in two different moments, both paintings employ deep green and blue colors that allude to tropical regions inhabited by pre-Hispanic ruins: *Paisaje de Papantla [Papantla Landscape]* (1955) recalls the encounter with Tajin's ruins, near Papantla, Veracruz, and *Muro azul [Blue wall] (Chiapas)* (1977), alludes to Mayan antique cities hidden in Chiapas' tropical forest. Both paintings still form part of the collection of the Galería de Arte Mexicano, the first gallery that exhibited Gerzo's work in the 1950s.

20.2.1 General Analytical Procedure

A two-step workflow was set, with first a global examination by imaging techniques under several lighting conditions: visible, ultraviolet and infrared light combined with optical microscopy in various regions of each painting. Afterwards, a spectroscopic non-invasive molecular analysis was performed by Fourier-transform Infrared spectroscopy (FTIR), Raman spectroscopy and Fiber Optics Reflectance Spectroscopy (FORS), as well as an elemental analysis by X-ray Fluorescence spectroscopy (XRF).

A set of acrylic painting references was prepared in our laboratory using known pigments and commercial paintings. This set was analyzed under the

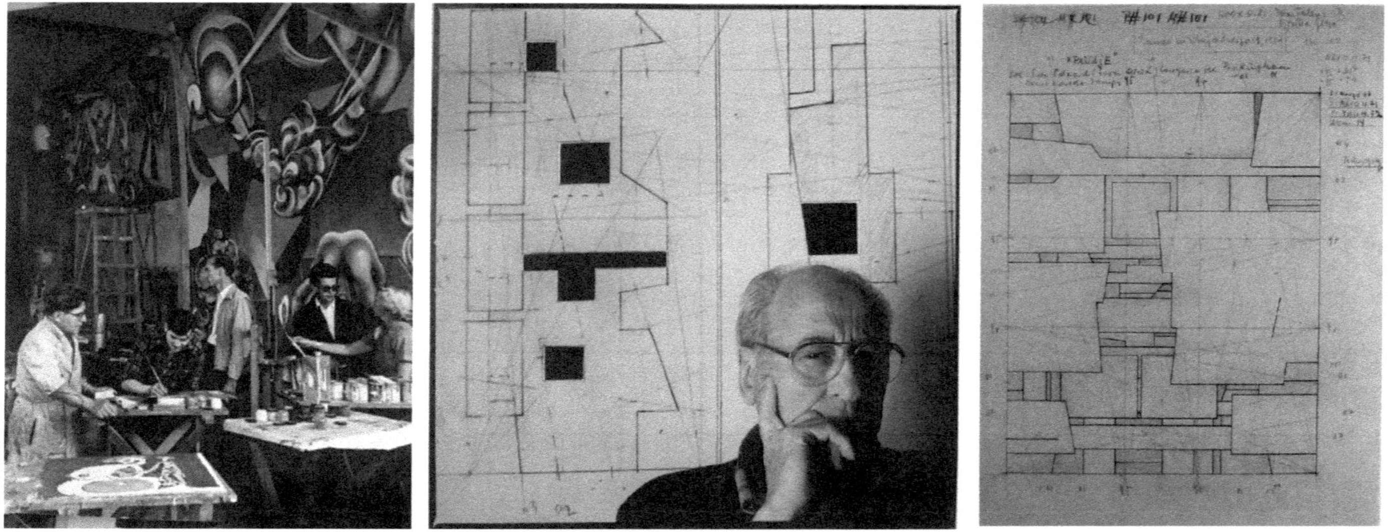

Figure 20.1 *Left*: José Gutiérrez (on the left) at the Taller Ensaye y Materiales Plásticos [Workshop for the Research of Plastic Materials] from Instituto Politécnico Nacional, Mexico City, *ca.* 1948. LANIC IIE UNAM Archive. *Center*: Gunther Gerzso in front of one of his golden section compositions, *ca.* 1980. Diana C. Dupont. *Right*: Gunther Gerzso golden section sketch of *Paisaje*, 1977. Diana C. Dupont.

same conditions as the paintings, generating a database that aided in the identification of the materials present in the paintings. Spectra of acrylic and pyroxylin paints and alkyd resin references were also acquired for comparison purposes.

20.2.2 Analytical Techniques

Visible photographs were captured with a Phase One® XF camera with an IQ3100 back, and ultraviolet images were acquired with a Nikon® D800E camera using a 50 mm FD macro lens. Visible images were illuminated with Elinchrom® flashlights, while ultraviolet fluorescence was achieved with UVP lamps, using both 254 and 366 nm wavelengths simultaneously.

Infrared imaging was carried out using two devices. For the 1000–1650 nm range a Xenics® camera model XEVA-1.7-640 with an InGaAs detector was used with no filter and Lowel® continuous halogen lighting. A second set of photographs were obtained with a Phase One® digital back attached to a Mamiya camera, with filters at 750 nm, 870 nm and 1000 nm.

FTIR was performed with the external reflection mode from a Bruker® Alpha II FTIR spectrometer. Spectra have a 4 cm^{-1} resolution in the near infrared range (400–4000 cm^{-1}) and a 50 mm^2 analysis area. 52 regions were analyzed on *Paisaje de Papantla* and 26 on *Muro azul*.

Raman spectra were generated using a B&W TeK™ spectrometer with a 785 nm laser (20 mW power and 50 μm dia). Data are in the 200–2000 cm^{-1} range with a 15 cm^{-1} resolution. Spectra were averaged on each analysis point, with an acquisition time in the 1–10 s range. All data were processed with the BWSpec® software (B&W Tek™).

Raman spectroscopy was only applied to *Muro azul*, as *Paisaje de Papantla* was very sensitive to the laser beam even in the lowest power settings, so it was decided not to use this method to avoid any beam damage to the painting. 55 regions were analyzed on *Muro azul*.

Fiber Optics Reflectance Spectroscopy (FORS) was carried out by a portable FieldSpec 4® from ASD Inc., allowing the acquisition of visible, NIR and shortwave near infrared (SWNIR) reflectance and absorbance (log[1/R]) spectra. A handheld probe was used which was placed in contact with the sample. A D65 illuminant provided illumination over the whole spectral range. The analysis area was 1 cm^2 and spectra were obtained with a 0.2 s integration time. Calibration was performed using a certified reflectance standard (AS-02035-000CSTM-SRM-990-362, ASD Inc). The visible and NIR regions of FORS spectra were presented together in a zone named visible near-infrared (VNIR). VNIR ranges from 300 nm to 1000 nm and SWNIR ranges from 1000 nm to 2500 nm. 81 measurements were done on *Muro azul* and 105 measurements were performed on *Paisaje de Papantla*.

Elemental analysis was carried out with SANDRA, XRF equipment built at our laboratory.[21] This equipment has a Mo tube and an Amptek® silicon drift detector (SDD) with a 2.5 mm^2 active area, 12.5 mm thick Be window, and a full-width half-maximum (FWHM) ¼ 130 eV at 5.9 keV. X-ray fluorescence conditions were 45 kV, 0.3 mA and 180 s using a collimator for 1 mm diameter spot size. Spectra were analyzed by PyMca software.[22] 76 regions were analyzed on *Paisaje de Papantla* and 55 on *Muro azul*.

20.2.3 Paisaje de Papantla, 1955

In *Paisaje de Papantla* Gerzso combined numerous applications of painting layers and glazes, some of which were sanded to generate a surface that presents the complexity of a quetzal feather, a blue-green color with reddish and yellow undertones that is almost impossible to reproduce photographically (see Figure 20.2). The color of the surface was generated by a process of light refraction that constitutes the sum of the vibration of several colors mixed optically over a white ground.

Gerzso used the smooth side of a Masonite panel; the back of the fiberboard has the characteristic pattern, the wire screen impression of the William Mason hot press process. The painter prepared it in the manner of the Old Masters: he applied several layers of gesso (possibly Plastilita, the acrylic gesso by Politec® – see discussion below). He carefully polished each application with the help of sandpaper or a knife, as can be detected by the lineal traces or

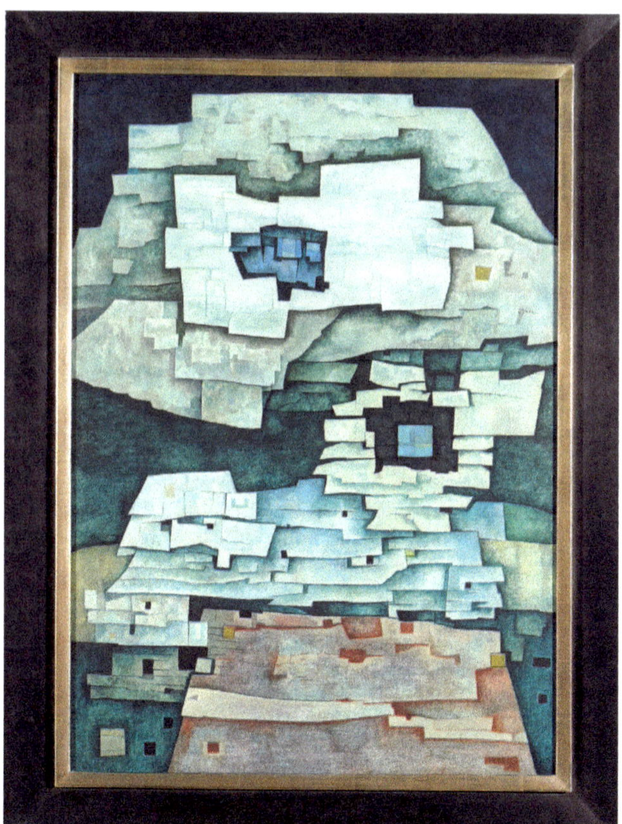

Figure 20.2 Gunther Gerzso, *Paisaje de Papantla*, 1955, copal, oil, acrylic, acrylic gesso on Masonite, 100 × 73 cm, Galería de Arte Mexicano, Mexico City. Photo: ©Eumelia Hernández Vázquez, LDOA–LANCIC IIE UNAM.

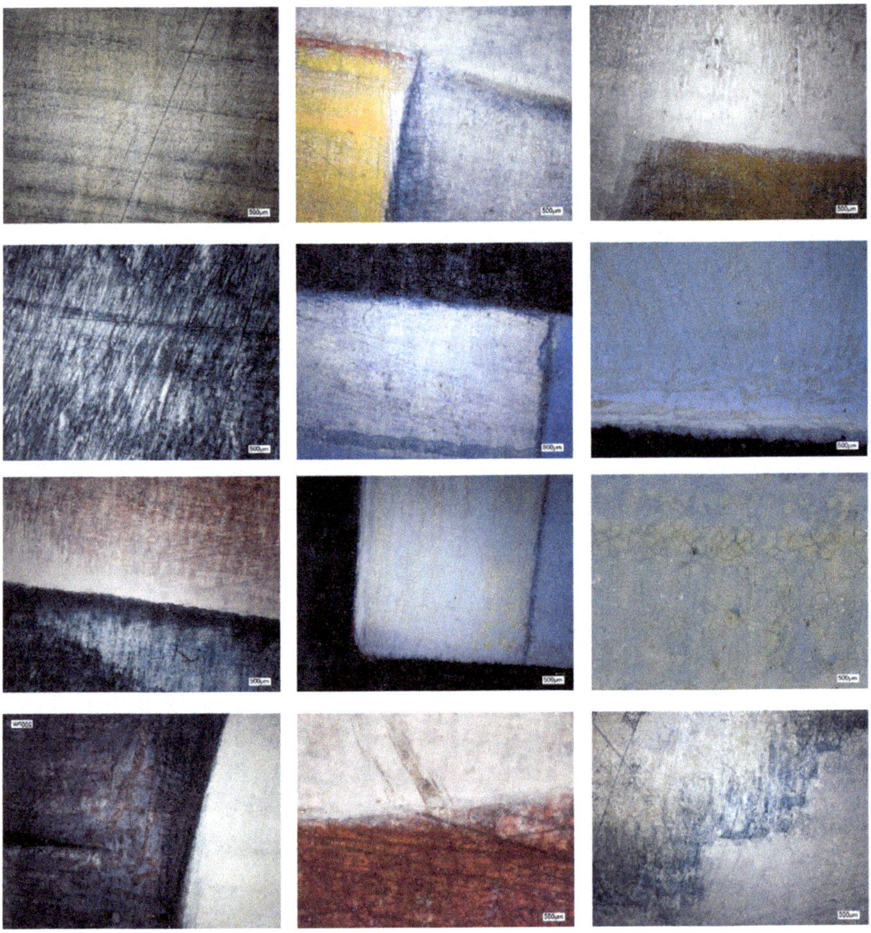

Figure 20.3 Microscopy of the surface of *Paisaje de Papantla*. *Left images*: linear marks of sanding process of the ground and acrylic underpainting. This method created transparencies that were overlapped. *Center*: the delimitation of regions was achieved with thin brushstrokes, also the oil glazes of bright yellow, red and whitish-blue colors is contraposed to sanded areas. *Right*: final touches and impastos were applied with oil with copal and a dryer; some cracks were observed in those areas. Photo: ©Eumelia Hernández Vázquez, LDOA–LANCIC IIE UNAM.

grooves observed all over with a microscope (see Figure 20.3). In many areas, the white ground can be seen throughout the painting, providing luminosity to the surface, even in the darkest areas. Then, Gerzso traced the preparatory drawing with a soft pencil; a few of those traces are still visible on the surface with microscopy and infrared photography (see Figures 20.3 and 20.4).

In *Painting with Acrylics*, Gutiérrez described the technique of *Paisaje de Papantla* as follows: "Underpainting was done with acrylic colors using a scumble technique. The artist used photo floodlights to hasten drying: he then applied thin glazes of oil paint over the acrylic painting. For final glazing, oil paints were thinned with copal painting medium, thinned with five parts of

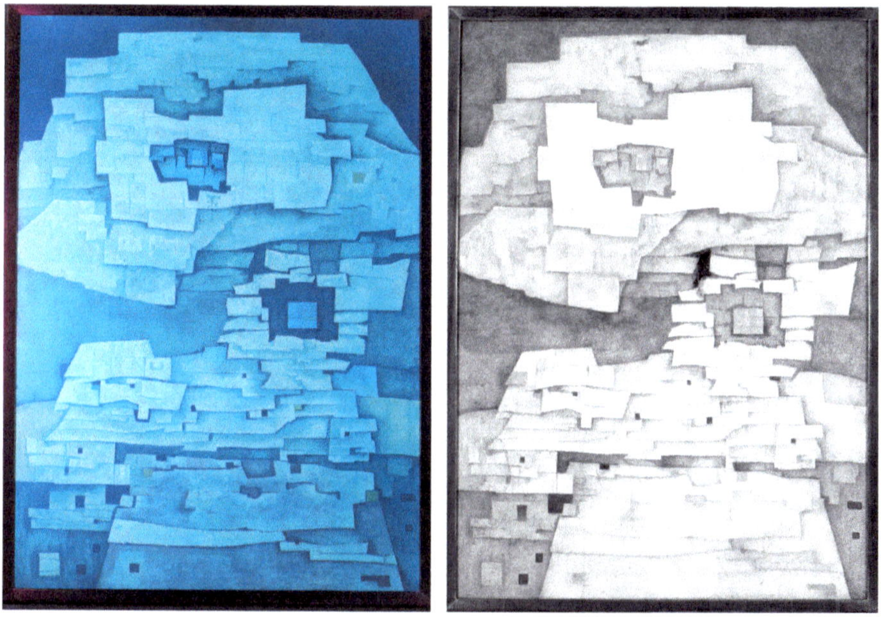

Figure 20.4 Gunther Gerzso, *Paisaje de Papantla*, 1955, copal, oil, acrylic, acrylic gesso on Masonite, 100 × 73 cm, Galería de Arte Mexicano, Mexico City. *Left*: UV image 254 nm. *Right*: IR image 750 nm. Photo: ©Eumelia Hernández Vázquez, LDOA–LANCIC IIE UNAM, 2016.

turpentine and five drops of dryer. Sable brushes were used for smooth, level brush strokes. As many as twenty thin coats of oil glaze were applied. Each coat was allowed to dry thoroughly before the next glaze was applied." It is interesting to notice the application of Gerzso's knowledge as set designer in the idea of using floodlights for drying and mixing colors optically.

The identification of these two binders, acrylic as foundation and oil as finishing, was possible through the spectroscopic studies. With FORS spectroscopy, the absorption bands in the NIR region (1000–2500 nm) of linseed oil, acrylic paint and pyroxylin references and blue colors in *Paisaje de Papantla* are shown in Figure 20.5. Alkyd resins and acrylics display similar spectra, with small differences in the resins' bands at 2300 and 2345 nm, which are present at 2312 and 2375 nm in the acrylics. Pyroxylin displays a different characteristic spectrum in both the 1675–1740 nm and 2050–2200 nm regions. On *Paisaje de Papantla*, independently of the color analyzed, the painting spectra present similar features to acrylic paint and the bands detected for the 2250–2400 nm range are similar to those of linseed oil.[24] This agrees with Gutiérrez' description of an oil finishing.

On the other hand, FTIR spectra from different areas in *Paisaje de Papantla* were similar, showing no apparent relationship to the color analyzed (see Figure 20.6). Although the spectra are noisy and some of the signals are weak and not well defined, the bands around 2917, 2846, 1731, 1471, 1460 and 1362 cm^{-1} have been previously associated with acrylic resins[23,24] (see Figure

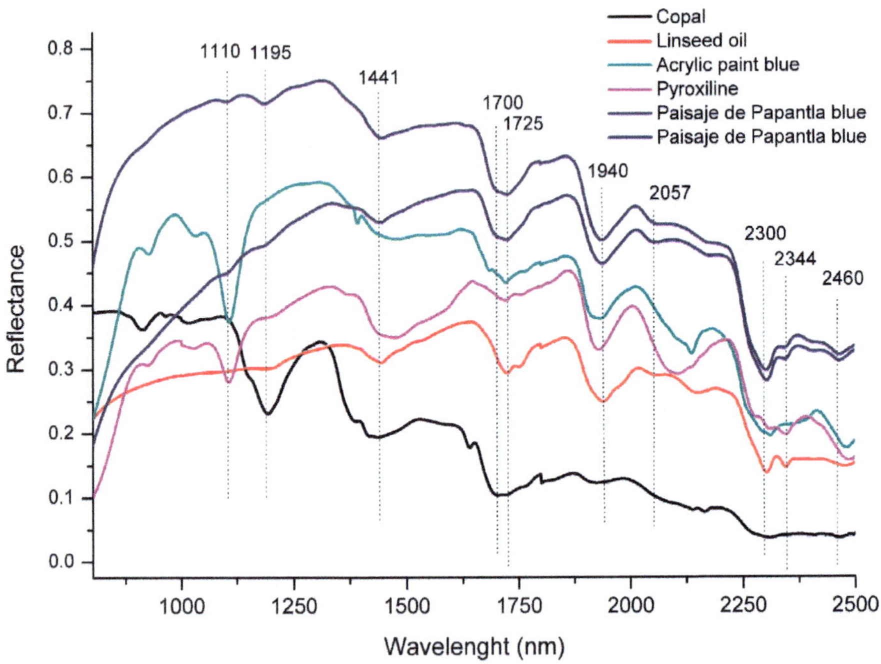

Figure 20.5 Characteristic FORS spectra from *Paisaje de Papantla* by Gunther Gerzso.

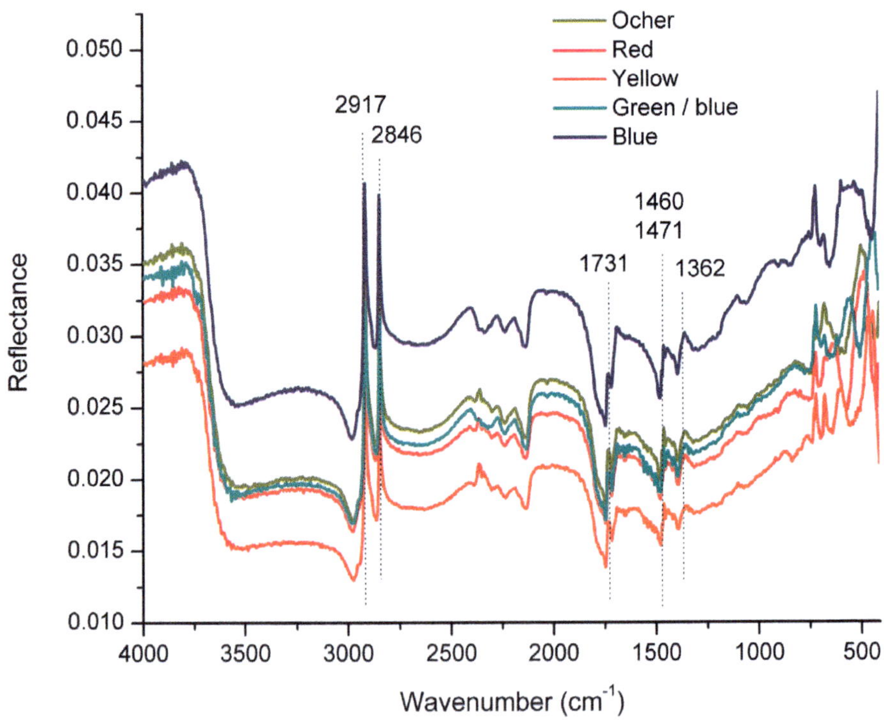

Figure 20.6 FTIR spectra acquired on *Paisaje de Papantla* for various colors.

20.6 and Table 20.1). The bands observed at lower wavenumbers (700–400 cm^{-1}) may be related to other components added to the mixture to obtain the desired texture, but their assignment is unclear.

Gutiérrez' account explains the technique that Gerzso developed for *Paisaje de Papantla*, where he used acrylics in the same manner that Flemish painters used tempera, as an underpainting to be covered by oil glazes and final touches. Over the polished acrylic gesso ground he applied several layers of acrylic glazes intermingled with thicker painting layers, sanded the layers between applications and in some areas applied impastos, opaque painting layers and glazes with commercial oil and a final glazing of oil mixed with copal and a dryer. This system gave profundity to the colors and a rich variable texture that intermingles opaque and translucent regions. Under shortwave ultraviolet, the painting has a bright appearance (see Figure 20.4), possibly due to the presence of copal, although it was not possible to identify this material, because it may be in a very small proportion. In microscopy some of the bright areas show signs of shrinking by a fast-drying process (see Figure 20.3 right).

The repeated action of sanding the painting layers disrupted the preparatory drawing, so Gerzso added delicate lines with very thin brushes to define the contours of the ruins. This process can be identified in the red area that projects a multiplicity of colors. In the end, on the upper part of the painting

Table 20.1 Comparison of FTIR bands observed on *Paisaje de Papantla*, and from pyroxylin, alkyd and acrylic paints and other acrylic polymer ref. 24.

Paisaje de Papantla observed bands (cm^{-1})	Observed bands in references (cm^{-1})			Reported bands[24] (cm^{-1})	
	Pyroxylin paint	Alkyd paint	Acrylic paint	Paraloid F-10 pnBMA 1	Paraloid-B72 p(EMA-MA) 1
2917				2962	2988
				2939	
2846				2876	
1731	1712	1715	1728	1729	1732
	1635				
		1575	1538		
			1508		
1471				1468	
1460	1443	1445	1420		1449
				1387	1388
1362	1370	1373	1358	1370	
	1274			1270	
	1252	1251	1227	1244	1238
		1162		1177	1170
				1155	
	1116	1113			
		1065	1096	1069	
		1037	1012	1022	1028
		956		967	971
				949	
	823			847	862
	728	703		750	756
			616		
		470	447		

he added another plane of profundity and brighter shades of blue, yellow and white, as well as dark blue-green colors to create the sensation of nooks and small niches. On the central area the painter acted as if he were a minuscule microorganism degrading a limestone surface: he invented fissures, patinas and lines that seem like cracked borders.

Some pigments and inorganic aggregates were identified by means of XRF, although because of the complex technique, pigments are overlaid in very thin transparent layers all over and it is not easy to ascribe its composition (see Figure 20.7). Ca, Ti, Zn, Ba and Pb were detected in all the analyzed regions. Pb does not show a pattern, while Ca has a relative high intensity in most of the surface of the painting. We consider that Ca and Ti may be related to the ground layer, thicker than all the other painting layers, which were polished, exposing partially the acrylic gesso, probably Politec® Plastilita, composed by gypsum or calcium carbonate and titanium white. Zn presented higher intensities in the most saturated color regions, and it could be related to the white oil glazes added as finishing touches, as well as barium, that may be related to commercial artist oil colors, because it is present in several color regions.

Concerning the color areas (see **Figure 20.7**), Cr is present throughout the painting with higher intensities in some green, blue, yellow and red areas, probably due to chrome yellow ($PbCrO_4$) acrylic glazes, because it is not always related to the green color. This fact suggests the presence of a green organic color, and moreover a green of chromium pigment. Cu is present in the dark areas of green, as well in some dark mixtures with blue, probably composed of phthalocyanine blues and greens (chlorinated copper phthalocyanine), usually found in Politec® acrylic colours. Also As is present in higher intensity in the greenish-blue regions.

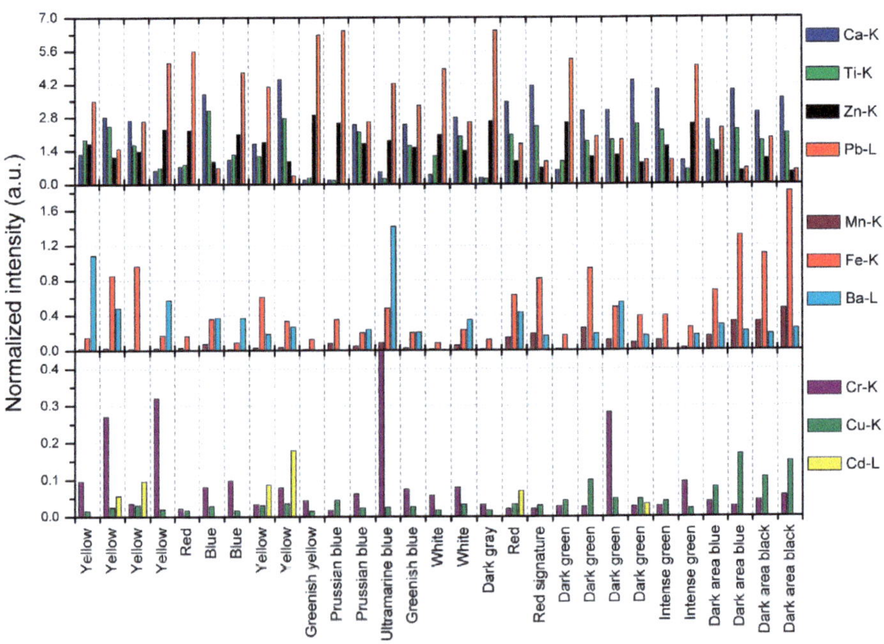

Figure 20.7 XRF results of *Paisaje de Papantla* for various color regions.

On the other hand, Cd is mainly related to some yellow areas and two regions of red, but it is not present in all the yellow regions. These results indicate the use of a yellow and organic red, and perhaps that the finishing touches could be oil commercial painting with Cd yellow and red.

The presence of Mn and Fe in the darker green regions indicates the use of iron oxide earths. Also, Fe and earths were used in ochre and red colors. Ni is also present in the yellow and red regions but also in some dark green regions.

There are several kinds of blue pigments. The light blue color may be related to ultramarine blue, while the darker blue regions could be related to the aforementioned phthalocyanine blue or Prussian blue, since Fe is present all over the surface. Co is present in low concentrations with a peak in the signature of the painter and in some darker regions; perhaps cobalt blue was used to prepare these darker colors.

20.2.4 *Muro Azul (Chiapas)*, 1977

Muro azul (Chiapas) was painted over a thick cotton canvas, a commercially prepared duck canvas, but he flipped it and used the back to take advantage of the texture of the taffeta weave. For *Muro azul* Gerzso created two types of surfaces (see Figure 20.8). The first was a translucent emerald

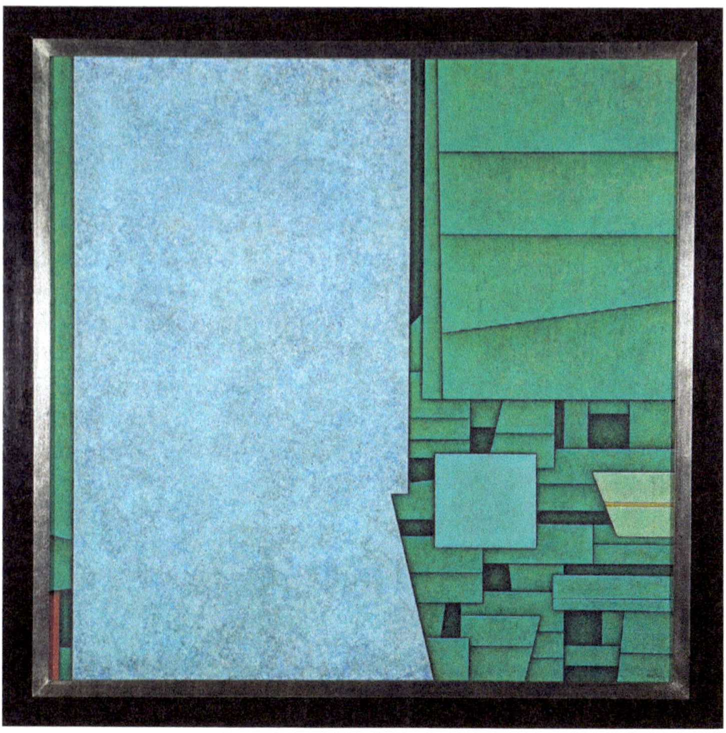

Figure 20.8 Gunther Gerzso, *Muro azul [Blue wall] (Chiapas)*, 1977, acrylic on canvas, 95 × 99 cm, Galería de Arte Mexicano, Mexico City. Photo: ©Eumelia Hernández Vázquez, LDOA–LANCIC IIE UNAM.

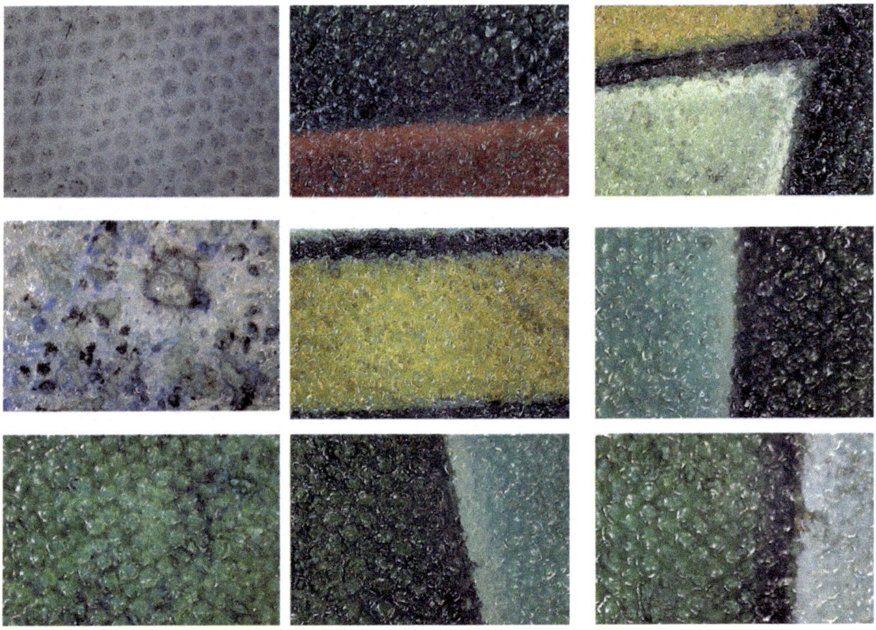

Figure 20.9 Gunther Gerzso, *Muro azul [Blue wall] (Chiapas)*, surface microscopy. The contrast of textures was important for Gerzso, the back of the canvas has a commercially prepared ground which he avoided, using instead the coarseness of the weave. He added middle-grained marble powder to the blue wall, while the greener areas are quite flat.

green achieved by applying diverse green shades, obtained in palette, sometimes mixed with white using an extremely thin brush. He contrasted these areas with a blue square block that recalls a stony surface. For this region he mixed marble dust or sand with a blue acrylic painting; the aggregates are visible through microscopy (see Figure 20.9). At 1000–1650 nm most of the colors seem to be transparent to the IR, even the darker shadows that give profundity to the blocks. The underdrawing seems to be traced with a ruling pen and carbon ink because of its opacity under IR; some black splattering and the marble powder or sand also are visible with this technique, and in raking light (see Figure 20.10). It is a previously established composition, which puts into play profiles like those of pre-Hispanic architecture, with the panel-slope (*tablero-talud*) inclinations that create almost flat solids of blue-green colors (at the center, one is almost a Maya blue shade). The black contours resemble the frame line of the Mesoamerican codex painting tradition. Gerzso tested these kinds of drawings based on the golden section: he created sketches in paper and meditated about them for a long time before turning them into paintings (see Figure 20.11).

The FORS and Raman spectroscopic studies suggested the presence of acrylic polymers as binders in *Muro azul*; the roughness of the painted surface did not allow the acquisition of any useful frontal reflection FTIR

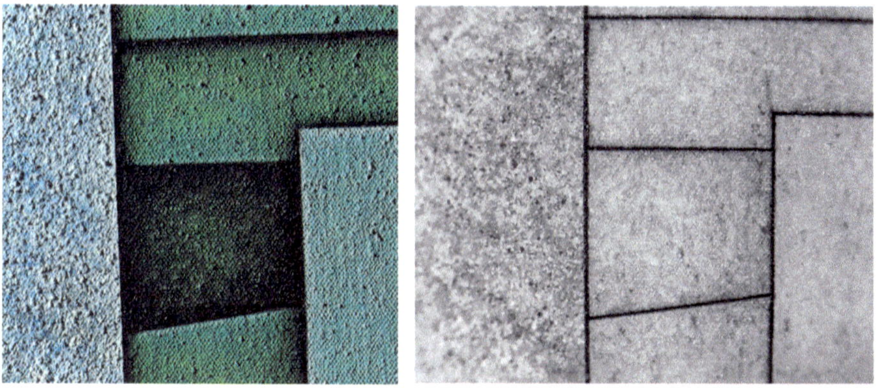

Figure 20.10 *Left*: detail of *Muro azul*, raking light. *Right*: the same region by IR reflectography, 1000–1650 nm.

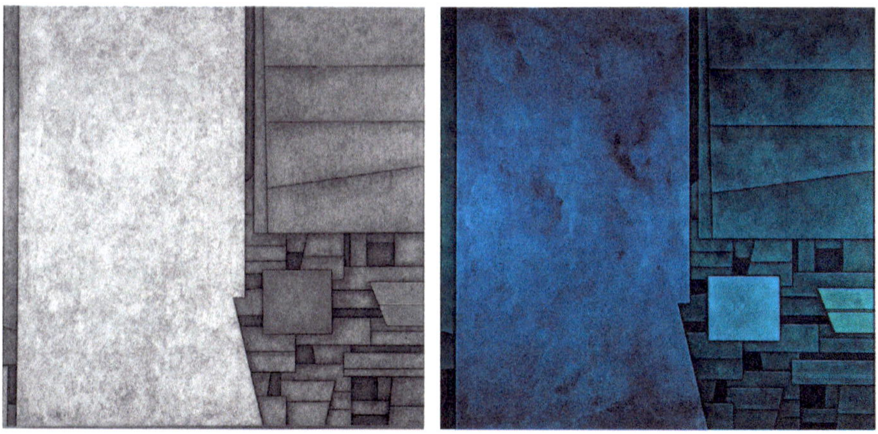

Figure 20.11 Gunther Gerzso, *Muro azul [Blue wall] (Chiapas)*. *Left*: IR imaging 870 nm. *Right*: UV image 365 nm.

spectra. The representative FORS spectra of green and blue colors acquired from *Muro azul* presented great similarity with the acrylic references in the 1600–2500 nm region (see Figure 20.12). The spectra from *Muro azul* blue areas do not match the analyzed blue references in the 900–1200 nm region; these bands are typical of chlorinated copper phthalocyanine. The *Muro azul* blue spectra correspond, moreover, to cobalt blue,[25] while the green spectra match the spectra of Politec® light green acrylic (see Figure 20.13). This coincidence with Politec®'s references suggests the presence of acrylic paints.

Regardless of the color analyzed, all of the Raman spectra acquired on *Muro azul* displayed similar characteristics (Figure 20.14). A comparison of the observed Raman bands and those reported for different resins and

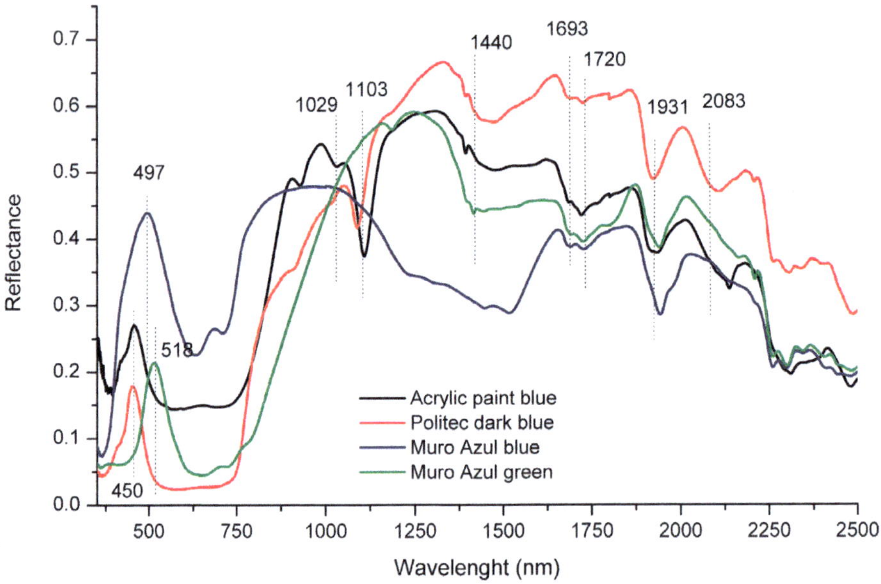

Figure 20.12 FORS reflectance spectra of blue acrylic references and *Muro azul* representative blue and green colors.

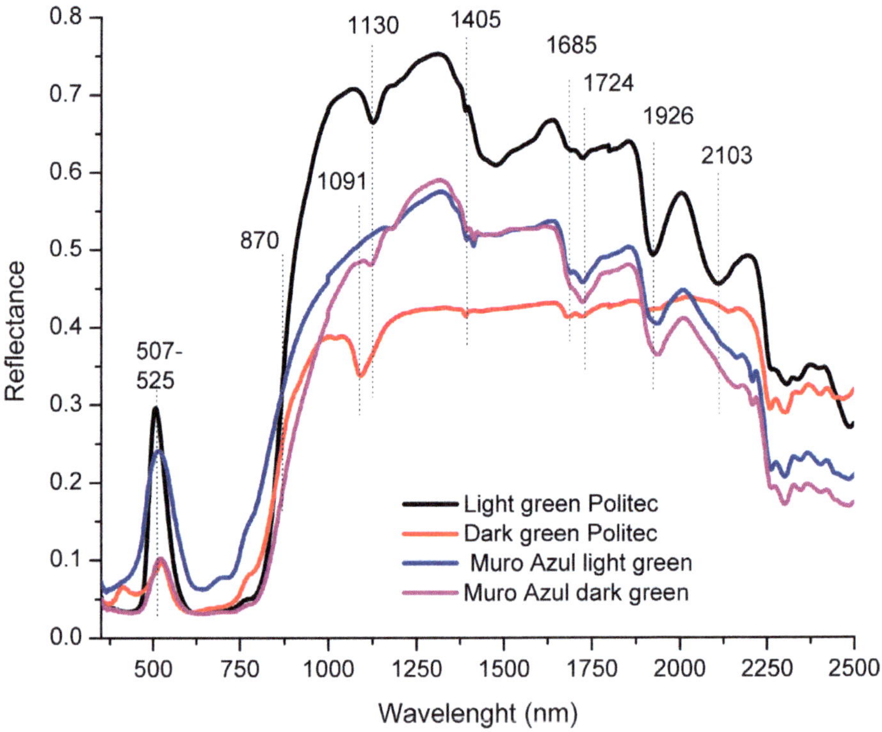

Figure 20.13 FORS reflectance spectra of green acrylic references and *Muro azul* representative light and dark green colors.

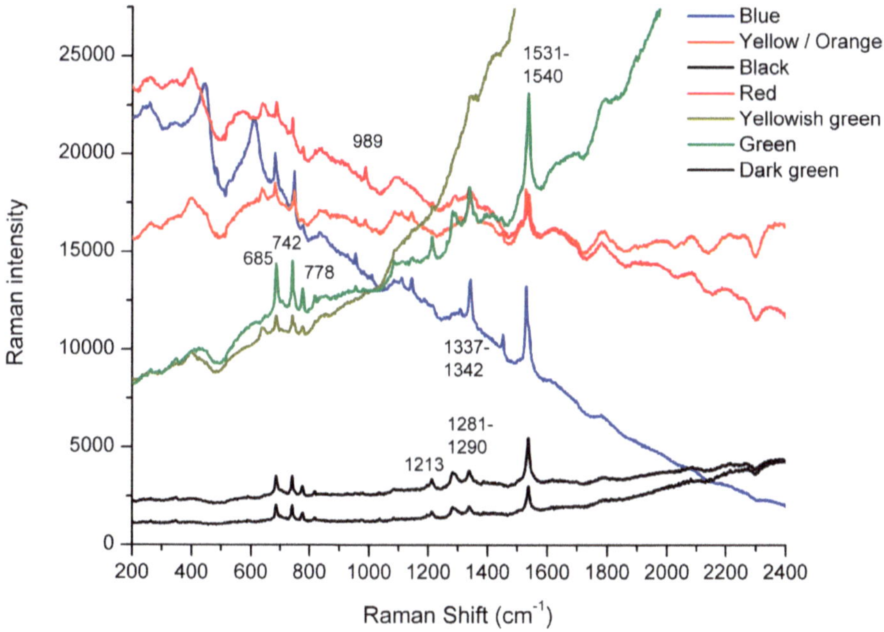

Figure 20.14 Raman spectra acquired on *Muro azul*.

polymers[26–30] is presented in Table 20.2. The acquired spectra suggest the presence of an acrylic polymer. Although the experimental data from Raman are not sufficient to ascertain the exact polymer present in the artwork, the observed bands greatly coincide with those reported for acrylic resin. However, the Raman bands of this resin reported in Table 20.2 are approximate, since some works provide the spectra without indicating the Raman shift value.[30]

In *Muro azul,* Ca seems to be related to the aqua blue and the most intense blue regions, probably due to the calcium carbonate of the marble powder aggregates added by the painter (see Figure 20.15). In the small red area Ca has the lowest intensity because there are no aggregates in there. Possibly titanium white was used for the brighter finishing touches; the intensity of Ti rises on those brushstrokes and also in the blue regions. Pb and Zn do not present a pattern. The presence of S is related to the gypsum preparation layer of the back of the painting.

The red and orange tones were made using Cd red and orange (see Figure 20.15) and Se is also present in this color (some cadmium reds are CdSe). The blue color is cobalt blue, as FORS studies previously indicated. Also some cobalt blue has been added to the red color to modify the tone. Cu is practically constant with some increments in the dark green regions but chlorinated copper phthalocyanine green bands are not clear in the FORS spectra. Moreover, the absence of Cr and other characteristic elements in the green color suggests the use of another organic pigment. It is interesting

Table 20.2 Comparison of Raman bands observed on *Muro azul* and reported bands for acrylics and other polymers[26–30].

Muro azul observed bands (cm^{-1})	Reported bands (cm^{-1})						
	Poly(acrylic acid)[26]	Acrylic acid[27]	Sodium acrylate[28]	Accura 60[29]	Poly(AN-*co*-VA) and Poly(AN-co-MA)[a,30]	Alkyd resin (*o*-phthalic)[a,30]	Acrylic resin[a,30]
	1698			1723–1730	1739		
		1660	1640	1620–1634			
		1637					
1531–1540			1550–1575			1540	1540
1452	1457	1434	1453	1445	1452	1450	1460
				1407		1400	
1337–1342	1333		1370		1345	1320	1340
		1395					
1309		1295					1310
1282–1290		1280	1283	1261–1262		1290	
				1234–1250			
1213				1214		1210	1220
	1191			1173–1186			
1145							1150
							1110
1085		1069	1053		1080	1090	
989		977	988				1010
956			950			960	960
			902	914			
830	846	861					
816		820	836				
775–778				760–788		780	
742						750	750
685						690	690
635–638		643	661	601		650	

[a]approximate values, from the spectra figures.[30]

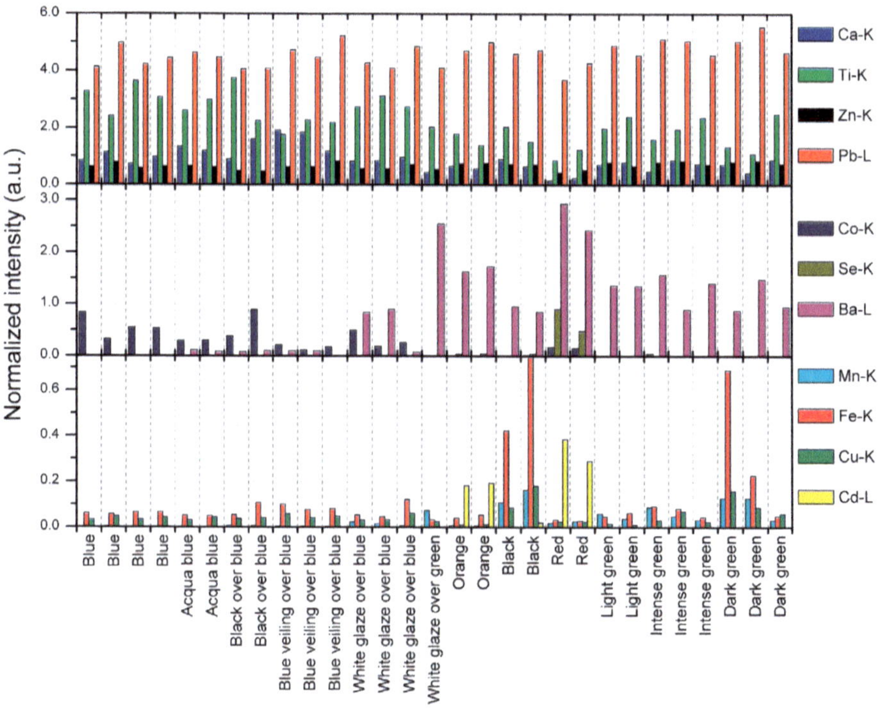

Figure 20.15 XRF results of *Muro azul* for various color regions.

to observe that Ba is related to red, orange and green colors but it has a lower intensity for the blue color, perhaps due to the addition of rich Ca aggregates in this color. On the other hand, Mn and Fe have higher intensities in the darker regions, suggesting that the shadows were achieved with pirolusite dark earth color. The presence of Al, Si and K in all the colored areas can be related to the aggregates from the acrylic paint to produce the rough surface.

20.3 Discussion and Conclusions

Siqueiros initiated a movement towards the use of modern industrial materials and tools with the idea that only with new means of production was a new modern art possible. José Gutiérrez created a laboratory to research and test new materials; he also advised Mexican artists to apply synthetic media and create new techniques. During the 1950s, Gutiérrez developed Politec® acrylic paintings and they were soon employed to create paintings that explored the creation of complex surface effects. Through the material study of Gunther Gerzso, it was possible to identify *Paisaje de Papantla* as one of the first Mexican paintings that employed acrylics, probably Politec®. Gerzso was interested in achieving optical effects to create a multiplicity of refractions, colors that have several dimensions due to a system of sanded painting layers, glazes and opaque brushstrokes, possibly mixing also the media, using acrylic paint as underpainting and oil with copal as finishing. Such a complex, multi-layered system makes difficult the precise characterization of each pigment and medium through non-invasive techniques.

Later, during the 1970s, Gerzso continued exploring the role of color and refraction, but instead of producing a smooth multi-layered and polished surface like *Paisaje de Papantla*, in *Muro azul* he used the texture of the canvas, added aggregates and played with well-defined compositions, structured with the golden section and deep dark profile. In both paintings he explored the emotional sensations and the experiences of Mexican pre-Hispanic ruins, with a very subtle understanding of color.

The use of a combined methodology of spectroscopic techniques, including FORS, Raman, FTIR and XRF with surface microscopy, was useful for determining the use of acrylic paintings and characterizing the grounds and fillers present in the paintings and also the aggregates added by the painter. Nevertheless, due to the complexity of Gerzso's technique, it was not always possible to identify precisely all the materials. The use of references was fundamental for this kind of study, although more research is needed on the chemistry of Politec® paintings and on the so-called "mixed media": the use of commercial oil paints over acrylic underpaintings.

Acknowledgements

This research has been supported by grants from CONACYT LANCIC LN 279749, LN 293904, LN 299076 and CB 239609, as well as PAPIIT UNAM IN402417 and IN112018. We thank James Oles and Julio García Murillo of the Centro Cultural Universitario Tlatelolco for the opportunity of doing this research and the Galería de Arte Mexicano, especially Mariana Pérez Amor and Alejandra Yturbe, for the support in studying Gunther Gerzso's paintings. Also to the UNAM Program of Doctoral and Postdoctoral Grants and to Clara Bargellini, advisory at IIE UNAM.

References

1. D. Alfaro Siqueiros, Los vehículos de la pintura dialéctico subversiva, conference presented at the John Reed Club, California, 2 September 1932, Published in *Palabras de Siqueiros*, ed. R. Tibol, Fondo de Cultura Económica, Mexico, 1996, pp. 62–80.
2. D. Alfaro Siqueiros, Qué es Ejericio plástico y cómo fue realizado, pamphlet published in Buenos Aires, December, 1932, Published in *Palabras de Siqueiros*, ed. R. Tibol, Fondo de Cultura Económica, México, 1996, pp. 103–110.
3. L. P. Hurlburt, *Art J.*, 1976, **35**(3), 237.
4. J. Harten, *Siqueiros/Pollock, Pollock/Siqueiros*, Dumont Verlag Kunsthalle Düsseldorf, Düsseldorf, 1995.

5. E. Arroyo, A. Aviram, M. A. Fernández Delgado, R. González Mello, A. Juárez, C. Mcglinchey and S. Zetina, *Low Viscosity: The Birth of Fascism and Other Solutions*, Instituto de Investigaciones Estéticas, Universidad Nacional Autónoma de México, Mexico, 2013.
6. S. Zetina Ocaña, MA thesis, Universidad Nacional Autónoma de México, Mexico, 2012.
7. J. L. A. Marontate, PhD thesis, Université de Montréal, 1996.
8. J. B. Lynch, *Américas*, 1970, **22**(7), 35.
9. Letter from M. Jiménez Rueda to the Head of the Administrative Department of the Instituto Politécnico Nacional, 11 March 1946, Sala de Arte Público Siqueiros Archive, folder 5-3-201, doc. 1.
10. J. L. Gutiérrez and M. Jiménez Rueda, *Taller de Ensaye de Materiales de Pintura y Plásticos. Boletín de Información*, undated brochure printed by IPN (circa 1946–1947), Sala de Arte Público Siqueiros Archive, folder 5-3-201, doc. 2.
11. E. Arroyo, A. Aviram, C. McGlinchey and S. Zetina, David Alfaro Siqueiros and the domain of industrial materials 1931–1945, in *Low Viscosity: The Birth of Fascism and Other Solutions*, ed. E. Arroyo *et al.*, Instituto de Investigaciones Estéticas, Universidad Nacional Autónoma de México, Mexico, 2013.
12. J. L. Gutiérrez, *From Fresco to Plastics: New Materials for Easel and Mural Paintings*, National Gallery of Canada, Ottawa, 1959.
13. J. López Gutiérrez and F. Pego Moscoso, Letter to Eng. L. Aguilar Álvarez, Technical Sub director of the Instituto Politécnico Nacional, Undated document, Sala de Arte Público Siqueiros Archive, folder 5-3-201, doc. 2.
14. Memoria del Subinstituto de Ensaye de Materiales de Pintura Mural, Decorativa e Industrial, dependiente del IPN para su publicación en los Anales del Instituto Politécnico Nacional, Mexico, Sala de Arte Publico Siqueiros Archive.
15. J. Gutiérrez and N. Roukes, *Painting with Acrylics*, Watson-Guptill, New York, 1966.
16. J. L. Gutiérrez, Techniques for mural and easel painting with instructions for the application and use of Politec acrylic artist colors, extract from *De la pintura rupestre a la contemporánea*, translated and distributed by The Politec Company, San Francisco California, undated printed document, Fondo José Gutiérrez, IIE Archive, UNAM.
17. A. Mejía-González, S. Zetina, M. Espinosa-Pesqueira and N. Esturau-Escofet, *Int. J. Polym. Anal. Charact.*, 2017, **22**, 473.
18. R. Eder, *Defying Stability, Artistic Processes in Mexico, 1952–1967*, Universidad Nacional Autonoma de Mexico and Turner, Mexico, 2014.
19. D. A. Siqueiros, *No hay mas ruta que la nuestra: Importancia nacional e internacional de la pintura mexicana moderna*, Talleres Gráficos de la Secretaría de Educación Pública, Mexico, 1945.
20. R. Eder, *Gunther Gerzo, El esplendor de la muralla*, CONACULTA, México, 1994, pp. 11–14.
21. J. L. Ruvalcaba Sil, D. Ramirez Miranda, V. Aguilar Melo and F. Picazo, *X-Ray Spectrom.*, 2010, **39**, 338.

22. A. Solé, E. Papillon, M. Cotte, Ph. Walter and J. Susini, *Spectrochim. Acta, Part B*, 2007, **62**, 63–68.
23. P. Ricciardi, J. K. Delaney, M. Facini, J. G. Zeibel, M. Picollo, S. Lomax and M. Loew, *Angew. Chem., Int. Ed.*, 2012, **51**, 5607.
24. T. J. S. Learner, *Analysis of Modern Paints*, The Getty Conservation Institute, Los Angeles, 2005.
25. J. K. Delaney, E. Walmsley, B. H. Berrie and C. F. Fletcher, Multi-Spectral Imaging of Paintings in the Infrared to Detect and Map Blue Pigments, in National Academy of Sciences, *Scientific Examination of Art: Modern Techniques in Conservation and Analysis*, The National Academies Press, Washington, DC, 2005, pp. 120–136, DOI: 10.17226/11413.
26. J. Dong, Y. Ozaki and K. Nakashima, *Macromolecules*, 1997, **30**, 1111.
27. W. R. Feairheller and J. E. Katon, *Spectrochim. Acta, Part A*, 1967, **23A**, 2225.
28. B. Martin, J. Puentes, L. Wruck and T. A. Osswald, *Polym. Eng. Sci.*, 2017, **58**, 228.
29. I. Rosales-Candelas, J. J. Soto-Bernal, R. Gonzalez-Mota and C. Frausto-Reyes, *Opt. Express*, 2013, **21**, 25544.
30. J. Zieba-Palus and R. Borusiewicz, *J. Mol. Struct.*, 2006, **792–793**, 286–292.

CHAPTER 21

Industrial Design Objects in the Museum Environment

L. TONIOLO*[a] AND A. NEVIN[b]

[a]Politecnico di Milano, Department of Chemistry, Materials and Chemical Engineering "Giulio Natta", Piazza Leonardo da Vinci 32, 20133, Milano, Italy; [b]Consiglio Nazionale delle Ricerche, Institute for Photonics and Nanotechnologies (IFN-CNR), Piazza Leonardo da Vinci 32, 20133, Milano, Italy
*E-mail: lucia.toniolo@polimi.it

21.1 Italian Industrial Design

Writing on the relationship between art and design, Gillo Dorfles cites the phenomenon of industrial design, recognising that "The concept of art and design will become increasingly interchangeable: it will be assigned an aesthetic value in some technological and scientific sectors, several artistic movements will decline, and a type of artisanal manufacturing will gain importance and value."[1]

Design developed in Italy, and particularly in Milan, starting shortly after the Second World War; it was not only functional, but grew as a way of expression with a strong symbolic dimension. The "function" of objects became progressively less significant.[1] Industrial design was born from the solid relationship among important architects who had been educated in schools of architecture, the productive culture of the Milan bourgeoisie and a class of inspired entrepreneurs. Design objects were able to unite both the creative process typical of artistic invention and technological advances that are necessary for industrial, serial production.[2]

Science and Art: The Contemporary Painted Surface
Edited by Antonio Sgamellotti, Brunetto Giovanni Brunetti and Costanza Miliani

Published by the Royal Society of Chemistry, www.rsc.org

Important Italian architects including Piero Portaluppi, Gio Ponti and Franco Albini designed furniture and small objects for the interiors of their buildings well before the Second World War.[1,2] In the two decades after the war other important groups of architects that included the Castiglioni brothers, Ignazio Gardella, Enzo Mari, Vittoriano Viganò, Marco Zanuso and the young designer Vico Magistretti[2,3] developed a new concept of design that was intended to become part of reconstructed life and new homes: the beginnings of Italian design in the 1950s were seen as a "new renaissance". During the 1950s and 1960s the idea of creating spaces and design extended into all aspects of everyday life – home interiors and domestic appliances, transportation and automobiles, working space – gaining popularity and accessibility through high-quality serial production.[3]

Plastic materials were engineered for industrial production during the first half of the 20th century, starting from the early semi-synthetic plastics, like cellulose acetate and nitrate, and later the synthetic polymers like Bakelite, a phenol-formaldehyde polymer that became the standard material for household goods.[4,5] In the second half of the 20th century, the novel formulation of polymeric materials and the introduction of additives like plasticizers and stabilizers paved the way for the production of new and versatile objects in important materials such as polyvinyl-chloride, polyethylene, polystyrene and acrylonitrile butadiene styrene (ABS).[5,6]

21.2 Italian Design Icons From the 1960s

Many objects that emerged from the Italian design movement in the 1960s have become iconic. They were widely adopted as part of everyday life in Western civilization, and they represent the lifestyle, the habits and the attitude of the distinct and sophisticated manufacturing that defines high-quality European design. Indeed Italian products designed during the 1960s (and early 1970s) remain in our collective imagination today as objects that have essentially redefined the concept and appearance of traditional things: from chairs, lamps and telephones to coffee machines, pots, bowls and cups.[3,7] In the same period, many appliances and office machines, traditionally in metal housings, such as radios, tape-recorders, cameras and televisions, were transformed into small portable products with simple shapes. This significant revolution was fostered through the adoption of specifically engineered polymeric materials. For example, thermoplastic resins were used for the emerging production technology of injection molding and extrusion, yielding a new and wider scope for design, in different shapes, colours and patterns.

In 1949 the visionary young chemist Giulio Castelli founded the company Kartell, that focused on plastic materials and their formulation and serial production with injection molding, and on strong collaboration with a team of well-known architects. In 1963 Kartell opened the Habitat division that produced plastic furniture. Among the most iconic objects to emerge was the "Componibili" system in ABS designed by Anna Castelli Ferrieri (Kartell, 1967) that entered homes worldwide (Figure 21.1). Other extraordinary objects from 1960s include: the "4006 lamp" by Achille and Piergiacomo Castiglioni (Kartell, 1959); the "Selene" chair by Vico Magistretti (Artemide, 1961); the "Acrylic Lamp" by Joe Colombo (Oluce, 1962); the "Perpetual Calendar" by Enzo Mari (Danese, 1963); the "Cube" radio by Marco Zanuso (Brionvega, 1963); and the "Valentine" typewriter by Ettore Sottsass (Olivetti, 1968). These iconic objects in plastic have both a distinct shape and functionality, and have entered important private and public collections because of their intrinsic cultural value.

The conservation of design objects within collections poses distinct challenges due to the deterioration of polymeric materials, which are relatively

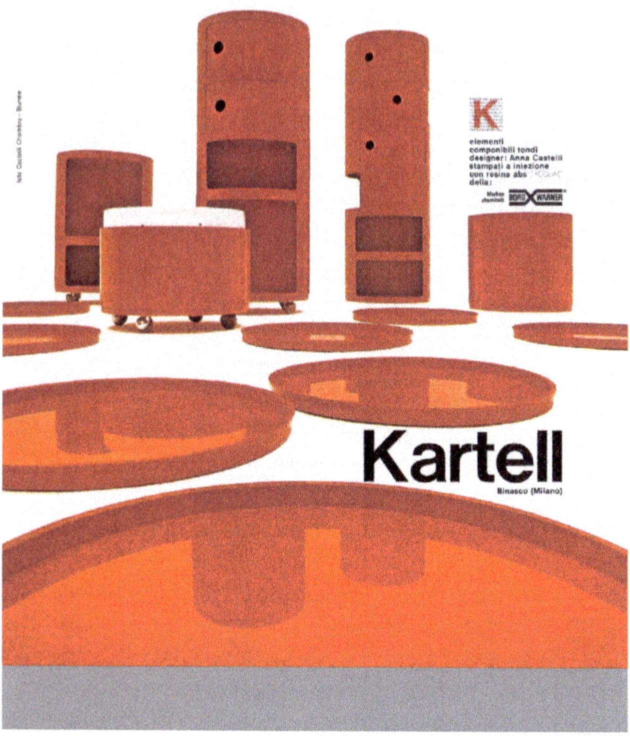

Figure 21.1 The iconic "Componibili" round shape, designed by Anna Castelli Ferrieri, 1967. In 2017, 50 years of continuous production was celebrated by the Kartell company. Kartell Museum, Binasco, Italy. Reproduced with permission.

sensitive compared to many other classes of materials including ceramics and many metals. In uncontrolled conditions (unfiltered lighting, heat and relative humidity) synthetic polymers may rapidly alter their main chemical and mechanical properties, and may change in appearance.[5,8] At the same time, limited resources are dedicated for research to preserve plastics in a changing world that is becoming increasingly aware of the need to mitigate and address the persistence of plastic residues in the environment (soils, water and air).

21.3 A General Overview of the Deterioration of Plastic

The environment, here understood as the combination of relative humidity, temperature and light, has a major influence on the stability of plastics. Exposure to ultraviolet (UV) radiation can lead to rapid deterioration of plastics and therefore suitable lighting, without UV emission, and appropriate UV-filters on windows are standard. More generally, within any collection, the conservation, storage and display of plastic objects needs to be considered in the context of their stability and may imply less stringent environmental control than may be recommended for natural organic materials, but more costly storage and long-term monitoring due to their inherent instability. Important publications and research projects have highlighted these issues in great depth, and broad classes of plastics and plastic-based objects have been reported with specific conservation dilemmas.[5,8,9] Dedicated research focusing on design object collections is scarcer, and has sparked interest in curators and registrars only recently.

In general, plastics show an induction period following production, during which no physical and chemical changes are apparent and no exterior signs of deterioration are perceived but during which chemical modifications may occur slowly. After the induction period chemical changes can be very rapid and degradation processes can lead to severe physical and mechanical damage, such as crumbling, powdering and collapse. Sometimes plastic objects in collections that are not deemed at risk exhibit sudden important deterioration.[5]

Photo-oxidative degradation is the process of decomposition of polymeric materials which is considered as one of the primary sources of damage of plastic collections in the indoor environment. The time over which significant deterioration processes can occur is relatively short and often curators are obliged to change or replace items in displayed collections. The duration and conservation of design objects in modern collections is now a major concern and the debate about preventive conservation and possible restoration is ongoing.[8–11]

While the causes of deterioration of design objects containing plastic depend chiefly on the chemistry and stability of the materials used to fabricate them, the preservation of multi-material plastic-based collections requires careful decisions regarding suitable environments for display cases or open spaces, the length of exhibitions and the type of storage. Analogies can be drawn between the preservation of plastics and design objects with other areas of conservation – for example that of photographic collections, where the type of photograph will determine the storage and display requirements; the well-recognized instability of cellulose acetate and cellulose nitrate negatives and prints and the instability of dyes requires careful balance between display, often in low lighting conditions, and long-term cold storage.[12,13]

The mechanism of the degradation and oxidation reactions (Figure 21.2) occurring in polymeric materials exposed to UV irradiation is initiated by some extraneous groups and/or impurities in the polymer that absorb UV photons and form excited states in radical species that can react by the extraction of hydrogen radicals from the polymer. Direct photolysis of C–C and C–H bonds is also possible as an initiation phase that leads to the production of polymeric radicals. The propagation and termination phases change depending on the functional groups and characteristic chemistry of the different polymers.[14]

The thermal degradation of polymers consists of two distinct reactions, which can occur simultaneously. One is a random scission of bonds causing a molecular weight reduction of the polymer, and the other is a chain-end scission of C–C bonds, generating volatile products such as monomers. In both cases it accounts for a depolymerisation process that affect severely the mechanical properties of the material. Generally, this kind of deterioration requires heat absorption and therefore high temperatures. Although polymers are strong insulating materials, heat absorption due to light irradiation (in case of solar light exposure) can play an important role in determining depolymerisation effect. Thermal degradation is particularly effective for

Initiation

$$I^{\bullet} + R\text{-}H \longrightarrow R^{\bullet} + I\text{-}H$$

Propagation

$$R^{\bullet} + O_2 \longrightarrow ROO^{\bullet}$$

$$ROO^{\bullet} + R\text{-}H \longrightarrow ROOH + R^{\bullet}$$

Termination

$$ROO^{\bullet} + ROO^{\bullet} \longrightarrow ROOR + O_2$$

$$RO^{\bullet} + R^{\bullet} \longrightarrow ROR$$

$$ROO^{\bullet} + R^{\bullet} \longrightarrow \text{stable products}$$

$$R^{\bullet} + R^{\bullet} \longrightarrow R\text{-}R$$

Figure 21.2 Summary of photo-oxidation reactions in polymers.

elastomers and low T_g thermoplastic polymers, constituting an important concern also in the museum and storage environment. Actually, temperature plays a major role in the stability of many plastic objects made of unstable polymers, and their degradation can be slowed only by lowering temperature. The use of cold storage for collection items made, for example, of nitrocellulose or cellulose acetate and polyvinyl-chloride, is the only approach to prolong their lifecycle.

One of the greatest challenges for polymer-based collections is the determination of the composition of materials present, and the degree to which they have already degraded or may continue to degrade in the future – a staggering necessity, since different materials require different storage or more or less stringent display.[15] Although they are industrial and recent artefacts, producers or manufacturers do not indicate precisely the materials they employ or use trademarks; serially produced design objects may have changed in terms of the materials and formulations (polymers, plasticizers and additives), hence composition should be checked to implement suitable conservation strategies. For this purpose a number of *in situ* (also known as non-invasive) techniques have been developed for the characterization of materials and their condition.[16,17] Molecular spectroscopies, for example Fourier Transform Infrared Spectroscopy (FTIR) in either Attenuated Total Reflectance (ATR) or Reflectance or Near Infrared Spectroscopy may be used to classify objects based on their chemical composition, and hence susceptibility, as part of a survey or monitoring.[17–19] Portable FTIR and Raman instruments are the basic equipment which can be used to survey polymeric collections for the classification of plastic materials and can complement data acquired with multi-spectral imaging, fluorescence imaging and Fibre Optic Reflectance Spectroscopy (FORS).[20,21] Spectrofluorimetry is also useful for the monitoring of changes in plastics which occur as a result of the migration of plasticizers or photo-oxidation.[21]

An issue commonly found with plastic-based works is that their surface is susceptible to scratching and wear, and may become tacky with ageing. Hence dust accumulation may become problematic and in some cases there are few viable cleaning solutions; therefore, limiting dust (through display in cases) is the best preventive strategy.[22–24]

The bulk or original composition is not the only relevant information required when making decisions about conservation and the environment. For example, to overcome the inherent susceptibility of polymers to oxidation, producers of plastics usually add suitable stabilizers to polymers in the formulation. While many additives to plastics are stable, others, including organic dyes and pigments, may fade following photo-oxidation. It is hence important to identify which materials are light sensitive, as well as to measure the colour of objects and to quantify colour changes over time as part of long-term monitoring. Approaches based on FORS or colorimetry are both valid, but may be challenging for semi-transparent or highly reflective surfaces.[19] In parallel, light sensitivity can be assessed through the identification of additives, a significant challenge as colourants and additives may be present in trace concentrations and may be identified non-invasively with Raman spectroscopy.[25] Similarly, if the materials used in an object have already degraded significantly, as will be illustrated in case studies in Section 21.4.2. For polyvinylacetate (PVAc), it is important to establish if the object is stable and whether or not degradation is on-going, if the object can be displayed, or – if there is active deterioration and the object continues to degrade, emitting for example acetic acid in the case of acetate-based materials – how the object can be stored without compromising the stability of other objects.

21.4 Case Studies: The Deterioration of ABS and PVAc in Design Icons of the 1960s

21.4.1 Mechanistic Investigations of the Photo-oxidation of ABS

ABS is a copolymer made of an acrylonitrile–styrene continuous phase (SAN) and partially grafted polybutadiene (PB), the mechanical properties of which can be fine-tuned by varying grafting conditions and the proportions of components in the polymer.

For over 50 years ABS has been widely used in industrial applications due to its excellent mechanical, thermal, chemical and electrical properties that can be modified by changing the grafting conditions and amount of monomer.[5,6] Nevertheless, ABS is susceptible to photo-oxidative degradation. This instability often leads to the alteration of its physical properties, chemical modifications such as crosslinking and chain scission, colour changes and worsening of mechanical performance.

The behaviour of ABS has been extensively studied in mild conditions, together with the changes following the artificial ageing of the material after solar irradiation at temperatures lower than 45 °C.[26] The susceptibility of ABS to photo-oxidative degradation is due to the butadiene phase and, in particular, to residual double bonds along the polymer chains, and photo-oxidation leads to molecular modifications that are localized preferentially on the material surface.[27] This degradation involves an autocatalytic free radical chain mechanism where the allylic radicals react with oxygen to form hydroperoxide products. Labile oxygen–oxygen bonds break to form RO· radicals that are precursors of typical oxidation products such as aldehydes, esters, carboxylic acids and ketones.[28] The penetration of radiation and oxygen into ABS will directly impact the depth of the degradation.[27]

The yellowing of ABS exposed to UV-A irradiation in the presence of oxygen is a consequence of surface photo-oxidation and it is particularly evident after visual inspection for white plastic samples.[29] In Table 21.1 the change of colour coordinates after UV ageing is shown: b increases dramatically (Δb +22) after 1000 h of exposure to solar irradiation. Radical species formed during oxidation of the PB component react with PS in the SAN macro-phase, leading to the formation of many coloured and highly absorbing products.[30]

One way of mapping molecular changes is through the use of micro-FTIR that is sensitive to modifications in ABS. The assignments of vibrations detected by FTIR spectroscopy, attributed to different functional groups in the ABS reference, are summarized in Table 21.2.

Bands from FTIR spectra are useful for following oxidation in samples of ABS and demonstrate that oxidation products develop in the top hundred micrometers in samples that can be followed by monitoring: the CN

Table 21.1 Colour changes represented as L^*, a^* and b^* values following photo-oxidation of ABS samples to accelerated ageing at 0, 100, 500 and 1000 hours of light.

Sample	L^*	a^*	b^*
ABS 0 h	85.6	−3.2	8.4
ABS 100 h	83.7	−3.8	17.5
ABS 500 h	79.1	−0.5	28.4
ABS 1000 h	74.5	1.8	30.5

Table 21.2 Assignment of the main bands' Fourier Transform Infrared Spectrum of ABS ascribed to different components in the blend Acrylonitrile (AN), Styrene (S), Polybutadiene (PB).

Band/cm^{-1}	Attribution	Intensity
3060	ν CH aromatic (S)	Weak
3025	ν CH aromatic (S)	Medium
2920	ν_a CH_2 (PB)	Strong
2850	ν_s CH_2 (PB)	Strong
2237	ν C≡N(AN)	Weak
1600	Aromatic ring skeletal vibration (S)	Weak
1495	ν_s ring skeleton in-plane (S)	Medium
1450	ν_s ring skeleton in-plane (S)	Medium
966	Trans 1.4 vinyl unit (PB)	Strong
911	1.2 vinyl unit (PB)	Medium
760	Ring out of plane deformation (S)	Medium
700	Ring out of plane deformation (S)	Strong

stretching at 2237 cm^{-1} for acrylonitrile; the stretching mode at 1600 cm^{-1} for styrene; the stretching mode of CO in carbonyl-containing groups around 1730 cm^{-1} (Table 21.3).[31–34]

In samples of ABS there is a limit in the depth of maximum oxidation corresponding to 70 μm regardless of the time of exposure to light (Figure 21.3). This strengthens the hypothesis that the layer of oxidized material acts as a protective shield against further photo-oxidative degradation processes (passivation) preventing oxygen diffusion in the material.[31] However, crazing and physical deterioration of the surface plays an important role in the formation of the layer of oxidized material.

Table 21.3 Oxidation products which contributed to the band between 1719 and 1735 cm^{-1} in the first 192 μm below the exposed surface of ABS.

Band/cm^{-1}	Attribution
1699	α,β-unsaturated ketone
1710	Carboxylic acids
1720	α,β-unsaturated aldehyde
1720	Saturated ketone
1731	Saturated aldehyde
1739	Ester group
1775	Anhydride/*g*-lactone/peracids

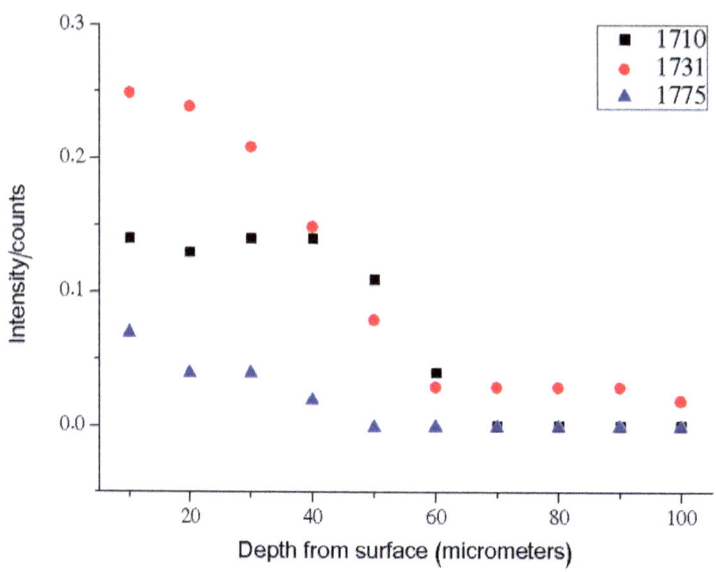

Figure 21.3 Gaussian fit of the intensity of three principal bands at 1710, 1731 and 1775 cm^{-1} in artificially aged ABS as a function of depth from the external irradiated surface.

21.4.2 An Iconic Object in ABS: The Grillo Folding Telephone (1965, Zanuso and Sapper)

An iconic design ABS object, the white Grillo folding telephone, was analysed.[21] The object, designed in 1965 by Marco Zanuso and Richard Sapper and produced by SIT Siemens® with a revolutionary and modern shape, is a recognisable precursor to the folding mobile telephone. Other telephones were also produced by Siemens® in different vibrant colours. The white Grillo telephone is from a private collection and is no longer in use. The phone is made of four separate parts: the external encasing parts, the internal part of the telephone with the microphone, the dial and the speaker inside (Figure 21.4).

The Grillo telephone is strongly luminescent under UV illumination and increases in surface fluorescence have been related to additives and oxidation products which develop on the surface of objects.[26] *In situ* analysis using spectrofluorimetry and excitation emission spectra with a fibre optic probe can rapidly detect that different parts of the phone have different luminescent signals, ascribed both to variation in the formulation of the components made in ABS and variations in the development of chromophores from photo-oxidation on exposed and unexposed surfaces (Figure 21.5). It is also noted that different pieces have different fluorescence lifetime signals, as measured with time-resolved fluorescence imaging.[21]

FTIR spectra acquired from the phone confirm that the material is ABS. The differences with the ABS reference spectrum are related to the degree of

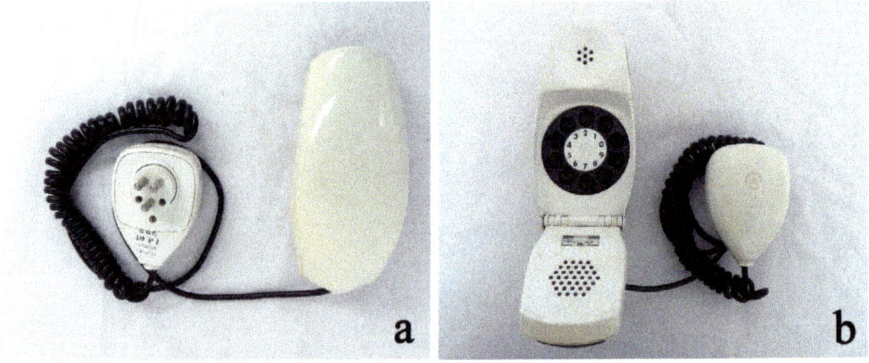

Figure 21.4 The Grillo telephone. Designed by Marco Zanuso and Richard Sapper, 1965. (a) Closed position, external yellowed encasing part; (b) open position, internal part with the microphone, the dial and the speaker. Private Collection. Reproduced with permission.

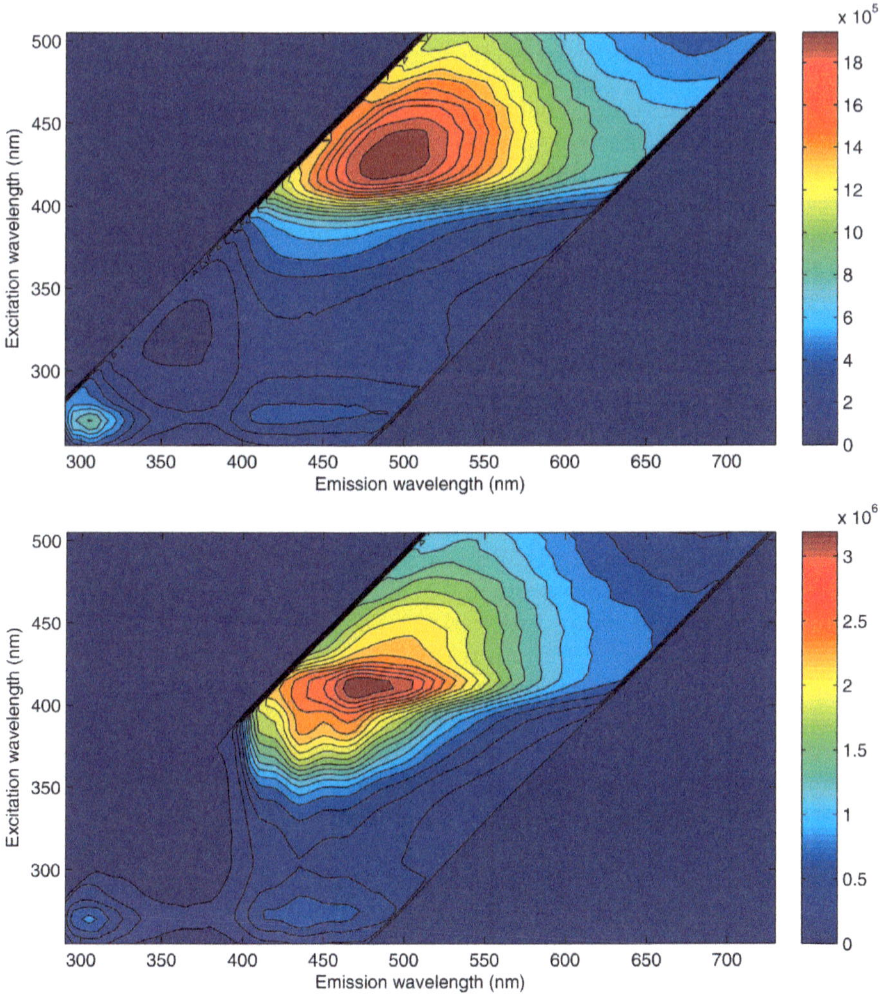

Figure 21.5 Fluorescence excitation emission spectra acquired from two different areas of the Grillo telephone, both in ABS, that highlight significant variation in the optical properties of different pieces of the phone, confirming that the example studied is a composite object that has been reassembled from different phones.

degradation of the polymer and to the presence of an additive (most probably a stabilizer) within the polymeric blend. Actually, two additional bands, one at 3300 cm^{-1} (assigned to a secondary amine) and a smaller one at 1565 cm^{-1} (assigned to a NH group) can be detected; the band at 1640 cm^{-1} is more intense than in reference ABS. Finally, a contribution from a carbonyl group, presumably an amide group, can be hypothesized, overlapping with the signal from the ABS vinyl group. These features can be ascribed to the presence of a stabilizer.

The photo-oxidative degradation of ABS in the Grillo phone is apparent by comparing FTIR spectra collected on samples taken from the exposed left and right sides of the ear piece and unexposed internal surface (Figure 21.6). The following changing can be observed: (a) decrease of the intensity of bands at 966 and 911 cm^{-1} related to PB; (b) formation of unsaturated aldehydes (1720 cm^{-1}) and anhydrides/lactones (1776 cm^{-1}); (c) increase of the broad 3400 cm^{-1}band due to hydroxyl group formation.

Photo-degradation and the accumulation of chromophores may compromise the surface cohesion and the appearance of the object and have significant implications for the display of design objects made in ABS in museum collections: photo-oxidation is indeed a significant problem for the conservation of historical objects which have not been stabilized with the effective UV stabilizers commonly used today, both in storage and on display. Indeed, the role of colourants and additives in ABS on the degradation of the material still requires further research. Methods for the monitoring of the degradation of design objects in ABS like Grillo are particularly useful, and have been based on an integration of results from point-like (FTIR) and imaging analyses.[21,35] Moreover, while practical solutions for the display of ABS objects without UV irradiation could limit the accumulation of photo-oxidation products, there are still no cleaning methods available to remove the yellowing effects, except for aggressive abrasion methods. Often curators choose to substitute the objects of the collection, and particularly those on display (serially produced objects may suffer particularly from substitution, which can be seen as an impediment for the development of adequate conservation strategies), rather than addressing risks from display conditions, sheltering the items from solar irradiation and heat sources. Preventive conservation in cool and dark environments is the only measure that can ensure a prolonged lifecycle for these kinds of collections.[11]

21.4.3 The Case of Cocoon® Resin: Plasticized Polyvinylacetate (PVAc) in the Fantasma Lamp (1959, Castiglioni Brothers)

We focus here on the iconic Italian lamp named “Fantasma”, designed by Piergiacomo and Achille Castiglioni in 1959 and produced by Flos® with a polymeric formulation known as “Cocoon®” (Figure 21.7).[36] Cocoon is a cobweb of white filaments of a resin which was extruded in threads and wrapped around a metallic structure forming the basis for the light diffuser; the same resin is sprayed all over this structure to form a continuous semi-transparent external surface. The effect is similar to that obtained in the past with parchment for light diffuser, with the advantage of creating free shape

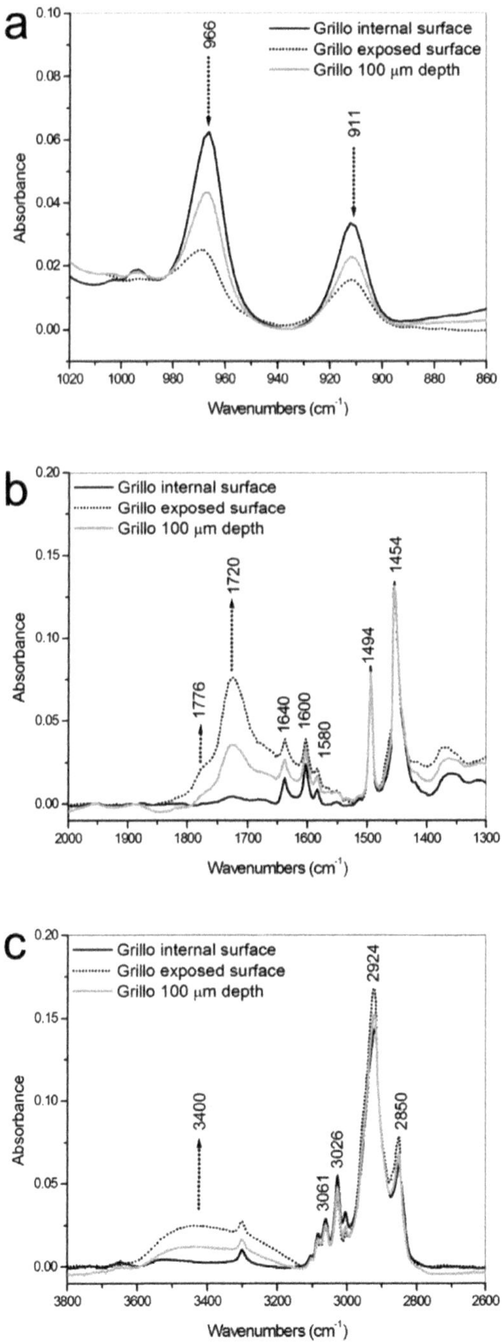

Figure 21.6 FTIR spectra acquired from samples taken from the Grillo telephone. Changes in intensity of bands in FTIR spectra acquired from differently deteriorated areas are observed in the finger print spectral region (a) and (b), and in the O–H and C–H stretching regions (c).

Figure 21.7 "Fantasma" lamp. Designed by Piergiacomo and Achille Castiglioni, 1959. Triennale Design Museum, Milan, Italy. Reproduced with permission.

and dimension. While the Cocoon® resin is still on the market, produced by Andek®, the chemical composition has changed significantly since the lamps were produced.

Polyvinyl-acetate (PVAc) is an unstable polymeric material that was nonetheless widely used in the 20th century, both as an adhesive and for the production of design objects. PVAc degrades *via* a well-known deacetylation autocatalytic process leading to the liberation of acetic acid and the formation of polyenes.[37,38] The degradation process is favoured by UV irradiation or heat absorption and may occur according to Norrish mechanisms either with (a) hydrogen abstraction by an acyl radical followed by aldehyde formation, or (b) an intramolecular hydrogen abstraction and the formation of a C=C double bond in the polymer backbone chain as a consequence of the loss of an acetic acid molecule. Furthermore, this process leads to the formation of ketone groups inside the main chain. The rate of loss of acetate side group was found to increase as degradation proceeds, so that an autocatalytic effect of the process should be inferred.

The typical deterioration patterns perceived on the objects are discoloration (mainly darkening and yellowing), and mechanical failure of the material that results in tears and embrittlement. Yellowed areas, deposition of particulate matter, micro-crazing and extended breakages are visible on the external and internal surfaces of the light diffuser. In Figures 21.8 and 21.9 micro-samples of Cocoon® fibres from the Fantasma lamp are yellowed and deteriorated.

In the lamp, the Cocoon® resin is composed of a PVAc matrix plasticized with a mixture of phthalate plasticizers (dibutyl-phthalate, DBP, or

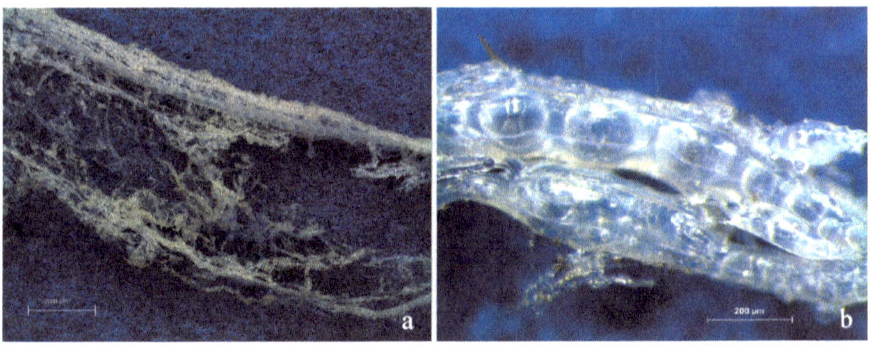

Figure 21.8 Images from the internal fibres at different magnification: a) extruded polymers grouped in bundles and b) magnification of a single fibre containing air bubbles.

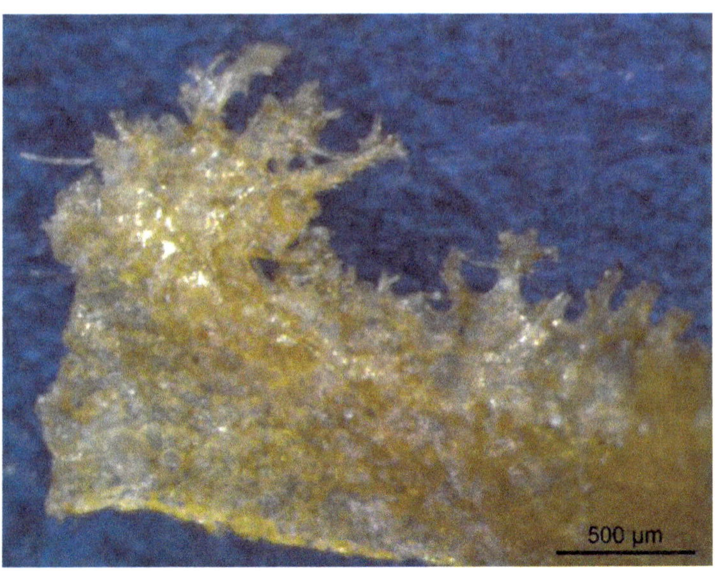

Figure 21.9 Magnification of a sample of Cocoon® resin: the widespread external yellowing of fibres is evident.

dioctyl-phthalate, DOP). The presence of plasticizer and the degradation phenomena in the polymer require a multi-analytical approach based on the integration of non-invasive analysis through spectrofluorimetry and micro-sampling of the objects, followed by analysis with Fourier Transform Infrared Spectroscopy (FTIR) and Pyrolysis–Gas Chromatography/Mass Spectrometry (Py-GC/MS).[35,39]

Spectrofluorimetry has been used to track the migration of phthalates in PVAc, and modifications in fluorescence are linked to oxidation of the Cocoon® formulation. In studies of thermal oxidation of PVAc emissions in the near-UV and blue region increase; this has been associated with the development of unsaturation following deacetylation.[34] Fluorescence excitation emission spectra acquired directly from the surface of Fantasma using a fibre optic probe map the presence of different fluorophores ascribed to additives, including plasticizers and optical brighteners. Spectra acquired can be compared with other lamps made of Cocoon® and highlight variations in the optical properties that can be linked to differences in the degree of oxidation of the polymer blends, the migration of plasticizers and the resulting surface composition. The broad band seen in Figure 21.10 at excitation at around 370 nm and emission between 430 and 510 nm in objects made of Cocoon® is associated with photo-oxidation and the accumulation of unsaturated and aromatic molecules which may also be responsible for the observed yellowing of the objects (Figure 21.9).[35]

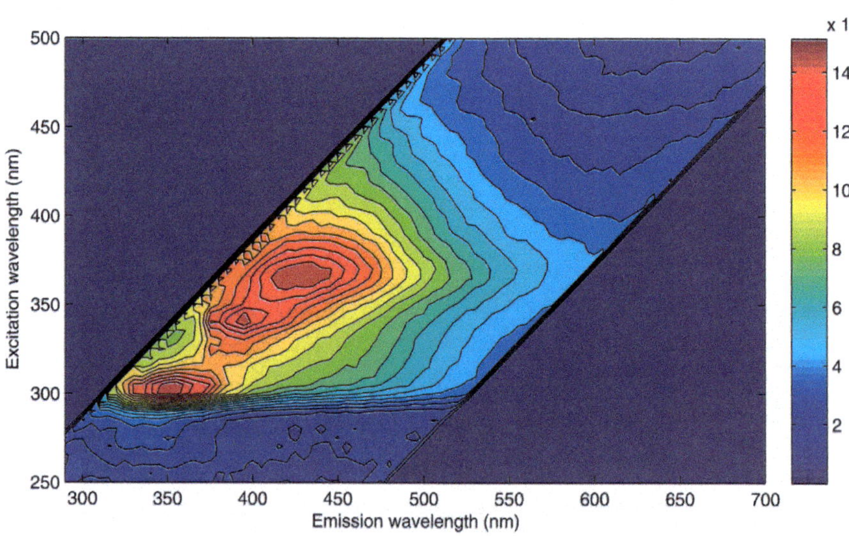

Figure 21.10 Fluorescence excitation emission spectrum acquired from the surface of Fantasma, using a fibre optic probe. The broad band excitation at 370 nm and emissions between 430 and 510 nm are ascribed to the accumulation of unsaturated and aromatic molecules.

Samples from Fantasma were extracted (solvent/non-solvent) to separate the polymer from additives and to characterize both separately with FTIR. The polymer was obtained by dissolving the sample in dichloromethane (solvent), precipitating it in hexane (non-solvent) and separating the precipitate by micro-filtration. The obtained solution was dried and the extract was analysed with FTIR to identify the other components of the blend (including plasticizer). Molecular spectra of samples allowed the detection of the PVAc and its deterioration, the presence of additives and the type of plasticizers.

In FTIR typical absorptions at 2961, 2917 and 2852 cm^{-1} (due to ν_a CH_3, ν_a CH_2 and ν_s CH_3, respectively), 1738 cm^{-1} (ν_s C=O ester), 1434 and 1372 cm^{-1} (due to ν_a and δ CH_3 groups associated with the acetate substituent, respectively), 1238 cm^{-1} (ν_a C–O–C) and 1024 cm^{-1} (ν_s CH–O) confirm PVAc as the polymeric matrix of both the internal fibres and the external layer (Table 21.4). Nevertheless, degradation phenomena of the polymer are evident for samples coming from the external coating of the light diffuser. The signals ascribed to the acetate side chain decrease (the ν_s C=O and the ν_a C–O–C at 1738 and 1238 cm^{-1} and ν_s CH–O at 1024 cm^{-1}), while, conversely, the OH stretching region (3355 cm^{-1}) and the C=C stretching signals (1640 cm^{-1}) show a decisive increase, in good agreement with published results assessing the ongoing deacetylation of the material.[36,37]

The FTIR spectrum of the extract showed absorptions at 2959 and 2858 cm^{-1} (due to ν_a CH_2 and ν_s CH_3, respectively), 1463 cm^{-1} (ν_a CH_3), 1730 cm^{-1} (ν_s C=O), 1288 cm^{-1} (ν_a C–O–C), 1072 cm^{-1} (ν_s CH–O), 1123 cm^{-1} (ν_s C–O–C), 1578 and 1601 cm^{-1} (aromatic ring skeletal vibration) that can be easily ascribed to a phthalate-based plasticizer. Given the high intensity of ν_a CH_2 absorption peak at 2959 cm^{-1}, it is possible to suggest the presence of a phthalate with a longer aliphatic substituent in the Cocoon® blend, probably a di-octyl phthalate (DOP). Actually, the correct identification of the plasticizer as

Table 21.4 The main functional groups and vibration bands in Fourier Transform Infrared Spectra in PVAc.

Band/cm−1	Attribution	Intensity
2980	ν_a CH_3	Weak
2931	ν_a CH_2	Weak
2870	ν_s CH_3	Weak
2860	ν_s CH_2	Weak
1740	ν_s C=O ester	Strong
1433	ν_a CH_3 group associated to the acetate substituent	Weak
1371	δ CH_3 group associated to the acetate substituent	Medium
1240	ν_a C–O–C	Strong
1130	ν_s C–O–C	Weak
1025	ν_s CH–O	Medium
606	δ C–CO	Weak

DOP was possible only by Py-GC/MS analysis. Specifically, dioctyl-phthalate, in addition to small amounts of dioctyl-adipate and dibutyl-phthalate, were identified as plasticizers.

The composition appears slightly different if the internal fibres and external coating are considered and reveals a heterogeneous plasticizer distribution between the two sides. This fact can be interpreted as a constituting original difference of the two types of Cocoon® material, given the different technological features necessary for extrusion polymeric fibres (internal fibres) and for spraying polymeric blend (external coating).

21.5 Conclusions

Industrial design objects deserve particular care in collections as they represent determinant cultural values of 20th century civilization. Unfortunately, a large number of these items are made of different plastics (formulations of polymeric materials and additives) that deteriorate rapidly in normal use, display and storage conditions. Even if design objects are serially produced in an industrial manner, the conservation of original objects is critical and curators and conservators have yet to develop adequate and standard protocols for their care. Photo-oxidation and thermal ageing are the main threats for plastic collections, but the impact of light and temperature changes with different polymeric substrates and global environmental conditions. Studies, such as those illustrated in this text, demonstrate that formulations, polymers and types of plasticizers and stabilizers are the factors that affect changes of chemical, physical and mechanical properties. Conservation scientists can help advance the field with innovative and non-invasive techniques to develop adequate monitoring tools for museum collections and it is necessary to foster collaborations to study real objects and gather experience on different polymeric materials. It is worth noting a growing interest in plastic heritage as part of focused scientific and conservation congresses, but open access literature and studies should be improved to build up a new scientific approach for the conservation of industrial design objects.

References

1. G. Dorfles, *Introduzione al Disegno Industriale*, Einaudi, Torino, 2001, p. 95.
2. E. Brigi, *Art e Dossier*, Giunti, Firenze, 2008, vol. 244, p. 7.
3. K. Fallan and G. Lees-Maffei, Introduction: The History of Italian Design, in *Made in Italy. Rethinking a Century of Italian Design*, ed. G. Lees-Maffei and K. Fallan, A&C Black, 2013, p. 1.

4. J. Morgan, *A Survey of Plastics in Historical Collections*, Plastics Historical Society and The Conservation Unit of the Museums & Galleries Commission, London, 1994.
5. Y. Shashoua, *Conservation of Plastics*, Routledge, London, 2012.
6. T. L. Richardson and E. Lokensgard, *Industrial Plastics: Theory and applications*, Cengage Learning, Boston, USA, 2004.
7. AA.VV, *Storie. Il Design Italiano*, Electa Mondadori, Milano, Italy, 2018.
8. *Preservation of Plastics*, 2010, Retrieved from The Getty Conservation Institute, http://www.getty.edu/conservation/our_projects/science/plastics/.
9. *IIC 2016 Saving the Now, Crossing Boundaries to Conserve Contemporary Works*, Conference Proceedings of IIC Los Angeles Congress, ed. A. Nevin, J. H. Townsend, J. K. Atkinson, D. Saunders and A. Brokerhoff, published in *Studies in Conservation*, 2016.
10. A. Hollander, Aging of plastics – what can we do about it?, in *Future Talks 009*, ed. T. Bechtold, Die Neue Sammlung, Munich, 2009.
11. Y. Shashoua, Inhibiting the inevitable; current approaches to slowing the deterioration of plastics, in *Macromolecular Symposia*, WILEY-VCH Verlag, Weinheim, 2006, pp. 67–77.
12. B. Lavédrine, S. Manod and J.-P. Gandolfo, *A Guide to the Preventive Conservation of Photograph Collections*, The Getty Conservation Institute, Los Angeles, California 2003.
13. Y. Shashoua, *Storage Strategies for Plastics*, Retrieved from Conservation Perspectives – GCI, 2014, Newsletter: http://www.getty.edu/conservation/publications_resources/newsletters/pdf/v29n1.pdf.
14. B. Singh and S. Nisha, *Polym. Degrad. Stab.*, 2008, **93**(3), 561.
15. D. Thickett and E. Richardson,*Preventive Conservation for Plastics on Open Display in Plastics. Looking at the Future and Learning from the Past*, ed. B. Keneghan and L. Egan, Archetype, London, 2008.
16. L. Toniolo, A. Dell'Acqua Bellavitis, S. Annicchiarico, D. Comelli, G. Galleani, M. Levi, A. Nevin and F. Toja, *Caring for Design Objects: Daily Practice and Advanced Analytical Tools in Future Talks 009*, ed. T. Bechtold, Die Neue Sammlung, Munich, 2009.
17. F. Rosi, C. Miliani, C. Clementi, K. Kahrim, F. Presciutti, M. Vagnini, V. Manuali, A. Daveri, L. Cartechini, B. G. Brunetti and A. Sgamellotti, *Appl. Phys. A*, 2010, **100**, 613.
18. E. Richardson, M. Graham and P. Wyeth, On-site collections management: NIR characterisation and condition monitoring of modern textiles, in *Conservation Science 2007: Papers from the Conference Held in Milan, Italy, 10–11 May 2007*, ed. J. H. Townsend, L. Toniolo and F. Cappitelli, Archetype Publications Ltd, London, United Kingdom, 2008.

19. C. Cucci, L. Bigazzi and M. Picollo, *J. Cult. Herit.*, 2013, **14**, 290.
20. D. Saviello, L. Toniolo, S. Goidanich and F. Casadio, *Microchem. J.*, 2016, **124**, 868.
21. D. Comelli, F. Toja, C. D'Andrea, L. Toniolo, G. Valentini, M. Lazzari and A. Nevin, *Polym. Degrad. Stab.*, 2014, **107**, 356.
22. B. Lavedrine, A. Fournier and G. Martin, *Preservation of Plastic Objects in Museum Collections*, Comité Des Travaux Historiques Et Scientifiques, Paris, 2012.
23. Y. Shashoua, K. Segel, T. van Oosten, A. Laganá, B. Keneghan, G. Barabant, C. Bollard and S. Kuperholc, in *ICOM-CC 16th Triennial Conference Lisbon 19–23 September 2011: Preprints*, ed J. Bridgland, ICOM Committee for Conservation (Corporate Author), Critério–Produção Grafica, Lda., Lisbon, Portugal, 2011.
24. T. Bechtold, Challenges related to plastics degradation in The Design Museum, in *Il futuro del contemporaneo: conservazione e restauro del design (The future of the contemporary: conservation and restoration of design.)*, ed. G. Cassese, Gangemi Editore, Rome, Italy, 2016, p. 146.
25. E. M. Angelin, S. Babo, J. L. Ferreira and M. J. Melo, *J. Raman Spectrosc.*, 2019, 232–241.
26. D. Saviello, E. Pouyet, L. Toniolo, M. Cotte and A. Nevin, *Anal. Chim. Acta*, 2014, **843**, 59.
27. R. M. Santos, G. L. Botelho and A. V. Machado, *J. Appl. Polym. Sci.*, 2010, **116**(4), 2005–2014.
28. D. Saviello, L. Andena, D. Gastaldi, L. Toniolo and S. Goidanich, *J. Appl. Polym. Sci.*, 2018, **135**(17), 46194.
29. M. Piton and A. Rivaton, *Polym. Degrad. Stab.*, 1997, **55**(2), 147.
30. X. Jouan and J. L. Gardette, *Polym. Degrad. Stab.*, 1992, **36**(1), 91.
31. J. G. Bokria and S. Schlick, *Polymer*, 2002, **43**(11), 3239.
32. D. J. Nagle, J. Dylan, M. Celina, L. Rintoul and P. M. Fredericks, *Polym. Degrad. Stab.*, 2007, **92**(8), 1446.
33. R. A. Nyquist, C. L. Putzig, M. A. Leugers, R. D. McLachlan and B. Thill, *Appl. Spectrosc.*, 1992, **46**(6), 981.
34. F. Toja, A. Nevin, D. Comelli and L. Toniolo, Novel Application of molecular spectroscopy and imaging for the analysis of design objects in ABS, in *Future Talks 2011*, ed. T. Bechtold, Die Neue Sammlung, Munich, 2011.
35. F. Toja, D. Saviello, A. Nevin, D. Comelli, M. Lazzari, M. Levi and L. Toniolo, *Polym. Degrad. Stab.*, 2012, **97**(11), 2441.
36. F. Toja, A. Nevin, D. Comelli, M. Levi, R. Cubeddu and L. Toniolo, *Anal. Bioanal. Chem.*, 2011, **399**(9), 2977.

37. I. C. McNeill, S. Ahmed and L. Memetea, *Polym. Degrad. Stab.*, 1995, **47**(3), 423.
38. M. F. Silva, M. T. Doménech-Carbò, L. Fuster-Lopéz, S. Martìn-Rey and M. F. Mecklenburg, *J. Anal. Appl. Pyrolysis*, 2009, **85**, 487.
39. M. Doménech-Carbó, G. Bitossi, L. Osete-Cortina, J. de la Cruz-Cañizares and D. J. Yusá-Marco, *Anal. Bioanal. Chem.*, 2008, **391**(4), 1371.

CHAPTER 22

From Context to Content: On the Preservation of Net-based Art

DOMENICO QUARANTA*

Accademia di Belle Arti di Carrara, Italy
*E-mail: quaranta@accademiacarrara.it

22.1 Form Art

Form Art is a net-based artwork made by Russian artist Alexei Shulgin during a residency at C3 in Budapest in May 1997 (Figure 22.1). Today, art critics would mostly agree in defining it as a piece of Net Art, or Internet Art. Back in those days, these labels didn't even exist. Alexei Shulgin self-identified using the label "net.art", first seen in the title of an exhibition curated by German artist and activist Pit Schultz in Berlin in 1995, and then mostly adopted and discussed by a small group of artists on *Nettime*, a mailing list co-founded by Schultz.[1] The label appealed to the group because it sounded like a joke on most artistic labels, identifying the "net" as its medium and context, while simultaneously turning art into a file extension, or a top-level domain.[†] This irony was important for the group that reconnected to the anti-art tradition – from the Avantgardes to Situationism to Neoism – and found in the Internet a platform where you could make artworks, discuss with friends and have an audience far from an hyper-formalized and hyper-commodified art world.

[†]Out of a joke, the .art domain became a reality in 2016, *e.g.* https://art.art/ (accessed January 2019).

Science and Art: The Contemporary Painted Surface
Edited by Antonio Sgamellotti, Brunetto Giovanni Brunetti and Costanza Miliani

Published by the Royal Society of Chemistry, www.rsc.org

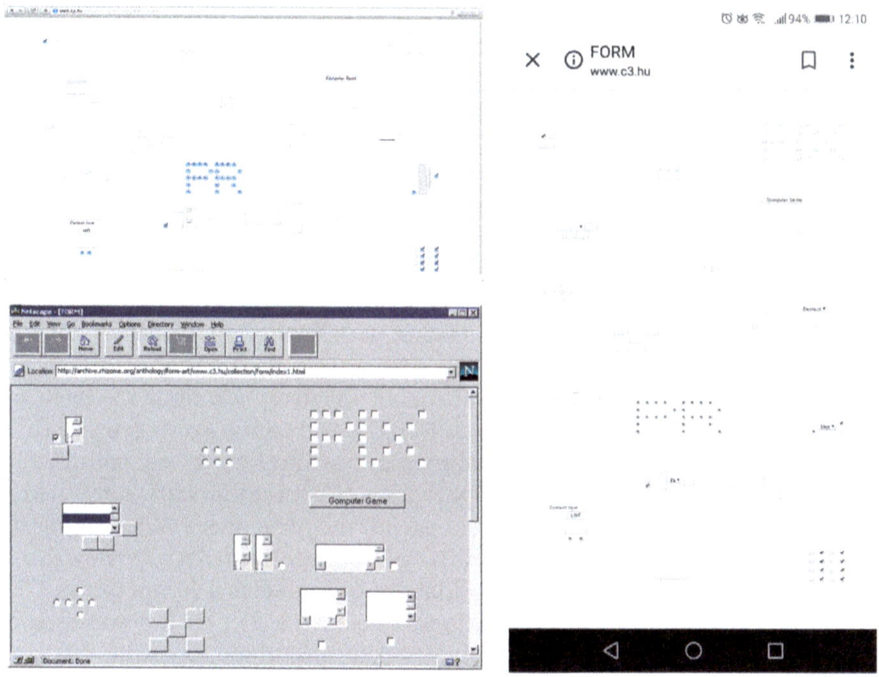

Figure 22.1 Alexei Shulgin, *Form Art*, 1997. Website, screenshots. Clockwise from top left: *Form Art* on Firefox® on a MacBook Air® with OS X (version 10.9.4); *Form Art* on Chrome® on a Huawei® P8 Lite 2017 with Android™ 8; *Form Art* emulated on remote browser, *via Rhizome*®. Reproduced with permission from Alexei Shulgin.

In pure Dadaist fashion, Alexei Shulgin contributed to shaping the myth of net.art as the last Avantgarde of the Twentieth Century, by writing an email to *Nettime* in which he explained that the term "net.art" was originated by a piece of malfunctioning software and "discovered" in a badly converted email by Slovenian artist Vuk Ćosić, who started to use it for his own work.[2] The story became viral, and it's still often reported to explain the origin of the label.

This beautiful, early example of Russian fake news is, to all effects, another manifestation of Shulgin's artistic practice. As a newborn platform where art did not exist yet, the web was essentially, for him and for most net.artists, a place where the entire game of art could be reinvented; as a disembodied communication platform where you are what you claim to be, the web was a place where you could play with identities and rules, define your practice and the criteria for evaluating it, write your own story, be the artist, the art critic, the curator, the institution, the gallery, all at once.[3]

Form Art is a good example of this way of approaching the web as a platform, and as a medium. It is, first of all, an artwork, a series of web pages made by the artist with the help of a Javascript® programmer. The piece uses

HTML forms – checkboxes, buttons, text boxes, menus, functional elements conceived to allow web users to enter data that is sent to a server for processing – depriving them of their functionality, and using them as the only visual elements available to design a web page. A basic feature of HTML forms is that they are simple textual tags in the HTML code, that are rendered graphically by your web browser of choice with a visual style which is embedded in the interface. For Shulgin, this meant HTML forms embodied a new, web-native, readymade aesthetics in the purest way. HTML forms allowed him to focus on "the unique and proper area of competence" of net.art, that is on what "was unique to the nature of its medium", to say it in Clement Greenberg's words.[4] He was making formalist art with forms, and he called it Form Art.

This reference to Modernist formalism is a way to reply, tongue-in-cheek, to a remark often made by critics of early Net Art, and Media Art in general: that its focus on the medium is, in post-media, postmodern times, completely outdated, and sets it outside of the paradigm of contemporary art. It's also a fundamental element, when you want to play the Avantgarde. With *Form Art*, Shulgin does not only make an artwork: he proposes Form Art as a "new form of artistic expression", and to consolidate it, he launches a competition. The announcement of the First International Form Art Competition was spread online in June 1997; as the only Form Art expert, Shulgin appointed himself as the only jury member for the competition, and after evaluating the submissions, he awarded two prizes and an honorary mention. The ceremony was celebrated in September at the Ars Electronica festival in Linz (Austria).[‡] In a short report written for Ars Electronica's online diary, and posted to *Nettime*, German journalist and critic Tilman Baumgaertel wrote:

> "While many observers agreed that they had witnessed the birth of an important new art movement, some felt that form art will mainly lead to more crashes of their browser. Shulgin refuses to serve as a leader to the new art movement, that he has started. Asked how he communicated with his followers, he replied: "Very formally...".[5]

In other words, in *Form Art* the formalist approach to the web page goes hand in hand with myth making, a performative play upon the social roles of the art world – from the art theorist to the curator and the juror – and a parasitic exploitation of existing institutions (Ars Electronica and its historical "Prix").

Made in 1997, both *Form Art* and *Form Art Competition* are still online and publicly enjoyable through a web browser. *Form Art* is part of the C3 online collection;[6] the *Form Art Competition* website, originally hosted by the Ars

[‡]Founded in 1979, Ars Electronica (https://ars.electronica.art, accessed January 2019) is one of the main events and institutions of reference for Media Art. While its annual festival is central in defining the trending topics of this field, the Prix Ars Electronica – with its evolving set of categories – plays a role in shaping the practice, underlying the importance of a given area of activity.

Electronica server, is currently available through Alexei Shulgin's online portfolio, easylife.org.[§] This domain change has some consequence upon the correct preservation of the piece, because we lose the parasitic implications that were made readable by the original domain name: "http://remote.aec.at/form/competition.html" is not just a location, it's also an endorsement. But apparently, both the work and the competition website are fully enjoyable in all their hyper-textual glory. I visited them with three mainstream current browsers – Firefox®, Chrome® and Safari® – on a MacBook Air® with OS X (version 10.9.4) and they look almost the same on all of them. This could be seen already as a small miracle: a code written in 1997, for a browser (Netscape® 3) and the operative systems of the time, still works properly in 2018 on interfaces and softwares that had yet to be conceived at the time. It even works on a device that in 1997 we could only dream of: the smartphone.

This miracle is, indeed, the consequence of a conscious choice: the effort – committed over the years by the World Wide Web inventor Tim Berners-Lee and the World Wide Web Consortium (W3C®) – to keep the HTML language an open standard; and, of course, of the persistent usefulness of HTML forms. Using just HTML and Javascript® code, *Form Art* can still be properly interpreted by current software, without any restoration.

And yet, this is not the work Alexei Shulgin designed in 1997. If we look at the work through the original software and hardware, we see that the colors are different; the design of the HTML form elements is different; their distribution on the web page is different; the "frame" provided by the web browser is different. And these are just the discrepancies that can be experienced at a surface level. Back in the days, this work offered an interactive, time-based experience that's totally different from the current one. The only way to see it at the time was from a desktop computer, with a CRT screen and a 800 × 600 pixel screen resolution. Netscape® was the main web browser. It didn't have tabs: the page you were visiting was the only web page open in front of you. Internet connection was served by a 56 K modem, so the time taken to visit the work, loading pages and interacting with them, moving from page to page, was slower than the current, hectic web experience. Professional web design didn't exist yet, and the graphic possibilities made available by HTML code were extremely basic and simple. In other words, these minimalistic abstract pages were floating in a context in which there was little difference between a mainstream media portal and an amateur personal homepage. Today, we can recover this experience by using obsolete machinery, or by faking it through an emulator:[¶] it feels archeological rather than alive, but it works. And yet, whatever we could do to artificially reconstruct the original experience of

---

[§]The original domain was http://remote.aec.at/form/competition.html; currently, the piece is available at http://easylife.org/form/competition/competition.html (accessed January 2019).

[¶]The piece can be experienced in emulated form through the *Net Art Anthology*: https://anthology.rhizome.org/form-art (accessed January 2019).

this piece, it would always be impossible to emulate the cultural shock generated among early web users by such an alternative use of HTML code.

In other words: even if not a single line of code was changed in the original piece, its experience changed completely, because the social and technical infrastructure it is part of evolved around it. To make an impossible comparison, it would be like – not because of deterioration, but through some weird law of our physical universe that causes this change to be perceived as natural and unavoidable – the colors in *Tableau 1*(1921) by Piet Mondriaan changing from primary yellow, red and blue to green, orange and purple, and the black lines turning 45° clockwise. Could we still understand it as the very same painting Mondriaan made in 1921?

22.2 Net-based Art and the Loss of Context

This is, nonetheless, the very basic condition of the existence of net-based art in an online, constantly evolving, shared space. Any online piece of art has to deal, sooner or later, with this form of obsolescence which is somehow inborn in the system it is part of. The Internet does not only provide, for a work of art, a location from which it can be experienced from any side of the world. It's not like a white, empty wall in an institutional space. It's, first of all, a *technological infrastructure* that evolves constantly at any level of its complex body: the physical network, the servers, the communication protocols, the languages used to build the interface, the content providers, the search engines and, on the client side, the devices, the operative systems, the softwares used to access it. It's, on a second level, *a social infrastructure*: an online society that grows in numbers (from 32 million users in 1995, to the 4 208 571 287 "netizens" estimated in June 2018)[7] and in social complexity, and that over the years shaped its social spaces, both mirroring existing institutional or commercial spaces and generating brand new ways of gathering, communication and exchange (*i.e.* social media platforms). This society, which was initially academic in terms of social composition, anarchic in terms of organization and amateurish in terms of design capabilities, became increasingly subject to economic interests and political interferences, which impacted both on its social and technological structure, and of course on our way of thinking about the Internet.

The net-based art that was produced over these two decades mirrors this evolution; but if the evolution of the social infrastructure of the Internet caused changes that, even when traumatic, affected only the story around the work, which can be effectively reconstructed with research, documentation and storytelling, the evolution of its technological infrastructure can affect, at various levels, the actual nature of the work – the lower level being the one we already discussed when talking about *Form Art*: the work remains the same, yet it has completely changed.

Let's consider early net.art again. Although jodi.org was registered in August 1995[||] and Pit Schultz's Berlin exhibition dates back to the same year, we can conventionally set a starting point in May 1996, when Slovenian artist Vuk Ćosić organized a meeting at Teatro Miela, Trieste, called *Net.art per se*. Ćosić celebrated the event with a webpage, a clone of the homepage of CNN Interactive, opening with the hot news: "Specific Net. art found possible." The piece is now recognized as one of the first net.art readymades.** Only three years later, everybody seemed to agree that net. art was over. Alexei Shulgin co-wrote "Introduction to net.art (1994–1999)", bearing an opening and closing date, and ending with a chapter titled: "Utopian Appendix (After Net.art)".[8] Other artists contributed to this narrative with symbolic gestures, like the infamous "fuck you" with which JODI greeted the Webby Award they received in the "Arts" category. Most artists who recognized themselves in this label stopped, for some time, making web-based projects, and moved on to software, music, video and installation. What had happened? How can an art movement expire in three years? Alexander Galloway, at the time editor at *Rhizome®*, a web-native institution and mailing list, explained it in this way:

> "Net-dot-art is dead.
>
> Or, as Tilman Baumgartel has noted, it's "the end of an era. The first formative period of net culture seems to be over."
>
> People want more than email. They want new interfaces. They want killer aps (*sic!*). They want to escape the offline. All art media involve constraints, and through these constraints creativity is born. Net.art (the jodi-vuk-shulgin-bunting style) was the product of a particular technological constraint: low bandwidth. Net.art is low bandwidth through and through. We see it in ASCII art, form art, HTML conceptualism – anything that can fit easily through a modem. As computers and bandwidth improve, the primary physical reality that governed the aesthetic space of net.art begins to fall away. Today, plug-ins and Java are good. And software trumps them both."[9]

If "the primary physical reality that governed the aesthetic space of net. art" was low bandwidth, its primary social reality was its existence in a space

[||]Jodi.org is the website of the Dutch/Belgian duo (JODI) Joan Heemskerk and Dirk Paesmans. Although the couple came out gradually with their names and faces, in the mid Nineties JODI was perceived as an opaque online entity – a place more than a subject, governed by some superhuman entity speaking a hybrid – half machinic half human – language on mailing lists and private emails, yet characterized by a signature style and a subversive take on the languages, aesthetics and behaviors of software. This opacity is still a distinguishing feature of their online presence: their website offers today random access to an increasing number of projects, with no index page, no captions, no information, no About page or artists' CVs. See, at your own risk, www.jodi.org (accessed January 2019).

**The work is available online at http://www.ljudmila.org/naps/cnn/cnn.htm (accessed January 2019).

with no institutions, no art professionals and no interest from institutions and art professionals working "outside of the Internet". Yet, this social reality was fading out as well. Documenta X introduced net.art in the institutional space already in 1997. In 1999, institutions as diverse as the Webby Awards, Ars Electronica, the Whitney Museum were honoring net.art. In 1999 ZKM, the Zentrum für Kunst und Medientechnologie in Karlsruhe, Germany[††] – at the time, the one and the only museum of media art – launched *Net_Condition*, a major exhibition of networked art: according to Galloway, the exhibition "was received as simultaneously an introduction to and a retrospective of net art (despite its relative youth as a medium)." And finally, a specialized network of critics, curators and institutions like *Rhizome*® was taking shape.

Of course, the end of net.art didn't mean the end of net-based art. On the one hand, the more neutral term Net Art had already been proposed as "the more generic term we use to identify work for which the network is a necessary and sufficient condition",[11] without necessarily identifying with the ethos and the narrative of net.art; on the other hand, other artists were starting working online, enjoying and experimenting with the new technological and social infrastructure. But if it's so difficult, today, to communicate to younger generations the excitement that those early works produced when first experienced in the late Nineties, this happens because, even if the pieces are still there, the Internet that made them possible has completely disappeared, to an extent that it's extremely hard to imagine (and even to recall for those who experienced it). Low bandwidth and 800 × 600 pixel CRT screens are not easy to imagine if you grew up in a world with WiFi connections and flat screens everywhere; the radical weirdness of jodi.org's first interface[12] is hard to understand if you don't know that, at the time, a mainstream website such as CNN Interactive had a graphical and technical complexity comparable to what can be achieved today by a web design student in ten minutes of coding. And of course, it's also hard to imagine a web inhabited almost only by nerds, Prof. Drs,[‡‡] amateurs and artists, with no commercial interests, no political tensions, no commodified spaces.

As we said, the problem of the loss of context does not concern only those early years, but any net-based artwork. Let's take another example. In 2000 in the US, Al Gore and George W. Bush were running for the Presidential Elections. James Baumgartner, then an MFA student at Rensselaer Polytechnic Institute (RPI) in Troy, New York, made a website called voteauction.com, inviting American voters to offer their vote for sale to the highest bidder. The website didn't allow people to sell their votes, it just invited them to register, either to sell or to bid. In other words, it was a work of political criticism, that by suggesting a paradoxical, imaginary solution made people think about

---

[††]ZKM was founded in 1989, but it opened its premises in Karlsruhe to the public only in 1997. Although a relatively young institution, in 1999 ZKM – defined by his Director Peter Weibel the "Mecca of Media Arts" – was already perceived in the net.art community as an established, old fashioned museum for its commitment in collecting and preserving huge interactive media installations from the previous decades – the so-called "ZKM art".[10]

[‡‡]Artist Olia Lialina has identified and described the "Prof. Dr Style" in her essay.[13]

the value of their vote and the relationships between business and politics. This brought it some attention, and the author understood that there was no way to run such a website from the US. He discussed this topic with a member of the ®™Ark collective, who was teaching at RPI. ®™Ark was, at the time, a newly established activist collective, which registered as a corporation with "corporate sabotage" as its core mission. The idea, rooted in culture jamming philosophy, was to adopt the privileges usually reserved to corporations (what is called the "corporate shield") to boycott them. Hiding behind a dotcom website, the faceless group was running email campaigns, organizing funds to support various causes and secretly registering domains where modified mirrors of corporate or political websites were uploaded. By running the gatt.org website, a modified version of the World Trade Organization (WTO), in 2000 they started receiving invitations to appear in public and on the media as spokespersons of the WTO. The two ®™Ark members who decided to accept the challenge started a new group identity, The Yes Men, under which they made a number of performances and interventions.[14]

®™Ark loved *Vote Auction*, so much that it was featured in one of the first presentations made by The Yes Men as WTO spokespersons. Speaking in Salzburg at a conference organized by the Center for International Legal Studies, Dr Andreas Bichlbauer (one of the many fake names used by The Yes Men's frontman) explained that "consumerism is the ultimate form of democracy", and presented *Vote Auction* as a positive example of how one could optimize democracy by directly connecting the corporations that financially support political campaigns with the "consuming voter".[§§] At that time, ®™Ark had already sold the website to UBERMORGEN, another mysterious dotcom just founded in Europe. UBERMORGEN was the brainchild of Hans Bernhard and his partner lizvlx – the latter working as an individual artist, the first as a co-founder and member of etoy.com[¶¶] until 1999. Got lost? The important information here is that if, on the one hand, UBERMORGEN was mostly unknown and invisible online, on the other hand it was supported by a strong awareness and experience of how the Internet actually works. Their opacity allowed UBERMORGEN to play the role of the aggressive Eastern-European dotcom trying to speculate on the American elections; their awareness helped them to anticipate the reactions and bring *Vote Auction* to an incredible level of visibility. UBERMORGEN re-designed the website, registered it under different domain names, bought server space on different servers around Europe and kept the project online for months, while government agencies started

[§§]The lecture, titled "Trade Regulation Relaxation from *ca.* 1970 to the present", is partially documented in the movie *Infowars* (J. F. Sebastian, Austria 2004).

[¶¶]Etoy is an online collective and a registered shareholder company founded in Switzerland in 1994 by a group of artists, musicians and activists who wanted to "leave reality behind" by abandoning individual identity and building up a strong online identity (http://www.etoy.com, accessed January 2019). Etoy was massively launched in March 1996 with the *Digital Hijack*, a huge online performance/promotional campaign that, by cheating search engines, brought around 1 600 000 unaware web users to their website (http://www.hijack.org/, accessed January 2019). The *Digital Hijack* won the Golden Nica at the Prix Ars Electronica, Linz in September 1996, the first artwork to be awarded in the "Web" category.

investigations, various American states took legal action against them – none of them following the correct procedure for delivering legal papers to non-US residents – and the media discussed the project widely, with over 2500 news features on television, radio and print magazines and a whole episode of CNN's legal format *Burden of Proof* dedicated to it.

Vote Auction can be understood as a media hack and a performance that takes its strength from the fact that it happened in the public space of the media and on the international platform of the Internet, without being understood as art throughout its development. The website is one part of it – the trigger of the whole thing. The background story, the media reactions, the legal prosecution are equally important to understand the work in its complexity. To present the project in the white cube, in later years UBERMORGEN designed an installation (Figure 22.2) usually including the printed seal of the project, a paper stack collecting all the press reactions and legal

Figure 22.2 UBERMORGEN, *[V]ote-Auction Seal*, 2000. Pigment print on fabric. *[V]ote-Auction 700 Kg*, 2005, A4 Paper. Reproduced with permission from Hans Bernhard, UBERMORGEN.

papers circulated around it and the screening of the *Burden of Proof* episode: documentation replaced the time-based, immaterial artwork, as often happens with performance. This is a good starting point, but a better understanding can only come by offering contextual information in the form of storytelling. This information shall not only reconstruct the networking dynamics that generated the piece, but also illustrate a moment in the history of the Internet that, again, belongs to the past. As Diana McCarty writes in a recent piece on *Rhizome*®:

> "Eighteen years ago, the internet was a different thing, in the process of casting off its utopic metallic cast of '90 s cyberspace and entering the popular imaginaries of people around the world (if not their homes). Dial-up modems were still a thing, website design was a viable career, the cool kids were surfing, Europeans were sniping at WIRED, and the whole world had just survived the impending doom of Y2K – the Millennium Bug – when the 20th century computer codes would reset the date to '00. There wasn't any social media, not a Friendster in sight. East Europe was another dangerous, exciting place (almost a real life second life in the American mind), where money could be made, wars could break out, and anything was possible. The masses mostly experienced this virtually, through the firmly established screen of the 24 hour cable news cycle. Like today, there was a deeply entrenched fear of election rigging, albeit for very different reasons. For the mainstream populations in Western Democracies, and especially in the US, the internet was a fantastical space and networked computers held a sexy, promising, and slightly dangerous allure. An oft-cited new wild west, similar to Eastern Europe, distant places where dreams could be realized – both offered gold, freedom, unlimited space, porn, perverts, and outlaws. Y2K and its attendant fear-mongering had given people some idea of just how networked and omnipresent computers were in the world. Al Gore's electronic superhighway was also a potential space for a different kind of gerrymandering and vote rigging. Cable news was rife with speculation about how the internet could be used to rig the vote and how computers could be infiltrated by those with bad intentions."
>
> [Reproduced with permission from ref. 15. Originally published on *Rhizome*®, November 6, 2018.]

It would be easy to object, to these considerations about the loss of context, that the same has always happened with any kind of art from the past. Time goes by, and the socio-cultural conditions that produced an artwork get lost and can only be reconstructed through research, historiography and storytelling. We do not live in the same world in which Botticelli painted his *Primavera*; the Paris in which Picasso painted the *Demoiselles d'Avignon* is not the same city we can visit today, and even Basquiat's New York is definitely lost. We can enjoy these works for what they still have to say to us, and then maybe learn to know them better and better through study and research. If I meet *Vote Auction* in the exhibition space, I should be open to approach it with the same, layered approach: check if it still engages me formally and conceptually and, if it does, be open to go deeper and learn more about it. So, what's the point here?

The point is that the Internet, as a platform, is both persistent and ever-evolving. We experience it as something that belongs to the present, to the here and now, even if it has been built over the years, and still encapsulates uncountable fragments of the past. If an artwork is kept online, and if it's not made dysfunctional by technological evolution, we experience it in the here and now, on our current devices, surrounded by the present web. On the Internet, the past is not just deleted, it's replaced. The way an old website or any media fragment (an animated GIF, a YouTube™ video) is experienced on the Internet is closer to the way an old building or a monument is experienced in the chaos of the city, rather than to the way an artwork is experienced in the museum, but with a difference: it doesn't belong to a respected past we can somehow connect it to, but to a continuous present to which it has little or no connection.

To clarify, let's make a last example. *YouTube Earthquake* is a one-minute video uploaded by Dutch artist Constant Dullaart on YouTube™ in June 2008, as part of a series titled *YouTube as a Subject*.[16] For the distant future reader, YouTube™ is a video sharing platform that was founded in February 2005 by three young guys, and bought one year later by Google for 1.65 billion dollars. The slogan of the platform was "Broadcast yourself", and the first video published, by one of the founders, was called *Me at the Zoo*. Amateur videos and pirated media content were, for a long time, what YouTube™ had to offer, and what actually made it grow exponentially. In 2008 it was three years old, and it still had the crappy interface design it was born with. The media player displayed a big grey 3D play button upon all videos, and a similarly invasive play bar under them.[||||] *YouTube as a Subject* was meant as a kind of instant, ironic, formalist take upon YouTube™, turning the medium into its own content: in the videos of the series, the 3D play button falls down, goes out of focus, bounces as the DVD player screensaver flickers, shivers, changes color upon a black background. The series responds to the platform, its influence over the content uploaded, its relationship with older media experiences. In 2008, experiencing these videos would mean clicking on the button and seeing *the very buttons* fall down, go out of focus, bounce, *etc.* People had fun with them, and Dullaart collected many video responses from other users.

In 2018, YouTube™ is a mainstream medium, with 5 billion videos shared to date, and 300 hours of videos uploaded every minute,[17] strongly regulated by copyright laws and commercial interests. Amateur videos are still uploaded, but are less and less visible if they are not by successful "YouTubers".

[||||]To have a feeling of how YouTube™ looked in 2008, check out this Reddit® thread: www.reddit.com/r/nostalgia/comments/20fcrw/2008_youtube_layout/ (accessed January 2019).

The interface design has changed, the slogan "Broadcast yourself" has disappeared from the interface, the default video ratio is 16:9, and *there is no 3D play button anymore*. Videos are controlled with a minimal, transparent play bar that disappears when you move the pointer away, and the only additional graphic that shows up on screen is the "loading" circle. For a user that was not there in 2008, the joke is lost, let alone the fact that it was not just a funny joke. Dullaart's videos are still there, in all their 4:3, 480 pixel wide majesty, encrusted like old gems in this functional interface, but they can speak only to older users' nostalgia.

22.3 Preserving Net-based Art: Saving the Context

Constant Dullaart was so aware of the fragility of net-based art that he started, together with Berlin-based curator Robert Sakrowski, the netart database: a platform where an online work is documented by two videos of the same length, a screen grab of the computer screen displaying the artwork and a video documenting the user while using the artwork, with the hardware and software of its time (Figure 22.3). The idea behind the project is that the very

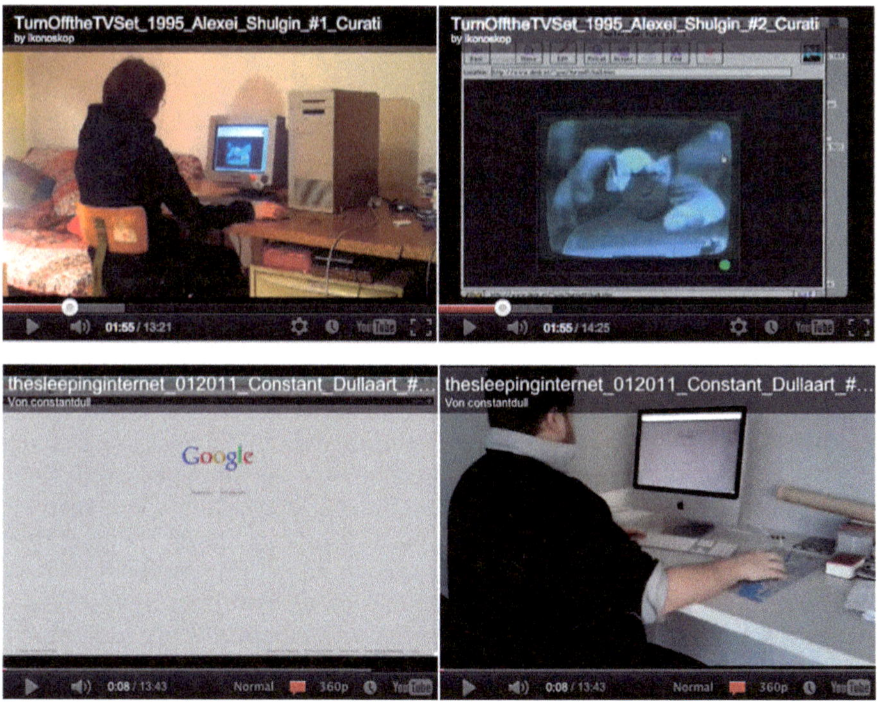

Figure 22.3 Netart database. Above: experiencing *Turn Off the TV Set*, 1995, by Alexei Shulgin. Below: experiencing *The Sleeping Internet*, 2011, by Constant Dullaart. Reproduced from netart database with permission from Constant Dullaart.

way in which we access a work of net-based art has changed, and will change in the near future, to such an extent that we might not be able to recall the situation in which an artwork from 1999 or 2018 was typically experienced. From here, the idea is to document this experience, and invite people to collaborate in this process of "subjective guerrilla archiving". According to Constant Dullaart:

> "There are so many different kinds of works, and only a very small percentage of the works are static enough to archive through copying or backing up the data. Also most of the works are very, if not fully, dependent on the context in which they are viewed if they are to function as intended. Think of works in domain names, works that exist on video or blogging services, generative works *etc*. To draw the comparison with Performance Art again, I felt as if people were trying to archive the body of the performer to be able to archive the work. As if you would freeze Marina Abramovic."[18]

Preserving, *via* storytelling, documentation, storage and emulation of original hardwares and softwares, *the original experience of a work of art*, is definitely a good starting point for a comprehensive strategy of preservation of net-based art, but only one of the many articulations of the larger problem of *preserving the context in which a work of art originated*. A provisional, inevitably incomplete list of them may include:

- *The preservation of the intellectual context, the discussions, the relational dynamics that originated the piece*. As for any other art form, this can be mostly achieved *via* storytelling and documentation: letters, memories, interviews, reviews, essays, photographs and videos. Without Man Ray's or Hans Richter's memories, the few physical artworks left by the Dadaists would mostly be lifeless leftovers of a game we can't understand anymore. Likewise, we would not understand *Form Art* if we look at it as a simple exercise with HTML coding, nor *Vote Auction* if we see it as a solo work by UBERMORGEN, developed online for a few months and later translated into an art installation. In this respect, the Internet apparently allows us to preserve even more than in the past, as a post on a mailing list, a blog, a forum, a social network is more persistent than a phone call or word of mouth. It is, indeed, *if somebody cares*: keeping things online, archiving them on hard disks or printing them, taking screenshots, devising strategies to archive something as ephemeral as a Facebook™ comment or any contribution on a proprietary platform – that might be, for the contextualization of that specific artwork, much more important than a review on *Artforum*.[19]
- *The preservation of the responses the piece produced, when it took place and over time, in the art field and outside of it, on a range that goes from mainstream media to anonymous media producers.* A net-based artwork is a performative piece that takes place in a public space that is intrinsically participative; and, no matter how old it is,

it is perceived as "present" any time it is accessed. So, a net-based artwork is not limited to the enclosed piece the artist uploaded or delivered (a website, a picture, a video, an email), but it includes the way it was delivered and the responses it received over time. *Vote Auction* is not just a website: it's the performance of its existence online between 1999 and 2000, between servers and domains; and of the responses it got from the media, the legal system, the Internet service providers and the audience. Similarly, *YouTube Earthquake* is not just a digital video that could be easily preserved on a local storage device: it's a public intervention on a commercial video sharing platform that responds to this platform and includes its own metadata (title, description, tags) and the reactions posted by the audience in the comments section of the video or on their own accounts. *YouTube Earthquake* is like a joke in a foreign language we don't understand; we need to recall what YouTube™ was in 2008 to understand it, but we need to snap into the conversation it generated to get its nuances, see how its concerns were shared by a diversified audience, and finally, maybe, be able to smile.

The problem with reactions – especially when they are user-generated – is that they are even more ephemeral than the artwork itself. They may be removed by the author, or censored by the platform; they may be dispersed in various context and conversations that require different levels of access; and it may take a long time before the one who's working on the preservation of a given piece recognizes their importance. The artist is probably the best archivist for them – the first to understand their importance, as well as their fragility. A fitting example is provided by Jon Rafman's video piece *Still Life (Betamale)*. Released in 2013, the video is part of a trilogy based on the materials collected over years of exploration of image boards, social networks and the so-called deep web, researching the visual debris of online subcultures. In a way, the video is itself an archive: an unsystematic, idiosyncratic archive with no labels and footnotes, arranged around the artist's stream of consciousness, preserving the visual outputs of obscure subcultures in a given moment in time. When it was published, the video immediately sparked a long discussion on 4chan®, one of the image boards it drew upon. One of the features, and problems, of 4chan® is that it isn't archived: a given discussion is up there while it is participated in; when nobody seems to care about it anymore, it disappears from the Internet, forever. Jon Rafman saved the conversation in a 43-page pdf that is made available from his website, on the artwork's page.[20] This document is relevant not just because it saves the public discussion generated by an artwork whose reach went far beyond the realm of art; but also because it provides many insights into the visual material featured in the video that only the community that produced, or first circulated, that material could offer. In the future, Rafman's iconographers – as well as Internet culture researchers – will find in this document a unique source of primary information about Internet cultures *ca.* 2013.

- Finally, and obviously, the preservation of the context in which a net-based artwork is born should imply *the preservation of the state of the Internet at the time in which it was conceived and published*. By all means: historiography and storytelling, documentation and actual archiving. In this titanic effort, of course, the art conservator is not alone, but is supported by all kinds of professionals: historians, researchers, media archeologists, cultural analysts, media theorists, hackers, archivists and so forth.

One impressive initiative that should be mentioned here in terms of actual archiving of the Internet, is the Internet Archive. "A non-profit library of millions of free books, movies, software, music, websites, and more", the initiative started in 1996 by archiving captures of web pages, that were later made available through the Wayback Machine®, a tool that allows the user to see, in the browser, how a specific website evolved over time. Today, the Internet Archive provides access to more than 345 billion web pages.*** Yet, although monumental, the Internet Archive is not enough – for various reasons: its spiders are not saving the whole Internet – they only take captures, and are often unable to save embedded content, like images, videos, sound files, *etc.* It has many problems with storing the content of most sharing platforms, like YouTube™ and Flickr®, and only saves the public side of password-protected social media accounts, such as your Facebook™ page.

Of course, no single initiative could be expected to archive "the whole Internet". And, at the same time, everybody can be of help. In 2009, Yahoo!™ announced that it was going to shut down Geocities®, a free hosting service that was launched in 1995 and bought by the company in 1999. On October 26, 2009, thousands of personal homepages – a big portion of what was left online of the web of the late Nineties – were taken down. Yet, before it happened, a hacker group called the Archive Team saved almost a terabyte of Geocities pages, that a year later were made available online as a huge Torrent file. Even if not on this scale, the "Save as" option is available to anybody; and as the netart database proved, we can all contribute to archiving the current web for the future by saving, or even printing out,††† the parts of the Internet we like the most, in a form of "subjective guerrilla archiving".

The history of the Internet is a big part of the history of our society and culture over the last twenty years, and its preservation doesn't matter only to art conservators. And yet, as we will see below, sometimes art preservation (and even art making) proved to be the ideal ground on which to test preservation solutions that could be used as well outside the field of art.

***For more information, visit https://archive.org (accessed January 2019).

†††*Printing Out the Internet* is a project launched in 2013 by New York based poet and artist Kenneth Goldsmith, on the occasion of an exhibition at LABOR Gallery in Mexico City. With an announcement, Goldsmith asked for people to print out pages from the Internet and send them to the gallery. The space was filled with ten tons of paper during the exhibit, sent by more than 20 000 people.[21]

22.4 Preserving Net-based Art: Saving the Artwork

So far, we have only considered works that, despite the evolution of the technological and social infrastructure of the Internet, are still up and running, without any restoration and with no need to change a single line of code. This is not to say that the preservation of the actual piece of art is easy, or less important: but to stress the importance of the preservation of the context that makes that piece of software, or social software, intelligible and enjoyable. Even so-called "single-purpose websites", one-page websites with no links to the outside, typically displaying a simple thing such as an image or a full-screen animation on a dedicated domain, do not exist in a vacuum; they may use the net as a canvas, but are not just like paintings; they are woven within the web and they change with it.[‡‡‡]

The preservation of net-based artworks includes so many variables that it would be better considered on a case-by-case basis. Also, the fast evolution of hardwares and softwares made the problem emerge very quickly, and the early interest of institutions in the practice made discussions about the preservation of net-based art take shape as early as the late Nineties. It would be impossible to sum up, here, more than two decades of debates and efforts, but a few references should be enough to outline some key questions.

In 1998 äda'web, a "digital foundry" and a website commissioning net-based projects, founded four years before by curator Benjamin Weil and entrepreneur John Borthwick, stopped its activities. The website, featuring works by respected artists such as Jenny Holzer, Julia Scher, Doug Aitken and Lawrence Weiner, was about to disappear from the web, but thanks to an agreement with the Walker Art Center, Minneapolis, which accepted to host it on its own servers, it's still online in its "frozen", final state.[§§§] Most works are still working, but Lawrence Weiner's *Homeport* (1996) was already inaccessible before the end of äda'web. The conceptual artists opened a chat room decorated with statements in his signature style within The Palace, a popular graphical chat software produced by Time Warner Interactive and first opened to the public in November 1995. The Palace required an external client, which first made the maintenance of *Homeport* problematic, and although it had its popularity peak around 1999, in the early 2000s it was already considered "abandonware".

Although storing äda'web in its frozen state, the Walker Art Center supported a discussion about its preservation by commissioning the artistic

---

[‡‡‡]The first "single-purpose websites" have been made within the framework of the NEEN art movement, launched in 2000 by Greek artist Miltos Manetas. NEEN artists such as Rafael Rozendaal and Angelo Plessas were mostly using dedicated domain names to display colorful, sometimes interactive animations made with Macromedia Flash™. So, *whywashesad.com* (2002) by Rafael Rozendaal is a blue sky filled by white clouds that you can delete by moving your browser cursor. In the following decade, this solution was widely adopted by net artists. For more information about NEEN, http://afterneen.com/ (accessed January 2019).

[§§§]äda'web is available at www.adaweb.com (accessed January 2019). For more information about the history of the website and its archiving strategies, see D. Quaranta, *Net art 1994–1998. La vicenda di äda'web*, Vita e Pensiero, Milan, 2004 [in Italian].

collaborative Janet Cohen, Keith Frank and Jon Ippolito to produce an artwork, *The Unreliable Archivist* (1998).¶¶¶ At the time, Ippolito was already working as a curator at the Guggenheim Museum, New York, with a focus on collecting and displaying New Media Art. Ippolito followed the integration in the museum's collection of some seminal pieces of net-based art, such as Mark Napier's *Net.Flag* and John F. Simon's *Unfolding Object*;‖‖‖ he was among the initiators of the Variable Media Initiative (since 1999), the first consistent institutional attempt to develop a theory and a practice for new media preservation.**** He was also the curator of *Seeing Double. Emulation in Theory and Practice* (2004), a groundbreaking exhibition that first tested the promises of emulation as an experimental preservation model, by pairing original art installations with their emulated versions.††††

At the time, the Guggenheim and the Walker were not the only institutional subjects dealing with collecting and preserving net-based art. Thanks to its Adjunct Curator of New Media Arts, Christiane Paul, in 2001 the Whitney Museum of American Art established artport,[25] a platform for commissioning works of net-based and New Media Art. But the Whitney's first huge effort in restoring a net-based artwork came about only in 2012, when the museum undertook the restoration of *The World's First Collaborative Sentence*, created by Douglas Davis in 1994 and donated to the Whitney in 1995.‡‡‡‡ The *Sentence* employed a CGI script to allow the audience to infinitely contribute to a sentence started by the artist, in a sort of endless, collaborative, online *cadavre exquis*, until it stopped working properly around 2005. The project was presented as a case study, adopting different strategies to preserve the piece. First, the historic online version was restored, in order to be fully readable. Second, a new live version was launched, using a new code to make the online experience of the *Sentence* fully interactive again, and to allow people to keep contributing (which is no longer possible in the historic version).

¶¶¶Ironically, neither the project nor the essays related to it are on the museum website anymore. Gallery9, the major output of the "New Media Initiatives" of the Walker Art Center, was stopped in 2003, and it gradually disappeared from the Internet. *The Unreliable Archivist* was restored by Jon Ippolito in 2010, and is now available at www.three.org/z/UA/ (accessed January 2019).

‖‖‖Both works have been commissioned by the Guggenheim in 2002 and made available through their website, respectively at http://netflag.guggenheim.org/ and http://unfoldingobject.guggenheim.org/ (accessed January 2019). *Unfolding Object* was restored in 2018, while *Net.flag* doesn't seem to work with the latest versions of Java®.

****The Variable Media Initiative, which included a number of institutions among its founding members, is responsible for a seminal publication[22] and for establishing the Variable Media Questionnaire: a tool for collecting information from artists that could be useful in establishing best practices for the preservation of their work.[23]

††††*Seeing Double* took place at the Solomon R. Guggenheim Museum, New York from March 19 to May 16, 2004.[24]

‡‡‡‡The *Sentence* is available online at https://whitney.org/artport/douglas-davis (accessed January 2019). The project team included Carol Mancusi-Ungaro, Associate Director of Conservation and Research; Christiane Paul, Adjunct Curator of New Media Arts; Farris Wahbeh, Manager, Cataloguing and Documentation; Ben Fino-Radin, Digital Conservator at *Rhizome*®; and the Museum's Digital Media department.

Third, documentary material was published, including photo documentation of the way it was presented in physical spaces, screenshots of how it looked on Netscape® in 1994, and an artist's essay about the work.

In 1999 *Rhizome®*, founded in 1996 as a mailing list and after a short-lived attempt to exist as a dotcom (1996–1998), launched the ArtBase, an online repository of born digital artworks. The ArtBase accepted submissions from artists, which soon turned it into the largest collection of digital artworks ever. According to Mark Tribe, *Rhizome®*'s founder:

> "We were talking about this stuff, but often the stuff was vanishing. Somebody could stop paying the bills on their internet hosting service and a systems administrator could delete art history with the click of a mouse. We wanted to give artists a place to put their work for safekeeping."[26,§§§§]

While archiving and storing the works, *Rhizome®* had to envision strategies for their preservation, which made this topic increasingly crucial for its institutional strategies. After becoming, in 2003, an affiliate of the New Museum, New York, in 2005 *Rhizome®* hired Lauren Cornell as Executive Director. In 2006, Cornell appointed Ben Fino-Radin as the institution's first Digital Conservator. Trained as an artist, Fino-Radin – who would get his Master of Science in Library and Information Science at Pratt Institute only in 2014 – somehow had to invent his discipline and methodology based on previous experiments and theories.[¶¶¶¶]

This brief, incomplete history of the establishment of media art conservation on an institutional level already shows a few things, when it comes to net-based art preservation. First of all, *it needs somebody who cares for the artwork*: for keeping it online and for maintaining it when some code becomes obsolete. It may seem a pretty obvious statement, but it isn't: physical objects have a resilience that online digital files don't have. If somebody doesn't keep them online, they become digital files on a local hard disk, or even less if we are talking about, for example, a social media account with no local backup that is disabled or censored for whatever reason. This could have been the fate of äda'web, and has been the fate of uncountable artworks so far. If somebody doesn't maintain their code, they can survive online as ghost artworks that work only partially or don't work at all: this happened to Douglas Davis' *Sentence* – a work collected by a museum – between 2005 and 2012, when the museum undertook restoration. The "somebody" can be the artist, a collector or an institution. While for museums and institutions the preservation of the art they own is somehow part of their missions (even if it

[§§§§]Ref. 26 is also a good reference for a detailed history of *Rhizome®* as an institution.

[¶¶¶¶]Ben Fino-Radin left *Rhizome®* in 2013 to become Associate Media Conservator at MoMA, New York. In 2017 he left MoMA to establish his own venture, Small Data Industries: "a consultancy providing services to support the collection, exhibition, preservation, and storage of time-based media art" (http://smalldata.industries/, accessed January 2019). The company works with individual artists as well as with big institutions, such as MoMA itself, the Museum of the Moving Image, the Whitney Museum and the DIA Foundation.[27]

can take years), with artists things get complicated. While many of them take care of their work, only a few have the patience to document the versioning. Many, when updating their online presence, feel free to "rewrite" their story, *i.e.* changing titles of the works, or deleting/hiding works they don't consider relevant anymore (but that may have, in background, a historical relevance they don't recognize yet). Some artists, who may have had a central role in a specific time and community, may stop making art and completely delete their online presence. An example is offered by American artist Kevin Bewersdorf, who had a key role – as a producer, animator and writer – in the so-called surfing clubs generation between 2007 and 2009. At some point, right after his first solo show, he decided to delete his online presence and to stop making art. Only partially recovered by the Internet Archive, his websites are now lost. In the following years, Bewersdorf occasionally resurfaced online with new net-based projects related to his daily practice of Taoist meditation, but this presence proved to be, again, temporary.[||||||]

If kept online, as we already noticed, a net-based piece can be put at risk by *various degrees of obsolescence*. The simplest one is link rot: the disappearance of pages or content that is linked from or embedded into the main site without changing its location. An extreme example of this is provided by artworks that are literally scattered through various servers and domains. Russian artist Olia Lialina often worked this way, to stress the fragility of a net-based artwork but also to entrust a network of friends, making them co-responsible for her own art. *Summer* (2013) is a full page animated GIF of the artist riding a swing (Figure 22.4). Yet, every single frame of the animation is "physically" located in a dedicated folder of a friend's domain: the second frame is at http://1x-upon.com/~despens/olia/summer/, the third frame at www.newrafael.com/olia/summer/, and so forth. Metaphorically, Olia Lialina rides her swing over the Internet, traveling from site to site and from country to country, seamlessly. But if one of her friends decides to let his domain expire, or just forgets to pay for it, then one frame is gone, and the swing stops running.

Other common problems may be: changes in the specifics of the browser between one version and the other; obsolescence or changes in the code of plug-ins and applets; changes in external services that are integrated into the artwork, as in the case of Google APIs. One famous example is provided by *Data Diaries*, a work by American artist Cory Arcangel in which he fooled Quicktime® into believing that his computer's memory file was actually a video file (Figure 22.5). So, a given day's computer RAM file was visually

[||||||] I was personally involved in the preservation of the work of Kevin Bewersdorf on various levels. I recovered most of his writings from the Internet Archive and other sources, and published them as a book.[28] I also launched an online project (*Share Your Sorrow*, 2012 – ongoing; http://shareyoursorrow.tumblr.com/, accessed January 2019) in which I archived and shared the fragments of his work I was able to collect over time – photographs, animated GIFs, performance documentation, songs, websites – and I invited people to do the same, turning their personal act of collecting and downloading from the Internet into a preservation strategy. In March 2020, Bewersdorf made an archive of his work available at the domain http://www.kevinbewersdorf.org (accessed March 2020).

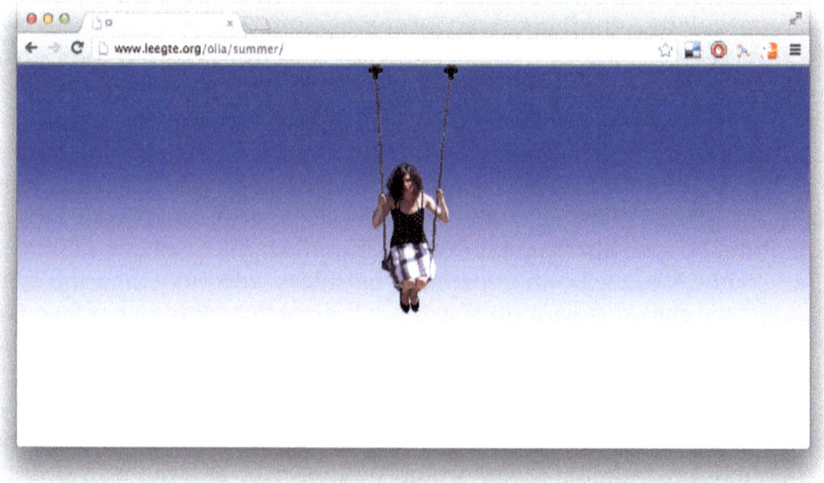

Figure 22.4 Olia Lialina, *Summer*, 2013. Website, screenshot. Reproduced with permission from Olia Lialina.

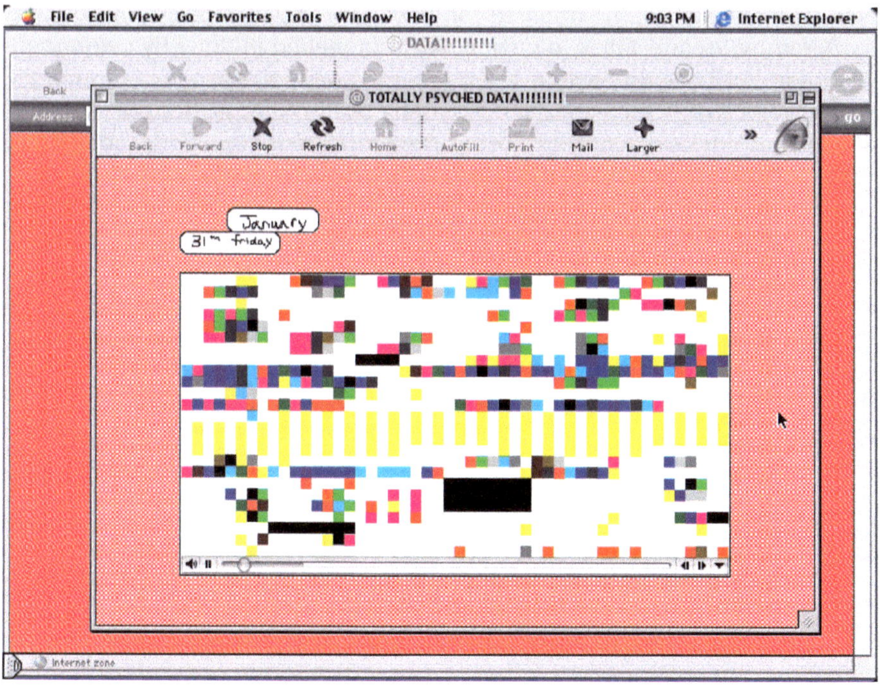

Figure 22.5 Cory Arcangel, *Data Diaries*, 2003. Commissioned by Turbulence.org. Website, screenshot, courtesy Cory Arcangel Studio. Reproduced with permission.

resolved as a glitchy, abstract geometric video consisting of black and white or colored squares – until it worked. At some point, Quicktime® changed its code and the edgy, pixelated aesthetics of the original work left room to an unwanted, blurry abstraction.

Furthermore, net-based works can rely on external platforms and services that change over time their design and ways of working, but also their internal regulations and terms and conditions. This, as we saw speaking about Constant Dullaart's videos, can change the context of an artwork beyond recognition, but can also make it disappear in the blink of an eye. Petra Cortright's *VVEBCAM* (2007), an apparently harmless video in which the artist stares into a webcam as cartoonish clip art figures float around her face, was removed from the platform three years later when YouTube™ realized that, by using spam keywords as tags to spread the video as much as possible, she was violating its "terms and conditions". Last December, Tumblr® – a popular micro-blogging platform widely used by artists – announced that posts containing "adult content" would no longer be allowed on Tumblr®, and that posts that had been flagged as mature would disappear from the platform. How many artworks have become fragmentary or completely disappeared in this process?[29]

Finally, many net-based artworks may not be browser-based, but rely, as in the case of *Homeport*, on external clients. This can open up another set of problems as these softwares evolve, or are abandoned by the community that used them, or stop being updated for newer operative systems. An example is offered by Second Life®, a virtual world launched in 2003 by a company called Linden Labs. In Second Life®, a large and active artistic community was active between 2005 and 2009 – staging performances, designing installations and environments, exploring the world and recording footage for travel logs and narrative videos. One of these artists was Gazira Babeli, the code performer: an artist that never revealed her real identity, consolidating the one of the avatar artist. Babeli made performances and scripted coded environments that for some time have been collected onto her studio/island, called Locus Solus. She also made exhibitions in real venues, showing prints and videos documenting her interventions. Before stopping being active in Second Life®, she exported most of her environments as standalone simulations that can be explored locally on a dedicated client. These are occasionally exhibited as part of the archive of Gazira Babeli, but nobody can meet the artist and her works in her native environment anymore.****

22.5 *Rhizome*®, Webrecorder and the *Net Art Anthology*

"Art is, in general, a very good field to test tools in, because art has never followed the rules and this generates edge cases all the time."[26]

****The online archive of Gazira Babeli can be found at http://gazirababeli.com/ (accessed January 2019).

Rhizome®'s recent preservation strategies are trying to take into account all the problems we have been enumerating so far. In 2014, *Rhizome*® appointed Dragan Espenschied to lead its Digital Art Conservation Program. An artist himself, Espenschied started working as a musician, co-founding the eight-piece band Bodenständig 2000 together with Bernd Kirsch, and accompanying his musical activity with a variety of other projects, mostly dealing with obsolete hardwares and low resolution graphics.[30] In 2003 he started working, together with net.art pioneer Olia Lialina, on various online projects. Their interest in the vernacular creativity of the early web brought them to publish essays and visual essays online, and to compile the seminal book *Digital Folklore Reader*.[31] As, in 2010, the Archive Team started seeding the Geocities torrent, they started downloading it and analyzing its content. The output of this long-term project is *One Terabyte of Kilobyte Age* (2011–present), a blog in which Olia and Dragan share screenshots, discoveries, researches, graphic elements, comparative studies and essays about what they have found in the Geocities archive.[32,33]

This short presentation outlines a keen interest in obsolescence, media preservation and Internet history, that easily translated from Espenschied's artistic activity into his new institutional role at *Rhizome*®.[34] One of the first things he focused on at *Rhizome*® was the development of a tool able to faithfully capture websites and webpages, but also dynamic social media pages, which are controlled server-side and do not belong to the artists, but can nevertheless be a legitimate place of art making. So, he reached out to developer Ilya Kreymer, who had worked on the Internet Archive's Wayback Machine®, and together they started developing the first prototype of Webrecorder. Made publicly available in 2016, Webrecorder was first tested by producing an archival copy of *VVORK*, a group blog about contemporary art active from 2006 to 2012,[†††††] and of Amalia Ulman's performance *Excellences and Perfections*, that took place over Instagram® in 2014.[‡‡‡‡‡] Apart from the fact it can "create and share high-fidelity, fully interactive copies of *almost* any website"

[†††††]Over the time of its activity, *VVORK* became a reference for many artists, and somehow "designed" the forms of art dissemination online that would be later adopted by other online platforms. *VVORK* is still available online on its own domain www.vvork.com, which outlines one of the main functions of Webrecorder: to capture what's online when it's alive and properly functioning, and keep it in this state.[35]

[‡‡‡‡‡]From April to September 2014, artist Amalia Ulman used her social media accounts – mostly Instagram® – to perform a radical "makeover" of herself, interpreting a story of self-degradation and progressive self-design according to the stereotype of the sugar baby: she started taking selfies in expensive hotel rooms, often in expensive sexy lingerie; she underwent breast surgery, dancing, crying and being sexy in front of the camera, expressing values such as "Stay pretty. Be educated. Dress well. Get money." People believed in the story and reacted to it in the comments. Webrecorder was able to capture Ulman's Instagram® account – the segment dedicated to the performance – as it was in 2014, with all posts and comments, the original design of the page, the number of followers, *etc*. This capture will stay as it is, regardless of future changes on the platform.[36,37]

(it still has some problems with Facebook™), what's really outstanding about Webrecorder is that it's available to be used by anybody, for any kind of website. This, on the one hand, shows *Rhizome*®'s belief that net-based art preservation is not just a task for institutions, but a project in which any of us can play an active role; on the other hand, it makes it not just a tool for net-based art preservation, but a truly public service, on a scale and with an ambition comparable to that of the Internet Archive's Wayback Machine®. This has been clearly understood by the Andrew W. Mellon Foundation, which in 2018 awarded *Rhizome*® $1 million for the further development of Webrecorder.[38]

Other tools developed by *Rhizome*® in recent years include oldweb.today, which allows one to explore web archives through emulated legacy browsers, choosing the browser and the version that better fits to the project we want to visit, and pywb, an "open-source, high-fidelity web archive replay and recording system (a more advanced version of the Wayback Machine®) that serves as the core of *Rhizome*®'s web archiving toolset."[39]

Currently, pywb is available online as a Github® project, not as a usable interface. But it is widely used by *Rhizome*® for its *Net Art Anthology* program, an ongoing exhibition charting the history of net art through 100 selected artworks.[§§§§§] If Webrecorder and pywb help to preserve or restore the artworks, and the use of emulators allow us to view the artworks in their native environments, the *Net Art Anthology* combines these approaches with an effort to reconstruct, around the project, the context that makes it interesting and intelligible: using captions and quotes, commissioning essays and interviews with the artists, providing documentation. Incidentally, it's the *Net Art Anthology* that allowed me to show you *Form Art* in its original form; that commissioned Diana McCarty's text about *Vote Auction*'s current topicality; that restored Cory Arcangel's *Data Diaries* and reconstructed the original context of Petra Cortright's *VVEBCAM*. And the same was done for dozens of other works we couldn't mention, including, as a very last example, Eva and Franco Mattes' *Life Sharing* (2001–2003): a three-year performance in which the artists shared, on their own domain, 0100101110101101.org, the entire content of their computer, updated live in an enduring exercise of data transparency – hiding in plain sight a few years before the explosion of social networking, and a decade before the Edward Snowden revelations. Once interrupted, the project was replaced by a portfolio website and never shown publicly.

Until 2017, when *Rhizome*® made it available again.

[§§§§§]Launched in 2016 and completed in June 2019, the *Net Art Anthology* (https://anthology.rhizome.org/, accessed January 2019) consists of five chapters collecting net-based projects from the emergence of the practice up to the present day. It's a curatorial project, aiming "to represent net art as an expansive, hybrid set of artistic practices that overlap with many media and disciplines", and dealing with preservation strategies only when it's required by the project.

References

1. *Nettime*, https://nettime.org, accessed January 2019.
2. A. Shulgin, Net.Art – the origin, in *Nettime*, March 18, 1997, http://www.nettime.org/Lists-Archives/nettime-l-9703/msg00094.html (accessed January 2019).
3. A. Shulgin and N. Bookchin, *Introduction to net.art (1994–1999)*, 1999, http://easylife.org/netart/ (accessed January 2019).
4. C. Greenberg, Modernist Painting, in *Forum Lectures*, Voice of America, Washington, DC, 1960.
5. T. Baumgaertel, Very formally, in *Nettime*, September 11, 1997, https://nettime.org/Lists-Archives/nettime-l-9709/msg00020.html (accessed January 2019).
6. http://www.c3.hu/collection/form/ (accessed January 2019).
7. https://www.internetworldstats.com/stats.htm (accessed January 2019).
8. http://www.easylife.org/netart/ (accessed January 2019).
9. A. Galloway, net.art Year in Review: State of net.art 99, in Switch, 1999, http://switch.sjsu.edu/archive/web/v5n3/D-1.html (accessed January 2019).
10. https://zkm.de/en/the-zkm (accessed January 2019).
11. S. Dietz, Beyond.interface: net art and Art on the Net II, in *Museum and the Web*, 1998, https://www.museumsandtheweb.com/mw98/beyond_interface/bi_frdietz.html (accessed January 2019).
12. http://wwwwwwwww.jodi.org/ (accessed January 2019).
13. O. Lialina, Prof. Dr. Style. Top 10 Web Design Styles of 1993 (Vernacular Web 3), in *Contemporary Home Computing*, July 2010, http://contemporary-home-computing.org/prof-dr-style/ (accessed January 2019).
14. www.theyesmen.org (accessed January 2019).
15. D. McCarty, Vote Auction 3.0: Digital déja vu. Rewind, news fake, fast forward, in *Rhizome*, November 2018, https://rhizome.org/editorial/2018/nov/06/vote-auction-news-fake/ (accessed January 2019).
16. www.constantdullaart.com/site/html/new/youtubeasasubject.html (accessed January 2019).
17. www.omnicoreagency.com/youtube-statistics/ (accessed January 2019).
18. F. Thalmair, Re:Interview #015: Ever-Changing Chains of Work | Constant Dullaart, in *Cont3xt.net*, 2011, http://net.artdatabase.org/about/ (accessed January 2019).
19. Z. Kaplan, The Accidental Archivist: Criticism on Facebook, and How to Preserve It, in *Rhizome*, May 2014, https://rhizome.org/editorial/2014/may/29/preserving-facebook-criticism/ (accessed January 2019).
20. http://jonrafman.com/betamale/ (accessed January 2019).
21. O. Gat, To Bind and to Liberate: Printing Out the Internet, in *Rhizome*, May 2014, http://rhizome.org/editorial/2014/may/01/printing-out-internet/ (accessed January 2019).

22. Permanence Through Change: The Variable media Approach, ed. A. Depocas, J. Ippolito and C. Jones, The Solomon R. Guggenheim Foundation, New York and The Daniel Langlois Foundation for Art, Science and Technology, Montreal, 2003, www.variablemedia.net/e/preserving/html/var_pub_index.html.
23. http://variablemediaquestionnaire.net/ (accessed January 2019).
24. www.variablemedia.net/e/seeingdouble/index.html (accessed January 2019).
25. https://whitney.org/artport/ (accessed January 2019).
26. M. Durón, A Net Art Pioneer Evolves With the Digital Age: Rhizome Turns 20, in *Artnews*, September 2016, www.artnews.com/2016/09/01/a-net-art-pioneer-evolves-with-the-digital-age-rhizome-turns-20/ (accessed January 2019).
27. T. Owens, ArtBase and the Conservation and Exhibition of Born Digital Art: An Interview with Ben Fino-Radin, in *The Signal*, May 2012, https://blogs.loc.gov/thesignal/2012/05/artbase-and-the-conservation-and-exhibition-of-born-digital-art-an-interview-with-ben-fino-radin/ (accessed January 2019).
28. K. Bewersdorf, *Spirit Surfing*, Link Editions, Brescia, 2012.
29. J. D'Onofrio, A better, more positive Tumblr, in *Tumblr Staff*, December 2018, https://staff.tumblr.com/post/180758987165/a-better-more-positive-tumblr (accessed January 2019).
30. http://1x-upon.com/~despens/ (accessed January 2019).
31. *Digital Folklore Reader*, ed. O. Lialina and D. Espenschied, merz & solitude, Stuttgart 2009.
32. http://blog.geocities.institute/ (accessed January 2019).
33. M. Connor, I (No Longer) Have a Web Site: Access, Authenticity, and the Restoration of GeoCities, in *Rhizome*, February 2014, http://rhizome.org/editorial/2014/feb/10/authenticity-access-digital-preservation-geocities/ (accessed January 2019).
34. T. Owens, "Digital Culture is Mass Culture": An interview with Digital Conservator Dragan Espenschied, *The Signal*, March 2014, https://blogs.loc.gov/thesignal/2014/03/digital-culture-is-mass-culture-an-interview-with-digital-conservator-dragan-espenschied/ (accessed January 2019).
35. M. Connor, After VVORK: How (and why) we archived a contemporary art blog, *Rhizome*, February 2015, http://rhizome.org/editorial/2015/feb/09/archiving-vvork/ (accessed January 2019).
36. http://webenact.rhizome.org/excellences-and-perfections (accessed January 2019).
37. M. Connor, First Look: Amalia Ulman—Excellences & Perfections, *Rhizome*, October 2014, http://rhizome.org/editorial/2014/oct/20/first-look-amalia-ulmanexcellences-perfections/ (accessed January 2019).
38. Z. Kaplan, Andrew W. Mellon Foundation Awards Rhizome $1 Million for Further Development of Webrecorder, *Rhizome*, January 2018, http://rhizome.org/editorial/2018/jan/11/rhizome-awarded-1000000/ (accessed January 2019).
39. http://rhizome.org/software/ (accessed January 2019).

CHAPTER 23

Painted Surfaces in Contemporary Architecture†

MOSÈ RICCI[a]*, ALESSANDRA BATTISTI[b], ALDO AYMONINO[c], PIO BALDI[d], FRANCO SUMMA[e] AND KAY BEA JONES[f]

[a]Trento University, Italy; [b]La Sapienza University, Rome, Italy; [c]IUAV, Venice, Italy; [d]Accademia dei Beati del Pantheon, Rome, Italy; [e]Franco Summa, Pescara, Italy; [f]Ohio State University, Columbus, Ohio, USA
*E-mail: mose.ricci@unitn.it, alessandra.battisti@uniroma1.it, aldoaym@iuav.it, baldipio@libero.it, summaars@alice.it, jones.76@osu.edu

23.1 Framework

All architectures are made of solid materials that have colors like wood, concrete, stone or bricks. Color is an integral part of architecture and also a fundamental element of its decorum. This is how it has always been. The theme of colored surfaces in contemporary architecture is so vast that in treating it one risks stating the obvious. Western historical buildings typically have the colors of their construction materials. Even when the facade is painted, in most cases the paint is selected to imitate the underlying building material. Many frescoed facades or those painted in the color of the air (*e.g.* the Quirinale Palace in Rome or the Sant'Agnese church in Agone and the Collegio by Borromini in Piazza Navona) were the exceptions, to celebrate the value of the construction or to limit its impact on the urban scene.

†The authors contributed to this chapter as follows: Alessandra Battisti: Section 23.3.1, Aldo Aymonino: Section 23.3.2, Pio Baldi: 23.3.3, Franco Summa: Section 23.4.1, Kay Bea Jones: Section 23.5.1.

Science and Art: The Contemporary Painted Surface
Edited by Antonio Sgamellotti, Brunetto Giovanni Brunetti and Costanza Miliani

Published by the Royal Society of Chemistry, www.rsc.org

As in the theory of Paolo Marconi (1933–2013),[†] still one of the greatest experts in the restoration of the chromatism of Italian historical architecture, painted stucco is a sacrificial surface. It serves to protect the building and its constructive elements from external agents, but not only. It imitates the noblest building materials that make up the building in other parts. It provides a unitary and coherent element to an architecture that would have been unnecessarily more demanding from a constructive and economic point of view.

Often the columns of Roman monuments or the doorposts of Renaissance palaces were made of stone only at their bases, where they suffered the most impact of people and carriages. The upper parts of these columns were made of brick and covered with a plaster painted to imitate stone which gave continuity to the column, making it look like a single piece of marble. The main facade of Palazzo Massimo alle Colonne (by Baldassarre Peruzzi, 1532) in Rome is not composed of stone as it appears. It is made of painted plaster to simulate a travertine ashlar. One can identify hundreds of these examples that lead everyone to reaffirm the concept that for historical architecture, until the advent of the *modern age* in the twentieth century, color had the precise function of protecting and simulating construction materials, unless it provided pure decoration in the form of frescoed facades.

Modernity breaks once and for all the profound link between material and color in architecture, finally making this relationship free. Nevertheless, for modern architecture the color was not always a matter of design and was often denied. The architecture of modernity has been above all conceived as a composition of spaces, of (pure) volumes, and almost always has the color of its building materials or the non-color of the white plaster.

However, it cannot be said that the coloring of modern buildings and spaces is a direct consequence of constructive and technological choices. Construction materials were often chosen precisely for their characteristics and for their chromatic quality.

Just think, for example, of the veined marble walls at the base of the Neue Nationalgalerie by Ludwig Mies van der Rohe in Berlin, the green copper roofs by Renzo Piano in the Rome Auditorium or even the COR-TEN® steel guard-rails designed by Pietro Porcinai in the 60 s for the Brenner motorway to reduce their impact on the Alpine landscape between Italy and Austria.

The color manages to assert its autonomous space in the works of many masters of the twentieth century, even if minimalism and purism theorized the monochromatic nature of the construction material or the abstraction of white. The significant presence of the sense of color has been verifiable for modern buildings since the beginning of the last century. Theo Van Doesburg, in the fifteenth article of the Fundamental Principles of Neoplasticism (De Stijl Manifesto, 1918) claims that: "the realization of color in space time

[†]See: Paolo Marconi quoting Marcello Paribeni, *Colore e "colorito" in architettura: il cantiere storico, le tecniche storiche di manutenzione contributo al problema del "Colore di Roma"*, Bollettino d'Arte, Ministero dei Beni Culturali.[1]

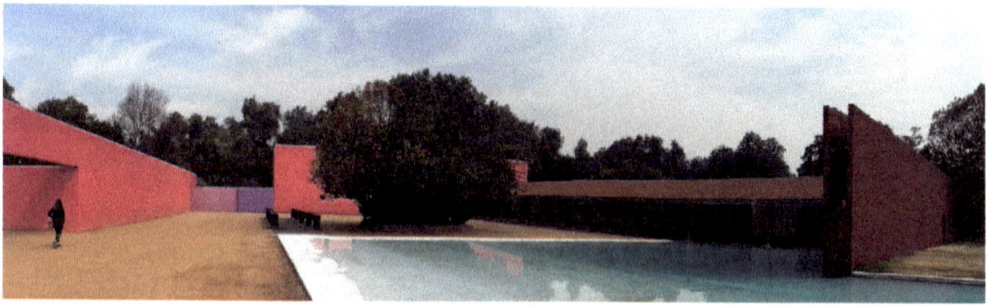

Figure 23.1 L. Barragán, *Cuadra San Cristóbal*, Atizapán, México, *A. Aymonino*, 2018.

makes a new dimension plastically visible". Le Corbusier painted the *brise-soleil* septa of the Brazilian Pavilion of the University of Paris (1930–1932)[2] and of the Unité d'habitation in Marseilles (1945) in cement to characterize the lodgings and in the algid Villa Savoye in Poissy (1930) he painted the interiors with more intimate nuances. Mies van der Rohe used colored marbles, as previously written, while Wright used the warm tones of earth and wood in the walls. Louis Barragan (Guadalajara, 1902) in Mexico imbue his solar architecture with color (Figure 23.1). With him, color becomes *solid* in the volumetric dimension of an emotional architecture[3] and *poetic*, with Giò Ponti using the *infinite blue* for the tiles at Sorrento Hotel Parco dei Principi (1960), and later Aldo Rossi using this for the *Blue of the Sky* of the Modena Cemetery.[‡]

From the mid-twentieth century the joy of color in architecture seems to have been successfully transferred to open spaces. It was a real turning point in the era of modernity and the purism of white buildings. Coloring architectural surfaces makes sense as much as they are visible and can be appreciated and experienced by more people. In modern cities since the end of the Second World War, open spaces are often more colorful than the buildings. Perhaps in this way a new need for happiness is projected into public spaces, creating a surprising and happy integration between architecture, art and society. In these squares and gardens, color becomes virtually volumetric, a pleasant habitable space.

In the years between 1960 and 1980 the Brazilian landscape architect Roberto Burle Marx created a series of public gardens as paintings, coloring their planimetric surface in free abstract forms, making popular modernity in the third dimension of open spaces(Figure 23.2).

In that same period and always in Brazil, Lina Bo Bardi painted with the tones of the four seasons the four sports courts in the concrete tower of SESC Pompéia, at once circumventing both international sports conventions and functionalist ideas of using colors in architecture.[4]

[‡]The Blue of the Sky is the title of an article by Aldo Rossi on his addition to the San Cataldo Cemetery, Modena, 1978.[32]

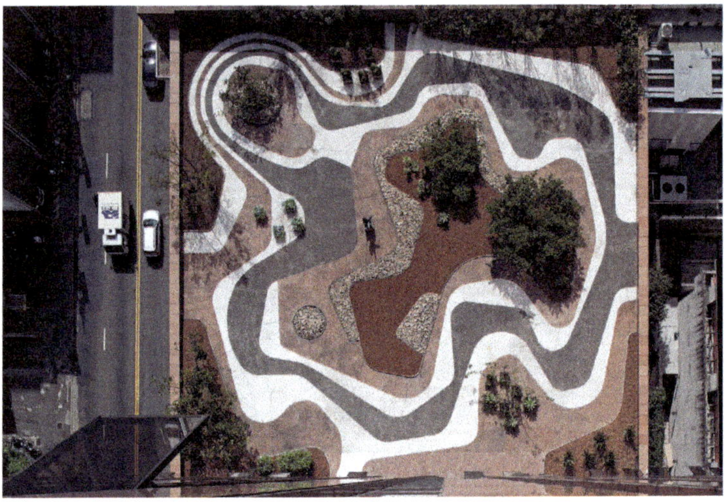

Figure 23.2 R. Burle Marx, *Banco Safra roof garden*, São Paulo (Br), *Courtesy of L. Finotti*, 1956.

23.2 Concepts

The field of interest of this synthetic overview refers to those contemporary artifacts that have been intentionally painted or composed with earth, metal, wood, ceramic or other applications expressly designed to give a color to the architecture, but also to the recovery of existing buildings, to landscape projects and public space.

After the end of the last century (after the end of modernity?) we could say that color, freed from the slavery of matter and also from that of construction, assumes an increasingly new and decisive importance for architecture. In particular, as we shall see, painted surfaces play a strategic role in the research and in the works of some authors and weaker social realities. Thus, the painting of architectural surfaces, often plastered, takes on new meanings as well as those of a performative nature that it has already had in the past for the simulation of materials. Color becomes a *milieu*, a landmark and a background for the recognition and the social emancipation of local communities, but also for the senses (aesthetic, psychological, relational) and narration in architectural and urban space. This chapter deals with colored surfaces by analyzing precisely those three non-opposing concepts – performative, social, narrative – which often characterize the new role of color in architecture.

23.3 Performative

The first concept is that of color as an architectural *performative device*.

It is the paradigm based on the use of color as a conceptual principle of an operational aesthetic and as capable of developing significant technological performances. The performative painting compared to the coloring itself means placing at the center of the architectural project of color not the use, but the appreciable innovative result in predominant ecological terms. But not only this. For many years, for example, the big hi-tech companies have studied the so-called "media facades",[5] those made of sensitive materials such as LED walls, which can become screens and project information, images, advertising, through connection networks and/or can simply take different colors at different times. In the restructuring of the Manifattura Tabacchi in Rome by Aldo Aymonino with the Seste Office (2012), color, *via* LED and painted surfaces, was the design device used to make the building less severe and sad (Figure 23.3).

An example that clearly highlights the performative role of the chromatic surface, even in a very simple and not technologically advanced way, is the famous *Kilometro Rosso* (2007) by Jean Nouvel,[6] built in Brembo's Science Park near Bergamo (I). It is a red wall along the motorway 10 meters high and one kilometer long. The red has an essential communicative value, because it is visible at high speeds and can clearly signal the presence of the industrial brand. It takes on the meaning of a cut, of a *caesura* between the chaotic environment of the highway and the calm and active world of research.

Also, we could say that performative, in an urban perspective, is the role of the color masterplans that in Italy have been designed since the end of the last century with the aim of ensuring a modification of the historical urban fabric consistent with its value, which the plan makes legible through its rules.

The principle of performance projects architecture onto the contemporary, making it the terminal or interface of a system of physical or immaterial relationships that can substantiate its existence. It is the re-contextualization of the idea of color within a new and not necessarily material intervention space.

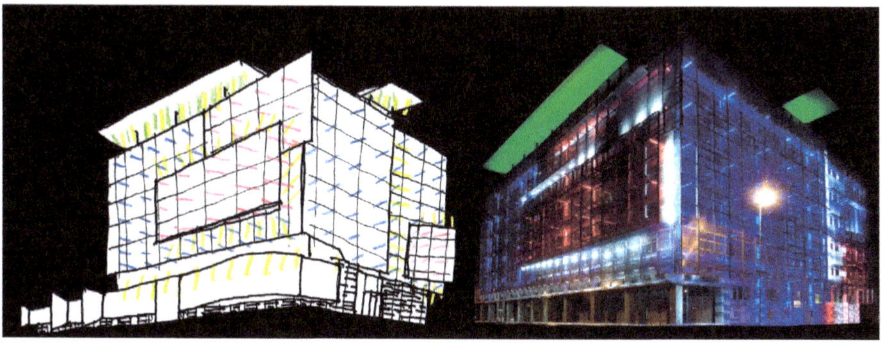

Figure 23.3 A. Aymonino with Seste Engineering, *Ex Manifattura Tabacchi*, Roma, sketch and photo, *A. Aymonino*, 2016.

23.3.1 Innovative Colors for Climate Mitigation in Contemporary Architecture

When we delve into the role of color innovation in architecture as a passive tool that can meet the needs generated by climate change – and therefore mitigate extreme phenomena such as overheating or desert climates – we find ourselves exploring a function that dates back to ancient Greek and Roman traditions, not only in response to the aesthetic and protective aspects found in Daniele Donghi's teachings,[§] but also to those cultural and functional aspects of a physiological/perceptive type linked to the physical and mental well-being of inhabitants, affected by the brilliance, gradation and *albedo* of colored materials. In contemporary architecture, color has now taken on myriad different meanings, dictated by a hyper-technological market and by speculative operations that force architecture to adapt to various different lifestyles, controlling the quality of surfaces and their response to new performance-related needs by using materials where the decisive interface between technology and new ways of living often depends on the role of color. "Our environment is filled with color, which, day and night, in public and in private, sometimes loud and sometimes quiet, demands our attention. This explosion of colors means something. We are exposed to a constant stream of colors. We are programmed by colors. They are an aspect of the codified world in which we have to live."[7]

Take a few of the thermochromic materials used as temperature sensors thanks to their color fluctuations.[8] As far back as 1987, at the World Teleport Association Conference entitled *Design for the Information Age*, Mike Davies announced the appearance of future buildings with these words: "Look up at a spectrum-washed envelope whose surface is a map of its instantaneous performance, stealing energy from the air with an iridescent shrug, rippling its photogrids as a cloud runs across the sun, a wall which, as the night chill falls, fluffs up its feathers and turning white on its north face and blue on the south, closes its eyes but not without remembering to pump a little glow down to the night porter, clear a view-patch for the lovers of the south side of level 22 and to turn 12% silver just before dawn." While this statement may have seemed fanciful thirty years ago, today, thanks to smart materials that can change their color and transparency, as well as computerised control systems, some buildings can respond to changes in climate and environment, turning what was once science fiction into reality, almost as if to prove Arthur C. Clarke's Third Law: "Any sufficiently advanced technology is indistinguishable from magic."[9]

[§]"The coloring and painting of a building's walls is done both to preserve them and to improve their appearance."

23.3.1.1 *Nanotechnologies and Smart Materials: Cool Colored Materials for Building Envelopes*

Among the geo-engineering-based solutions used in architecture to mitigate climate change phenomena, the most relevant and consolidated category is that of highly reflective materials, also referred to as 'cool materials' due to their low surface temperatures, followed by phase change materials (PCM) and thermochromic materials. The category of cool materials is vast and comprises: a first generation, including natural materials with high reflectivity; a second generation, based on the development of artificial white materials designed to have a high *albedo* equal to or higher than 0.85; a third generation of cool materials, known as 'cool colored materials', including colored coatings that feature high reflectivity in the infrared spectrum of solar radiation (NIR), thus appearing like traditional dark materials but with a low surface temperature; and a fourth generation, based on nanotechnological additives like thermochromic paints or PCM-doped cool materials. As has already been observed,[10] previous research into highly reflective materials primarily focused on their effects on outdoor temperatures, thus focusing only on microclimate parameters such as air temperature and surface temperature variations and, as regards that particular aspect, a consistent review has been carried out.[11] Nevertheless, recent studies have highlighted the importance of including other parameters as well, such as mean radiant temperature (T_{mrt}) and, above all, comfort analysis (in terms of PET or UTCI), since it appears that 'cool colored materials' have different side-effects on architectural thermal comfort, depending on the surface of application, on their exposure to solar radiation and on the combination with other elements.[12] In general, increasing reflectivity facilitates the reduction of daytime surface temperatures during the day, whereas during the night the *albedo* of materials becomes ineffective due to the absence of solar radiation. However, since more radiation is reflected during the day, less heat is stored in reflective materials compared to conventional ones, thus leading to small reductions in night-time surface temperatures. Cool colored materials can be applied on roofs (*i.e.* cool roofs), walls (*i.e.* cool walls) and pavements (*i.e.* cool pavements). A field study conducted in Athens that specifically involved the replacement of an urban park's traditional pavements with 'cool pavements'[13] showed that reflective pavements could reduce surface temperature in an urban park by up to 7.6 °C under non-shaded conditions. As regards thermochromic materials for building envelopes, such innovative materials can change their optical and thermal properties in a dynamic way, responding to their environment, changing their color and solar properties: moving from darker to lighter shades as the temperature rises. This change is reversible, which means that, as temperature decreases, these materials turn back to their original color. A multidisciplinary team developed eleven thermochromic coatings for building envelopes using thermochromic pigments in an appropriate binder system and investigated their performance compared to similarly colored highly reflective (cool) and common coatings.[14] The results demonstrated that the surface temperatures of thermochromic samples were lower than the temperatures of color-matched cool and common coatings by a maximum of 7 °C and 11 °C, respectively. (However, problems to do with durability and stability arise when applying thermochromic materials to buildings and urban structures, even with limited outdoor exposure, and thus further research is required).

23.3.1.2 High-performance Pigments in Contemporary Architecture: Case Studies

The thermochromic furniture designed by Jürgen Mayer changes its color in response to the heat released by its user. In recent years, there have been significant developments as regards materials and coatings that react to 'stimuli' in the external environment. In particular, thermochromic pigments, which were developed in response to the need for control and safety (*e.g.* so as to ensure that frozen products were not overheating), are now used in everyday products and they open up infinite possibilities for experimentation in interior design and fashion. Some examples include wallpapers that change color by touch and chairs that record the form of whoever sits on them. Systems based on plastics that react to temperature changes are currently being researched at the Fraunhofer Institute. According to experts, the demand for intelligent polymer sheets for use as thermal sensors is expected to increase. They are applied to the facades of buildings, for example, so as to optimise the use of air conditioning inside.¶

Another high-profile 'demonstration' project is Diller and Scofidio's Brasserie Restaurant on the ground floor of Mies van der Rohe's seminal Seagram Building, where the architects clad the interior facades with smart materials, enhancing their design from a box to an interactive place of flexibility and transformability. Transformability is the key word of the United Network Studio's (UNS) facade project for the La Defense office complex in Almere too, where the facades are clad with glass panels in which tinted sheets reflect different colors depending on the time of day and the angle of incidence, thanks to the material employed: a film generally used for packaging. This material turns interior facades into highly reflective, prismatic parts that appear to change color when the sun goes past the building and crosses the office.

¶A material develops thermochromic qualities, *i.e.* the capacity to vary color in response to a change in temperature, when we add thermochromic pigments in a coating. The thermochromic effect is caused by changes in crystal structure. Typical thermochromic materials include inorganic metal oxides (*e.g.* zinc oxide, vanadium oxide), polymer blends or liquid crystals, which, on warming, do not change directly from the crystalline to the liquid state. In particular, thermosensitive hydrogels for sun protection are the subject of ongoing research. Thermal lacquers are already available on the market for many different applications and can easily be applied to most material surfaces. The reaction temperature can be adjusted and the color changes defined. Thermochromic resins can be mixed with conventional moulding materials and used for the production of plastic components.

Innovative pigments have made it possible to formulate red/pink-colored surface coatings that reduce the heating effect in sunlight compared to surfaces with the same hue in the nursery school on Boulevard de Grenelle in the heart of the 15th *arrondissement* of Paris, where the French group Périphériques, in partnership with the Terreal paint factory, has developed a specific range of nine enamelled colors to meet aesthetic expectations and create a brightly colored facade. 'Grâce à une mise en oeuvre à claire-voie et à une pose sur la diagonale, la perception des couleurs est enrichie et offre à chaque élément deux faces aux teintes différentes: l'une dans la lumière, l'autre dans l'ombre', in the words of the architects themselves.[15] As well as a cooling effect, there are other benefits that come with using these pigments, such as increased service life and improved comfort levels for building occupants.

23.3.2 Ephemeral, Changeable, Unstable

In the last three decades the tools of the architectural language have developed a wealth of devices able to change the spatial understanding of architecture. The translucent, the opalescent, the meshes, the wraps created by polymer sprays, glass that shows different degrees of transparency according to the angle from which it is observed, or is able to pass from complete transparency to opacity in fractions of a second. The perception of the volumes in which we daily move is changing.

LEDs have changed the night figure of the cities, but not only this. The traditional building volume is often reduced to support an 'other' language that connotes the surrounding area much beyond its actual physical and dimensional consistency.

Today color, with the double meaning of fixed and mutable *cromia*, serves to draw precise points of reference in the isotropic mesh of the territory-city and at the same time to give local identity. This semantic device is used with paradoxically similar objectives from opposing economic social tribes. It is difficult to imagine two more culturally and financially different worlds than those represented by the skyscraper of the Rogier Tower in Bruxelles of Samyn and Partners and by Vila Brasilândia in São Paulo by Boa Mistura. Yet both use chromatism (though interpreting it with techniques that are the opposite of technological sophistication) to draw some landmarks in the urban landscape and 'to make place' for the communities of people that they live and work close to.

In the case of the old Manifattura Tabacchi in Rome, the recovery strategy is inspired by orders of problems of different scales, impacts and significance. The first of these refers to an urban/representative framework: the restructuring of the building in question should be the first step of a major urban redevelopment of institutional significance. The second concerns areas more related to the specific function and use of the building: proposing an 'exemplary' restructuring that can serve as a model for the future. The building is required, therefore, to look at the city and its specific methods of use, while at the same time satisfying the need for its urban and institutional roles and perfecting its internal operations as a place of work in the service of citizens. The choice of the colored LEDs and of the top yellow canopy enhance the characters of innovation that the whole project of the building recovery wants to express. It is conceived like a colored *urban lantern.*

23.3.3 The Color Masterplan of Historical Architecture

The restoration of elevations in historical centers, until not many years ago, was considered an intervention of ordinary maintenance that could be carried out in a silent-assent regime. The urban planning instrumentation at the local level often, as far as the coloring is concerned, limited itself to reiterating some simple prescriptions:

- replicate the pre-existing colors
- do not use colors that are dissonant with respect to the decoration of the surrounding historical context.

This last indication, which appeals to general good environmental sense, is in truth rather generic and certainly does not identify sure lines of behavior; the other prescription is only apparently significant. In fact, the replication of the pre-existing tint tends to provoke a sequence of variations that are generally quite predictable because the colors found on site are those - altered - of the last painting, surely different from the penultimate and the antepenultimate, not only for probable intentional variations, but also due to a phenomenon of the natural weighting of color with time, caused by atmospheric pollution.

A plan for the color and restoring of the historical elevations can constitute an important and effective urban planning tool, as long as it does not propose itself as a set of additional rules, or it would have the counterproductive effect of discouraging the object of its goal.

It should be clarified that attention should also be paid to all the elements of the facade: window frames, signs, balconies, railings, tiles, canals, gutters, lanterns, must be restored so as not to constitute disturbing or dissonant elements for materials and shapes. The decisions to be made on the color finish must therefore be based on a series of historical, technical and analytical data, analyses and surveys that are in agreement, to make it possible to draw up the general restoration project.

The facade is the surface on which all the deterioration processes induced by the environment on the building itself (atmospheric pollution, meteoclimatic factors, biological deterioration, *etc.*) take place in the initial phase. It follows that the facade, subjected to incessant deterioration and weakening, is affected, with a certain frequency, also by reparations, maintenance renewals and protections.

Traces of such interventions are often identifiable, albeit in a small part and in a situation of stratified superimpositions, and can constitute precious technical indications and historical traces to be preserved. The facade is therefore a privileged place for the accumulation of information, a sort of limestone archive to be protected and treated with care.

In any case, even with different methodological commitment, the choice of colors and materials to be programmed for restoration work is subordinated to the determination of what can be called the chromatic history of a building (parallel to the history of restoration and maintenance), and that is the sequence of all the coloring applied on it, from the original one to the most recent one.

The methods, now well known, used for the determination of previous painting, are substantially of three types: historical research (bibliographic, iconographic, archival, *etc.*), direct analysis on samples of plaster appropriately taken from the building, affinity and typological criteria. In this regard, it should be noted that such methodologies of analysis preparatory to the interventions to be carried out should not be applied separately, but that, on the contrary, precisely their complementarity and interrelation can lead to more reliable results, since documentary evidence, verified on site and related to the type of building, can provide decisive results.

23.4 Social

The painted architectural surface as *social action device* is the second concept that I explore in this chapter.

This quite radical idea is the focus of many contemporary works on buildings and public spaces. Almost always the authors are not architects, but artists who use the architectural surface as a medium in the city to make their artworks visible beyond the commercial circuit of galleries or exhibitions, freely communicated to everyone.

Frescoed facades have always existed in historical architecture, but the aims of the artistic intervention were the decoration of the architectural apparatus, the simulation of a different and "nobler" material or the historical celebration. For contemporary architecture the use of painting or graffiti on the facade is a drastic choice, often with radical meaning, that can form a dialogue with the architecture that supports it, as in the case of Franco Summa's works, or using the facade as canvas, only with people watching. Even when the work is conceived autonomously by the support, then becomes an integral part of it, this gives the hosting architecture a different symbolic and economic value. Who does not remember the murals of Diego Rivera, David Alfaro Siqueiros and José Clemente Orozco that started to appear on the walls of Mexico City after the 1910 revolution?

The mural is a form of collective art that belongs to the viewer. It is a social action that can contribute to the growth of a country and its culture. This idea affirmed itself all over the world at the end of the last century as street art or urban art through the works of renowned writers: Keith Haring, Paul Morrison, Gordon Matta-Clark, Jean-Michel Basquiat and Bansky,[16] just to name a few of them. Today these artworks are taking on a different role in

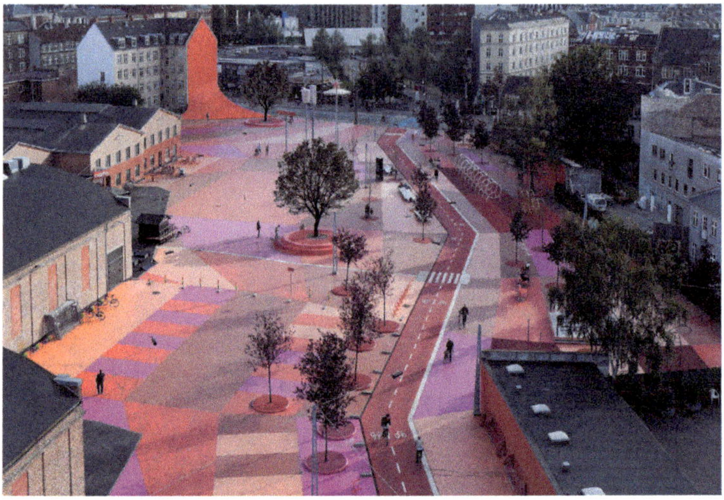

Figure 23.4 Topotek1, BIG, Superflex; *Superkilen* SUK, Copenhagen, Denmark. *Iwan Baan*, 2012, Bjarke Ingels Group, 2012. Reproduced under the terms of the Creative Commons Attribution License, CC BY 2.0, [https://creativecommons.org/licenses/by/2.0/].

their respective cities. In recent years, architecture has become more aware of the value of urban art as a tool for regeneration, often through processes of shared planning (made in co-authorship with citizens) rather than only participated[17] (made with social approval).

Today, artists conceive of the urban environment no longer as merely a background or a stage, but the object of the art work itself. Artists are giving public evidence to the citizens' requests, transforming urban space and abandoned or degraded buildings into places of value, intercepting the desires of beauty of the inhabitants, creating relationships between public actors, clients, institutions and citizens. Also many times it is the same citizens or artists who spread their stories about buildings, such as the intervention of Dan Pitera with the Detroit Collaborative Design Center.[18]

Often for urban art or graffiti works, the traditional concept of authorship, as previously stated, could be questioned by the sharing of the creative process and/or the project implementation is self-managed and *hic et nunc*, exceeding the long bureaucratic times of approvals and public permissions.

Gravalos–Di Monte in Zaragoza[19] use with citizens' collaboration graffiti, colors and urban tattoos to regenerate the abandoned spaces of the suburbs as boxes of the goose game, triggering a new cycle of social life in the neighborhood. Topotek1 and BIG in Copenhagen, Denmark have invented the Superkilen (Figure 23.4) by elevating a road to a park almost exclusively with color (black, white, red). The transition of the colors from the road to the facades is the manifestation of the belonging of the color to the public space domain. Among many others, the work of Boa Mistura is a leading proponent of architecture in this field, such as 'Luz Na Vielas' in São Paolo,[20] where the color on the buildings redeems the public space, filling it with happiness and social pride (Figure 23.5).

The basic idea underlying the interventions of Boa Mistura is that of positivity: changing the aesthetics of a site does not just mean giving it an acceptable face, but, through a visual change, helping a process of social and cultural re-evaluation of a place which in many cases was previously negatively connoted (crime, poverty, abandonment, discomfort) to create aggregation. For example, this is what underlies the idea of interventions in the already quoted favela Vila Brasilândia of São Paolo in Brazil, as well as the intervention in the Chorrillo district of Panama.

Urban art is a social project, it is the projection of an idea in the environmental context for a transfiguration that aims to illuminate significant social values. Urban art is always dealing with the solid context of intervention, transforming it with colors into a work of art that is at the same time profoundly autonomous and deeply linked to the architectural characteristic of the place. It's like giving a second chance to a building or to a place by turning it into a landmark just by re-inventing its physical space with color.

23.4.1 Color as a Social Tool for Understanding the Different Dimensions of Architecture and Urban Sites

The habit of painting facades of buildings can remind us of the parietal interventions, currently in great vogue, of the so-called street art. It could appear as a recovery of that frescoed architecture of the past centuries,

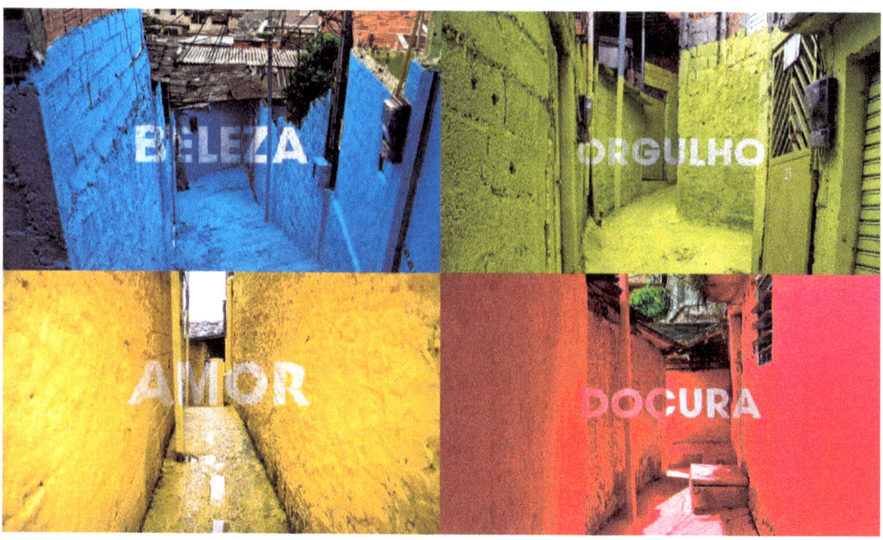

Figure 23.5 Boa Mistura, *Luz Nas Vielas*, São Paulo (Br), Favela of Vila Brâsilandia. Reproduced with permission from Boa Mistura, 2015.

but there are substantial differences. In particular, the lack of relationship between the contents and the figuration forms and the architectural, social and urban context. In the street-art interventions, in general, any environmental reference which might make them contextually significant is missed. Instead, the possibility of environmental requalification through the hypothetical cathartic ability of the masterpiece is relied upon.

The assumption of the city as the artistic working area implies a necessary interpenetration with the environment, aiming at the knowledge and at the conceptual assimilation of the customs and the senses of the context. That considers its cultural, memorial, symbolic, social, psychological, historical dimensions and of course, the spatial and environmental ones. This is, in other terms, a way to enter into 'consonance', to absorb, almost through an osmotic process, every dimension and sense of the city – obvious or not.

In the finalisation of an idea, of a 'draw' that has to be realised, every constitutive element of the environmental dimension must be considered and valued. It must be found in its specificity the possible diapason, the reference, the element or the elements to start and generate the elaboration of a work that 'rewrites' the sense of the context, powering up new dimensions, new perspectives, new meanings. To make, therefore, an 'invention', in the Latin sense of *invenire*, 'to find' but also 'to create'. The environmental artwork, when it really is the case, never appears shut down; it is not presented as an artistic object that is completed and simply located in a determinate urban place, but it is given as sign–form–color that consciously and dynamically affects the context, transfiguring and giving it a new meaning, becoming, at the end, an essential constructive part of the environment.

In 1975 I conceived an installation to exalt a specific urban aspect of Città Sant'Angelo, a historical village of the hill site in Abruzzo: the main street; a road axis that passes through the entire extension of the historical centre following the hill's profile up to the visual and physical conclusion on the scene of an eighteenth-century church elevated on a stairway.

This particularity is assumed as appropriate at the elaboration of a substantial modification, not only in the urban environment, but also in the way the citizens perceive and live the daily paths.

The work *A Rainbow at the End of the Street* (F. Summa, Città Sant'Angelo, 2013) highlights the poetic sense of the transfiguration realized through the coloration of the twenty-four steps of the stairway (Figure 23.6). The church, suspended on the colors, proposes a rainbow that can be climbed color by color as a metaphor for a possible rise from the material to the spiritual. *Il Giardino Incantato*,[21] an urban place overlooking a Mario Botta building, is interpreted as a garden-square with a malleable-pictorial visionary presence.

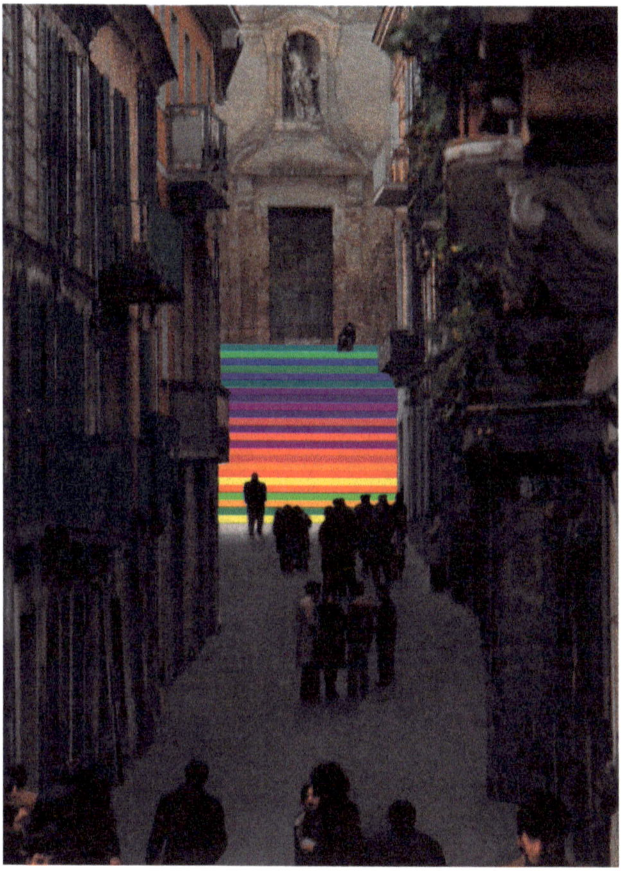

Figure 23.6 F. Summa, *Un arcobaleno in fondo alla via*, Città S. Angelo, Italy, *F. Summa*, 1975.

It is rhythmically divided by sixteen colored slabs, and appears to be a physically and emotionally accessible place inside a garden-square. Pescara, 2017 (Figure 23.7).

23.5 Narrative

The third concept is that of *color as a narrative device.*

It expresses the need to give sense to the architectural project, to discover with different eyes and through stronger figures what architecture in other ways already expresses or implies. Rem Koolhaas describes it in a superb way in the project for the Prada Foundation in Milan (2008–2018)[22] when he paints the old industrial warehouse with a golden patina, as if to say that

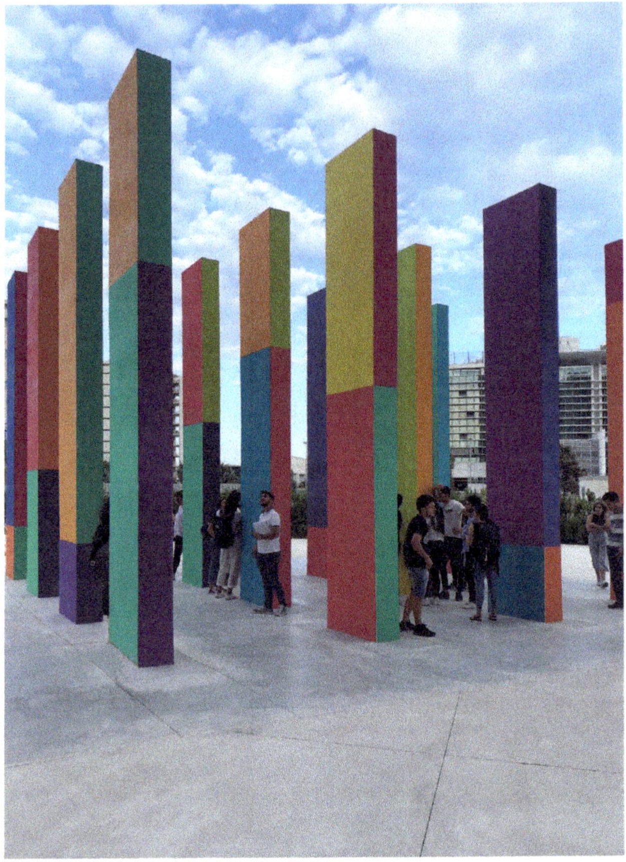

Figure 23.7 F. Summa, *Il giardino incantato*, Pescara, Italy, *F. Summa*, 2018.

the existing is the value to be taken into account in an epoch and in a city where the rampant tendency in urban planning is to demolish and rebuild everywhere, a city that seems unaware of its history and disinterested in its heritage.

> "A narrative architecture," as Giancarlo De Carlo wrote, "is capable of listening, welcoming and annexing the tensions of the city and its inhabitants. An architecture that must become a 'process', undermining the consolidated vision of the building as a perfect *and concluded* unicum."[22]

Even in a strict etymological sense, the use of color as a narrative device reveals different meanings. It tells the story of the building and of those who live in it and reveals a new dimension that can only be perceived for example through color, as in the research of the German study founded in 1989 by Matthias Sauerbruch (Constance, 1955) and Louisa Hutton (Norwich, 1957). Their work is characterized by the importance given to the color, used not as a simple decorative element but as a fundamental part in the definition of the significance of spaces.[23]

Their book *Sauerbruch Hutton: Color in Architecture*[24] presents nine colored projects, but also introduces the complexities inherent in the practice of designing and thinking about color: how to materialize color; how to use color to manipulate or to address impressions of space and form; how to perceive color in the built environment; and how to connect the color with the place. In the project for the Brandhorst Museum of Modern and Contemporary Art in Munich, the facade has been managed as a large abstract mosaic of twenty-three colors, made vibrant by the superimposition of narrow colored modules and horizontal and vertical bands. The effect achieved through the use of colors by the German office is that of integration with the context with an almost chameleonic mimicry that gives a fascinating and pointillist three-dimensionality to the new architecture. Their latest Italian Museo del Novecento in Mestre (Figure 23.8) is conceptually very similar to the previous one. It consists of the recovery of an old convent and is characterized by the polychrome ceramic coating made of strips placed diagonally across the two new additions that at

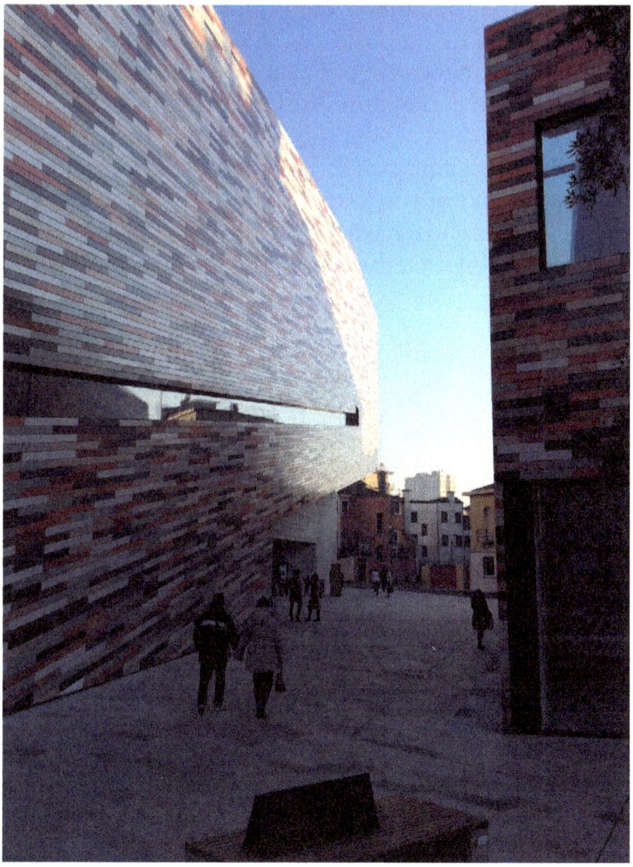

Figure 23.8 Sauerbruch & Hutton, *M9 museum in Mestre*, Venice. *P. Baldi*, 2018.

the same time appear abstract and integrated to the existing urban fabric through the use of color that decomposes in the facade the colors of the architectural materials of the surrounding historical city.

A similar approach can also be found in the work of the Spanish Mansilla + Tuñón[25] with the MUSAC de Castilla y Léon of 2005, and the Town Hall in Lalin (Spain) of 2011.

Still in the early 2000s for Enric Miralles and Benedetta Tagliabue, the large colored tapestry on the roof of the Santa Caterina Market in Barcelona (2003) is the giant image of a photo taken from a Mediterranean fruit and vegetable counter printed as a puzzle of 325 000 hexagonal tiles of 67 colors. This is in reference and homage to the works of Gaudì, clearly one of the most representative architects of Catalan modernism.

The same type of theme, with the painted surfaces that tell the joy of the market, is developed with great quality in the internal vault of the Rotterdam market by the Dutch architect group MVRDV in 2014 (Figure 23.9).[26] In the 'Didden Village' (Rotterdam, 2009), a project to increase the height of a traditional nineteenth-century house, MVRDV uses symbolically the colors that become the hallmark of the two simple volumes superimposed on the existing building. In the designer's concept it is like a 'crown at the top of the monument'. Didden Village can represent a prototype for a new type of expansion of historical cities where color plays a strategic role.

Jean Nouvel uses the same bright color of the Kilometro Rosso for the tenth temporary pavilion of the Serpentine Gallery in 2010, at Kensington Gardens, London. Red is chosen with a narrative intent. It tells about the city. Red like the telephone booths, the double-decker buses and the main doors

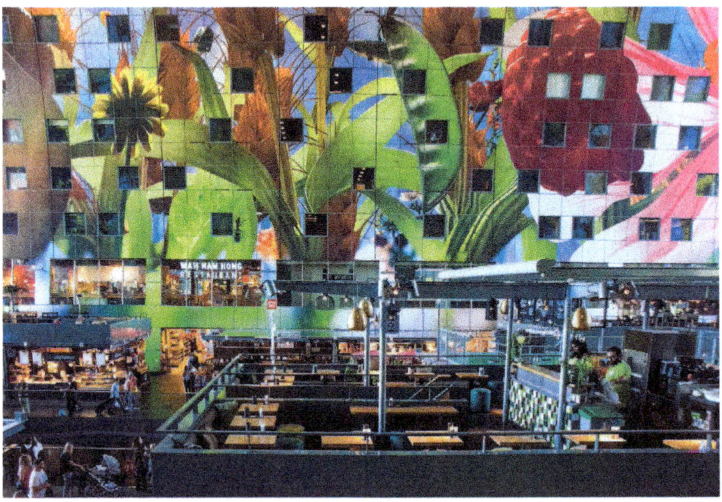

Figure 23.9 MVRDV, *Markthal*, Rotterdam, *Courtesy of M. Mantelli*, 2018. Reproduced with permission.

of the houses of London. The red pavilion, so lively and intense, creates a strong contrast with the green lawn where it is inserted to pay homage to the city.

23.5.1 Prodromes of Neo-Fauvism

Color is architecture's most troublesome material. Even as the medium was abstracted, theorized for relativity and phenomenologically re-perceived as its own subject during modernism, chroma was then sanitized from the built environment. Instead of producing color-coded environments, material ethics and truth served as an aesthetic. The predominance of form for form's sake continued the hierarchies of Plato's legacy in building design. In contemporary architecture, transcending the clarity and simplicity of black and white is typically a foray into shades of gray. Only daylight is more seductive, uncontrollable, sensuous and unstable than color. No wonder modern architects – employed to manage and characterize so much in the way of international nation-building, Western hegemony and global construction – became chromophobes.

In the last decade, NMDA has proposed some daring color insertions in projects and buildings all over the world. The firm's website even mentions relationships of color to culture and locus in the portrayed work. Perhaps Neil Denari is poised to lead a Neo-Fauvist revolution in urban form. Denari's radical color is not immediately apparent. His singular chromatic preference has been for a bright eye-grabbing accent to the tectonically assertive semi-transparent massing of torqued or tilted rectangular forms with silvery finishes.[27] In this way, he perpetuates the minimalist tradition of singular color accents to neutrally toned architecture. Urban insertions are his forte, and each artifact speaks boldly to its unique surroundings. Beyond formal objects or surfaces, spaces are often rendered as colored light. It is in these sweeping ethereal suggestions that the problem of color's embrace begs some attention. For the preceding generation of 'white' architects, the minimalists of the senses and those seduced by formal mind games, color got in the way. Color as a foreign body acts as something out of control, more often associated with the exotic, the feminine, oriental, infantile or vulgar. When color is supplemental, superficial or cosmetic, it may also be conceived of as trivial or – alternately – dangerous.[28]

Yet in the context of these challenges to color's troublesome nature in architecture, Denari's New Keelung Center in Taiwan is of the greatest interest (Figure 23.10). The grand public project is also distinct from Denari's other predominant uses of red hues. "Since Keelung is a rainy city, often shrouded in wet, heavy air, the chartreuse sections of the complex, in combination with specially designed lighting, will push through the gauzy atmosphere providing vitality and life."

Denari's portrayals of the massive metallic form in the harbor essentially feature lime yellow light emitted from within and reflected in that water. In some depictions, the light appears in competition with the rising sun and

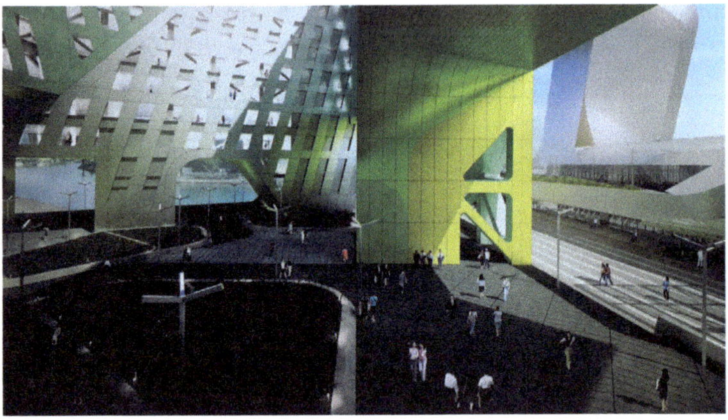

Figure 23.10 Neil M. Denari/NMDA, *Keelung Harbor Terminal*, Taiwan, in construction, NMDA, 2012. Reproduced with permission from Neil M. Denari.

focuses viewers' attention to the building near grade with both planted landscapes and hardscapes. So luminous is the ambience that the building seems to be in its own timeless glow. As more of his projects come to fruition, we will watch his spectra of atmospheric light and color evolve and come to life. This is the promise of a richer, luminous, deep sensual architecture of its time, of its place and for its colorful citizens.

23.5.2 Some Italian Tales

In Italy, in Rome, the Palazzina Papanice by Paolo Portoghesi in the early 70 s refers to the Baroque tradition with an organic form that affirms its identity in color: the blue of the sky, the green of the plants, the brown of the trunks, the gold of the sun. "In the vertical band design of the majolica, the allusion to the colors of nature, overlaps the 'musical' rhythm of their weaving, while inside, where the colored bands run horizontally, the ceilings of the living room are continually rippled by a series of concentric degrading cylinders."[29]

The linear park between Caltagirone and Piazza Armerina, 1999, by Marco Navarra makes the landscape narrative with a continuous strip of color: "This is a landscape project and it is about, in details, the acknowledgement of ancient agricultural, natural and historical landscapes in an area of Sicily in the threshold between Erei and Iblei, through the building of a new landscape (...). "

The *Black House* by Tierstudio (Alberto Bertagna and Sara Marini) in Urbania, 2012,[30] completely covered with slate slabs, is significant in this text for the precision and the intentionality of the narrative use of color (Figure 23.11). The blackboard changes over time and from black to very clear; it records the context. The black house is camouflaged in the winter sky, but inside it is white to mark the relationship with the inner intimacy.

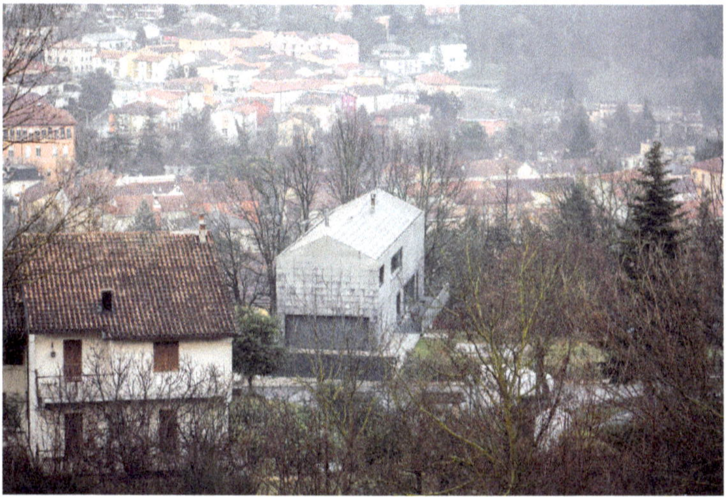

Figure 23.11 S. Marini, A. Bertagna, *Casa Nera*, Urbania, *Courtesy of F. Mantovani*, 2018. Reproduced with permission.

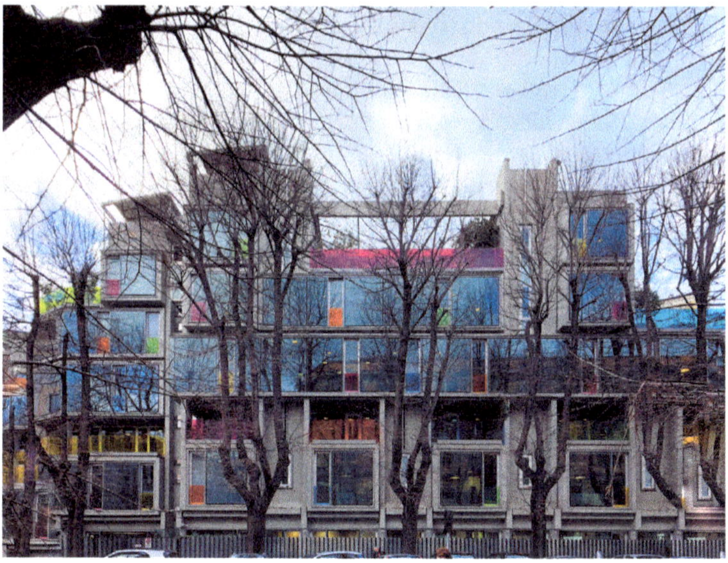

Figure 23.12 M. Ricci with RicciSpaini AA, *Ghella EcOffice*, Roma, Italy, *Courtesy of M. Mantelli*, 2017. Reproduced with permission.

In the brutalist building by Alvaro Ciaramaglia recovery intervention in Rome, designed with the RicciSpaini AA office, color is a plot of the inner history of the regeneration of the building itself wishing a cheerful architecture made of cement, transparencies and colors even at night, and the internal feeling of an almost non-urban work environment between the hanging gardens and the linden trees (Figure 23.12).

"When they asked Goethe what his favorite color was, he said: *I like rainbows*. Here's what I love about architecture: if it's good, it contains all the colors of the spectrum of life, if it's bad, the colors will fade to fade away from the ruins of Byzantium to the streets of New York, from the roof of a Chinese pagoda to the top of the Eiffel Tower, each building tells a story, even better, different stories."[31]

The use of painted surfaces to support and enhance the narrative value of the architectural work is a design device capable – as in art and fashion – of describing and interpreting the resistance of contemporary architecture to the digital revolution when it drastically cuts the sphere of meanings and emotions by pushing them in a non-material world. Reporting these values to the reality of physical space, the painted surfaces seem to express not only a return of importance of color in architecture and even more in public spaces, but also a new desire to make cities happier.

References

1. http://www.bollettinodarte.beniculturali.it/opencms/multimedia/BollettinoArteIt/documents/1371726293991_04_-_P._Marconi.pdf.
2. W. Boesiger, *Le Corbusier*, Zanichelli, Bologna, 1991.
3. E. Ambasz, *The Architecture of Luis Barragan*, The Museum of Modern Art, New York, 1976.
4. *Lina Bo Bardi*, ed. M. Carvalho Ferraz, Imprensa Oficial do Estado de São Paulo/Instituto Lina Bo e P.M Bardi, 2008.
5. K. Gasparini, *Schermi Urbani. Tecnologia e Innovazione. Nuovi sistemi per le facciate mediatiche*, Wolters Kluver Italia, Milan, 2012.
6. http://www.jeannouvel.com/en/projects/the-kilometro-rosso-science-and-technology-park/.
7. V. Flusser, *La Cultura dei Media*, Mondadori, Milan, 2004.
8. D. M. Addington and D. L. Schodek, *Smart Materials and New Technologies: for the architecture and design professions*, Elsevier, Oxford, 2005.
9. A. C. Clarke, *Profiles of the Future*, Harper & Row, 1958.
10. J. Yang, Z.-H. Wang and K. E. Kaloush, *Renewable Sustainable Energy Rev.*, 2015, **47**, 830–843.
11. H. Akbari, C. Cartalis, M. Santamouris, A. Synnefa, D. Kolokotsa, A. Muscio, A. L. Pisello, F. Rossi, N. H. Wong and M. Zinzi, *J. Civ. Eng. Manage.*, 2016, **22**, 1–16.
12. J. Yang, Z.-H. Wang and K. E. Kaloush, Environmental impacts of reflective materials: Is high albedo a 'silver bullet' for mitigating urban heat island?, *Renew. Sustain. Energy Rev.*, 2015, **47**, 830–843.
13. M. Santamouris, N. Gaitani, A. Spanou, M. Saliari, K. Giannopoulou, K. Vasilakopoulou and T. Kardomateas, *BAE Build. Environ.*, 2012, **53**, 128–136.

14. T. Karlessi, M. Santamouris, K. Apostolakis, A. Synnefa, and I. Livada, *Sol. Energy*, 2009, **83**, 538–551.
15. https://wellcom.fr/presse/terreal/2014/05/terreal-orne-creche-parisienne/.
16. C. McCormik, M. Schiller, S. Schiller and E. Seno, *Trespass. A History of Uncommissioned Urban Art*, Taschen, Los Angeles, 2010.
17. P. Bishop and L. Williams, *The Temporary City*, Routledge, New York, 2012.
18. http://www.dcdc-udm.org.
19. www.plataformaarquitectura.cl/cl/02-349303/esto-no-es-un-solar-reconvirtiendo-parcelas-vacias-en-espacio-publico-parte-ii.
20. https://www.designboom.com/art/boa-mistura-luz-nas-vielas-favela-anamorphic-street-art-03-23-2017/.
21. Il Giardino Incantato, Urban Art Installation, Court of Pescara (Italy), Franco Summa, 2018.
22. http://www.fondazioneprada.org.
23. M. Sauerbruch and L. Hutton, in *Lessico del XXI secolo*, Enciclopedia Treccani, 2013.
24. M. Sauerbruch and L. Hutton, *Color in Architecture*, Distanz Verlag, Berlin, 2012.
25. Mansilla and Tunon, 1992–2012, in *el Croquis* no. 161, Edicion Impresa, Madrid, 2012.
26. https://www.archdaily.com/13370/didden-village.
27. www.denari.co/projects.
28. D. Batchelor, *Color*, The MIT Press, Cambridge, 2008.
29. P. O. Rossi, *Roma. Guida all'architettura Moderna 1909–2011*, Laterza, Roma, 2012.
30. S. Marini and A. Bertagna, Casa Nera, in *Arcipelago Italia. Progetti per il futuro dei territori interni del Paese*, La Biennale di Venezia-Quodlibet, 2018.
31. D. Libeskind, *Breaking Ground*, Sperling & Kupfer, 2004.
32. A. Rossi, The Blue of the Sky, *Architectural Design*, 1982, **52**.

Subject Index

References to figures are given in *italic* type. References to tables are given in **bold** type. Notes are indicated by a 'n' following the page number.